Applied Mathematical Sciences | Volume 31

William T. Reid

Sturmian Theory for Ordinary Differential Equations

Springer-Verlag
New York Heidelberg Berlin

William T. Reid
formerly of the
Department of Mathematics
University of Oklahoma

Prepared for publication by
John Burns *and* Terry Herdman
Department of Mathematics
Virginia Polytechnic Institute
and State University
Blacksburg, Virginia 24061/USA

Calvin Ahlbrandt
Department of Mathematics
University of Missouri
Columbia, Missouri 65201/USA

AMS Subject Classifications: 34-01, 34B25

Library of Congress Cataloging in Publication Data

Reid, William Thomas, 1907 (Oct. 4)—1977.
Sturmian theory for ordinary differential equations.

(Applied mathematical sciences; v. 31)
Bibliography: p.
Includes indexes.
1. Differential equations. I. Title. II. Series.
QA1.A647 vol. 31a [QA372] 510s [515.3′52] 80-23012

Printed in the United States of America.

9 8 7 6 5 4 3 2 1

ISBN 0-387-90542-1 Springer-Verlag New York Heidelberg Berlin
ISBN 3-540-90542-1 Springer-Verlag Berlin Heidelberg New York

Dedicated to

DR. HYMAN J. ETTLINGER

Inspiring teacher, who introduced the author as a graduate student to the wonderful world of differential equations.

PREFACE

A major portion of the study of the qualitative nature of solutions of differential equations may be traced to the famous 1836 paper of Sturm [1], (here, as elsewhere throughout this manuscript, numbers in square brackets refer to the bibliography at the end of this volume), dealing with oscillation and comparison theorems for linear homogeneous second order ordinary differential equations. The associated work of Liouville introduced a type of boundary problem known as a "Sturm-Liouville problem", involving, in particular, an introduction to the study of the asymptotic behavior of solutions of linear second order differential equations by the use of integral equations.

In the quarter century following the 1891 Göttingen dissertation [1] of Maxime Bôcher (1867-1918), he was instrumental in the elaboration and extension of the oscillation, separation, and comparison theorems of Sturm, both in his many papers on the subject and his lectures at the Sorbonne in 1913-1914, which were subsequently published as his famous *Lecons sur les méthodes de Sturm* [7].

The basic work [1] of Hilbert (1862-1941) in the first decade of the twentieth century was fundamental for the study of boundary problems associated with self-adjoint differential systems, both in regard to the development of the theory of integral equations and in connection with the interrelations between the calculus of variations and the characterization of eigenvalues and eigensolutions of these systems. Moreover, in subsequent years the significance of the calculus of variations for such boundary problems was emphasized by Gilbert A. Bliss (1876-1951) and Marston Morse (1892-1977). In particular, Morse showed in his basic 1930 paper [1] in the *Mathematische Annalen* that variational principles provided an appropriate environment for the extension to self-adjoint differential systems of the classical Sturmian theory.

The prime purpose of the present monograph is the presentation of a historical and comprehensive survey of the Sturmian theory for self-adjoint differential systems, and for this purpose the classical Sturmian theory is but an important special instance. On the other hand, it is felt that the Sturmian theory for a single real self-adjoint linear homogeneous ordinary differential equation must be given individual survey, for over the years it has continued to grow and continually provide impetus to the expansion of the subject for differential systems. There are many treatments of the classical Sturmian theory, with varied methods of consideration, and in addition to Bôcher [7] attention is directed to Ince [1-Chs. X, XI], Bieberbach [1-Ch. III, §§1-4], Kamke [7-§6, especially Art. 25], Sansone [1, I-Ch. IV], Coddington and Levinson [1-Chs. 7,8,11,12], Hartman [13-Ch. XI], Hille [2-Ch. 8], and Reid [35-Chs. 5,6].

In the present treatment there has been excluded work on the extension of Sturmian theory to the areas of partial differential equations, and functional differential equations with delayed argument. Also, for ordinary differential equations the discussion and references on the asymptotic behavior of solutions has been limited to a very small aspect that is most intimately related to the oscillation theorems of the classical Sturmian theory.

For older literature on the subject the reader is referred to the 1900 Enzyklopadie article by Bôcher [4], and his 1912 report to the Fifth International Congress of Mathematicians on one-dimensional boundary problems [5]. For discussions of Bôcher's work and his influence on this subject, attention is directed to the review of R. G. D. Richardson [5] of Bôcher's *Lecons sur les méthodes de Sturm*, and the article by G. D. Birkhoff [4] on the scientific work of Bôcher. The account of subsequent literature prior to 1937 has been materially aided by the author's old report [6]; in particular, not all of the Bibliography of that paper has been reproduced in the set of references at the end of this volume. For more recent literature the author has been greatly helped by the survey articles in 1969 by Barrett [10] and Willett [2]. Also, of special aid has been the report of Buckley [1], which presented brief abstracts of many papers dealing with the oscillation of solutions of scalar linear homogeneous second order differential equations, and which appeared in a number of journals, largely in the decade ending with 1966.

Although the appended Bibliography is extensive, undoubtedly the author has overlooked some very relevant papers of which he is cognizant, and unfortunately others of which he is not aware. To the authors of all such papers, regrets are extended herewith and the hope expressed that they will inform the author of the omission. Special regrets are extended to the authors of papers written in the Russian language, for the author's inability to read the original papers has necessitated his reliance upon translations and reviews.

In organization, most of the chapters contain a body of material which might be described as textual, and which presents concepts and/or methods that the author feels are central for the considered topic. Such material is then usually followed by a section with more detailed comments and references to pertinent literature, and finally there is a section on Topics and Exercises devoted to a variety of examples of related results with references, and sometimes comments on the principal ideas involved in derivation or proof. Clearly such a selection involves a high order of subjectivity on the part of the author, for which he assumes full responsibility.

References to numbered theorems and formulas in a chapter other than the one in which the statement appears include an adjoined Roman numeral indicating the chapter of reference, while references to such items in the current chapter do not contain the designating Roman numeral. For example, in Chapter VI a reference to Theorem 6.4 or formula (4.6) of Chapter V would be made by citing Theorem V.6.4 or formula (V.4.6), whereas a reference to Theorem 1.2 or formula (3.15) would mean the designated theorem or formula in Chapter VI.

Profound thanks are extended to the Administration of the University of Oklahoma for support in providing secretarial help. The author is also deeply grateful to Mrs. Debbie Franke for her typing of preliminary working papers and the final version of this manuscript.

W. T. Reid
Norman, Oklahoma
September, 1975

ADDITION TO THE PREFACE

As indicated above, the main text of this book was completed in September 1975. However, at the time of Professor Reid's death (October 14, 1977) the manuscript was still in the review process. In 1979 Calvin Ahlbrandt and I accepted the responsibility for having the manuscript reviewed by several publishers and an agreement for publication by Springer-Verlag was completed. I agreed to undertake the usual author's responsibility concerning proofreading, etc. Therefore, I accept all responsibility for errors in the final copy. I am certain that these errors would have been corrected by Professor Reid had he lived to complete the publication of the book. The main text of the present book is essentially a faithful copy of Professor Reid's final manuscript except for minor corrections and a few additions to the bibliography. Many of the references were published after 1975 and these references were updated wherever possible. However, we have made no attempt to add references beyond those available to Professor Reid in 1975.

I wish to express my sincere appreciation to Calvin Ahlbrandt and Terry Herdman for their assistance in completing, proofreading and publishing the manuscript. They devoted considerable time to the project and without their efforts it would have been impossible to complete the book within any reasonable time period. Also, I wish to thank Mrs. Kate MacDougall for her excellent typing of the final camera-ready copy for this volume.

Calvin, Terry and I were fortunate to have studied under Professor Reid and we feel privileged to have helped in the completion of this book.

John Burns
Blacksburg, Virginia
July 22, 1980

CONTENTS

Page

CHAPTER V. STURMIAN THEORY FOR DIFFERENTIAL SYSTEMS

CHAPTER VI. SELF-ADJOINT BOUNDARY PROBLEMS

CHAPTER VII. A CLASS OF DEFINITE BOUNDARY PROBLEMS

CHAPTER VIII. GENERALIZATIONS OF STURMIAN THEORY

CHAPTER I. HISTORICAL PROLOGUE

1. Introduction

The present volume is concerned with the Sturmian theory for differential equations, and the interrelations of this theory with the calculus of variations. A remarkable historical fact is the almost simultaneous occurrence in the fourth decade of the nineteenth century of basic works in various facets of this subject.

Firstly, in 1836 there appeared the classic paper of Sturm [1], dealing with oscillation and comparison theorems for linear homogeneous second-order ordinary differential equations. Closely allied to Sturm's work was that of Liouville [1] in the period 1835-1841 dealing with the asymptotic form of solutions of ordinary differential equations of the second order linear in a characteristic parameter, and the ensuing type of boundary problem known as the Sturm-Liouville problem.

Although the interrelations were not appreciated at the time, almost simultaneous with Sturm's work Jacobi [1] presented in 1837 some fundamental results on the non-negativeness and the positiveness of the second variation for a non-parametric simple integral variational problem with fixed end-points, and results of the type formulated by Jacobi form one of the basic ingredients of the extensive discipline now known as the Sturmian theory for differential equations. Moreover, although some fifty years elapsed before there was appreciable research in the area opened by Sturm and Liouville, a separation theorem of Sturm was used by Weierstrass in the late 1870's to provide a rigorous proof for a result stated by Jacobi, and concerning which more details will be presented in the following section.

The early 1890's were marked by two phenomena which gave decided impetus to the study of problems of the type introduced by Sturm and Liouville. In 1890, Picard [1] determined for the special differential system

$$u''(t) = f(t), \quad u(a) = 0 = u(b) \tag{1.1}$$

the kernel function $g(t,s)$ which provides the solution for this problem in the integral form

$$u(t) = \int_a^b g(t,s)f(s)ds. \tag{1.2}$$

Shortly thereafter, Burkhardt [1] considered boundary problems involving the differential equation of (1.1) and more general end-conditions at $t = a$ and $t = b$, and introduced the term "Green's function", since for the ordinary differential systems studied this function assumes a role similar to that played by a function introduced much earlier (1828), by G. Green for certain boundary problems involving partial differential equations.

The second phenomenon of the early 1890's that vitally affected the subject under consideration was the thesis by Maxime Bôcher [1], written under the direction of Felix Klein. In his study of problems in potential theory, Klein was led to the question of when certain linear homogeneous second order differential equations involving two parameters had for two given non-overlapping intervals on the real line a pair of solutions which possessed on the respective intervals a prescribed number of zeros. In particular, Klein was concerned with the Lamé differential equation to which the partial differential equation $\Delta u + k^2 u = 0$ is reduced through the introduction of elliptic coordinates, (see, for example, Kamke [7, p. 500]). The dissertation of Bôcher dealt with this and related problems, and thus introduced Bôcher to the work of Sturm. The major portion of Bôcher's subsequent research may be described as giving rigorous and accessible form to the work of Sturm and Klein on the real solutions of ordinary differential equations, and extensive elaboration of these theories.

In particular, the sequence of three papers [2] of Bôcher in the Bulletin of the American Mathematical Society in the years 1897-99 did much to present the principal results of Sturm in a more rigorous form, and to call the attention of mathematicians to this area of research. As formulated in the first of this sequence of papers, the First Comparison Theorem of Sturm was presented in the following context:
If $p_1(t)$ *and* $p_2(t)$ *are continuous, real-valued functions satisfying on* $[a,b]$ *the inequality* $p_1(t) \geq p_2(t)$, *with the inequality sign not holding for all values of* t *in a*

neighborhood of $t = a$, *and* $u_1(t)$, $u_2(t)$ *are solutions of the respective differential equations*

$$u_1''(t) - p_1(t)u_1(t) = 0, \quad u_2''(t) - p_2(t)u_2(t) = 0, \tag{1.3}$$

satisfying the initial conditions $u_1(a) = u_2(a) = \alpha$, $u_1'(a) = u_2'(a) = \alpha'$, $|\alpha| + |\alpha'| > 0$, *and if* $u_1(t)$ *has* n *zeros* $a < t_1 < \ldots < t_n < b$, *then* $u_2(t)$ *has at least* n *zeros between* a *and* b, *and the* i-th *zero measured from* a *is less than* t_i, $(i = 1,\ldots,n)$.

The second paper of this sequence dealt with the now familiar "Sturmian boundary problem" involving a differential system

$$\begin{aligned} &u''(t) - p(t,\lambda)u(t) = 0, \\ &\alpha'u(a) - \alpha u'(a) = 0 = \beta'u(b) - \beta u'(b), \end{aligned} \tag{1.4}$$

with α, α', β, β' real constants satisfying $|\alpha'| + |\alpha| > 0$, $|\beta'| + |\beta| > 0$, and p a continuous function of its two arguments satisfying monotoneity conditions which imply the existence of a sequence of eigenvalues and eigenfunctions $(\lambda_j, u_j(t))$, $(j = 1,2,\ldots)$, with $u_j(t)$ possessing exactly j zeros on the open interval (a,b). In the third of these papers Bôcher showed that results of the earlier papers remained valid when $p(t)$ and $p(t,\lambda)$ were allowed as functions of t to have a finite number of points of discontinuity $t = c_j$, $(j = 1,\ldots,k)$, with these functions in absolute value dominated by respective functions $[(t-c_1)\ldots(t-c_k)]^{-\alpha}\psi(t)$ and $[(t-c_1)\ldots(t-c_k)]^{-\alpha}\psi(t,\lambda)$, where $\psi(t)$ and $\psi(t,\lambda)$ are continuous in the designated arguments, while α is a value satisfying $0 < \alpha < 1$. Prefatory to discussing the extension of the Sturmian results of the preceding two papers, Bôcher established for such equations the analogues of the usual initial value theorems for equations with continuous coefficients. In particular, this third paper is of historical interest, for it antedated the work of Lebesgue on integration by a few years and was some twenty years before Carathéodory [1, Ch. XI], generalized the concept of a solution of a differential equation to that of an absolutely continuous function satisfying almost everywhere the equation.

Historically, Riccati differential equations have occurred in various manners in the theory of linear differential equations and associated boundary problems. In his classical memoir [1, pp. 158-160], Sturm notes that if $u(t)$ is a solution of an equation of the form

$$[r(t)u'(t)]' - p(t)u(t) = 0 \tag{1.5}$$

then $w(t) = r(t)u'(t)/u(t)$ is a solution of the Riccati differential equation

$$w'(t) + \frac{1}{r^2(t)}w^2(t) - p(t) = 0, \tag{1.6}$$

and also $z(t) = 1/w(t)$ is a solution of the Riccati differential equation

$$z'(t) + p(t)z^2(t) - \frac{1}{r(t)} = 0, \tag{1.7}$$

where clearly it should be stipulated that in the respective cases the functions $u(t)$ and $u'(t)$ are assumed to be non-vanishing on an interval of consideration. In essence, however, Sturm employed the relationship between linear equations and Riccati equations only for the simple result that if $r(t)$ and $-p(t)$ are both positive on a given interval then the function $w(t)$ and $z(t)$ are, respectively, strictly monotone decreasing and increasing on such an interval.

For the present treatment, a basic reference in regard to the use of Riccati equations is the paper [3] of Bôcher, in which he derives the comparison theorems of Sturm as consequences of comparison theorems for a pair of Riccati equations

$$w'(t) = A_1(t) + C_1(t)w^2(t), \tag{1.8$_1$}$$

$$w'(t) = A_2(t) + C_2(t)w^2(t), \tag{1.8$_2$}$$

where A_α, C_α, $(\alpha = 1,2)$, are supposed to be real-valued continuous functions, satisfying on a given compact interval $[a,b]$ the inequalities

$$A_2(t) \geq A_1(t), \quad C_2(t) \geq C_1(t), \quad \text{for} \quad t \in [a,b].$$

Because of its historical interest, the Introduction to Bôcher's paper is repeated here, the only alterations being changes to incorporate references to the accompanying Bibliography.

"Of the many theorems contained in Sturm's famous memoir in the first volume of Liouville's Journal (1836), p. 106, two, which I have called the Theorems of Comparison, may be regarded as most fundamental. I have recently shown [2] how the methods which Sturm used for establishing these results can be thrown into rigorous form. In the present paper I propose to prove these theorems by a simpler (Footnote: Simpler, at least if we wish to establish the theorems in all their generality) and more direct method. This method was suggested to me by a passage, to which Professor H. Burkhardt kindly called to my attention, in one of d'Alembert's papers on the vibration of strings, [Memoirs of the Berlin Academy,

Vol. 70 (1763), p. 242]. D'Alembert's fundamental idea, and indeed all that I here preserve of his method, consists in replacing the linear differential equations by Riccati's equations, (Footnote: Sturm also in the paper quoted (p. 159) uses Riccati's equations, but only incidentally, and for quite a different purpose)."

Although at that point in time, (1900), Bôcher felt that the use of Riccati equations afforded the simpler proof of the Sturm comparison theorems in full generality, his opinion was evidently altered by the time of his report [5] to the Fifth International Congress of Mathematicians in 1912, as in that paper his reference to the method is limited to a brief footnote, and in his subsequent *Lecons* [7] there is seemingly no reference to the method. Undoubtedly this change in feeling was occasioned by the intervening work of Picone [1,2], as in [5] Bôcher emphasized the "extremely elegant" method given by Picone, and employed the "Picone identity". As a whole, the treatment of Picone [1,2] of boundary problems and associated theorems of oscillation and comparison was neither in the spirit of Sturm, nor in the context of a variational problem, even though a major portion of the proof of his "identity" may be considered as a particular instance of the Jacobi transformation of a second variation problem. Rather, Picone's basic existence proofs were established by a method of successive approximations that had been employed earlier by Picard [2, Ch. VI] for a special boundary problem of the form

$$u''(t) + \lambda k(t)u(t) = 0, \quad u(a) = 0 = u(b), \tag{1.9}$$

where $k(t)$ is a positive continuous function on $[a,b]$, and utilizing the "Schwarz constants" in the derivation of a test for the range of convergence of the procedure. This method, introduced originally by Schwarz in the consideration of a boundary problem involving a partial differential equation, may be characterized as the determination of the circular domain of convergence in the complex plane of the Maclaurin series expansion for the resolvent of an associated functional problem linear in a complex parameter. Previous to the work of Picone the method had also been used by E. Schmidt [1] in his theory of linear integral equations with real symmetric kernel. In later years it has been used by Bliss [4;6] and Reid [9; 35 Ch. IV, 6] in the study of so-called definitely self-adjoint differential systems.

Aside from the personal contributions of Bôcher which culminated in his *Lecons* [7], the major papers of the period 1909-21 using essentially the methods of Sturm-Bôcher, and which contributed to the extension of Bôcher's results to general self-adjoint boundary problems associated with a real, linear, homogeneous differential equation of the second order involving non-linearity a parameter, are those of G. D. Birkhoff [3], R. G. D. Richardson [4], and H. J. Ettlinger [1,2,3].

2. Methods Based Upon Variational Principles

To illustrate the interrelations that exist between the theory of boundary problems for ordinary differential equations and variational principles, attention will be restricted to the simplest type of problem of the calculus of variations. Consider the problem of determining an arc

$$C: x = x(t), \quad t \in [t_1, t_2], \tag{2.1}$$

in the real (t,x)-plane joining two given points $P_1 = (t_1, x_1)$ and $P_2 = (t_2, x_2)$, and minimizing an integral functional of C of the form

$$J[C] = \int_{t_1}^{t_2} f(t, x(t), x'(t))dt. \tag{2.2}$$

For definiteness, the real-valued integrand function $f = f(t,x,r)$ is supposed to be continuous and to have continuous partial derivatives of the first two orders in an open region $\mathscr{R}$ of real (t,x,r)-space. Moreover, the class of differentially admissible arcs (2.1) to be considered consists of those $x(t)$ which are continuous, piecewise continuously differentiable on $[t_1, t_2]$, and for which all elements $(t, x(t), x'(t)) \in \mathscr{R}$ for $t \in [t_1, t_2]$. In this connection, it is to be emphasized that if $t_0 \in (t_1, t_2)$ is a value at which the derivative $x'(t_0)$ does not exist, then each of the unilateral values $(t_0, x(t_0), x'(t_0^+))$, $(t_0, x(t_0), x'(t_0^-))$ is required to belong to $\mathscr{R}$. Also, it is supposed that the class of *admissible arcs* consisting of those differentially admissible arcs joining the end-points P_1 and P_2 is non-empty. Assume that $\mathscr{R}$ is convex in r.

A function $\eta: [t_1, t_2] \to \mathbb{R}$ which is continuous, piecewise continuously differentiable, and satisfies $\eta(t_1) = 0 = \eta(t_2)$, will for brevity be called an *admissible variation*, and the set of all such admissible variations denoted by $\mathscr{V}_0$. In particular, if

$$C_0: x = x_0(t), \quad t \in [t_1, t_2],$$

is an admissible arc and η is an admissible variation, then for s a real value sufficiently small in absolute value the arc

$$C_s : x = x_0(t) + s\eta(t), \quad t \in [t_1, t_2], \tag{2.3}$$

is also an admissible arc, and

$$J[C_s] = \int_{t_1}^{t_2} f(t, x_0(t) + s\eta(t), x_0'(t) + s\eta'(t))dt. \tag{2.4}$$

Consequently, if C_0 minimizes $J[C]$ in the class of admissible arcs, then $J[C_s]$ has a relative minimum at $s = 0$, and in view of the above stated assumptions on the character of f on $\mathscr{R}$ it follows that

$$J_1[\eta] \equiv \int_{t_1}^{t_2} [f_r^o(t)\eta'(t) + f_y^o(t)\eta(t)]dt = 0, \textit{ for } \eta \in \mathscr{V}_0, \tag{2.5}$$

$$J_2[\eta] \equiv \int_{t_1}^{t_2} 2\omega^o(t, \eta(t), \eta'(t))dt \geq 0, \quad \textit{for } \eta \in \mathscr{V}_0, \tag{2.6}$$

with $f_r^o(t) = f_r(t, x_0(t), x_0'(t))$, $f_y^o(t) = f_y(t, x_0(t), x_0'(t))$, and $2\omega^o(t, \eta, \rho)$ is the quadratic functional

$$2\omega^o(t, \eta, \rho) = f_{rr}^o(t)\rho^2 + 2f_{ry}^o(t)\rho y + f_{yy}^o y^2, \tag{2.7}$$

and where f_{rr}^o, f_{ry}^o, f_{yy}^o denote the respective second order partial derivatives of f with arguments $(t, x_0(t), x_0'(t))$.

From (2.5) it follows, (see, for example, Bliss [7, Ch. I] or Bolza [2, Ch. I]), that there exists a constant c such that

$$f_r^o(t) = \int_{t_1}^{t} f_y^o(s)ds + c, \quad t \in [t_1, t_2], \tag{2.8}$$

which implies that between values of t defining corners on C_0 the function f_r^o is differentiable with derivative f_y^o; that is, on such subintervals $x_0(t)$ satisfies the (Euler) differential equation

$$\frac{d}{dt} f_r(t,x_o(t),x_o'(t)) - f_y(t,x_o(t),x_o'(t)) = 0. \qquad (2.9)$$

Indeed, under the hypotheses stated above for f, with the aid of an implicit function theorem it follows that on a sub-interval between corners throughout which the non-singularity conditions $f^o_{rr} \neq 0$ holds the function $x_o(t)$ has continuous derivatives of the first two orders. If an arc C_o is without corners and satisfies (2.9) on $[t_1,t_2]$ then C_o is called an *extremal* arc for the considered variational problem, the term resulting from the fact that as a candidate for being a minimizing arc for $J[C]$ in the class of admissible arcs such an arc C_o satisfies the "first necessary condition" (2.5) for such an extremizing arc.

The connecting link between the calculus of variations and the theory of linear differential equations is the condition (2.6). At an early stage in the development of variational theory there was considered the problem of reducing the second variation $J_2[\eta]$ to a form from which it could be readily verified that condition (2.6) holds, or the stronger positive definite property that $J_2[\eta] > 0$ for all non-identically vanishing admissible variations η. In 1786, Legendre [1] noted that if η is an admissible variation, and w is an arbitrary function of t, (which from the standpoint of mathematical rigor should obviously be assumed to possess certain properties of differentiability), the integral of $(w\eta^2)' = 2w\eta\eta' + w'\eta^2$ over $[t_1,t_2]$ is zero, and hence the value of $J_2[\eta]$ is unaltered when $2w\eta\eta' + w'\eta^2$ is added to the integrand $2\omega^o$ of (1.6). Moreover, if w is a solution of the first order non-linear differential equation

$$[f^o_{ry}(t) + w(t)]^2 - f^o_{rr}(t)[f^o_{yy}(t) + w'] = 0. \qquad (2.10)$$

then upon such adjunction the integrand of the modified expression for $J_2[\eta]$ would be non-negative and condition (2.6) would hold. The equation (2.10), which is known as "Legendre's differential equation" for the variational integral (2.2), is a special instance of a so-called "Riccati differential equation".

Employing such a w, Legendre proceeded to write $J_2[\eta]$ in the form

$$J_2[\eta] = \int_{t_1}^{t_2} f^o_{rr}\{\eta' + ([f^o_{ry}+w]/f^o_{rr})\eta\}^2 dt, \qquad (2.11)$$

from which he concluded that if $f^o_{rr} > 0$ along an admissible arc C_o then the second variation would be positive for η a non-identically vanishing admissible variation, and indeed in this case C_o would render a minimum to $J[C]$. Legendre's argument was in error, however, as there does not always exist a solution w of (2.10) on the interval $[t_1,t_2]$. Moreover, at that point in time there had not been a precise formulation of the class of arcs in which the minimization of $J[\eta]$ was to be considered. Actually, Legendre's conclusion that existence of the form (2.11) for the second variation with $f^o_{rr}(t) > 0$ on $[t_1,t_2]$ would insure a minimum, is true if "minimum" is interpreted as "weak relative minimum" in the present accepted sense of this latter term. Also, the condition $f^o_{rr}(t) \geq 0$ along an admissible arc C_o is necessary if C_o is to furnish a weak relative minimum to $J[C]$, but is not sufficient, and is known as the "Legendre necessary condition".

In 1837, some fifty years after Legendre's initial attack

on the problem, Jacobi [1] established the circumstances under which Legendre's transformation of $J_2[\eta]$ to the form (2.11) was valid, and characterized the cases in which it is not possible. Briefly, on an extremal arc with fixed initial point $P_1 = (t_1,x_1)$ there is in general a point $P' = (t_1',x_1')$ which is "conjugate to P_1," and such that Legendre's transformation is possible on any subarc of P_1P' excluding P', but not possible on any subarc P_1P including P'. In the derivation of this condition Jacobi introduced the linear second order differential equation

$$\begin{aligned} L[u](t) \equiv [f^o_{rr}(t)u'(t) + f^o_{ry}(t)u(t)]' \\ - [f^o_{yr}(t)u'(t) + f^o_{yy}(t)u(t)] = 0 \end{aligned} \tag{2.12}$$

in place of the Riccati equation (2.10) used by Legendre. Appropriately, (2.12) is called the "Jacobi equation" for the variational integral (2.2). In particular, if η is continuous and continuously differentiable throughout a subinterval $[s_1,s_2]$ of $[t_1,t_2]$ then $2\omega^o(t,\eta(t),\eta'(t)) = \{\eta[f^o_{rr}\eta' + f^o_{ry}\eta]\}' - \eta L[\eta]$ on this subinterval, and

$$\int_{s_1}^{s_2} 2\omega^o(t,\eta(t),\eta'(t))dt = \eta[f^o_{rr}\eta'+f^o_{yr}\eta]\Big|_{s_1}^{s_2} - \int_{s_1}^{s_2}\eta L[\eta]dt. \tag{2.13}$$

With the aid of this formula, which is known as the *Jacobi transformation of the second variation,* Jacobi then obtained the result that if there exists a non-identically vanishing solution $u = u_o(t)$ of (2.12) which is zero at distinct values s_1 and s_2, $(s_1 < s_2)$, on $[t_1,t_2]$, then η_o defined as $\eta_o(t) = u_o(t)$ on $[s_1,s_2]$, $\eta_o(t) = 0$ on $[t_1,s_1] \cup [s_2,t_2]$ is a particular admissible variation for which $J_2[\eta_o] = 0$. Also, Jacobi showed that if $f^o_{rr} \neq 0$ on

$[t_1,t_2]$, and there exists a solution $u = u(t)$ of (2.12) which is different from zero throughout $[t_1,t_2]$, then for η an arbitrary admissible variation $J_2[\eta]$ may be written as

$$J_2[\eta] = \int_{t_1}^{t_2} f^o_{rr}(t)[\eta'(t) - \frac{u'(t)}{u(t)}\eta(t)]^2 dt. \qquad (2.14)$$

The interrelations between the Jacobi equation (2.12) and the Legendre equation (2.10) are as follows:

(i) if on a subinterval $[s_1,s_2]$ of $[t_1,t_2]$ we have $f^o_{rr} \neq 0$ and $u = u(t)$ is a solution of (2.12) which is non-zero on $[s_1,s_2]$. Then $w(t) = -[f^o_{rr}(t)u'(t) + f^o_{ry}(t)u(t)]/u(t)$ is a solution of (2.10) on $[s_1,s_2]$;

(ii) if on a subinterval $[s_1,s_2]$ of $[t_1,t_2]$ we have $f^o_{rr} \neq 0$, and $w = w(t)$ is a solution of (2.10) on $[s_1,s_2]$, then a function $u = u(t)$ which satisfies the first order differential system

$$f^o_{rr}(t)u'(t) + [f^o_{ry}(t) + w(t)]u(t) = 0, \quad u(\tau) \neq 0 \quad \text{for some } \tau \in [s_1,s_2],$$

is a solution of (2.12) which is non-zero on $[s_1,s_2]$.

The reader is referred to Bolza [2, Ch. I] and Bliss [1,5,7] for more detailed discussions of the contributions of Legendre and Jacobi to variational theory, both in regard to the overall extent of results and their shortcomings as to mathematical rigor by today's standards. The study of the analogue of the Jacobi transformation of the second variation of more complicated "Lagrange type problems" involving auxiliary differential equations as restraints was initiated by Clebsch [1,2] in 1858, followed by a simplifying treatment by A. Mayer [1] in 1868, and a much more thorough discussion by von Escherich [1] in a sequence of five papers in the years 1898-1901.

The conversion of a two-point boundary problem to an integral equation through the use of a Green's function provided for Hilbert [1] a ready application of his theory of integral equations in the first decade of the twentieth century. In the development of his theory of integral equations with real symmetric kernels, and its application to self-adjoint boundary problems, Hilbert pointed out the interrelations with variational theory. In particular, from the extremizing properties of the eigenvalues as deducible from the expansion theory associated with such integral equations, it follows that for an extremal arc C_0 of (2.2) which is non-singular in the sense that $f^o_{rr}(t) \neq 0$ on $[t_1,t_2]$ the non-negativeness, {positive definiteness}, of the second variation $J_2[\eta]$ on the class of admissible variations η is equivalent to the positiveness of f^o_{rr} on $[t_1,t_2]$ and the non-negativeness, {positiveness}, of the smallest eigenvalue λ_1 of the associated boundary problem

$$L[u;\lambda](t) \equiv L[u](t) + \lambda u(t) = 0, \ u(t_1) = 0 = u(t_2). \quad (2.15)$$

In particular, in 1906 Hilbert suggested that for Sturm-Liouville boundary problems the oscillation properties of the eigenfunctions were consequences of the Jacobi condition imposed on these functions by the fact that they afforded a minimum to certain associated problems of isoperimetric type in the calculus of variations. This question was considered by Robert König [1] in his 1907 Göttingen dissertation, but König's treatment contained errors. Consequelty, Hilbert proposed to R. G. D. Richardson the reconsideration of the problem, and Richardson [1] established the validity of Hilbert's conjecture for the particular boundary problem

$$[r(t)u'(t)]' - p(t)u(t) + \lambda k(t)u(t) = 0, \; u(0)=0=u(1), \tag{2.16}$$

wherein r,p,k were supposed to be real-valued, analytic, with $r(t) > 0$, $p(t) \geq 0$ and $k(t) \neq 0$ for $t \in [0,1]$.

For self-adjoint boundary problems involving a real second-order differential equation linear in the parameter, and two-point boundary conditions independent of the parameter, Mason [2] gave a proof of the existence of eigenvalues, each of which was characterized as the minimizing function for a quadratic functional in a suitable class of functions satisfying the set of boundary conditions of the problem and associated integral orthogonality conditions. His proof was the first such treatment that did not rely upon integral equation theory. In nature his proof was indirect, in that with the aid of the solvability theorems for such differential systems he was able to show that if the infimum of the quadratic functional on an appropriate class of functions was not an eigenvalue, then there would exist functions of the class which negated the definitive property of the infimum. For the case of a self-adjoint differential equation of the second order whose coefficient functions were periodic of common period ω, his results implied the existence of the sequence of eigenvalues for which the equation has solutions that are of period ω, and by supplementary argument using results of Sturm he established the oscillation properties characteristic of the individual eigenfunctions.

Historically, the 1920 survey paper [1] of Bliss is noteworthy for his comments on the ability of viewing the problem of non-negativeness of the second variation functional as a minimum problem within a minimum problem, together with remarks on his success only shortly before that date of verifying the validity of this approach for the more complicated Lagrange type problem, as well as for the simpler problems of the calculus of variations. For a variable end-point problem involving an integral functional of the form (2.2), but with end-values allowed to vary on two given curves in the (t,x)-plane, this program was carried out in Bliss [2]. For this particular calculus of variations problem the second variational functional along an extremal arc $C_o : x = x_o(t)$, $t \in [t_1,t_2]$, is of the form

$$J_2[\eta] = Q_2\eta^2(t_2) - Q_1\eta^2(t_1) + \int_{t_1}^{t_2} 2\omega^o(t,\eta(t),\eta'(t))dt, \quad (2.17)$$

where $2\omega^o$ is as in (2.7), Q_1 and Q_2 are constants determined by the elements of C_o and the points of contact with the prescribed end-curves, and the "Jacobi condition" becomes that of determining conditions for the non-negativeness and the positive definiteness of (2.17) in the class of arcs η which are piecewise continuously differentiable on $[t_1,t_2]$. Under the assumption that $f^o_{rr}(t) > 0$ on $[t_1,t_2]$, Bliss used methods of the calculus of variations to show that the non-negativeness, {positive definiteness}, of (2.17) on this class of arcs η was equivalent to the non-negativeness, {positiveness}, of the smallest eigenvalue λ_1 of the Sturm-Liouville boundary problem

$$\begin{aligned} L[u](t) + \lambda u(t) &= 0, \\ Q_1u(t_1) + f^o_{rr}(t_1)u'(t_1) + f^o_{ry}(t_1)u(t_1) &= 0, \\ Q_2u(t_2) + f^o_{rr}(t_2)u'(t_2) + f^o_{ry}(t_2)u(t_2) &= 0. \end{aligned} \quad (2.18)$$

In view of the comments in the latter portion of Bliss [1], at the time of its writing he evidently considered the Clebsch-von-Escherich transformation theory of the second variation for Lagrange type problems to be elaborate, if not excessively laborious. It is of interest to note that less than five years later, in Bliss [3], he presented a proof of this transformation that may truly be termed elegant.

In accord with the above-expressed view of Hilbert on possible interrelations between the calculus of variations and the theory of boundary problems, in the Introduction of [2] Bliss writes, "It seems likely that a complete theory of self-adjoint boundary value problems for ordinary differential equations, with end conditions of very general type, can be deduced from theorems already well known in the calculus of variations." For the accessory boundary problem associated with a fixed end-point non-parametric variational problem in (n+1)-space this procedure was first carried out by Hickson [1] in his 1928 Chicago dissertation under Bliss; it is to be commented that for the proof of higher order eigenvalues characterized by the property of minimizing the second variation functional in a class of arcs satisfying appropriate integral orthogonality conditions, the thus formulated problem of isoperimetric type was transformed into a Lagrange type problem, to which he applied results established for such problems by Bliss in a set of 1925 lectures, and which subsequently was published as Bliss [5]. Undoubtedly the above mentioned elegant proof of Bliss of the transformation theory of the second variation was a non-trivial aid in the writing of Hickson [1], and other Chicago dissertations of the same general period dealing with similar problems for various of the more sophisticated simple integral problems of the calculus of variations, (see, for example, M. E. Stark [1]; T. F. Cope [1]; R. L. Jackson [1]).

Near the end of the Introduction to Hickson [1], and subsequent to his comment that his results provided a generalization of the existence theorem of R. G. Richardson [1], he adds, "One would not expect to find a direct generalization of his (Richardson's) oscillation theorem, but it is very probable that there exist results for a system of ordinary linear differential equations of the second order with a parameter, which would be somewhat analogous to his oscillation theorem". This prophecy was almost immediately realized in the paper [1] of Morse, wherein for self-adjoint second order linear differential systems there were presented generalizations of the separation, comparison, and oscillation theorems in the Sturmian theory for real second-order linear differential equations. Indeed, this paper of Morse and the related discussion of Chapter IV of Morse [4], may truly be said to form the corner stone for subsequent extensions and elaborations of the classical Sturmian theory.

Briefly, the basic concepts of the Morse theory are illustrated by the following example. If E is an extremal arc along which $f^{o}_{rr}(t) > 0$, then the number of values on the open interval (t_1,t_2) defining points on E conjugate to the initial point $P_1 = (t_1,x_1)$ is equal to the number of negative eigenvalues of the boundary problem (2.15), and this number is the greatest integer k for which there is a k-dimensional subspace of $\mathcal{V}_0$ on which $J[\eta]$ is negative definite. Specifically, with the aid of "broken extremal arcs" Morse presented a finite dimensional space S_N such that if $\eta \in \mathscr{V}_0$ then there is a unique $v = v(\ ;\eta) \in S_N$ such that $J_2[v(\ ,\eta)]$ is a real quadratic form $Q[v] \equiv \sum_{i,j=1}^{N} Q_{ij} v_i v_j$ on $S_N \times S_N$, while $v(\ ;v(\ ;\eta)) = v(\ ;\eta)$ and $J_2[\eta] \geq J_2[v(\ ;\eta)]$ with the equality sign holding only if $\eta = v(\ ;\eta)$. Consequently, the problem of determining the number of values on (t_1,t_2) defining points on E conjugate to P_1 is reduced to finding the number of negative eigenvalues of the algebraic problem $\sum_{j=1}^{n} Q_{ij} v_j - \lambda v_i = 0$, $(i = 1,\ldots,n)$. As Morse [1, p. 61] notes, in the consideration of variational problems Hahn [1,2] and Rozenberg [1] had employed broken extremal arcs with a single intermediate vertex. In this connection, it is to be remarked that Hahn [1, p. 110] notes that his use of broken extremals was prompted by the earlier use of this device by Scheeffer [1] in studying the second variation. Thus for the linear boundary problems of the Sturmian theory, the cited paper of Scheeffer is indeed a direct precursor of Morse's work.

As indicated above for a simple problem of the calculus of variations the Jacobi conditions on the non-negativeness and positive definiteness of the second variation functional may be phrased in terms of a boundary problem. For a detailed discussion of this approach for a simple problem the reader is referred to Lovitt [1; Ch. VI in particular]; it is to be remarked that this volume is actually the published form of Lovitt's notes on the lectures of Bolza on integral equations in the summer of 1913 at the University of Chicago. Lichtenstein ([1], [3], [4]), following the methods of Hilbert [1], utilized boundary problems in deriving sufficient conditions for a weak relative minimum in certain problems of the calculus of variations. Boerner ([2], [3]) applied the procedures of Lichtenstein to problems involving higher derivatives and to the parametric problem in the plane with fixed end-points. As indicated above, Bliss [2] expressed the Jacobi condition for a plane calculus of variations problem in terms of the eigenvalues of a boundary problem, and several of his studies continued this program for more sophisticated variational problems. For the use of this formulation of the Jacobi condition in connection with sufficiency theorems for more general problems of the calculus of variations, the reader is referred to Morse ([3], [4], [5]), Myers [1], Birkhoff and Hestenes [1], Bliss ([5], [7]), and Reid ([4], [5], [33]). A comparison of different formulations of the Jacobi condition is presented in Reid [3]. It is of interest to note that the boundary problem used by Reid ([4], [33]) in treating discontinuous solutions in the non-parametric problem involves boundary conditions at more than two points. By a simple transformation, however, this problem is reducible to one involving boundary conditions at two points.

3. Historical Comments on Terminology

The following remarks concern the classification of certain points relative to a given linear homogeneous second order differential equation

$$L[u](t) \equiv u''(t) + p(t)u'(t) + q(t)u(t) = 0 \tag{3.1}$$

where for definiteness it will be assumed that the coefficient functions $p(t)$ and $q(t)$ are real-valued and continuous on a given interval I of the real line. For a given

$t = a$ on I, let $u_0(t)$ and $u_1(t)$ be solutions of $L[u] = 0$ satisfying the initial conditions

$$\begin{aligned} &\text{(i)} \quad u_0(a) = 0, \quad u_0'(a) \neq 0, \\ &\text{(ii)} \quad u_1(a) \neq 0, \quad u_1'(a) = 0. \end{aligned} \tag{3.2}$$

A value $t = b$ on I is called a *right-hand*, {*left-hand*} *conjugate point* to $t = a$ if $b > a$, $\{b < a\}$, and $u_0(b) = 0$. This concept of conjugate point seemingly goes back to Weierstrass, who in his 1879 lectures introduced the term in a corresponding situation for the Jacobi equation in a plane parametric variational problem. Correspondingly, a value $t = b$ is called a *right-hand* {*left-hand*} *focal point* to $t = a$ if $b > a$, $\{b < a\}$, and $u_1(b) = 0$. The English term focal appeared in the early 1900's, and dates from A. Kneser's use of *Brennpunkte* in German for the consideration of the Jacobi equation for a parametric variational problem wherein an endpoint was restricted to lie on a given surface. These remarks are in agreement with comments by Kneser in his 1900 Enzyklopädie article on the calculus of variations; see also, Bolza [1, footnote on p. 109].

Picone [1; Sec. 1] introduced further terms for a more extended classification of points that has not been widely used. Specifically, in Picone's terminology, a value $t = b$ is called a *right-hand*, {*left-hand*} *pseudoconjugate*, (pseudo-coniugato), of $t = a$ if $b > a$, $\{b < a\}$, and $u_0'(b) = 0$. Also, a value $t = b$ is called a *right-hand* {*left-hand*} *hemiconjugate*, (emiconiugato) to $t = a$ if $b > a$, $\{b < a\}$, and $u_1(b) = 0$. Finally, Picone called $t = b$ a *right-hand* {*left-hand*} *deconjugate* (deconiugato) to $t = a$ if $b > a$,

$\{b < a\}$, and $u_1'(b) = 0$. Thus in Picone's terminology, a focal point to a is a hemiconjugate to a. So far as the author is aware, subsequent to Picone's papers [1,2] there was no use of this extended classification. Indeed, it appears that the only subsequent comprehensive terminology corresponding to that of Picone is due to Borůvka, ([4, §3 of Ch. I]), who uses the term "conjugate point of the first, second, third and fourth class", with the associated right- or left-hand designations, for the respective Picone conjugate, deconjugate, pseudoconjugate, and hemiconjugate. Within recent years there has been some application of specific terms to the values Picone termed pseudoconjugates. Certain authors have referred to these values as "focal to a", but this is definitely in contradiction to the long-time usage of "focal", and should be discouraged. Leighton [13] has used the term σ-point for this concept. Also, Leighton [10] uses the term "f-point" for the value which in Picone's terminology is the first right-hand deconjugate to a.

CHAPTER II. STURMIAN THEORY FOR REAL LINEAR HOMOGENEOUS SECOND ORDER ORDINARY DIFFERENTIAL EQUATIONS ON A COMPACT INTERVAL

1. Introduction

The differential equations to be considered in this chapter are of the form

$$\ell[u](t) \equiv [r(t)u'(t) + q(t)u(t)]' - [q(t)u'(t) + p(t)u(t)] = 0, \quad t \in I, \tag{1.1}$$

where on a given non-degenerate interval I on the real line the coefficient functions r, p, q satisfy one of the following conditions.

($\mathscr{A}_C$) r, p, q *are real-valued and continuous on* I, *with* $r(t) \neq 0$ *on this interval.*

($\mathscr{A}_L$) r, p, q *are real-valued, Lebesgue measurable on* I *with* $r(t) \neq 0$ *on this interval, while* $1/r$, q/r *and* $p - q^2/r$ *are locally Lebesgue integrable, i.e., they are integrable on arbitrary compact subintervals of* I.

The conditions of ($\mathscr{A}_C$) imply those of ($\mathscr{A}_L$), and in the following discussion, unless stated otherwise, the presented results hold under ($\mathscr{A}_L$). Indeed, it is to be emphasized that when the concept of solution is made clear the details of proof under ($\mathscr{A}_L$) vary little, if any, from those employed when the stronger hypothesis ($\mathscr{A}_C$) holds, so the reader unfamiliar with the Lebesgue integral need feel no handicap in this regard since it may be assumed that hypothesis ($\mathscr{A}_C$) is satisfied.

Under either of the above stated hypotheses, by a "solution u of (1.1)" is meant a continuous function u for which there is an associated function v such that

$$r(t)u'(t)+q(t)u(t) = v(t), \quad v'(t) = q(t)u'(t)+p(t)u(t), \quad t \in I.$$

That is, u is a solution of (1.1) if and only if there is an associated function v such that $(u;v)$ is a solution of the first order system

$$\begin{aligned} \ell_1[u,v](t) &\equiv -v'(t) + c(t)u(t) - a(t)v(t) = 0, \\ \ell_2[u,v](t) &\equiv u'(t) - a(t)u(t) - b(t)v(t) = 0, \end{aligned} \quad t \in I. \tag{1.2}$$

where the coefficient functions in (1.2) are

$$a = -q/r, \quad b = 1/r, \quad c = p - q^2/r. \tag{1.3}$$

Under hypothesis ($\mathscr{A}_C$) these functions are all continuous with $b(t) \neq 0$ on I, and $(u;v)$ is a solution of (1.2) if and only if these functions are continuously differentiable and satisfy equations (1.2) throughout I. Under hypothesis ($\mathscr{A}_L$) these coefficient functions are locally Lebesgue integrable on I, and by definition a solution of (1.2) is a pair of locally a.c., (absolutely continuous), functions satisfying (1.2) a.e., (almost everywhere), on I; equivalently, u and v are continuous functions which for arbitrary $t_0 \in I$ satisfy the integral equations

$$\begin{aligned} v(t) &= v(t_0) + \int_{t_0}^{t} \{c(s)u(s) - a(s)v(s)\}ds, \\ u(t) &= u(t_0) + \int_{t_0}^{t} \{a(s)u(s) + b(s)v(s)\}ds, \end{aligned} \qquad t \in I. \tag{1.4}$$

This concept of solution is due to Carathéodory, [1; Ch. XI], and for a treatment in English of differential equations employing this concept of solution the reader is referred to such texts as Coddington and Levinson [1, Ch. II], or Reid [35, Ch. II]. Under either hypothesis, for $t_0 \in I$ and arbitrary real u_0, v_0 there is a unique solution $(u;v)$ of (1.2) satisfying $u(t_0) = u_0$, $v(t_0) = v_0$. Under hypothesis ($\mathscr{A}_C$) such initial conditions are clearly equivalent to corresponding initial conditions $u(t_0) = u_0$, $u'(t_0) = u_0'$, where $r(t_0)u_0' + q(t_0)u_0 = v_0$.

An intermediate case, which is subsumed under hypothesis ($\mathscr{A}_L$), is that wherein the following hypothesis is satisfied.

($\mathscr{A}_{PC}$) r, p, q *are real-valued and piecewise continuous on* I, *with* $r(t) \neq 0$ *and* $1/r(t)$ *locally bounded on* I.

Whenever ($\mathscr{A}_{PC}$) holds and $(u;v)$ is a solution of (1.2), then for each $t \in I$ these functions have right- and left-hand derivatives which with the corresponding unilateral limit values of the coefficient functions satisfy the equations of (1.2). In this case, the initial values of a solution (u,v) at a $t_0 \in I$ may equally well be phrased as initial values of u and a unilateral derivative $u'(t_0^+)$ or $u'(t_0^-)$.

It is to be noted that hypothesis ($\mathscr{A}_C$) implies that $r(t)$ is of constant sign on I, whereas ($\mathscr{A}_L$) and ($\mathscr{A}_{PC}$) do not imply this restriction. For brevity, the notations ($\mathscr{A}_C^+$), ($\mathscr{A}_L^+$), ($\mathscr{A}_{PC}^+$) will denote the corresponding hypothesis ($\mathscr{A}_C$),

$(\mathscr{A}_L)$, $(\mathscr{A}_{PC})$ with the added restriction that $r(t) > 0$ for $t \in I$.

Let $y = (y_\alpha)$, $(\alpha = 1,2)$, denote the two-dimensional vector function with $y_1 = u$ and $y_2 = v$. Then (1.2) may be written as the vector equation

$$\mathscr{J}y'(t) + \mathscr{A}(t)y(t) = 0, \quad t \in I, \tag{1.2}$$

where the constant matrix J and the 2×2 real symmetric matrix function $\mathscr{A}(t)$ are given by

$$\mathscr{J} = \begin{bmatrix} 0 & -1 \\ 1 & 0 \end{bmatrix}, \quad \mathscr{A}(t) = \begin{bmatrix} c(t) & -a(t) \\ -a(t) & -b(t) \end{bmatrix}. \tag{1.5}$$

It is to be emphasized that the theory of an equation (1.1) is no more general, or no less general, than that for an equation

$$\ell^0[u](t) \equiv [r(t)u'(t)]' - p(t)u(t) = 0, \quad t \in I, \tag{1.1^0}$$

wherein $q(t) \equiv 0$, in which case $a(t) \equiv 0$ and the system (1.2) becomes

$$\begin{aligned} \ell_1^0[u,v](t) &\equiv -v'(t) + c(t)u(t) = 0, \\ \ell_2^0[u,v](t) &\equiv u'(t) - b(t)v(t) = 0, \end{aligned} \quad t \in I, \tag{1.2^0}$$

Indeed, if either $(\mathscr{A}_C)$ or $(\mathscr{A}_L)$ is satisfied, and $f(t) = \exp\{-\int_\tau^t [q(s)/r(s)]ds\}$, then for $\tilde{u} = (1/f)u$ we have

$$f\ell[u] = \tilde{\ell}[\tilde{u}], \text{ where } \tilde{\ell}[\tilde{u}](t) = [\tilde{r}(t)\tilde{u}'(t)]'-\tilde{p}(t)\tilde{u}(t), \tag{1.6$_1$}$$

with

$$\tilde{r} = rf^2, \; \tilde{p} = [p-q^2/r]f^2 = cf^2. \tag{1.7$_1$}$$

Correspondingly, under the transformation

$$\tilde{u} = (1/f)u, \quad \tilde{v} = fv, \tag{1.8}$$

we have $f\ell_1[u,v] = \tilde{\ell}_1[\tilde{u},\tilde{v}]$, $(1/r)\ell_2[u,v] = \tilde{\ell}_2[\tilde{u},\tilde{v}]$, where

$$\begin{aligned} \tilde{\ell}_1[\tilde{u},\tilde{v}](t) &= -\tilde{v}'(t) + \tilde{c}(t)\tilde{u}(t), \\ \tilde{\ell}_2[\tilde{u},\tilde{v}](t) &= \tilde{u}'(t) - \tilde{b}(t)\tilde{v}(t) \end{aligned} \tag{1.6$_2$}$$

with

$$\tilde{b} = b/f^2, \quad \tilde{c} = cf^2 = p. \tag{1.7$_2$}$$

In particular, it is to be noted that whenever r, p, q satisfy hypothesis $(\mathscr{A}_C)$, $(\mathscr{A}_L)$, $(\mathscr{A}_{PC})$, $(\mathscr{A}_C^+)$, $(\mathscr{A}_L^+)$, or $(\mathscr{A}_{PC}^+)$, the new coefficient functions $r, p, q \equiv 0$ satisfy the same hypothesis.

The above reduction of (1.1) to (1.1^0) has been accomplished by a "change of the dependent variable u." Using a "change of the independent variable" an equation of the form (1.1^0) may be reduced further to one of the form

$$\ell^{\#}[u](t) \equiv u''(t) - p(t)u(t) = 0, \quad t \in I^{\#}, \tag{1.1$^{\#}$}$$

wherein $r(t) \equiv 1$, in which case the system (1.2^0) becomes

$$\begin{aligned} \ell_1^{\#}[u,v](t) &\equiv -v'(t) + c(t)u(t) = 0, \\ \ell_2^{\#}[u,v](t) &\equiv u'(t) - v(t) = 0. \end{aligned} \quad t \in I^{\#}, \tag{1.2$^{\#}$}$$

For $\tau \in I$ the integral

$$s = \int_\tau^t \frac{d\sigma}{r(\sigma)} = \int_\tau^t b(\sigma)d\sigma, \quad t \in I \tag{1.9}$$

defines a function $s: I \to \mathbf{R}$ that is strictly monotone increasing, and which maps I onto an interval $I^{\#}$. Let $T: I^{\#} \to \mathbf{R}$ denote the inverse function of s. If r is positive and continuous on I then T is continuously differentiable and $T'(s) = r(T(s))$ on $I^{\#}$. If r merely satisfies the condition of $(\mathscr{A}_L)$, then T is locally a.c. and $T'(s) = r(T(s))$ a.e. on $I^\#$. Moreover, $u^\#: I^\# \to \mathbf{R}$ defined by $u^\#(s) = u(T(s))$ is such that

$$r(t)\ell^0[u(t)]\Big|^{t=T(s)} = D^2u^\#(s) - p^\#(s)u^\#(s) \tag{1.10}$$

where D^2 denotes the second order derivative with respect to s, and

$$p^\#(s) = r(T(s))p(T(s)), \quad s \in I^\#. \tag{1.11}$$

In particular, if the coefficient functions r and p in (1.1^0) are continuous on I then $p^\#$ of (1.9) is continuous on $I^\#$. If r and p merely satisfy the conditions of $(\mathscr{A}_L)$ then $p^\#$ of (1.9) is locally integrable on $I^\#$; indeed, if $[s_1,s_2]$ is a compact subinterval of $I^\#$ then

$$\int_{s_1}^{s_2} p^\#(s)ds = \int_{s_1}^{s_2} p(T(s))T'(s)ds = \int_{T(s_1)}^{T(s_2)} p(t)dt.$$

It is to be remarked that if the coefficient functions of (1.1) satisfy the intermediate hypothesis $(\mathscr{A}_{PC})$ then the coefficients of the above associated equations (1.1^0) and $(1.1^\#)$ also satisfy this hypothesis.

Consequently results on solutions of a differential equation of the form (1.1), (1.1^0) or $(1.1^\#)$ with coefficient functions satisfying either $(\mathscr{A}_C)$, $(\mathscr{A}_L)$, or $(\mathscr{A}_{PC})$ are readily translatable into solutions of either of the other of these forms with coefficient functions satisfying the same hypothesis.

If in (1.1) the coefficient functions r and q are continuously differentiable on I, then this equation may be written as

$$r(t)u''(t) + r'(t)u'(t) + [q'(t) - p(t)]u(t) = 0,$$

with continuous coefficients. Conversely, if real-valued functions $p_j(t)$, $(j = 0,1,2)$ are continuous and $p_2(t) \neq 0$ for $t \in I$, then for $\tau \in I$ the function $\mu(t) = [1/p_2(t)]\exp\{\int_\tau^t [p_1(s)/p_2(s)]ds\}$ is such that

$$\mu[p_2u'' + p_1u' + p_ou] = [ru']' - pu \tag{1.12}$$

where

$$r(t) = \exp\{\int_\tau^t [p_1(s)/p_2(s)]ds\},\ p(t) = -r(t)p_o(t)/p_2(t), \tag{1.13}$$

and thus the differential equation

$$p_2(t)u''(t) + p_1(t)u'(t) + p_o(t)u(t) = 0,\ t \in I, \tag{1.14}$$

is transformable to the form (1.1^o) with coefficients satisfying hypothesis ($\mathscr{A}_C$). If the real-valued functions p_j are merely Lebesgue measurable with $p_2(t) \neq 0$ on I and the functions p_o/p_2, p_1/p_2 locally integrable on this interval, then the same transformation reduces (1.14) to an equation (1.1^o) with coefficients satisfying hypothesis ($\mathscr{A}_L$).

2. Preliminary Properties of Solutions of (1.1).

The following properties of solutions of (1.1) are readily derivable from the above definitions, where it is to be noted that in case of oscillation phenomena individual results for (1.1) or (1.2) are equivalent to the corresponding results for the equation $\tilde{\ell}[\tilde{u}] = 0$ defined by (1.6_1) or the system $\tilde{\ell}_1[\tilde{u},\tilde{v}] = 0$, $\tilde{\ell}_2[\tilde{u},\tilde{v}] = 0$ defined by (1.6_2) and satisfied by the associated u, v of (1.8). In particular, the equation

$$\tilde{u}(t_2) - \tilde{u}(t_1) = \int_{t_1}^{t_2} \tilde{b}(s)\tilde{v}(s)ds, \text{ for } t_1 \in I,\ t_2 \in I$$

implies immediately the following separation theorems.

THEOREM 2.1. *If* $r(t)$ *is of constant sign on* I *and* u *is a non-identically vanishing solution of* (1.1), *then the zeros of* u *are isolated; moreover, between any two zeros of* u *there is a zero of* $v = ru' + qu$.

The following results may be verified directly.

THEOREM 2.2. *If* u_α, $(\alpha = 1,2)$, *are solutions of* (1.1) *and* $y^{(\alpha)} = (u_\alpha, v_\alpha \equiv ru_\alpha' + qu_\alpha)^T$ (T is transpose), *then*

$$\{u_1,u_2\}(t) \equiv y^{(2)*}Jy^{(1)}(t) \equiv v_2(t)u_1(t) - u_2(t)v_1(t)$$

is constant on I; *moreover,* $\{u,u\}(t) \equiv 0$ *for* u *an arbitrary solution. Also, if* u_1 *and* u_2 *are solutions of* (1.1) *and* $u_1(t) \not\equiv 0$, *then* $\{u_1,u_2\} = 0$ *if and only if there is a constant* k *such that* $u_2 = ku_1$ *on* I.

In particular, if u_1 and u_2 are real linearly independent solutions of (1.1), with $v_\alpha = ru_\alpha' + qu_\alpha$ as in Theorem 2.2, then on a subinterval I_0 of I throughout which $u_1(t) \neq 0$ we have $\{u_1,u_2\} \neq 0$ and the function $\phi = u_2/u_1$ is such that $\phi' = \{u_1,u_2\}/[ru_1^2]$, so that ϕ is monotone on I_0. In particular, if I_0 is the open interval (s_1,s_2), with s_1 and s_2 consecutive zeros of u_1, then since $u_2(s_\alpha) \neq 0$, $(\alpha = 1,2)$, we have that $|\phi(t)| \to +\infty$ as $t \to s_\alpha$, $(\alpha = 1,2)$, and consequently one of these limiting values is $+\infty$, the other is $-\infty$, and there is a unique value $s \in (s_1,s_2)$ at which $\phi(s) = 0$ and therefore $u_2(s) = 0$. Thus we have the following preliminary separation theorem.

THEOREM 2.3. *If* u_1, u_2 *are real linearly independent solutions of* (1.1), *then the zeros of* u_1 *and* u_2 *separate each other.*

The following result may also be verified directly.

THEOREM 2.4. *If* u_1 *is a solution of* (1.1) *such that* $u_1(t) \neq 0$ *for* t *on a subinterval* I_o *of* I, *and* $t_o \in I_o$, *then* $u(t)$ *is a solution of* (1.1) *on* I_o *if and only if*

$$\begin{aligned} u(t) = u_1(t)h(t), \text{ with } h(t) &= k_o + k_1\int_{t_o}^{t} \frac{b(s)}{u_1^2(s)}ds \\ &= k_o + k_1\int_{t_o}^{t} \frac{ds}{r(s)u_1^2(s)}, \end{aligned} \tag{2.1}$$

where k_o *and* k_1 *are constants; in particular, the corresponding* $v = ru' + qu$ *is given by*

$$\begin{aligned} v(t) = v_1(t)h(t)+k_1/u_1(t), \text{ and } k_o &= u(t_o)/u_1(t_o), \\ k_1 &= \{u_1,u\}. \end{aligned} \tag{2.2}$$

In particular, if u_1, u are linearly independent solutions of (1.1) with $\tau \in I$ a zero of $u_1(t)$, then since $h(t)$ defined by (2.1) is such that $|h(t)| \to +\infty$ as $t \to \tau$, it follows that if $t_o \in I$ is such that $u_1(t) \neq 0$ for $t \in [t_o,\tau)$ or for $t \in (\tau,t_o]$ according as $t_o < \tau$ or $t_o > \tau$, then the integral $\int_{t_o}^{t} \frac{ds}{r(s)u_1^2(s)}$ diverges as $t \to \tau$.

If t_1 and t_2 are distinct values on I, then these values are said to be *conjugate*, {with respect to (1.1)}, if there exists a non-identically vanishing solution u of this equation such that $u(t_1) = 0 = u(t_2)$. If I_o is a subinterval of $[a,b]$ and there exist no pair of points of I_o which are conjugate with respect to (1.1), then this equation is said to be *nonoscillatory* on I_o, or *disconjugate* on

I_o. This latter designation, which is due to Wintner [7], has been widely adopted and will be used throughout the present discussion. In view of the result of Theorem 2.3, we have the following condition for disconjugacy.

THEOREM 2.5. *If* $[a,b]$ *is a compact subinterval of* I, *and* $(u;v) = (u_a;v_a)$ *is a solution of* (1.2) *satisfying* $u_a(a) = 0$, $v_a(a) \neq 0$, *then* (1.1) *is disconjugate on* $[a,b]$, $\{$*on* $(a,b)\}$, *if and only if* $u_a(t) \neq 0$ *on* $(a,b]$, $\{$*on* $(a,b)\}$. *Correspondingly, if* $(u;v) = (u_b;v_b)$ *is a solution of* (1.1) *satisfying the initial condition* $u_b(b) = 0$, $v_b(b) \neq 0$, *then* (1.1) *is disconjugate on* $[a,b]$, $\{$*on* $(a,b)\}$, *if and only if* $u_b(t) \neq 0$ *on* $[a,b)$, $\{$*on* $(a,b)\}$.

Now suppose that $u_a(t) \neq 0$ for $t \in (a,b]$, and $u_b(t) \neq 0$ for $t \in [a,b)$. Then $\{u_a,u_b\} \neq 0$, and upon selecting properly $v_a(a)$ and/or $v_b(b)$ the value of this constant function is negative. With this choice it follows with the aid of Theorem 2.4 that $u_a(t)$ and $u_b(t)$ are of the same algebraic sign on the open interval (a,b), and consequently $(u;v) = (u_a + u_b, v_a + v_b)$ is a solution of (1.2) with $u(t) \neq 0$ for all $t \in [a,b]$. Hence we have the following corollary to Theorem 2.5.

COROLLARY. *An equation* (1.1) *is disconjugate on a subinterval* $[a,b]$ *of* I *if and only if there exists a real solution* $u(t)$ *of this equation which is non-zero throughout* $[a,b]$.

If hypothesis $(\mathscr{U}_C)$ holds and $u(t)$, $v(t) = r(t)u'(t) + q(t)u(t)$ are continuously differentiable with $u(t) \neq 0$ on a subinterval I_o of I, then on this subinterval $w(t) = v(t)/u(t)$ is continuously differentiable and

$$u(t)\ell[u](t) = u^2(t)k[w](t), \tag{2.3}$$

where $k[w]$ is the Riccati formal differential operator

$$k[w](t) = w'(t) + 2a(t)w(t) + b(t)w^2(t) - c(t). \tag{2.4}$$

In particular, if $w(t)$ is a real solution of the Riccati differential equation

$$k[w](t) = 0 \tag{2.5}$$

on I_0, and $u(t) = \exp\{\int_\tau^t [a(s) + b(s)w(s)]ds\}$ for some $\tau \in I_0$, then $u(t), v(t) = w(t)u(t)$ is a real solution of (1.1) with $u(t) \neq 0$ on I_0. Whenever hypothesis $(\mathscr{A}_L)$ holds and $u(t)$, $v(t)$ are locally a.c. functions on a subinterval I_0 of I with $v(t) = r(t)u'(t) + q(t)u(t)$ a.e. on this subinterval, then the above stated results remain valid with statements on differentiability holding a.e. on I_0 and solutions of occurring differential equations interpreted in the Carathéodory sense. Consequently, in view of the above Corollary we have the following result.

THEOREM 2.6. *An equation* (1.1) *is disconjugate on a subinterval* $[a,b]$ *of* I *if and only if there exists on this subinterval a real solution of the Riccati differential equation* (2.5).

In regard to the general solution of the Riccati differential equation, we have the following basic result.

THEOREM 2.7. *If* $w_0(t)$ *is a solution of* (2.5) *on a subinterval* I_0 *of* I, *and for* $\tau \in I_0$ *the functions* $g = g(t,\tau:w_0)$, $\theta = \theta(t,\tau:w_0)$ *are defined as*

$$g(t,\tau:w_0) = \exp\{-\int_\tau^t \{a(s) + b(s)w_0(s)\}ds\}, \tag{2.6}$$

$$\theta(t,\tau:w_0) = \int_\tau^t b(s)g^2(s,\tau:w_0)ds \tag{2.7}$$

then w *is a solution of* (2.5) *on* I_0 *if and only if the constant* $\gamma = w(\tau) - w_0(\tau)$ *is such that* $1 + \gamma\theta(t,\tau:w_0) \neq 0$ *for* $t \in I_0$, *and*

$$w(t) = w_0(t) + \frac{g^2(t,\tau:w_0)}{1 + \gamma\theta(t,\tau:w_0)}, \quad t \in I_0. \tag{2.8}$$

In particular, $w(t) \neq w_0(t)$ *for all* $t \in I_0$ *if there exists a value* $\tau \in I_0$ *such that* $w(\tau) \neq w_0(\tau)$.

3. The Classical Oscillation and Comparison Theorems of Sturm

The major results of Sturm [1] are concerned with interrelations between the character of solutions of two differential equations

$$\ell_\alpha^0[u](t) \equiv [r_\alpha(t)u'(t)]' - p_\alpha(t)u(t) = 0, \; t \in I, \tag{3.1$_\alpha$}$$

where for $\alpha = 1,2$ the functions $r = r_\alpha$, $p = p_\alpha$, $q \equiv 0$ satisfy hypothesis ($\mathscr{H}_C$); that is, r_α and p_α are continuous and $r_\alpha(t) > 0$, $(\alpha = 1,2)$ for t on a given non-degenerate interval I. The following result in the case of $\ell_\alpha^0[u_\alpha] = 0$, $\alpha = 1,2$, was established initially by Picone [2, p. 20], and may be verified directly by differentiation.

LEMMA 3.1. {PICONE'S IDENTITY}. *If* u_α *and* $v_\alpha = r_\alpha u_\alpha'$, $(\alpha = 1,2)$, *are differentiable on a non-degenerate subinterval* I_0 *and* $u_2(t) \neq 0$ *for* $t \in I_0$, *then*

$$\left\{u_1^2 \frac{r_2u_2'}{u_2} - u_1r_1u_1'\right\}' + [r_1-r_2]u_1'^2 + [p_1-p_2]u_1^2 + r_2\left[u_1'-u_2'\frac{u_1}{u_2}\right]^2$$
$$= \frac{u_1}{u_2}\{u_1\ell_2^0[u_2] - u_2\ell_1^0[u_1]\}. \tag{3.2}$$

Now suppose that for $\alpha = 1,2$ the function u_α is a solution of $\ell_\alpha^o[u_\alpha] = 0$, with $t = c$ and $t = d$ consecutive zeros of u_1, while $u_2(t) \neq 0$ for $t \in (c,d)$. In this case, the right-hand member of (3.2) is zero on $[c,d]$. Also the function $u_1^2 r_2 u_2'/u_2 - u_1 r_1 u_1'$ has limiting values at $t = c$ and $t = d$ equal to zero. This result, which is obvious if u_2 is non-zero at each of these endpoints, is also valid in case u_2 vanishes at an end-point since in this case, the function has at such an end-point the non-zero finite limit of $u_1'(t)/u_2'(t)$ at this endpoint.

Let ($\mathscr{A}_1$) denote the following hypothesis.

($\mathscr{A}_1$) *For* $\alpha = 1,2$ *the functions* $r = r_\alpha$, $p = p_\alpha$, $q \equiv 0$ *satisfy* ($\mathscr{A}_C$) *on a non-degenerate interval* I, *and*

$$r_1(t) \geq r_2(t),\ p_1(t) \geq p_2(t) \text{ for } t \in I. \tag{3.3}$$

In view of the Picone identity and the above comments, we then have the following result.

LEMMA 3.2. *Suppose that hypothesis* ($\mathscr{A}_1$) *is satisfied, and* u_1 *is a real solution of* (3.1_1) *with consecutive zeros at* $t = c$ *and* $t = d$. *If* u_2 *is a real solution of* (3.1_2), *then either:*

(i) *there exists a value* $s \in (c,d)$ *such that* $u_2(s) = 0$, *or,*

(ii) $[r_1(t) - r_2(t)]u_1'(t) \equiv 0$, $[p_1(t) - p_2(t)] \equiv 0$ *on* (c,d), *and there exists a non-zero constant* κ *such that* $u_1(t) \equiv \kappa u_2(t)$ *on this interval.* (3.4)

Indeed, if conclusion (3.4i) does not hold then $u_2(t) \neq 0$ for $t \in (c,d)$, and upon integrating the left-hand member of (3.2) over the compact subinterval $[c+\varepsilon,d-\varepsilon]$ of (c,d) and letting $\varepsilon \to 0$ it follows that

$$\int_c^d \{[r_1 - r_2]u_1'^2 + [p_1 - p_2]u_1^2 + r_2(u_2[\frac{u_1}{u_2}]')^2\}dt = 0, \quad (3.5)$$

and conclusion (3.4ii) is a consequence of the non-negativeness of each of the three terms in the integrand of (3.5).

It is desirable to determine conditions which, together with $(\mathcal{A}_1)$, exclude the result of (3.4ii) and thus insure the existence of a value on the open interval (c,d) at which u_2 vanishes. In his consideration of this problem, Bôcher [7, Ch. III, 14] employed hypotheses of the following forms.

$(\mathcal{A}_2)$ *If* I_0 *is a non-degenerate subinterval of* I, *then either* $r_1(t) \not\equiv r_2(t)$ *or* $p_1(t) \not\equiv p_2(t)$ *on* I_o.

$(\mathcal{A}_3)$ *If* I_0 *is a non-degenerate subinterval of* I, *then the condition* $p_1(t) \equiv p_2(t) \equiv 0$ *does not hold on* I_0.

THEOREM 3.1. *Suppose that hypothesis* $(\mathcal{A}_1)$ *is satisfied, and* u_1 *is a real solution of* (3.1_1) *with consecutive zeros at* $t = c$ *and* $t = d$. *If* u_2 *is a real solution of* (3.1_2) *then* u_2 *has at least one zero on the open interval* (c,d) *if either:*

(i) *hypothesis* $(\mathcal{A}_2)$ *is satisfied, or*

(ii) *hypothesis* $(\mathcal{A}_3)$ *is satisfied, and the conditions* $r_1(t) \equiv r_2(t)$, $p_1(t) \equiv p_2(t)$ *do not hold throughout* (c,d).

Indeed, suppose that hypotheses $(\mathcal{A}_1)$ and $(\mathcal{A}_2)$ hold. Since $u_1(t) \neq 0$ for $t \in (c,d)$, and $u_1'(c) \neq 0$, it follows that there exists a value $t_0 \in (c,d)$ such that $u_1'(t) \neq 0$ and $u_1(t) \neq 0$ for $t \in (c,t_0]$, and the first two relations of (3.4ii) imply that $r_1(t) \equiv r_2(t)$ and $p_1(t) \equiv p_2(t)$ on $(c,t_0]$, contrary to $(\mathcal{A}_2)$. Consequently, whenever hypotheses

$(\mathscr{A}_1)$ and $(\mathscr{A}_2)$ hold the assumption that u_2 is non-zero on the open interval (c,d) has led to a contradiction, and thus conclusion (i) is established.

Now suppose that hypotheses $(\mathscr{A}_1)$ and $(\mathscr{A}_3)$ are satisfied, and there exists a value $t_0 \in (c,d)$ such that $r_1(t_0) > r_2(t_0)$. Then there is a subinterval $[t_1,t_2]$ of (c,d) such that $r_1(t) - r_2(t) > 0$ for $t \in [t_1,t_2]$. The first identity of (3.4ii) then implies that $u_1'(t) \equiv 0$ on $[t_1,t_2]$, and consequently $p_1(t)u_1(t) = [r_1(t)u_1'(t)]' \equiv 0$ on $[t_1,t_2]$. As the zeros of u_1 are isolated it then follows that $p_1(t) \equiv 0$ on $[t_1,t_2]$, and from the second identity of (3.4ii) it then follows that both p_1 and p_2 are identically zero on $[t_1,t_2]$, contrary to $(\mathscr{A}_3)$. That is, whenever hypotheses $(\mathscr{A}_1)$ and $(\mathscr{A}_3)$ hold the assumption that u_2 is non-zero on (c,d) implies a contradiction unless both $r_1(t) \equiv r_2(t)$ and $p_1(t) \equiv p_2(t)$ on (c,d), thus establishing conclusion (ii) of the theorem.

If $r(t)$ and $p(t)$ are constant functions $r(t) \equiv r_0 > 0$ and $p(t) \equiv p_0$ on I, then for $\tau \in I$ the solution of $\ell^0[u] = 0$ satisfying $u(\tau) = 0$, $u'(\tau) = 1$ is given by $u(t) = t - \tau$ if $p_0 = 0$, $u(t) = \sinh(\sqrt{p_0/r_0}\,[t-\tau])$ if $p_0 > 0$, and $u(t) = \sin(\sqrt{-p_0/r_0}\,[t-\tau])$ if $p_0 < 0$. Consequently, application of results of Theorem 3.1 to the equation $\ell^0[u] = 0$ and the equation $r_0 u'' - p_0 u = 0$ for appropriate values of r_0, p_0 yields the following result.

COROLLARY. *If hypothesis* $(\mathscr{A}_1)$ *is satisfied, and* $[c,d]$ *is a non-degenerate subinterval of* I, *then:*

(a) (1.1^0) *is disconjugate on* $[c,d]$ *if*

$$\frac{\text{Min}\{p(t) : t \in [c,d]\}}{\text{Min}\{r(t) : t \in [c,d]\}} > -\frac{\pi^2}{(d-c)^2};$$

(b) *an arbitrary solution of* (1.1^o) *has at least* m *zeros on* (c,d) *if*

$$\frac{\text{Max}\{p(t) : t \in [c,d]\}}{\text{Max}\{r(t) : t \in [c,d]\}} < -\frac{m^2\pi^2}{(d-c)^2}.$$

It is to be noted that in the above proof of Theorem 3.1 the effect of the two hypotheses $(\mathcal{A}_1)$ and $(\mathcal{A}_2)$ could be interpreted as the following property.

$(\mathcal{A}_2^*)$ *If* u_1 *is a non-identically vanishing solution of* (3.1_1) *and* $\tau \in I$ *is such that* $u_1(\tau) = 0$, *then for any non-degenerate subinterval* I_o *of* I *with* τ *an end-point of* I_o *we have*

$$J_{1,2}[u_1;I_o] \equiv \int_{I_o} \{[r_1(t) - r_2(t)]u_1'^2(t) + [p_1(t) - p_2(t)]u_1^2(t)\}dt > 0. \tag{3.6}$$

A stronger form of this condition is the following hypothesis $(\mathcal{A}_3^*)$, and the latter portion of the proof of Theorem 3.1 may be used to show that if hypotheses $(\mathcal{A}_1)$ and $(\mathcal{A}_3)$ hold then $(\mathcal{A}_3^*)$ also holds.

$(\mathcal{A}_3^*)$ *If* u_1 *is a non-identically vanishing solution of* (3.1), *then for any non-degenerate subinterval* I_o *of* I *inequality* (3.6) *holds.*

These conditions are mentioned here, because they are intimately related to conditions which appear in subsequent sections concerned with variational methods. Also from the standpoint of generality, they will be employed in the following classical Sturmian comparison theorems.

THEOREM 3.2. *Suppose that hypotheses* $(\mathscr{A}_1)$ *and* $(\mathscr{A}_2^*)$ *hold on* I, *and for* $\alpha = 1,2$ *the function* $u = u_\alpha$ *is a non-identically vanishing solution of* $\ell_\alpha^0[u] = 0$ *on a compact subinterval* $[a,b]$ *such that* $u_\alpha(t)$, $v_\alpha(t) = r_\alpha(t)u'_\alpha(t)$ *satisfy the following initial condition:*

$$\text{If } u_1(a) \neq 0 \text{ then } u_2(a) \neq 0, \text{ and } \frac{v_1(a)}{u_1(a)} \geq \frac{v_2(a)}{u_2(a)}. \tag{3.7}$$

Then we have the following results.

(a) (FIRST COMPARISON THEOREM OF STURM). *If* u_1 *has exactly* $m \geq 1$ *zeros* $t = t_j^1$, $(j = 1,\ldots,m)$, *with* $a < t_1^1 < \ldots < t_m^1 < b$, *then* u_2 *has* $r \geq m$ *zeros* $t = t_k^2$, $(k = 1,\ldots,r)$, *with* $a < t_1^2 < \ldots < t_r^2 < b$ *and* $t_j^2 < t_j^1$ *for* $j = 1,\ldots,m$.

(b) (SECOND COMPARISON THEOREM OF STURM). *If there exists a value* $c \in (a,b]$ *such that* $u_1(c) \neq 0$, $u_2(c) \neq 0$, *while* u_1 *and* u_2 *have the same number* $m \geq 1$ *of zeros on the open interval* (a,c) *then*

$$\frac{v_1(c)}{u_1(c)} > \frac{v_2(c)}{u_2(c)} . \tag{3.8}$$

If $u_1(t) \neq 0$ *and* $u_2(t) \neq 0$ *on* $(a,c]$, *then*

$$u_1^2(c)\left[\frac{v_1(c)}{u_1(c)} - \frac{v_2(c)}{u_1(c)}\right] \geq J_{1,2}[u_1;[a,c]],$$

so that $v_1(c)/u_1(c) \geq v_2(c)/u_2(c)$, *and the strict inequality* (3.8) *persists if hypotheses* $(\mathscr{A}_1)$ *and* $(\mathscr{A}_3^*)$ *hold.*

No details of proof of the results of this theorem will be given, as they are of the same type as those appearing in the proof of Theorem 3.1. Also, in essentially this form they are to be found in many references, (for example; Bôcher [7, Ch. III, Secs. 13, 14], Ince [1, Ch. X, Secs. 10.3, 10.4]).

4. Related Oscillation and Comparison Theorems

For a real solution u of a linear homogeneous second order differential equation Sturm [1] also considered oscillation and comparison properties of linear forms in u and u'. Such results are discussed in Bôcher [5; 6; 9, Ch. III, Sec. 12] and presented in a problem set in Ince [1, pp. 251-252]; also, an important reference for such problems is Whyburn [1, Sec. 3]. This topic will be discussed here briefly, with specific details restricted to the case of solutions u of equations (1.1^0), or equivalently to solutions $u(t)$, $v(t) = r(t)u'(t)$ of the system (1.2^0). That is, we now consider the behavior of certain functions $\Phi = \Phi(\;;u,v)$ of the form

$$\Phi(\;;u,v) = \phi_2(t)u - \phi_1(t)v, \tag{4.1}$$

where ϕ_1 and ϕ_2 are real-valued continuously differentiable functions on an interval I throughout which the functions $r, p, q \equiv 0$ satisfy hypothesis $(\mathscr{A}_C)$. It is to be remarked that the employed notation differs slightly from that of either Bôcher or Whyburn.

If $u(t)$ is a solution of (1.1^0), or equivalently $u(t)$, $v(t) = r(t)u'(t)$ is a solution of (1.2^0), we have

$$\Phi'(\;;u,v) = -u\ell_1^0[\phi_1,\phi_2] - v\ell_2^0[\phi_1,\phi_2]. \tag{4.2}$$

For brevity, we set $\langle\phi_1,\phi_2;u,v\rangle = u\ell_1^0[\phi_1,\phi_2] + v\ell_2^0[\phi_1,\phi_2]$, and abbreviate $\langle\phi_1,\phi_2;\phi_1,\phi_2\rangle = \langle\phi_1,\phi_2\rangle$.

Now if Φ possesses an infinite number of zeros on a compact subinterval, and τ is a point of this interval which is a limit point of the set of zeros of Φ, then it follows that $\Phi(\tau;u,v) = 0$ and $\Phi'(\tau;u,v) = 0$. As the determinant of coefficients of the pair of linear forms $\Phi(\ ;u,v)$ and $\Phi'(\ ;u,v)$ in u,v is equal to $-\phi_1\ell_1^o[\phi_1,\phi_2] - \phi_2\ell_2^o[\phi_1,\phi_2] = -\langle\phi_1,\phi_2\rangle$, it follows that if $\langle\phi_1,\phi_2\rangle(t) \neq 0$ for t on a non-degenerate subinterval I_o of I then for a non-identically vanishing solution u of (1.1^o) the corresponding function Φ of (4.1) has at most a finite number of zeros on any compact subinterval of I_o, and each zero is isolated. The assumption that $\langle\phi_1,\phi_2\rangle(t) \neq 0$ on a subinterval I_o clearly implies that ϕ_1 and ϕ_2 are not both zero at any point of this subinterval. Moreover, if u_1 and u_2 are linearly independent solutions of (1.1^o), one may verify readily that the Wronskian $\Phi(\ ;u_1,v_1)\Phi'(\ ;u_2,v_2) - \Phi(\ ;u_2,v_2)\Phi'(\ ;u_1,v_1)$ is equal to the product of $\langle\phi_1,\phi_2\rangle$ and the non-zero constant $\{u_1,u_2\} = v_2u_1 - u_1v_2$, so that $\Phi(\ ;u_1,v_1)$ and $\Phi(\ ;u_2,v_2)$ are linearly independent functions throughout a subinterval I_o on which $\langle\phi_1,\phi_2\rangle(t) \neq 0$. In particular, if t_1 and t_2 are successive zeros of $\Phi(\ ;u_1,v_1)$ on such a subinterval I_o, then on the open interval (t_1,t_2) the function $\Phi(\ ;u_2,v_2)/\Phi(\ ;u_1,v_1)$ has derivative equal to $\{u_1,u_2\}\langle\phi_1,\phi_2\rangle/\Phi^2(\ ;u_1,v_1)$, and consequently this function is strictly monotone, and tends to opposite infinite values at the endpoints of this interval. Consequently, there is a unique value $s \in (t_1,t_2)$ at which $\Phi(s;u_2,v_2) = 0$. That is, if u_1, u_2 are linearly independent solutions of (1.1^o) then on any subinterval I_o throughout which $\langle\phi_1,\phi_2\rangle$ is nonvanishing the zeros of $\Phi(\ ;u_1,v_1)$ and $\Phi(\ ;u_2,v_2)$ separate each other.

Now consider together with $\Phi(\ ;u,v)$ a second similar linear form

$$\Psi(\ ;u,v) = \psi_2(t)u - \psi_1(t)v, \tag{4.3}$$

where $\psi_1(t)$ and $\psi_2(t)$ are real-valued continuously differentiable functions on I such that $\delta(\phi_1,\phi_2;\psi_1,\psi_2) \equiv \phi_1\psi_2 - \phi_2\psi_1$ is different from zero on a non-degenerate subinterval I_o. Direct computation, which is greatly facilitated by the use of matrix notation, then yields the result that if u is a solution of (1.1^o) on I, then on I_o the pair of functions $(\Phi;\Psi) = (\Phi(\ ;u,v);\Psi(\ ;u,v))$ is a solution of the differential system

$$\begin{aligned} -\Psi'(t) + [\langle\psi_1,\psi_2\rangle/\delta](t)\Phi(t) - [\langle\psi_1,\psi_2;\phi_1,\phi_2\rangle/\delta](t)\Psi(t) &= 0, \\ \Phi'(t) - [\langle\phi_1,\phi_2;\psi_1,\psi_2\rangle/\delta](t)\Phi(t) + [\langle\phi_1,\phi_2\rangle/\delta](t)\Psi(t) &= 0. \end{aligned} \tag{4.4}$$

Therefore, if

$$\begin{aligned} \omega_1(t) &= -\int_a^t [\langle\phi_1,\phi_2;\psi_1,\psi_2\rangle/\delta](s)ds, \\ \omega_2(t) &= \int_a^t [\langle\psi_1,\psi_2;\phi_1,\phi_2\rangle/\delta](s)ds, \end{aligned} \tag{4.5}$$

it follows that $(\Phi(t);\Psi(t))$ is a solution of (4.4) if and only if

$$\Phi_1(t) = \Phi(t)\exp\omega_1(t), \quad \Psi_1(t) = \Psi(t)\exp\omega_2(t) \tag{4.6}$$

is such that $(\Phi_1(t);\Psi_1(t))$ is a solution of the differential system

$$-\Psi_1'(t) + c_o(t)\Phi_1(t) = 0, \quad \Phi_1'(t) - b_o(t)\Psi_1(t) = 0, \tag{4.7}$$

where

$$\begin{aligned} b_o(t) &= -\frac{\langle\phi_1,\phi_2\rangle(t)}{\delta(t)}\exp\{\omega_1(t) - \omega_2(t)\}, \\ c_o(t) &= \frac{\langle\psi_1,\psi_2\rangle(t)}{\delta(t)}\exp\{\omega_2(t) - \omega_1(t)\}. \end{aligned} \tag{4.8}$$

Now if $(\Phi_1(t);\Psi_1(t))$ is a solution of (4.7) the function $\Omega(t) = \Phi_1(t)\Psi_1(t)$ satisfies the equation

$$\Omega'(t) = b_0(t)\Psi_1^2(t) + c_0(t)\Phi_1^2(t). \tag{4.9}$$

In particular, if $b_0(t)$ and $c_0(t)$ are both positive or both negative on a subinterval of I_0, then on this subinterval $\Omega(t)$ has at most one zero. Translated in terms of the initial functions $\Phi(\ ;u,v)$ and $\Psi(\ ;u,v)$ we have the following results.

THEOREM 4.1. *Suppose that* $\phi_1,\phi_2,\psi_1,\psi_2$ *are continuously differentiable real-valued functions on* I *such that on a non-degenerate subinterval* I_0 *of* I *we have*

$$\delta = \delta(\phi_1,\phi_2;\psi_1,\psi_2) \neq 0, \ \langle\phi_1,\phi_2\rangle \neq 0, \ \langle\psi_1,\psi_2\rangle \neq 0 \quad \textit{for} \ t \in I_0. \tag{4.10}$$

If u(t) *is a non-identically vanishing solution of* (1.1^0) *on* I, *then:*

(i) *on* I_0 *the zeros of* $\Phi(\ ;u,v)$ *and* $\Psi(\ ;u,v)$ *separate;*

(ii) *if* $\langle\phi_1,\phi_2\rangle(t)$ *and* $\langle\psi_1,\psi_2\rangle(t)$ *are of opposite signs for* $t \in I_0$, *then on this subinterval each of the functions* $\Phi(\ ;u,v)$ *and* $\Psi(\ ;u,v)$ *can have at most one zero, and if one of these functions is zero for some value on* I_0 *then the other function is different from zero throughout this subinterval.*

In particular, if $\psi_1(t) \equiv 0$, $\psi_2(t) \equiv 1$, $\phi_1(t) \equiv 1$, then for arbitrary real-valued continuously differentiable functions on a non-degenerate subinterval I_0 of I we have

$$\begin{aligned} \langle\phi_1,\phi_2\rangle(t) &\equiv -\phi_2'(t) + p(t) - [1/r(t)]\phi_2^2(t), \\ \langle\psi_1,\psi_2\rangle(t) &\equiv -[1/r(t)], \quad \delta(\phi_1,\phi_2;\psi_1,\psi_2) \equiv 1, \end{aligned} \tag{4.11}$$

and therefore as a corollary to conclusion (ii) of the above theorem we have the following result.

COROLLARY. *If the functions* $r, p, q \equiv 0$ *satisfy* $(\mathscr{A}_C)$ *on* I, *and on a non-degenerate subinterval* I_0 *there exists a solution* $\phi(t)$ *of the Riccati differential inequality*

$$\phi'(t) + [1/r(t)]\phi^2(t) - p(t) < 0 \tag{4.12}$$

then for a non-identically vanishing solution $u(t)$ *of* (1.1^0) *neither* $u(t)$ *nor* $\phi(t)u(t) - r(t)u'(t)$ *can have more than one zero on* I_0, *and if one of these functions is zero for some value on* I_0 *then the other function is different from zero throughout this subinterval.*

5. Sturmian Differential Systems

Suppose that for open intervals $I = (a^0, b^0)$ and $\Delta = \{\lambda : \Lambda_1 < \lambda < \Lambda_2\}$ on the real line the functions $r(t,\lambda)$, $p(t,\lambda)$, $\alpha_a(\lambda)$ and $\alpha_b(\lambda)$ satisfy the following conditions:

$(\mathscr{C}_0)$ $r(t,\lambda)$ *and* $p(t,\lambda)$ *are real-valued continuous functions with* $r(t,\lambda) > 0$ *for* $(t,\lambda) \in I \times \Delta$;

$(\mathscr{C}_\alpha)$ $\alpha_a(\lambda)$ *and* $\beta_a(\lambda)$ *are real-valued continuous functions such that* $\alpha_a^2(\lambda) + \beta_a^2(\lambda) > 0$ *for* $\lambda \in \Delta$, *and either* $\alpha_a(\lambda) \equiv 0$ *on* Δ, *or* $\alpha_a(\lambda) \neq 0$ *for all* $\lambda \in \Delta$.

Moreover, let $u = u(t,\lambda)$ be the solution of the differential equation

$$\ell^0[u:\lambda](t) \equiv [r(t,\lambda)u'(t)]' - p(t,\lambda)u(t) = 0, \ t \in I, \tag{5.1}$$

determined by the initial conditions

$$u(a,\lambda) = \alpha_a(\lambda), \quad v(a,\lambda) = \beta_a(\lambda), \quad \lambda \in \Delta, \tag{5.2}$$

where $t = a$ is a point in I, which remains fixed throughout the following discussion, and $v(t,\lambda)$ denotes the function $r(t,\lambda)u(t,\lambda)$.

By well-known existence theorems, (see, for example, Coddington and Levinson [1, Ch. I, Sec. 7], Hartman [13, Ch. V, Sec. 2], Reid [35, Ch. I, Sec. 5]), the functions $u(t,\lambda)$ and $v(t,\lambda)$ are continuous in (t,λ) on $I \times \Delta$. Now suppose that conditions $(\mathscr{C}_o)$ and $(\mathscr{C}_\alpha)$ hold, and denote by $t_j(\lambda)$ the j-th zero of $u(t,\lambda)$ on (a,b^0), with $a < t_1(\lambda) < t_2(\lambda) < \dots < b^0$ whenever this zero exists. Since $u'(t,a) = \partial u(t,a)/\partial t$ is non-zero for $t = t_j(\lambda)$, by an elementary continuity argument it follows that if for $\lambda = \lambda_0$ the j-th zero $t_j(\lambda_0)$ of $u(t,\lambda_0)$ exists then there is a neighborhood $N(\lambda_0) = \{\lambda : \lambda \in \Delta, |\lambda - \lambda| < \delta\}$ such that $t_j(\lambda)$ exists and is a continuous function of λ on $N(\lambda_0)$.

It will now be supposed that in addition to the above conditions the following hypotheses are also satisfied.

$(\mathscr{C}_a^+)$ *The functions* $\alpha_a(\lambda)$, $\beta_a(\lambda)$ *satisfy* $(\mathscr{C}_\alpha)$; *moreover, if* $\alpha_a(\lambda) \neq 0$ *for* $\lambda \in \Delta$ *then* $\beta_a(\lambda)/\alpha_a(\lambda)$ *is a monotone non-increasing function on* Δ.

$(\mathscr{C}_1)$ *For* $t \in I$ *the functions* $r(t,\lambda)$ *and* $p(t,\lambda)$ *are monotone non-increasing functions of* λ *on* Δ, *and if* $\Lambda_1 < \lambda' < \lambda'' < \Lambda_2$ *then for* I_0 *a non-degenerate subinterval of* I *either* $r(t,\lambda') \not\equiv r(t,\lambda'')$ *or* $p(t,\lambda') \not\equiv p(t,\lambda'')$ *on* I_0.

As a direct consequence of conclusion (i) of Theorem 3.1 and the above comments on the continuity of the zeros of $u(t,\lambda)$, it follows that whenever conditions $(\mathscr{C}_o)$, $(\mathscr{C}_a^+)$, $(\mathscr{C}_1)$ are satisfied and for $\lambda_0 \in \Delta$ the j-th zero $t_j(\lambda_0)$ of $u(t,\lambda_0)$ exists, then $t_j(\lambda)$ exists for $\lambda \in [\lambda_0,\Lambda_2)$, and is a strictly monotone decreasing function of this interval.

The basic results of this section concern the differential system

$$\ell^o[u:\lambda](t) \equiv [r(t,\lambda)u'(t)]' - p(t,\lambda)u(t) = 0,$$
$$B_a(\lambda) \equiv \beta_a(\lambda)u(a) - \alpha_a(\lambda)v(a) = 0, \qquad (5.3)$$
$$B_b(\lambda) \equiv \beta_b(\lambda)u(b) + \alpha_b(\lambda)v(b) = 0,$$

involving boundary conditions in the values of u and $v = ru'$ at the end-points of a compact non-degenerate subinterval $[a,b]$ of I. A value for which there exists a non-identically vanishing solution of (5.3) is called an *eigenvalue, (proper value or characteristic value)* of this system, and a corresponding non-identically vanishing solution is called an *eigenfunction, (proper function or characteristic function)*. For the consideration of (5.3) it will be understood that the coefficient functions $r(t,\lambda)$, $p(t,\lambda)$, $\alpha_a(\lambda)$ and $\beta_a(\lambda)$ satisfy the conditions $(\mathscr{C}_o)$, $(\mathscr{C}_a^+)$, $(\mathscr{C}_1)$, and that $\alpha_b(\lambda)$, $\beta_b(\lambda)$ satisfy the following condition.

$(\mathscr{C}_b^+)$ $\alpha_b(\lambda)$ *and* $\beta_b(\lambda)$ *are real-valued continuous functions such that* $\alpha_b^2(\lambda) + \beta_b^2(\lambda) > 0$ *for* $\lambda \in \Delta$, *and either* $\alpha_b(\lambda) \equiv 0$ *on* Δ *or* $\alpha_b(\lambda) \neq 0$ *for all* $\lambda \in \Delta$; *moreover, if* $\alpha_b(\lambda) \neq 0$ *on* Δ *then* $\beta_b(\lambda)/\alpha_b(\lambda)$ *is a monotone non-increasing function on* Δ.

Also, the following discussion will involve the following conditions for the functions $r(t,\lambda)$ and $p(t,\lambda)$.

$(\mathscr{C}_2)$ *There exists a non-degenerate compact subinterval* $[a_o,b_o]$ *of* $[a,b]$ *such that the continuous functions* $r(t,\lambda)$, $p(t,\lambda)$ *satisfy the condition*

$$\frac{\operatorname{Max}\{p(t,\lambda):t \in [a_0,b_0]\}}{\operatorname{Max}\{r(t,\lambda):t \in [a_0,b_0]\}} \to -\infty \text{ as } \lambda \to \Lambda_2. \tag{5.4}$$

($\mathscr{C}_3$) *The continuous functions* $r(t,\lambda)$, $p(t,\lambda)$ *satisfy on the compact subinterval* $[a,b]$ *of* I *the condition*

$$\frac{\operatorname{Min}\{p(t,\lambda):t \in [a,b]\}}{\operatorname{Min}\{r(t,\lambda):t \in [a,b]\}} \to \infty \text{ as } \lambda \to \Lambda_1. \tag{5.5}$$

The basic existence theorem for the boundary problem (5.3) is as follows.

THEOREM 5.1. *If hypotheses* ($\mathscr{C}_0$), ($\mathscr{C}_1$), ($\mathscr{C}_2$), ($\mathscr{C}_a^+$), *and* ($\mathscr{C}_b^+$) *are satisfied, then there exists a positive integer* m *such that the eigenvalues of* (5.3) *on* Δ *may be written as an increasing sequence* $\lambda_m < \lambda_{m+1} < \ldots$, *and for* $j \geq m$ *an eigenfunction* $u_j(t)$ *of* (5.3) *for* $\lambda = \lambda_j$ *has exactly* $j-1$ *zeros on the open interval* (a,b); *moreover,* $\{\lambda_j\} \to \Lambda_2$ *as* $j \to \infty$. *If the further condition* ($\mathscr{C}_3$) *is satisfied, then* $m = 1$.

A detailed proof of the above theorem will not be presented as like theorems, with little or no change in hypotheses may be found in various places, (see, for example, Bôcher [7, Ch. III, Sec. 15], Ince [1, Ch. X, Sec. 10.6], Reid [35, Ch. V, Sec. 7]. In view of interrelations with other methods to be presented later, it is to be noted that if $u(t,\lambda)$ is the solution of (5.1) satisfying the initial conditions (5.2) with associated $v(t,\lambda) = r(t,\lambda)u'(t,\lambda)$ then in view of hypothesis ($\mathscr{C}_2$) and conclusion (b) of the Corollary to Theorem 4.1, then for each positive integer j the j-th zero $t_j(\lambda)$ of $u(t,\lambda)$ exists on $[a,b^0)$, and satisfies $t_j(\lambda) < b$ for λ sufficiently large. Consequently, there exists a smallest positive integer k such that $t_k(\lambda) < b$ is not satisfied for λ in a suitably small neighborhood of Λ_1, and hence for $j = k, k + 1, \ldots$, there is a unique value $\lambda = \mu_j$ such that $t_j(\mu_j) = b$; moreover, $\mu_k < \mu_{k+1} < \ldots$ and $\mu_j \to \Lambda_2$ as $j \to \infty$. If $\alpha_b(\lambda) \equiv 0$, so that the boundary condition $B_b(\lambda)$ of (5.3) is $u(b) = 0$, the first conclusion of the theorem regarding the existence of eigenvalues and eigenfunctions holds for $m = k$ and $\lambda_j = \mu_j$, $(j = k, k+1, \ldots)$. In case $\alpha_b(\lambda) \neq 0$ for $\lambda \in \Delta$, conclusion (b) of Theorem 4.1 implies that for $j \geq k$ the function $v(b,\lambda)/u(b,\lambda)$ is strictly monotone decreasing on (μ_j, μ_{j+1}) and tends to ∞ and $-\infty$ as $\lambda \to \mu_j^+$ and $\lambda \to \mu_{j+1}^-$, respectively. Consequently, in view of condition $(\mathscr{C}_b^+)$, for $j \geq k$ there exists a unique value $\lambda = \lambda_{j+1}$ on (μ_j, μ_{j+1}) such that the second boundary condition of (5.3) holds for $\lambda = \lambda_{j+1}$, $u = u(t, \lambda_{j+1})$, $v = v(t, \lambda_{j+1})$, and $u(t, \lambda_{j+1})$ has exactly j zeros on (a,b). Now the function $-\beta_b(\lambda)/\alpha_b(\lambda)$ is monotone nondecreasing on (Λ_1, μ_k) in view of $(\mathscr{C}_b^+)$ and if

$$\lim_{\lambda \to \Lambda_1} [-\beta_b(\lambda)/\alpha_b(\lambda)] < \lim_{\lambda \to \Lambda_1} [v(b,\lambda)/u(b,\lambda)] \quad (5.6)$$

then (5.3) has a single eigenvalue λ_k on (Λ_1, μ_k), the corresponding eigenfunction $u(t, \lambda_k)$ has exactly $k - 1$ zeros on (a,b), and the first part of the theorem holds for $m = k$. On the other hand, if the inequality (5.6) does not hold then there is no eigenvalue of (5.3) on (Λ_1, μ_k), and the first part of the theorem holds for $m = k + 1$. Finally, whenever condition $(\mathscr{C}_3)$ holds the result that $m = 1$ is obtained by a comparison of the solutions of the differential equation of (5.3) with the solutions of the equation

$$\rho(\lambda)u''(t) - \tau(\lambda)u(t) = 0,$$

where $\rho(\lambda)$ and $\tau(\lambda)$ are for given $\lambda \in \Delta$ the minima on $[a,b]$ of $r(t,\lambda)$ and $p(t,\lambda)$, respectively.

Of particular interest are differential systems (5.3) in which $r(t,\lambda)$ is independent of λ, $p(t,\lambda)$ is linear in λ, while the $\alpha_a(\lambda)$, $\beta_a(\lambda)$, $\alpha_b(\lambda)$, $\beta_b(\lambda)$ are constants. Such a system, which is called a Sturm-Liouville system, will be written as

$$\begin{aligned} \ell^{oo}[u:\lambda](t) \equiv [r(t)u'(t)]' - [p(t)-\lambda k(t)]u(t) &= 0, \\ \beta_a u(a)-\alpha_a v(a) &= 0, \qquad (5.7) \\ \beta_b u(b)+\alpha_b v(b) &= 0, \end{aligned}$$

with the usual understanding that $v(t) = r(t)u'(t)$. Our consideration of this system will be under the following hypothesis.

($\mathscr{A}^o$) *On the compact interval* $[a,b]$ *the functions* r,p,k *are real-valued and continuous, with* $r(t) > 0$, $k(t) \not\equiv 0$, *while* α_a, β_a, α_b, β_b *are real constants such that* $\alpha_a^2 + \beta_a^2 > 0$, $\alpha_b^2 + \beta_b^2 > 0$.

In constrast to system (5.3), wherein the coefficient functions $r(t,\lambda)$ and $p(t,\lambda)$ were assumed to be defined only for $(t,\lambda) \in I \times \Delta$, where I and $\Delta = (\Lambda_1,\Lambda_2)$ were intervals on the real line, for (5.7) the function $r(t,\lambda) \equiv r(t)$ and $p(t,\lambda) = p(t)-\lambda k(t)$ are well-defined for complex values λ. Consequently, in (5.7) the parameter λ is allowed to have complex values, and any λ, real or complex, for which there is a non-identically vanishing solution $u(t)$ is called an *eigenvalue, (proper value or characteristic value)*, of this system and any non-identically vanishing solution is called a corresponding *eigenfunction, (proper function, or characteristic function)*.

If $u = u_0(t)$ is an eigenfunction of (5.7) for an eigenvalue $\lambda = \lambda_0$, then

$$\begin{aligned}\lambda_o\int_a^b k(t)|u_0(t)|^2dt &= \int_a^b u_0\{-[ru_0']' + pu_0\}dt \\ &= -u_0 ru_0'\Big|_a^b + \int_a^b \{r|u_0'|^2 + p|u_1|^2\}dt \qquad (5.8) \\ &= \gamma_a|u_0(a)|^2 + \gamma_b|u_0(b)|^2 + \int_a^b \{r|u_0'|^2+p|u_0|^2\}dt\end{aligned}$$

where $\gamma_a = 0$ if $\alpha_a = 0$, $\gamma_a = \beta_a/\alpha_a$ if $\alpha_a \neq 0$, $\gamma_b = 0$ if $\alpha_b = 0$, $\gamma_b = \beta_b/\alpha_b$ if $\alpha_b \neq 0$. The following result is a ready consequence of the relation (5.2).

THEOREM 5.2. *For a Sturm-Liouville system* (5.7) *all eigenvalues are real, and the eigenfunctions may be chosen real, in each of the following cases:*

(a) $k(t)$ *is of constant sign on* $[a,b]$;

(b) $k(t)$ *changes sign on* $[a,b]$, *while* $p(t) > 0$ *for* $t \in [a,b]$, *and* $\alpha_a\beta_a \geq 0$, $\alpha_b\beta_b \geq 0$.

If the coefficient functions r,p,k satisfy the conditions of ($\mathscr{A}^0$), then there exist real-valued extensions of these functions on an interval (a^0,b^0) containing $[a,b]$, and to which one might be able to apply the type of analysis used in considering (5.3). For example, if $r(t) > 0$ and $k(t) > 0$ for $t \in [a,b]$, let $r(t) = r(a)$, $k(t) = k(a)$, $p(t) = p(a)$ for $t \in (-\infty,a]$, and $r(t) = r(b)$, $k(t) = k(b)$, $p(t) = p(b)$ for $t \in [b,\infty)$. Then $r(t,\lambda) = r(t)$, $p(t,\lambda) = p(t) - \lambda k(t)$ are functions of (t,λ) on $(-\infty,\infty) \times (-\infty,\infty)$

which satisfy the conditions $(\mathscr{C}_0)$, $(\mathscr{C}_1)$, $(\mathscr{C}_2)$, $(\mathscr{C}_3)$, $(\mathscr{C}_a^+)$, $(\mathscr{C}_b^+)$ whenever hypothesis $(\mathscr{A}^0)$ holds, and hence we have the following result.

THEOREM 5.3. *Suppose that system* (5.7) *satisfies hypothesis* $(\mathscr{A}^0)$, *and* $k(t) > 0$ *for* $t \in [a,b]$. *Then all eigenvalues of this system are real, the totality of eigenvalues may be written as a sequence* $\{\lambda_j\}$, *where* $\lambda_1 < \lambda_2 < \dots$, $\{\lambda_j\} \to \infty$ *as* $j \to \infty$, *and an eigenfunction* $u = u_j(t)$ *of* (5.1) *for* $\lambda = \lambda_j$ *has exactly* $j - 1$ *zeros on the open interval* (a,b).

In case the function $k(t)$ changes sign on $[a,b]$ the results of Theorem 5.1 cannot be directly applied to (5.7). As pointed out initially by Bôcher [5, p. 173], however, there is a modification of the form of this system to which the above results are applicable to establish the following result.

THEOREM 5.4. *Suppose that system* (5.7) *satisfies hypothesis* $(\mathscr{A}^0)$, *the function* $k(t)$ *changes sign, while* $p(t) > 0$ *for* $t \in [a,b]$ *and* $\alpha_a\beta_a \geq 0$, $\alpha_b\beta_b \geq 0$. *Then all eigenvalues of* (5.7) *are real, the totality of proper values may be written as two sequences* $\{\lambda_j^+\}$, $\{\lambda_j^-\}$, *with* $0 < \lambda_1^+ < \lambda_2^+ < \dots$, $0 > \lambda_1^- > \lambda_2^- > \dots$, $\{\lambda_j^+\} \to \infty$ *and* $\{\lambda_j^-\} \to -\infty$ *as* $j \to \infty$, *and an eigenfunction* $u = u_j^+(t)$ *or* $u = u_j^-(t)$ *for respectively* $\lambda = \lambda_j^+$ *or* $\lambda = \lambda_j^-$, *has exactly* $j = 1$ *zeros on the open interval* (a,b).

Under the hypotheses of the theorem it follows from Theorem 5.2 that all eigenvalues λ of the system (5.7) are real, and the stated results on $\{\lambda_j^+, u_j^+(t)\}$, $j = 1,2,\dots$, may be deduced from Theorem 5.1 by considering for $\lambda \in (0,\infty)$ the system

$$\left[\frac{r(t)}{\lambda} u'(t)\right]' - \left[\frac{p(t)}{\lambda} - k(t)\right]u(t) = 0,$$
$$\frac{1}{\lambda} \beta_a u(a) - \alpha_a v(a) = 0, \qquad (5.9)$$
$$\frac{1}{\lambda} \beta_b u(b) + \alpha_b v(b) = 0,$$

where it is to be emphasized that now $v(t) = v(t,\lambda)$ is given by $v(t,\lambda) = (1/\lambda)r(t)u'(t,\lambda)$. Finally, the stated results on $\{\lambda_j^-, u_j^-(t)\}$, $j = 1,2,\ldots$, follow from the preceding case of positive eigenvalues for the associated system obtained from (5.7) upon replacing $k(t)$ by $-k(t)$.

The following result follows from Theorem 5.4 and a direct verification that in the application of Theorem 5.1 to (5.9) the conclusion $m = 1$ remains valid under the altered hypothesis.

COROLLARY. *If either* $\alpha_b = 0$ *or* $\alpha_a = 0$, *the result of Theorem 5.4 remains true when the condition on* $p(t)$ *is weakened to* $p(t) \geq 0$ *on* $[a,b]$. *In particular, if hypothesis* $(\mathscr{A}^0)$ *holds and* $p(t) \geq 0$ *on* $[a,b]$, *then the equation* $\ell^{00}[u:1] = 0$ *is disconjugate on* $[a,b]$, $\{$*on* $(a,b)\}$, *if and only if there is no positive eigenvalue* λ *of the boundary problem* $\ell^{00}[u:\lambda] = 0$, $u(a) = 0 = u(b)$ *satisfying* $\lambda \leq 1$, $\{\lambda < 1\}$.

6. Polar Coordinate Transformations

We return to the consideration of the general equation (1.1), or the associated system (1.2), with coefficients satisfying either $(\mathscr{A}_C)$ or $(\mathscr{A}_L)$.

Under the polar coordinate transformation

$$u(t) = \rho(t) \sin \theta(t), \quad v(t) = \rho(t) \cos \theta(t), \qquad (6.1)$$

the system (1.2) is equivalent to the differential system

$$\begin{aligned} \theta'(t) &= a(t)\sin 2\theta(t) + b(t)\cos^2\theta(t) - c(t)\sin^2\theta(t), \\ \rho'(t) &= \{\tfrac{1}{2}[b(t)+c(t)]\sin 2\theta(t) - a(t)\cos 2\theta(t)\}\rho(t). \end{aligned} \tag{6.2}$$

In particular, for a differential equation (1.1°), the system (6.2) becomes

$$\begin{aligned} \theta'(t) &= \frac{1}{r(t)}\cos^2\theta(t) - p(t)\sin^2\theta(t), \\ \rho'(t) &= \{[\frac{1}{r(t)} + p(t)]\sin\theta(t)\cos\theta(t)\}\rho(t). \end{aligned} \tag{6.2'}$$

Although the first equation of (6.2) is non-linear, for t on a compact subinterval $[a,b]$ of I, and $-\infty < \theta < \infty$, the right-hand member of this equation is bounded and satisfies a Lipschitz condition in θ. Consequently, for $t_o \in I$ and θ_o an arbitrary real value, there exists a unique solution $\theta(t) = \theta(t;t_o,\theta_o)$ of this equation satisfying the initial condition $\theta(t_o) = \theta_o$, and the maximal interval of existence of this solution is I. With the function $\theta(t)$ thus determined, the second equation of (6.2) has the unique solution $\rho(t) = \rho_o\{\exp\int_{t_o}^{t} h(s)ds\}$, where

$$h(t) = \frac{1}{2}[b(t) + c(t)]\sin 2\theta(t) - a(t)\cos 2\theta(t),$$
$$\text{and } \rho_o = \rho(t_o).$$

Moreover, either $\rho(t) \equiv 0$ or $\rho(t) \neq 0$ for all $t \in I$, and $\rho^2(t) = u^2(t) + v^2(t)$. In particular, if u is a non-identically vanishing solution of (1.1) then whenever $\rho(t)$, $\theta(t)$ are related to $u(t)$, $v(t) = r(t)u'(t) + q(t)u(t)$ by (6.1) we have $u(t_o) = 0$ if and only if $\theta(t_o)$ is a multiple of π. Moreover, if $\theta(t_o)$ is a multiple of π then

$\theta'(t_0) = b(t)$, and since $b(t)$ is of constant sign on $[a,b]$ it follows that a given multiple of π is attained by $\theta(t)$ for at most one value of t. In particular, the condition $r(t) > 0$ implies that $b(t) > 0$ on I and $\theta(t)$ is increasing at each value where this function is a multiple of π.

Let $q(t;c,s)$ denote the real quadratic form in s,c defined by

$$\begin{aligned} q(t;c,s) &= b(t)c^2 + 2a(t)cs - c(t)s^2, \\ &= \frac{1}{r(t)}\,c^2 - \frac{2q(t)}{r(t)}\,cs + \left[\frac{q^2(t)}{r(t)} - p(t)\right]s^2. \end{aligned} \tag{6.3}$$

Then the first equation of (6.2) may be written as

$$\theta'(t) = q(t;\cos\theta(t), \sin\theta(t)); \tag{6.4}$$

in particular, if the quadratic form (6.3) is non-negative, {positive definite}, then $\theta(t)$ is a non-decreasing, {strictly increasing function}, on I. Also, as a prototype of systems occurring in the later discussion of higher dimensional problems, it is to be noted that (6.4) implies that

$$y^{(1)}(t) = (\sin\theta(t);\cos\theta(t)),\ y^{(2)}(t) = (\cos\theta(t);-\sin\theta(t))$$

are individually solutions of the two-dimensional linear differential system

$$\begin{bmatrix} 0 & -1 \\ 1 & 0 \end{bmatrix} y'(t)$$

$$= \begin{bmatrix} -q(t;\cos\theta(t), \sin\theta(t)) & 0 \\ 0 & -q(t;\cos\theta(t), \sin\theta(t)) \end{bmatrix} y(t).$$

Moreover, $w^{(1)}(t) = \operatorname{ctn}\theta(t)$ and $w^{(2)}(t) = -\tan\theta(t)$ are solutions of the corresponding Riccati differential equation (2.5), which is now

$$w'(t) + q(t;\cos\theta(t), \sin\theta(t))\{w^2(t) + 1\} = 0.$$

It appears impossible to ascribe the introduction of the polar coordinate transformation (6.1) to any specific person, as the use of polar coordinates in the study of differential systems is of long standing, appearing in particular in the perturbation theory of two-dimensional real autonomous dynamical systems. In particular, for a more general linear system of the form (6.5) below, wherein the coefficient functions are periodic with a common period, the corresponding equation (6.5a) is to be found in Levi-Civita [1, p. 352].

The first published use of this substitution in the derivation of certain results of the Sturmian theory for a linear homogeneous second order ordinary differential equations appears to have been by Prüfer [1], and in the literature this substitution is widely known as the "Prüfer transformation". Other authors (see, in particular, Whyburn [2] and Reid [32]) for references to work of H. J. Ettlinger), have also used it estensively, and we shall refer to it as a "polar coordinate transformation".

For an equation of the form (1.1^o) Prüfer [1] derived the oscillation theorems of Sturm and expansion theorems for certain Sturm-Liouville type boundary problems, including some related problems wherein the end-points of the interval of consideration were singular points for the involved differential equation. Such an equation has also been considered by Sturdivant [1]. In Kamke [3] and [4], respectively, the method was used to establish separation and oscillation theorems of Sturmian type. Also, in Kamke [5] the method was employed to establish comparison theorems of Sturmian type for a first order system

$$\begin{aligned} -v'(t) + c(t)u(t) - d(t)v(t) &= 0, \\ u'(t) - a(t)u(t) - b(t)v(t) &= 0, \end{aligned} \tag{6.5}$$

which reduces to (1.2) the case $d(t) \equiv a(t)$. Under the substitution (6.1) system (6.2) is replaced by

$$\begin{aligned} &\text{(a)} \quad \theta'(t) = q(t;\cos\theta(t), \sin\theta(t)), \\ &\text{(b)} \quad \rho'(t) = \{\tfrac{1}{2}[b(t) + c(t)]\sin 2\theta(t) \\ &\qquad\qquad - d(t)\cos^2\theta(t) + a(t)\sin^2\theta(t)\}\rho(t), \end{aligned} \tag{6.6}$$

where

$$q(t;c,s) = [b(t) + c(t)]cs + a(t)s^2 - d(t)c^2. \tag{6.7}$$

In particular, for two systems (6.5) with respective continuous coefficients $a_\alpha(t)$, $b_\alpha(t)$, $c_\alpha(t)$, $d_\alpha(t)$ for $\alpha = 1,2$, and respective quadratic forms $q_\alpha(t;c,s)$, whenever $q_2(t;c,s) - q_1(t;c,s)$ is a non-negative quadratic form in c,s for arbitrary $t \in I$, and

$$\theta_\alpha'(t) = q_\alpha(t;\cos\theta_\alpha(t), \sin\theta_\alpha(t)), \ \alpha = 1,2,$$

with $0 \le \theta_1(a) \le \theta_2(a) < \pi$, then $\theta_2(t) \ge \theta_1(t)$ for $t \ge a$, and if for a positive integer k there exists a value $t_k^1 > a$ such that $\theta_1(t_k^1) = k\pi$ then on $(0,t_k^1]$ there exists a value t_k^2 such that $\theta_2(t_k^2) = k\pi$.

It is to be remarked that the first equation of (6.2') is intimately related to the Riccati differential equation

$$w'(t) + \frac{1}{r(t)}w^2(t) - p(t) = 0. \tag{6.8}$$

Indeed, if $u(t)$ is a real solution of (1.1^{o}) which is non-zero throughout a subinterval I_0 of I, then under the substitution (6.1) we have that $\theta(t)$ satisfies the first equation of (6.2') if and only if $w(t) = \text{ctn}\,\theta(t)$ is a solution of (6.8) on I_0.

Various modifications of the polar coordinate transformation (6.1) have been introduced. In general, if m_1 and m_2 are non-zero functions which are continuously differentiable, then under the modified polar coordinate transformation

$$m_1(t)u(t) = \rho(t)\sin\theta(t), \quad m_2(t)v(t) = \rho(t)\cos\theta(t), \tag{6.9}$$

the real second-order differential system (1.2) with coefficients satisfying either $(\mathscr{A}_C)$ or $(\mathscr{A}_L)$ is equivalent to the system

$$\begin{aligned}
\text{(a)}\quad \theta'(t) &= \left\{a + \frac{1}{2}\left[\frac{m_1'}{m_1} - \frac{m_2'}{m_2}\right]\right\}\sin 2\theta + \frac{1}{2}\left\{b\,\frac{m_1}{m_2} + c\,\frac{m_2}{m_1}\right\}\cos 2\theta \\
&\quad + \frac{1}{2}\left\{b\,\frac{m_1}{m_2} - c\,\frac{m_2}{m_1}\right\}, \qquad (6.10) \\
\text{(b)}\quad \rho'(t) &= \left\{\left[a + \frac{m_1'}{m_1}\right]\sin^2\theta + \frac{1}{2}\left[b\,\frac{m_1}{m_2} + c\,\frac{m_2}{m_1}\right]\sin 2\theta \right. \\
&\quad \left. - \left[a - \frac{m_2'}{m_2}\right]\cos^2\theta\right\}\rho(t),
\end{aligned}$$

which clearly reduces to (6.2) when $m_1 \equiv m_2 \equiv 1$.

In particular, for a differential equation (1.1^{o}), the system (6.10) becomes

$$\begin{aligned}
\text{(a)}\quad \theta' &= \frac{1}{2}\,\frac{m_1'}{m_1}\sin 2\theta + \frac{1}{2}\left\{\frac{1}{r}\,\frac{m_1}{m_2} + p\,\frac{m_2}{m_1}\right\}\cos 2\theta + \frac{1}{2}\left\{\frac{1}{r}\,\frac{m_1}{m_2} - p\,\frac{m_2}{m_1}\right\}, \\
&\qquad\qquad (6.11) \\
\text{(b)}\quad \rho' &= \left\{\frac{m_1'}{m_1}\sin^2\theta + \frac{1}{2}\left[\frac{1}{r}\,\frac{m_1}{m_2} + p\,\frac{m_2}{m_1}\right]\sin 2\theta + \frac{m_2'}{m_2}\cos^2\theta\right\}\rho.
\end{aligned}$$

In both the polar coordinate transformation (6.1) and its modification (6.9), the differential equation in $\theta(t)$ is useful in the study of oscillation properties of the functions u,v, while the equation in $\rho(t)$ may be employed in the study of the behavior of the modulus $[m_1^2(t)u^2(t) + m_2^2(t)v^2(t)]^{\frac{1}{2}}$ of a solution (u,v).

For a differential equation of the form

$$[r(t)u'(t)]' + g(t)u(t) = 0 \tag{6.12}$$

where $r(t)$ and $g(t)$ are positive functions of class $\mathscr{C}^1[a,b]$, the substitution

$$\begin{aligned} &[g(t)/r(t)]^{\frac{1}{4}}u(t) = r(t)\sin\theta(t), \\ &[g(t)/r(t)]^{-\frac{1}{4}}u'(t) = r(t)\cos\theta(t) \end{aligned} \tag{6.13}$$

is of the form (6.9) in $u(t)$, $v(t) = r(t)u'(t)$ with $m_1(t) = [g(t)/r(t)]^{\frac{1}{4}}$, $m_2(t) = [g(t)r^3(t)]^{-\frac{1}{4}}$, and the equations (6.11) are

$$\begin{aligned} &(a)\ \theta'(t) = \left[\frac{g(t)}{r(t)}\right]^{\frac{1}{2}} + \frac{1}{4}\left[\frac{g'(t)}{g(t)} + \frac{r'(t)}{r(t)}\right]\sin 2\theta(t), \\ &(b)\ \rho'(t) = -\frac{1}{4}\left\{\frac{g'(t)}{g(t)}\cos 2\theta(t) + \frac{r'(t)}{r(t)}[1+2\cos^2\theta(t)]\right\}\rho(t). \end{aligned} \tag{6.14}$$

Correspondingly, for the differential equation

$$[r(t)u'(t)]' - g(t)u(t) = 0, \tag{6.15}$$

where $r(t)$ and $g(t)$ are positive functions of class $\mathscr{C}^1[a,b]$, under the substitution (6.13) the equations (6.11) are

$$\text{(a)}\ \theta'(t) = \left[\frac{g(t)}{r(t)}\right]^{\frac{1}{2}}\cos 2\theta(t) + \frac{1}{4}\left[\frac{g'(t)}{g(t)} + \frac{r'(t)}{r(t)}\right]\sin 2\theta(t),$$

$$\text{(b)}\ \rho'(t) = \left\{\left[\frac{g(t)}{r(t)}\right]^{\frac{1}{2}}\sin 2\theta(t) - \frac{1}{4}\left\{\frac{g'(t)}{g(t)}\cos 2\theta(t) + \frac{r'(t)}{r(t)}[1 + 2\cos^2\theta(t)]\right\}\right\}\rho(t). \tag{6.16}$$

For a Sturm-Liouville differential system

$$\begin{aligned} &[r(t)u'(t)]' + [\lambda - p(t)]u(t) = 0, \\ &u(a)\cos\theta_a - r(a)u'(a)\sin\theta_a = 0, \\ &u(b)\cos\theta_b - r(b)u'(b)\sin\theta_b = 0, \end{aligned} \tag{6.17}$$

wherein r,p are functions of class $\mathscr{C}^1[a,b]$, $\theta_a \in [0,\pi)$, $\theta_b \in (0,\pi]$, while λ is real and so large that $\lambda - p(t) > 0$ on $[a,b]$, the above substitution (6.13) and the ensuing equations (6.14) may be used to establish that for $j = 1,2,\ldots$ the j-th eigenvalue of (6.17) satisfies the condition

$$\lambda_j^{1/2} = \left[\int_a^b [r(\xi)]^{-\frac{1}{2}}d\xi\right]^{-1}(j + \nu) + 0(\tfrac{1}{j}), \tag{6.18}$$

where $\nu = 0, \frac{1}{2}$ or 1 according as to the values of certain combinations of the θ_a, θ_b appearing in the boundary conditions of (6.17), (see, for example, Hille [2, Section 8.4]). For example, if $\theta_a = 0 = \theta_b$, so that the boundary conditions of (6.17) are $u(a) = 0 = u(b)$, then $\nu = 1$; also, if $\theta_a = 0$ and $\theta_b = \pi/2$ then $\nu = 1/2$, while if both θ_a and θ_b are distinct from θ and $\pi/2$ then $\nu = 0$.

Another type of "polar coordinate transformation" associated with a differential equation (1.1) involves a basis for the vector space of all solutions of this equation; that is, a pair of real-valued solutions u_1, u_2 which are linearly independent on I. For such a pair of solutions the function $\{u_1,u_2\}(t) \equiv v_2'(t)u_1(t) - u_2(t)v_1(t)$ is equal to a non-zero constant k on I, and by direct computation it follows that the positive function

$$\mu(t) = \sqrt{u_1^2(t) + u_2^2(t)} \tag{6.19}$$

is a solution of the non-linear differential equation

$$\ell[\mu](t) = \frac{k^2}{r(t)\mu^3(t)}, \quad t \in I. \tag{6.20}$$

Conversely, if $\mu(t)$ is any solution of (6.20), then one may show that the general solution of (1.1) is given by

$$u(t) = C_1\mu(t) \sin \{|k| \int_\tau^t \frac{ds}{r(s)\mu^2(s)} + C_2\}, \quad t \in I, \tag{6.21}$$

For the case of an equation of the form (1.1^o) the described interrelations between this equation and the corresponding equation (6.20) is known as the Bohl transformation, (see Bohl [1,2]). The result for the general equation (1.1) is a direct composition of the transformation (1.8) for this equation and the Bohl transformation for the equation (1.1^o) equivalent to the system (1.6_2). Clearly the basic results of the Sturmian theory for equation (1.1) are derivable from (6.21); in this connection the reader is referred to Willett [2, §3]. In particular, for $a \in I$ the j-th right-hand conjugate point to $t = a$, if existent, is the value $t = \tau_j$ of I such that

$$\int_a^{\tau_j} \frac{ds}{r(s)[u_1^2(s) + u_2^2(s)]} = \frac{j\pi}{|k|}, \tag{6.22}$$

where, as before, u_1 and u_2 are linearly independent real-valued solutions of (1.1) and $\{u_1,u_2\} = k$.

7. Transformations for Differential Equations and Systems

In Section 1 of this Chapter there were introduced certain transformations for differential equations (1.1) and equivalent systems (1.2). At this time there will be discussed further transformations for such equations and systems. The list is definitely limited and selective, and even with additional examples occurring in the Exercises appended to this Chapter the coverage is far from comprehensive. For brevity, definite statements will be limited to cases wherein involved functions and certain derivative functions are assumed to be continuous. In all cases, however, for equations and systems whose coefficients satisfy ($\mathscr{A}_L$) the stated conditions may be weakened to the extent that certain of the occurring derivative relations hold only a.e. and solutions of certain appearing differential equations are in the Carathéodory sense.

1°. If the real-valued coefficient functions $a(t)$, $b(t)$, $c(t)$, $d(t)$ are continuous on a non-degenerate interval and $\mu(t)$, $\nu(t)$ are continuously differentiable functions which are non-zero on this interval, then under the transformation

$$u(t) = \mu(t)u_0(t), \quad v(t) = \nu(t)v_0(t) \tag{7.1}$$

the first order differential expressions

$$\begin{aligned} \ell_1[u,v](t) &\equiv u'(t) - a(t)u(t) - b(t)v(t), \\ \ell_2[u,v](t) &\equiv v'(t) - c(t)u(t) - d(t)v(t), \end{aligned} \tag{7.2}$$

satisfy the identities

$$\ell_1[u,v](t) \equiv \mu(t)\ell_1^0[u_0,v_0](t), \quad \ell_2[u,v](t) \equiv \nu(t)\ell_2^0[u_0,v_0](t), \tag{7.3}$$

where

$$\begin{aligned} \ell_1^0[u_0,v_0] &\equiv u_0'(t) - a_0(t)u_0(t) - b_0(t)v_0(t), \\ \ell_2^0[u_0,v_0] &\equiv v_0'(t) - c_0(t)u_0(t) - d_0(t)v_0(t), \end{aligned} \tag{7.2^0}$$

with

$$\begin{aligned} a_0(t) &= [a(t)\mu(t)-\mu'(t)]/\mu(t), \quad b_0(t) = b(t)\nu(t)/\nu(t) \\ c_0(t) &= c(t)\mu(t)/\nu(t), \quad d_0(t) = [d(t)\nu(t) - \nu'(t)]/\nu(t). \end{aligned} \tag{7.4}$$

In particular, if $a_1(t) = \mu'(t)/\mu(t)$ and $d_1(t) = \nu'(t)/\nu(t)$, so that

$$\mu(t) = \mu_0 \exp\left\{\int_{t_0}^{t} a_1(s)ds\right\}, \qquad \nu(t) = \nu_0 \exp\left\{\int_{0}^{t} d_1(s)ds\right\}, \tag{7.5}$$

then $a_0(t) = a(t) - a_1(t)$, $d_0(t) = d(t) - d_1(t)$; moreover, $d_0(t) \equiv -a_0(t)$ if and only if $a_1(t) + d_1(t) \equiv a(t) + d(t)$. In particular, if $a_1(t) \equiv a(t)$ and $d_1(t) \equiv d(t)$, then

$$\ell_1^0[u_0,v_0](t) = u_0' - b(t)v_0, \quad \ell_2^0[u_0,v_0](t) = v_0' - c(t)u_0,$$

where

$$b(t) = [\nu_0/\mu_0]b(t)\,\exp\left\{\int_{t_0}^{t} [d(s) - a(s)]ds\right\},$$

$$c(t) = [\mu_0/\nu_0]c(t)\,\exp\left\{\int_{t_0}^{t} [a(s) - d(s)]ds\right\}.$$

Consequently, if $b(t) \neq 0$ for $t \in I$ the differential system

$$\ell_1[u,v](t) = 0, \quad \ell_2[u,v](t) = 0 \tag{7.6}$$

is equivalent to the differential equation

$$\left[\frac{\delta(t)}{b(t)}\, u_0'(t)\right]' - \delta(t)c(t)u_0(t) = 0, \tag{7.7}$$

where $\delta(t) = \exp\left\{\int_{t_0}^{t} [a(s) - d(s)]ds\right\}$.

2^0. If $r(t)$ and $p(t)$ are continuous with $r(t) > 0$ and $p(t) \neq 0$ for $t \in [a,b]$, then the equation

$$s = s_1 + \int_a^t \sqrt{\frac{|p(\xi)|}{r(\xi)}}\, d\xi, \quad t \in [a,b], \tag{7.8}$$

defines a strictly monotone increasing continuously differentiable function $t = T(s)$ for $s_1 \leq s \leq s_2 = s_1 + \int_a^b \sqrt{|p(\xi)|/r(\xi)}\, d\xi$, and for $t = T(s)$, $s_1 \leq s \leq s_2$, the solution of (7.8), we then have that $dT/ds = \sqrt{r(T(s))/|p(T(s))|}$ on $[s_1,s_2]$. For a general function $\check{F}:[a,b] \to \mathbf{R}$, let $\check{F}:[s_1,s_2] \to \mathbf{R}$ denote the function defined by $F(s) = F(T(s))$, and let D signify the operation of differentiation with respect to s. We then have the identity

$$\ell^0[u](T(s)) = \sqrt{|\check{p}(s)|/\check{r}(s)}\; [D\{R(s)D\check{u}\} - P(s)\check{u}],$$
$$t \in [s_1,s_2]. \tag{7.9}$$

where

$$R(s) = \sqrt{\check{r}(s)|\check{p}(s)|}, \quad P(s) = (\check{p}(s)/|\check{p}(s)|)R(s). \tag{7.10}$$

In the case of positive functions $r(t)$, $p(t)$ this transformation has been used by Reid [13].

3^0. Suppose that $[a,b]$ is a compact interval and the coefficient functions $r(t)$, $p(t)$ of (1.1^0) are of class $\mathscr{C}^2[a,b]$ with $r(t) > 0$ on this interval, and for $m(t)$ and $n(t)$ positive functions which are respective of class $\mathscr{C}^2[a,b]$ and $\mathscr{C}^1[a,b]$ set

$$m(t)u(t) = w(t), \quad s = s_1 + \kappa^{-1}\int_a^t n(\xi)d\xi, \quad t \in [a,b]. \tag{7.11}$$

Let $t = T(s)$, $s_1 \leq s \leq s_2 = s_1 + \kappa^{-1}\int_a^b n(\xi)d\xi$, denote the function defined by the second equation of (7.11). Moreover, as in the preceding case 2^0, for a general function

$F:[a,b] \to \mathbf{R}$, let $\check{F}:[s_1,s_2] \to \mathbf{R}$ denote the function defined by $\check{F}(s) = F(T(s))$, and let D signify the operation of differentiation with respect to s. Then

$$\ell^o[u](T(s)) = \rho(s)[D^2w + h_1(s)Dw - h_2w](s) \tag{7.12}$$

where

$$\rho = \frac{\hat{r}\hat{n}^2}{\kappa^2\hat{m}}, \quad h_1 = \frac{\hat{m}^2}{\hat{r}\hat{n}} D\left(\frac{\hat{r}\hat{n}}{\hat{m}^2}\right)$$
$$h_2 = \frac{D^2\hat{m}}{\hat{m}} + \frac{\hat{m}}{\hat{n}\hat{r}} D\left(\frac{\hat{n}\hat{r}}{\hat{m}^2}\right) D\hat{m} + \frac{\kappa^2\hat{p}}{\hat{r}\hat{n}^2} \tag{7.13}$$

If m,n,r are chosen so that $nr/m^2 \equiv \kappa_1$, then

$$h_1 \equiv 0, \quad h_2 = (D^2\hat{m})/\hat{m} + (\kappa/\kappa_1)^2(rp/m^2). \tag{7.14}$$

In particular, if $p(t) = p_0(t) - \lambda k(t)$, where p_0 and k are of class $\mathscr{C}^2[a,b]$, and

$$m(t) = [k(t)r(t)]^{\frac{1}{4}}, \quad n(t) = [k(t)/r(t)]^{\frac{1}{2}}, \tag{7.15}$$

then $rn/m^2 \equiv 1$, so that $h_2 = (D^2\hat{m})/\hat{m} + \kappa^2\hat{p}_0/\hat{k} - \kappa^2\lambda$. In particular, with this choice of m and n the differential equation

$$[r(t)u'(t)]' - [p_0(t) - \lambda k(t)]u(t) = 0 \tag{7.16}$$

reduces to

$$D^2w + [\mu - P(s)]w(t) = 0 \tag{7.17}$$

where

$$P(s) = (D^2\hat{m})/\hat{m} + \kappa^2\hat{p}_0/\hat{k}, \quad \mu = \kappa^2\lambda. \tag{7.18}$$

In terms of derivatives with respect to t, we have $\dfrac{D^2\hat{m}}{\hat{m}} = \dfrac{\kappa^2}{n^2}\left[\dfrac{m''}{m} - \dfrac{m'}{m}\dfrac{n'}{n}\right]$, and, since the substitutions (7.15) imply $n = m^2/r$, by direct computation one obtains

$$P(s) = \kappa^2\left\{\frac{p_0}{k} + \frac{r}{4k}\left[\left(\frac{k'}{k}\right)' + \left(\frac{r'}{r}\right)' - \frac{1}{4}\left(\frac{k'}{k}\right)^2 + \frac{3}{4}\left(\frac{r'}{r}\right)^2 + \frac{1}{2}\left(\frac{k'}{k}\right)\left(\frac{r'}{r}\right)\right]\right\} \tag{7.19}$$

with argument $t = T(s)$.

The transformation (7.11) with m and n defined as in (7.15) for the differential equation (7.16) was introduced by Liouville [1] in 1837, and is known as the "Liouville transformation" for (7.16). In particular, if κ is chosen as

$$\kappa = \frac{1}{\pi}\int_a^b [k(\xi)/r(\xi)]^{\frac{1}{2}}d\xi,$$

then the second equation of (7.11) defines a mapping of the t-interval $[a,b]$ onto the s-interval $[s_1, s_1 + \pi]$.

In connection with the theory of Sturm-Liouville systems (5.7), it is to be noted that under a transformation (7.11) with m and n defined by (7.15) the boundary conditions are also transformed into conditions

$$\hat{\beta}_a w(s_1) - \hat{\alpha}_a (Dw)(s_1) = 0, \quad \hat{\beta}_b w(s_2) + \hat{\alpha}_b (Dw)(s_2) = 0,$$

where $\hat{\alpha}_a = 0$ if and only if $\alpha_a = 0$ while $\hat{\alpha}_a$ and α_a are of the same sign if different from zero, with similar statements concerning $\hat{\alpha}_b$ and α_b.

Under obvious change of variables in certain of the involved functions, the transformation (7.11) is equivalent to the so-called Kummer [1] transformation involving functions $\phi(t)$, $\psi(t)$ which are respectively of classes $\mathscr{C}^1[a,b]$ and $\mathscr{C}^2[a,b]$ with $\phi'(t) > 0$ and $\psi(t) \neq 0$ on this interval. Under the substitutions

$$\text{(a)} \quad s = \phi(t),\ t \in [a,b], \quad \text{(b)} \quad u(t) = \psi(t)z(s), \tag{7.20}$$

with again $T(s)$, $s_1 \le s \le s_2$ and $s_1 = \phi(a)$, $s_2 = \psi(b)$, the inverse function of ϕ, we have

$$(\frac{\psi}{\phi'} \ell^o[u])(T(s)) = D[R(s)Dz] - P(s)z(s), \quad s \in [s_1,s_2], \tag{7.21}$$

where

$$R(s) = (r\phi'\psi^2)(T(s)), \quad P(s) = ([\psi/\phi']\ell^o[\psi])(T(s)). \tag{7.22}$$

In particular, if ϕ and ψ are so chosen that

$$\phi(t) = \int_{t_0}^{t} [r(s)\psi^2(s)]^{-1}ds, \tag{7.23}$$

then $R(s) \equiv 1$ on $[s_1,s_2]$, and (7.21) reduces to

$$(r\psi^3\ell^o[u])(T(s)) = D^2z - P_o(s)z(s), \tag{7.24}$$

where

$$P_o(s) = (r\psi^3\ell^o[\psi])(T(s)). \tag{7.25}$$

8. Variational Properties of Solutions of (1.1)

For the further consideration of differential equations (1.1) we shall now consider the behavior of the functional

$$J[\eta_1,\eta_2;a,b] = \int_a^b \{\eta_2'[r\eta_1' + q\eta_1] + \eta_2[q\eta_1' + p\eta_1]\}dt \tag{8.1}$$

on a larger class of functions than solutions of this differential equation. If the coefficient functions satisfy hypothesis ($\mathscr{H}_L$) on a given interval I, then for a non-degenerate subinterval I_o of I the symbol $D'(I_o)$ will denote those real-valued functions $\eta: I_o \to \mathbf{R}$ which are locally a.c. and for which there exists a $\zeta: I_o \to \mathbf{R}$ that is (Lebesgue) measurable, locally bounded, and such that $r\eta' + q\eta = \zeta$ a.e. on I_o; that is, η is a solution in the Carathéodory sense of the differential equation

$\ell_2[\eta,\zeta](t) \equiv \eta'(t) - a(t)\eta(t) - b(t)\zeta(t) = 0$ with ζ locally of class $\mathscr{L}^\infty$ on I_0. This association of ζ with η is denoted by the symbol $\eta \in D'(I_0):\zeta$. The subclass of functions $\eta \in D'(I_0)$ for which the associated ζ is locally a.c. is designated $D''(I_0)$, and the association signified by the symbol $\eta \in D''(I_0):\zeta$. If hypothesis $(\mathscr{A}_C)$ is satisfied, then also $(\mathscr{A}_L)$ holds, and the above statements define associated classes $D'(I_0)$ and $D''(I_0)$, which however would involve concepts of (Lebesgue) measure and integrability in view of the above conditions stated for the associated functions ζ. In order to avoid this complication for the reader unfamiliar with the Lebesgue integral, in case hypothesis $(\mathscr{A}_C)$ holds the above definitions of the classes $D'(I_0)$ and $D''(I_0)$ are altered as follows: in the definition of $D'(I_0)$, the associated ζ is piecewise continuous, so that with this restriction under hypothesis $(\mathscr{A}_C)$ a function η belongs to $D'(I_0)$ if and only if it is continuous and has piecewise continuous derivatives on I_0. In the case of $(\mathscr{A}_C)$, for the class $D''(I_0)$ the associated ζ is supposed to be continuous and to have piecewise continuous derivatives, so that with this convention whenever $(\mathscr{A}_C)$ holds we have that $\eta \in D''(I_0):\zeta$ if and only if $\eta \in D'(I_0):\zeta$ and $\zeta \in D'(I_0)$. For further unity of notation in the separate cases of $(\mathscr{A}_C)$ and $(\mathscr{A}_L)$, let $D^0(I_0)$ denote the class of functions $\eta: I_0 \to \mathbf{R}$ which in case of $(\mathscr{A}_C)$ are required to be continuous and piecewise continuously differentiable on I_0, and which in case of $(\mathscr{A}_L)$ are required to be locally a.c. on this interval. With these definitions, we then have in either case that $\eta \in D''(I_0):\zeta$ if and only if

$\eta \in D'(I_0):\zeta$ and $\zeta \in D^0(I_0)$. It is to be noted that under either hypothesis ($\mathscr{A}_C$) or ($\mathscr{A}_L$), if (u,v) is a solution of system (1.2) then $u \in D''(I):v$.

In general, if I_0 is a compact subinterval $[a,b]$ of I, then the simplified notations $D'[a,b]$, $D''[a,b]$, $D^0[a,b]$ are employed for the respective precise symbols $D'([a,b])$, $D''([a,b])$, $D^0([a,b])$. Also, the subclasses of functions η belonging to $D'[a,b]$, $D''[a,b]$, $D^0[a,b]$ and satisfying the end-conditions $\eta(a) = 0 = \eta(b)$ are denoted by $D_0'[a,b]$, $D_0''[a,b]$, $D_0^0[a,b]$, respectively.

If $[a,b]$ is a compact subinterval of I, and $\eta_\alpha \in D'[a,b]:\zeta_\alpha$, it follows readily that the integral of (8.1) exists, and this functional may also be written as

$$J[\eta_1,\eta_2;a,b] = \int_a^b \{\zeta_2 b \zeta_1 + \eta_2 c \eta_1\} dt, \tag{8.2}$$

where the functions $b(t)$, $c(t)$ are defined by (1.3). Also, for $\eta \in D'[a,b]$ the symbol $J[\eta,\eta;a,b]$ is abbreviated to $J[\eta;a,b]$.

It follows readily that for $[a,b] \subset I$ the functional $J[\eta_1,\eta_2;a,b]$ is a real-valued symmetric functional on $D'[a,b] \times D'[a,b]$. That is, if $\eta_\alpha \in D'[a,b]$, $(\alpha = 1,2,3)$, then

(a) $J[\eta_1,\eta_2;a,b] = J[\eta_2,\eta_1;a,b]$,

(b) $J[c\eta_1,\eta_2;a,b] = cJ[\eta_1,\eta_2;a,b]$, for c a real constant,

(c) $J[\eta_1+\eta_2,\eta_3;a,b] = J[\eta_1,\eta_3;a,b] + J[\eta_2,\eta_3;a,b]$. (8.3)

Also, if $\eta_1 \in D''[a,b]:\zeta_1$ and $\eta_2 \in D'[a,b]:\zeta_2$, then

$$
\begin{aligned}
J[\eta_1,\eta_2;a,b] &= \eta_2\zeta_1\Big|_a^b - \int_a^b \eta_2\ell[\eta_1]dt \\
&= \eta_2\zeta_1\Big|_a^b + \int_a^b \eta_2\ell_1[\eta_1,\zeta_1]dt.
\end{aligned}
\tag{8.4}
$$

In particular, for $\eta \in D''[a,b]:\zeta$ we have

$$J[\eta;a,b] = \eta\zeta\Big|_a^b + \int_a^b \eta\ell_1[\eta,\zeta]dt; \tag{8.5}$$

moreover, if $\eta_\alpha \in D''[a,b]:\zeta_\alpha$, $(\alpha = 1,2)$, then

$$\int_a^b \eta_1\ell[\eta_2]dt - \int_a^b \eta_2\ell[\eta_1]dt = (\zeta_2\eta_1-\eta_2\zeta_1)\Big|_a^b = \{\eta_1,\eta_2\}\Big|_a^b \tag{8.6}$$

If $[a,b] \subset I$ and $u \in D'[a,b]$, then for arbitrary $\eta \in D_0'[a,b]$ an integration by parts in (8.1) yields the relation

$$J[u,\eta;a,b] = \int_a^b \eta'(t)g(t)dt, \tag{8.7}$$

where

$$g(t) = r(t)u'(t) + q(t)u(t) - \int_a^t \{q(s)u'(s)+p(s)u(s)\}ds. \tag{8.8}$$

Now if (u,v) is a solution of (1.2) on $[a,b]$ this function is equal a.e. to the constant value $v(a)$, and $J[u,\eta;a,b] = 0$ for arbitrary $\eta \in D_0'[a,b]$. Conversely, if $J[u,\eta;a,b] = 0$ for all $\eta \in D_0'[a,b]$ then from the so-called Fundamental Lemma of the calculus of variations, see, for example, Bolza [2, p. 25] or Bliss [7, p. 10] for the case of $(\mathscr{A}_C)$, and Reid [35, Problem III.2:1] for the case of $(\mathscr{A}_L)$, it follows that there exists a constant c such that $g(t) = c$ a.e. on $[a,b]$, and consequently $u(t)$, $v(t) = c + \int_a^b \{q(s)u'(s) + p(s)u(s)\}ds$ is a solution of the differential system (1.2) on $[a,b]$. Hence we have the following functional characterization of solutions of the differential equation (1.1) or the equivalent system (1.2).

THEOREM 8.1. *If* $[a,b] \subset I$ *then* u *is a solution of* (1.1) *on* $[a,b]$ *if and only if* $u \in D'[a,b]$, *and*

$$J[u,\eta;a,b] = 0 \quad \textit{for} \quad \eta \in D_0'[a,b]. \tag{8.9}$$

Now if $\eta_\alpha \in D_0'[a,b]$, $(\alpha = 1,2)$, then the quadratic character of the functional J implies

$$J[\eta_1+\eta_2;a,b] = J[\eta_1;a,b] + 2J[\eta_1,\eta_2;a,b] + J[\eta_2;a,b]. \tag{8.10}$$

In particular, for $\eta_1 = u$ and $\eta_2 = \sigma u$ where σ is a real constant, with the aid of this identity one obtains the following result.

COROLLARY 1. *If* $[a,b] \subset I$ *and* $J[\eta;a,b]$ *is non-negative on* $D_0'[a,b]$, *while* u *is an element of* $D_0'[a,b]$ *such that* $J[u;a,b] = 0$, *then* u *is a solution of* (1.1).

In essence the above theorem and corollary state that (1.1) is the Euler equation for the variational functional $J[\eta;a,b]$ in the class $D'[a,b]$. Also, if u is a solution of (1.1) on a subinterval $[a,b]$ of I, and $y \in D'[a,b]$ with $y(a) = u(a)$ and $y(b) = u(b)$, then the identity (8.8) for $\eta_1 = u$, $\eta_2 = y - u$ provides a ready proof of the following result.

COROLLARY 2. *Suppose that* $[a,b] \subset I$ *and* $J[\eta;a,b]$ *is non-negative on* $D_0'[a,b]$. *If* u *is a solution of* (1.1) *and* $y \in D'[a,b]$ *satisfying* $y(a) = u(a)$, $y(b) = u(b)$, *then* $J[y;a,b] \geq J[u;a,b]$; *moreover, if* $J[\eta;a,b]$ *is positive definite on* $D_0'[a,b]$ *then* $J[y;a,b] > J[u;a,b]$ *unless* $y(t) \equiv u(t)$ *on* $[a,b]$.

As the results of the above Corollaries are concerned with the non-negativeness of the functional $J[\eta;a,b]$ on the class $D_0[a,b]$, it is to be pointed out that this condition imposes a restriction on the algebraic sign of $r(t)$ on $[a,b]$. In particular, when hypothesis $(\mathscr{A}_C)$ holds this condition implies that $r(t) \geq 0$ on $[a,b]$, and when hypothesis $(\mathscr{A}_L)$ holds it implies that $r(t) \geq 0$ for t a.e. on $[a,b]$. Indeed, suppose that $(\mathscr{A}_C)$ holds, and $J[\eta;a,b] \geq 0$ for arbitrary $\eta \in D_0[a,b]$, while there exists a value $s \in [a,b]$ at which $r(s) < 0$; in view of the continuity of the function r it may be supposed that $s \in (a,b)$. For such a value s, and $0 < \varepsilon < \text{Min}\{s-a,b-s\}$ let $\eta_\varepsilon(t) = [\varepsilon - |t-s|]/\sqrt{\varepsilon}$ for $t \in [s-\varepsilon,s+\varepsilon]$ and $\eta_\varepsilon(t) = 0$ for $t \in [a,s-\varepsilon] \cup [s+\varepsilon,b]$. Then $\eta_\varepsilon \in D_0[a,b]$, and since $0 \leq \eta_\varepsilon(t) \leq \sqrt{\varepsilon}$ and $|\eta'(t)| \leq 1/\sqrt{\varepsilon}$, while the functions p,q are bounded on $[a,b]$, there exists a constant k such that

$$J[\eta_\varepsilon;a,b] = \frac{1}{\varepsilon}\int_{s-\varepsilon}^{s+\varepsilon} r(t)dt + p(\varepsilon), \text{ where } |p(\varepsilon)| \leq k\varepsilon.$$

It then follows that $J[\eta_\varepsilon;a,b] \to 2r(s)$ as $\varepsilon \to 0^+$, and consequently for ε sufficiently small in value we have a function $\eta_\varepsilon \in D_0[a,b]$ with $J[\eta_\varepsilon;a,b] < 0$, contrary to the assumption that $J[\eta;a,b]$ is non-negative on the class $D_0[a,b]$. Whenever hypothesis $(\mathscr{A}_L)$ holds, and $J[\eta;a,b] \geq 0$ for $\eta \in D_0[a,b]$, the assumption that there is a set $S \subset [a,b]$ of positive measure on which $r(t)$ is negative may be shown to lead to a contradiction by a somewhat similar argument involving a point $s \in (a,b)$ that is a point of density of S and the η_ε defined above are now replaced by particular solutions η of $\ell_2[\eta,\zeta] = 0$ with $\eta(t) \equiv 0$ outside an ε-neighborhood of s. Such a proof is to be found in Reid [35, pp. 325, 326], and will not be given here.

The identity of the following theorem may be verified directly.

THEOREM 8.2. *If* $[a,b] \subset I$, $u \in D''[a,b]:v$, *and* $h \in D^0[a,b]$ *is such that* $\eta = uh \in D'[a,b]$, *then*

$$\eta'[r\eta'+q\eta] + \eta[q\eta'+p\eta] = r[uh']^2 + [\eta vh]' + h^2u\ell_1[u,v]. \quad (8.11)$$

As an immediate corollary of this result, we have the following integral formula.

COROLLARY. *If* u,v *is a solution of* (1.2) *on a subinterval* $[a,b]$ *of* I, *and* $h \in D^0[a,b]$ *is such that* $\eta = uh \in D'[a,b]$, *then*

$$J[\eta;a,b] = \eta vh\Big|_a^b + \int_a^b r(t)[\eta'(t) - u'(t)h(t)]^2dt; \quad (8.12)$$

in particular, if $u(t) \neq 0$ *for* $t \in [a,b]$, *then* (8.12) *holds for arbitrary* $\eta \in D'[a,b]]$ *and* $h = \eta/u$.

Also, as a direct consequence of this corollary we have the following result.

THEOREM 8.3. *If* $[a,b] \subset I$, *and there exists a real-valued solution* u *of* (1.1) *such that* $u(t) \neq 0$ *for* $t \in [a,b]$, *then for* $\eta \in D_0'[a,b]$ *and* $h = \eta/u$ *we have*

$$\begin{aligned} J[\eta;a,b] &= \int_a^b r(t)[\eta'(t) - u'(t)h(t)]^2dt \\ &= \int_a^b r(t)[u(t)h'(t)]^2dt; \end{aligned} \quad (8.13)$$

in particular, if $r(t) > 0$ *for* $t \in [a,b]$, *then* $J[\eta;a,b]$ *is positive definite on* $D_0'[a,b]$.

It is to be pointed out that for $J[\eta;a,b]$ the second variational functional $J_2[\eta]$ of (I:2.6), relation (8.4) is the Jacobi transformation formula (I:2.13) and the result of Theorem 8.3 with the formula (8.13) embodies the Legendre or Clebsch transformation of the second variation functional.

THEOREM 8.4. *If* $[a,b] \subset I$, *and* $J[\eta;a,b]$ *is non-negative on* $D_0'[a,b]$, *then*

(i) *if* $u_a(t)$, $v_a(t)$ *is a solution of* (1.2) *satisfying* $u_a(a) = 0$, $v_a(a) \neq 0$, *then* $u_a(t) \neq 0$ *for* $t \in (a,b)$;

(ii) *if* $u_b(t)$, $v_b(t)$ *is a solution of* (1.2) *satisfying* $u_b(b) = 0$, $v_b(b) \neq 0$, *then* $u_b(t) \neq 0$ *for* $t \in (a,b)$.

Moreover, if $J[\eta;a,b]$ *is positive definite on* $D_0'[a,b]$, *then the* u_a, v_a *and* u_b, v_b *of the above* (i) *and* (ii) *are such that* $u_a(t) \neq 0$ *for* $t \in (a,b]$, $u_b(t) \neq 0$ *for* $t \in [a,b)$, *and there exists a real-valued solution* u,v *of* (1.2) *such that* $u(t) \neq 0$ *for* $t \in [a,b]$.

If $t_0 \in I$, and $u_0(t)$, $v_0(t)$ is the solution of (1.2) satisfying the initial conditions $u(t_0) = 0$, $v(t_0) = 1$, then any solution $u(t)$, $v(t)$ of this system for which $u(t_0) = 0$ is of the form $u(t) = cu_0(t)$, $v(t) = cv_0(t)$, with $c = v(t_0)$. Consequently, in considering solutions u_a,v_a and u_b,v_b satisfying the conditions of (i) and (ii) there is no loss of generality in restricting these solutions to be real-valued, and we shall make this restriction in the following argument.

In order to establish conclusion (i), suppose that there exists a $c \in (a,b)$ such that $u_a(c) = 0$. If η is defined as $\eta(t) = u_a(t)$ on $[a,c]$, $\eta(t) \equiv 0$ on $[c,b]$, then $\eta \in D_0'[a,b]$, and $J[\eta;a,b] = J[u_a;a,c] = u_a(c) - u_a(a)v_a(a) = 0$, and from Corollary 1 to Theorem 8.1 it follows that u is a solution of (1.1) on $[a,b]$. If v is the function such that u,v is a solution of (1.2) on $[a,b]$, then $v(t) \equiv 0$ on $(c,b]$, and hence also $v(c) = 0$ by continuity. From the initial conditions $u(c) = 0$, $v(c) = 0$ it then follows that $u(t) \equiv 0$, $v(t) \equiv 0$ on $[a,b]$, contradictory to the stated conditions $u_a(a) = 0$, $v_a(a) \neq 0$. That is, the assumption that u is equal to zero at a value on (a,b) has led to a contradiction. Conclusion (ii) is a consequence of conclusion (i) in view of the separation result of Theorem 2.3.

Now if $J[\eta;a,b]$ is positive definite on $D_0'[a,b]$, by an argument as in the proof of (i) it follows that $u_a(t) \neq 0$ for t on the interval $(a,b]$, and, similarly, $u_b(t) \neq 0$ for $t \in [a,b)$. Hence $u_a(b) \neq 0$ and $u_b(a) \neq 0$, and the constant function $\{u_a,u_b\} = v_b u_a - u_b v_a$ is equal to the non-zero value $v_b(b)u_a(b) = -u_b(a)v_b(a)$. Consequently, by a suitable choice of $v_a(a)$ or $v_b(b)$ one may attain the normalization $\{u_a,u_b\} = -1$, and with this choice we set $u = u_a + u_b$, $v = v_a + v_b$. Then u,v is a real solution of (1.2) on $[a,b]$, and it will be established that $u(t) \neq 0$ for $t \in [a,b]$. As $u(a) = u_b(a)$ and $u(b) = u_a(b)$, these values are different from zero. Now if $c \in (a,b)$ and $u(c) = 0$, define $\eta(t) = u_a(t)$ on $[a,c]$, and $\eta(t) = -u_b(t)$ on $[c,b]$. Then $\eta \in D_0'[a,b]$, and relation (8.5) for the individual intervals $[a,c]$ and $[c,b]$, with $(\eta,\zeta) = (u_a,v_a)$ and $(\eta,\zeta) = (u_b,v_b)$, respectively, yields the contradictory relation $0 \leq J[\eta;a,b] = J[u_a;a,c] + J[u_b;c,b] = u_a(c)v_a(c) - v_b(c)u_b(c) = -u_b(c)v_a(c) + v_b(c)u_a(c) = \{u_a,u_b\}(c) = -1$. Consequently, we also have $u(t) \neq 0$ for $t \in (a,b)$, and hence $u(t) \neq 0$ throughout the closed interval $[a,b]$.

A result complementary to that of Theorem 8.3 is presented in the following theorem.

THEOREM 8.5. *Suppose that* $[a,b] \subset I$, $r(t) > 0$ *on* $[a,b]$, *and there exists a solution* $u = u_0(t)$ *of* (1.1) *such that* $u_0(a) = 0 = u_0(b)$, *and* $u_0(t) \neq 0$ *for* $t \in (a,b)$. *then*

$$J[\eta;a,b] \geq 0, \quad \textit{for} \quad \eta \in D_0'[a,b], \tag{8.14}$$

and the equality sign holds in (8.14) *if and only if there is a constant* k *such that* $\eta(t) \equiv ku_0(t)$ *on* $[a,b]$.

As noted in the proof of Theorem 8.4, any solution of (2.1) which vanishes at a point of I is a multiple of a real-valued solution, so without loss of generality $u_0(t)$ may be chosen as a real-valued solution of (1.1), and in the following argument this choice will be made. Now for $0 \leq \varepsilon \leq \varepsilon_0 < (b-a)/2$, set $T_\varepsilon(t) = [(b-a)t-(b+a)\varepsilon]/[(b-a)-2\varepsilon]$, and for arbitrary $\eta \in D_0'[a,b]$, define $\eta_\varepsilon(t) = \eta(T_\varepsilon(t))$ for $t \in [a+\varepsilon,b-\varepsilon]$, $\eta_\varepsilon(t) = 0$ for $t \in [a,a+\varepsilon] \cup [b-\varepsilon,b]$. Then $\eta_\varepsilon \in D_0'[a+\varepsilon,b-\varepsilon]$, and also $\eta_\varepsilon \in D_0'[a,b]$; moreover, η_0, the function η_ε for $\varepsilon = 0$, is equal to η. The result of Theorem 8.3 then implies that $J[\eta_\varepsilon;a+\varepsilon,b-\varepsilon] \geq 0$, and conclusion (8.14) follows from the fact that a simple continuity argument yields the limit relation $J[\eta_\varepsilon;a+\varepsilon,b-\varepsilon] = J[\eta_\varepsilon;a,b] \to J[\eta;a,b]$ as $\varepsilon \to 0$. As $u_0(a) = 0 = u_0(b)$, from relation (8.5) for $\eta = u_0$ it follows that $J[u_0;a,b] = 0$, and the final conclusion of the theorem is a consequence of the result of Corollary 1 of Theorem 8.1.

In terms of the concept of disconjugacy, or lack of oscillation of solutions of (1.1), the results of Theorems 8.3, 8.4 and 8.5 yield the following theorem.

THEOREM 8.6. *If* $[a,b] \subset I$ *and* $r(t) > 0$ *for* $t \in [a,b]$, *then* (1.1) *is disconjugate on the open interval* (a,b) *if and only if* $J[\eta;a,b]$ *is non-negative on* $D_0'[a,b]$, *and* (1.1) *is disconjugate on* $[a,b]$ *if and only if one of the following conditions holds:*

(i) $J[\eta;a,b]$ *is positive definite on* $D_0'[a,b]$;

(ii) *if* u_a, v_a *is a solution of* (1.2) *with* $u_a(a) = 0$, $v_a(a) \neq 0$, *then* $u_a(t) \neq 0$ *for* $t \in (a,b]$;

(iii) *if* u_b, v_b *is a solution of* (1.2) *with* $u_b(b) = 0$, $v_b(b) \neq 0$, *then* $u_b(t) \neq 0$ *for* $t \in [a,b)$;

(iv) *there exists a real-valued solution* $u(t)$ *of* (1.1) *such that* $u(t) \neq 0$ *on* $[a,b]$;

(v) *there exists on* $[a,b]$ *a real-valued solution* w *of the Riccati differential equation* (2.5).

In particular, the above treatment provides an alternate proof of disconjugacy criteria already given in Theorem 2.5 with Corollary, and Theorem 2.6, and the fact that an equivalent criterion is the positive definiteness of $J[\eta;a,b]$ on $D_0'[a,b]$.

For a given subinterval $[a,b]$, the set of functions $\eta \in D'[a,b]$ satisfying $\eta(b) = 0$ is denoted by $D_{*0}'[a,b]$; similarly, $D_{0*}'[a,b]$ denotes the class $\{\eta : \eta \in D'[a,b], \eta(a) = 0\}$. In particular, we have $D_0'[a,b] = D_{*0}'[a,b] \cap D_{0*}'[a,b]$. Attention is now directed to functionals of the form

$$\begin{aligned} J_a[\eta_1,\eta_2;a,b] &= \gamma_a \eta_2(a)\eta_1(a) + \int_a^b \{\eta_2'[r\eta_1'+q\eta_1] \\ &\quad + \eta_2[q\eta_1'+p\eta_1]\}dt = \gamma_a\eta_2(a)\eta_1(a) + J[\eta_1,\eta_2;a,b], \end{aligned} \tag{8.15}$$

and

$$J_b[\eta_1,\eta_2;a,b] = \gamma_b\eta_2(b)\eta_1(b) + J[\eta_1,\eta_2;a,b], \qquad (8.16)$$

where the coefficient functions satisfy either hypothesis $(\mathscr{A}_C)$ or $(\mathscr{A}_L)$, and γ_a,γ_b are real constants. In accord with the previous notation, $J_a[\eta;a,b]$ and $J_b[\eta;a,b]$ will denote $J_a[\eta,\eta;a,b]$ and $J_b[\eta,\eta;a,b]$, respectively. Corresponding to the results of Theorem 8.1 and its Corollary 1 we have the following results.

THEOREM 8.7. *If* $[a,b] \subset I$ *then* u *is a solution of* (1.1) *on* $[a,b]$ *which satisfies with its associated function* $v = ru' + qu$ *the initial condition*

$$\gamma_a u(a) - v(a) = 0, \qquad (8.17a)$$

$$\gamma_b u(b) + v(b) = 0, \qquad (8.17b)$$

if and only if $u \in D'[a,b]$, *and* $J_a[u,\eta;a,b] = 0$ *for* $\eta \in D'_{*0}[a,b]$, $\{J_b[u,\eta;a,b] = 0$ *for* $\eta \in D'_{0*}[a,b]\}$. *Moreover, if* $J_a[\eta;a,b]$ *is non-negative on* $D'_{*0}[a,b]$, $\{J_b[\eta;a,b]$ *is non-negative on* $D'_{0*}[a,b]\}$, *and* u *is an element of* $D'_{*0}[a,b]$, $\{D'_{0*}[a,b]\}$, *such that* $J_a[u;a,b] = 0$, $\{J_b[u;a,b] = 0\}$, *then* u *is a solution of* (1.1) *which satisfies with its associated function* v *the boundary conditions*

$$\gamma_a u(a) - v(a) = 0, \quad u(b) = 0, \qquad (8.18a)$$

$$\gamma_b u(b) + v(b) = 0, \quad u(a) = 0. \qquad (8.18b)$$

For the functionals $J_a[\eta;a,b]$ and $J_b[\eta;a,b]$ we have the following results, corresponding to the combined conclusions of Theorems 8.3, 8.4 and 8.5 for the functional $J[\eta;a,b]$.

THEOREM 8.8. *If* $[a,b] \subset I$ *and* $r(t) > 0$ *on* $[a,b]$, *then* $J_a[\eta;a,b]$ *is non-negative on* $D'_{*0}[a,b]$, $\{J_b[\eta;a,b]$ *is non-negative on* $D'_{0*}[a,b]\}$ *if and only if the solution* $(u;v)$ *of* (1.2) *determined by the initial values*

$$u(a) = 1, \quad v(a) = \gamma_a, \tag{8.19a}$$

$$u(b) = 1, \quad v(b) = \gamma_b, \tag{8.19b}$$

is such that $u(t) \neq 0$ *for* $t \in [a,b)$, $\{t \in (a,b]\}$. *If* $u(b) \neq 0$, $\{u(a) \neq 0\}$, *then for* $\eta \in D'_{*0}[a,b]$, $\{\eta \in D'_{0*}[a,b]\}$, *and* $h = \eta/u$ *we have* $J_a[\eta;a,b]$, $\{J_b[\eta;a,b]\}$ *given by the integral*

$$\int_a^b r(t)[u(t)h'(t)]^2 dt, \tag{8.20}$$

and $J_a[\eta;a,b]$ *is positive definite on* $D'_{*0}[a,b]$, $\{J_b[\eta;a,b]$ *is positive definite on* $D'_{0*}[a,b]\}$. *If* $u(b) = 0$, $\{u(a) = 0\}$, *however, then* $J_a[\eta;a,b] \geq 0$ *for* $\eta \in D'_{*0}[a,b]$, $\{J_b[\eta;a,b] \geq 0$ *for* $\eta \in D'_{0*}[a,b]\}$, *and the equality sign holds if and only if there is a constant* k *such that* $\eta(t) = ku(t)$ *for* $t \in [a,b]$.

Relative to the functional (8.15), or relative to the differential equation (1.1) with initial condition (8.17a), a value $\tau > a$ on I is a *right-hand focal point* to $t = a$ if there exists a non-identically vanishing solution u of (1.1) which with its associated function v satisfies the initial condition (8.17a) and $u(b) = 0$. Correspondingly, relative to the functional (8.16), or relative to the differential equation (1.1) with initial condition (8.17b), a value $\tau < a$ on I is a *left-hand focal point* to $t = b$ in case there exists a non-identically vanishing solution u of (1.1) which with its associated function v satisfies the initial condition (8.17b) and $u(a) = 0$.

9. Comparison Theorems

Now consider two differential systems

$$\begin{aligned}\ell_\alpha[u](t) \equiv {} & [r_\alpha(t)u'(t) + q_\alpha(t)u(t)]' \\ & - [q_\alpha(t)u'(t) + p_\alpha(t)u(t)] = 0, \quad (\alpha = 1,2),\end{aligned} \tag{1.1^α}$$

where for $\alpha = 1,2$ the coefficient functions r_α, p_α, q_α are supposed to satisfy either hypothesis $(\mathscr{A}_C^+)$ or $(\mathscr{A}_L^+)$ on I. For subintervals I_o of I the respective classes of functions $D'(I_o)$ are denoted by $D'^\alpha(I_o)$, and if I_o is a non-degenerate compact subinterval then the corresponding classes $D'_o[a,b]$, $D'_{*o}[a,b]$, $D'_{o*}[a,b]$ for the two problems are designated by $D'^\alpha_o[a,b]$, $D'^\alpha_{*o}[a,b]$, $D'^\alpha_{o*}[a,b]$, respectively. In the following theorems we shall be concerned with equations (1.1^α) for which $D'^1(I_o) = D'^2(I_o)$ for arbitrary non-degenerate subintervals I_o of I, and hence for arbitrary non-degenerate compact subintervals the classes $D'^\alpha_o[a,b]$, $D'^\alpha_{*o}[a,b]$ and $D'^\alpha_{o*}[a,b]$ are the same for the two problems. This condition clearly holds in case the coefficient functions satisfy $(\mathscr{A}_C^+)$, since then for each problem the class $D'(I_o)$ consists of those functions $\eta: I_o \to \mathbf{R}$ which are continuous and piecewise continuously differentiable on I_o. In case $(\mathscr{A}_L^+)$ holds, then $D'^1(I_o)$ and $D'^2(I_o)$ are clearly equal whenever $a_1(t) = a_2(t)$ and $b_1(t) = b_2(t)$ on I_o, but are also equal under more general circumstances since whenever $\eta \in D'^1(I_o):\zeta_1$ and $\eta \in D'^2(I_o):\zeta_2$ it is not required that $\zeta_1 = \zeta_2$. For example, if each of the functions b_1/b_2, $(a_1-a_2)/b_2$, b_2/b_1 and $(a_2-a_1)/b_1$ is locally of class $\mathscr{L}^\infty$ on I, then $\eta \in D'^1(I_o):\zeta_1$ if and only if $\eta \in D'^2(I_o):\zeta_2$, where $\zeta_2 = (1/b_2)[b_1\zeta_1 + (a_1-a_2)\eta]$ and

$\zeta_1 = (1/b_1)[b_2\zeta_2 + (a_2-a_1)\eta]$.

For $\alpha = 1,2$, we set

$$2\omega_\alpha(t,\eta,\zeta) = r_\alpha(t)\zeta^2 + 2q_\alpha(t)\eta\zeta + p_\alpha(t)\eta^2, \tag{9.1}$$

and correspondingly for $[a,b] \subset I$ we write

$$J^\alpha[\eta;a,b] = \int_a^b 2\omega_\alpha(t,\eta(t),\eta'(t))dt, \quad (\alpha = 1,2) \tag{9.2}$$

which is well-defined for $\eta \in D'^\alpha[a,b]$. In particular, when $D'^1[a,b] = D'^2[a,b]$ the *difference functional*

$$J^{1,2}[\eta;a,b] = J^1[\eta;a,b] - J^2[\eta;a,b] \tag{9.3}$$

is well defined for $\eta \in D'^1[a,b] = D'^2[a,b]$. As a first comparison theorem we have the following result.

THEOREM 9.1. *Suppose that for* $\alpha = 1,2$ *the coefficient functions of* (1.1^α) *satisfy either hypothesis* $(\mathscr{A}_C^+)$ *or* $(\mathscr{A}_L^+)$, *and* $D'^1(I_o) = D'^2(I_o)$ *for arbitrary subintervals* I_o *of* I. *If* $[a,b] \subset I$ *and* $J^{1,2}[\eta;a,b] \geq 0$ *for arbitrary* $\eta \in D_o'^1[a,b] = D_o'^2[a,b]$, *and* $t = b$ *is the first right-hand conjugate point to* $t = a$ *for* (1.1^1), *then for a real-valued solution* $u_2(t)$ *of* (1.1^2) *exactly one of the following conditions holds:*

(i) *there exists an* $s \in (a,b)$ *such that* $u_2(s) = 0$;

(ii) $u_2(t) \neq 0$ *for* $t \in (a,b)$, *in which case* $t = b$ *is also the first right-hand conjugate point to* $t = a$ *for* (1.1^2), *and if* $u_1(t)$ *is a solution of* (1.1^1) *which determines* $t = b$ *as conjugate to* $t = a$ *with respect to this equation then there is a non-zero constant* κ *such that* $u_1(t) \equiv \kappa u_2(t)$.

In particular, if $J^{1,2}[\eta;a,b]$ *is positive definite on*

$D_0'^1[a,b] = D_0'^2[a,b]$, *then conclusion* (i) *holds.*

Let $(u_1;v_1)$ be a solution of (1.2^1) such that u_1 determines $t = b$ as conjugate to $t = a$ with respect to this equation. Then $u_1(t) \not\equiv 0$ on $[a,b]$, $u_1 \in D_0'^1[a,b]:v_1$ and $J^1[u_1;a,b] = 0$, so that in view of the hypotheses of the theorem we have that $u_1 \in D_0'^2[a,b]$ and $J^2[u_1;a,b] \leq 0$. If $J^2[u_1;a,b] < 0$ it then follows from the results of Theorems 8.5 and 8.6 that any real-valued solution u_2 of (1.1^2) must vanish at least once on the open interval (a,b); in particular, this situation holds when $J^{1,2}[\eta;a,b]$ is positive definite on $D_0'^1[a,b] = D_0'^2[a,b]$. On the other hand, if conclusion (i) does not hold and u_2 is a real-valued solution of (1.1^2) with $u_2(t) \neq 0$ on the open interval (a,b), then since $J^2[\eta;a,b]$ is not positive definite on $D_0'^2[a,b]$ it follows from the results of Theorems 8.5 and 8.6 for this functional that $u_2(a) = 0 = u_2(b)$ and $J^2[\eta;a,b] \geq 0$ for arbitrary $\eta \in D_0'^2[a,b]$. Moreover, in this case we have $0 = J^1[u_1;a,b] \geq J^2[u_1;a,b] \geq 0$, so that $J^2[u_1;a,b] = 0$ and there is a non-zero constant κ such that $u_1(t) = \kappa u_2(t)$ on $[a,b]$.

Associated with the functional $J^{1,2}[\eta;a,b]$ of (9.3) is the differential equation

$$\begin{aligned} \ell_{1,2}[u] \equiv {} & [(r_1 - r_2)u' + (q_1 - q_2)u]' \\ & - [(q_1 - q_2)u' + (p_1 - p_2)u] = 0, \end{aligned} \tag{9.4}$$

where in general the coefficient $r_1 - r_2$ may fail to be non-zero throughout a subinterval $[a,b]$ of I. However, if $r_1(t) - r_2(t) > 0$ on $[a,b]$, and the coefficients of (9.4) satisfy suitable conditions so that the class of functions

$D_0'[a,b]$ for this equation is the same as the corresponding class for each of the equations (1.1^α), this equation may be used to establish certain comparison criteria. Specifically, we have the following corollary to the above theorem.

COROLLARY. *Suppose that hypothesis* $(\mathscr{A}_C^+)$ *holds for each of the equations* (1.1^α), *while* $r_1(t) > r_2(t)$ *throughout a non-degenerate subinterval* $[a,b]$ *of* I, *and the equation* (9.4) *is disconjugate on the open interval* (a,b). *Then whenever* (1.1^2) *is disconjugate on* $[a,b]$ *the equation* (1.1^1) *is also disconjugate on this interval.*

The results of Theorem 9.1 also yield criteria for disconjugacy involving differential inequalities. Again, for simplicity details will be presented only in case of an equation with coefficients satisfying hypothesis $(\mathscr{A}_C^+)$, although a similar result is readily established for an equation with coefficients satisfying $(\mathscr{A}_L^+)$, {see, for example, Reid [35, Theorem VII:5.3, p. 340]}.

Indeed, for a differential equation (1.1) with coefficients r,p,q satisfying hypothesis $(\mathscr{A}_C^+)$, suppose that u is a real-valued function such that $u \in D''[a,b:\ell]:v$ with v continuously differentiable on $[a,b]$, while $u(t) \neq 0$ and $u(t)\ell[u](t) \leq 0$ for $t \in [a,b]$. Then $p_0(t) = u(t)\ell[u](t)$ is a non-positive continuous real-valued function on $[a,b]$, and for

$$r_2 = r, \quad q_2 = q, \quad p_2 = p + p_0/u^2 \tag{9.5}$$

it follows that u is a real-valued solution of $\ell_2[u] = 0$ with $u(t) \neq 0$ on $[a,b]$. This equation $\ell_2[u] = 0$ with the equation $\ell_1[u] \equiv \ell[u] = 0$ is such that $2\omega^1(t,\eta,\eta') - 2\omega^2(t,\eta,\eta') \equiv -p_0(t)/u^2(t) \geq 0$ on $[a,b]$, and hence the corresponding functional $J^{1,2}[\eta;a,b]$ is non-negative on $D_0'[a,b]$. These remarks, together with the relation between the existence of non-vanishing solutions of $\ell[u] = 0$ and the existence of solutions of the corresponding Riccati differential equation leads to the following criteria for disconjugacy.

THEOREM 9.2. *If hypothesis* $(\mathscr{A}_C^+)$ *holds and* $[a,b] \subset I$, *then* (1.1) *is disconjugate on* $[a,b]$ *if and only if one of the following conditions holds:*

(i) *there exists a real-valued function* $u \in D''[a,b:\ell]:v$ *with* v *continuously differentiable, such that* $u(t) \neq 0$ *and* $u(t)\ell[u](t) \leq 0$ *for* $t \in [a,b]$;

(ii) *there exists a real-valued continuously differentiable function* w *on* $[a,b]$, *such that* $k[w](t) \leq 0$ *for* $t \in [a,b]$, *where* $k[w]$ *is the associated Riccati differential operator* (2.4).

Now consider two functionals

$$J_a^\alpha[\eta;a,b] = \gamma_a^\alpha \eta^2(a) + J^\alpha[\eta;a,b], \quad (\alpha = 1,2), \tag{9.6$^\alpha$}$$

where the $J^\alpha[\eta;a,b]$ are as in (9.2) and the γ_a^α are real constants. Similar to (9.3), we write

$$J_a^{1,2}[\eta;a,b] = J_a^1[\eta;a,b] - J_a^2[\eta;a,b] = (\gamma_a^1 - \gamma_a^2)\eta^2(a) + J^{1,2}[\eta;a,b], \tag{9.7}$$

and corresponding to (8.17a) we consider sets of boundary conditions

$$\gamma_a^\alpha u(a) - v(a) = 0, \; u(b) = 0, \; \alpha = 1,2. \tag{9.8$^\alpha$}$$

Analogous to the result of Theorem 9.1 on conjugate points we now have the following result on focal points to $t = a$

relative to the functionals $J_a^1[\eta;a,b]$ and $J_a^2[\eta;a,b]$. No details of proof are given, however, as they parallel those of Theorem 9.1.

THEOREM 9.3. *Suppose that for* $\alpha = 1,2$ *the coefficient functions of* (1.1^α) *satisfy either* $(\mathscr{A}_C^+)$ *or* $(\mathscr{A}_L^+)$, *and* $D'^1(I_0) = D'^2(I_0)$ *for arbitrary subintervals* I_0 *of* I. *If* $[a,b] \subset I$ *and* $J_a^{1,2}[\eta;a,b] \geq 0$ *for arbitrary* $\eta \in D_{*0}'^1[a,b] = D_{*0}'^2[a,b]$, *and* $t = b$ *is the first right-hand focal point to* $t = a$ *relative to* $J_a^1[\eta;a,b]$, *then for a real solution* u_2 *of* (1.1^2) *satisfying* $\gamma_a u_2(a) - v_2(a) = 0$, $u_2(a) \neq 0$ *exactly one of the following conditions holds:*

(i) *there exists an* $s \in (a,b)$ *such that* $u_2(s) = 0$;

(ii) $u_2(t) \neq 0$ *for* $t \in (a,b)$, *in which case* $t = b$ *is also the first right-hand focal point to* $t = a$ *relative to* $J_a^2[\eta;a,b]$, *and if* $u_1(t)$ *is a solution of* (1.1^1) *which determines* $t = b$ *as a focal point to* $t = a$ *relative to* $J_a^1[\eta;a,b]$ *then there is a non-zero constant* κ *such that* $u_1(t) = \kappa u_2(t)$.

In particular, if $J_a^{1,2}[\eta;a,b]$ *is positive definite on* $D_{*0}'^1[a,b] = D_{*0}'^2[a,b]$, *then conclusion* (i) *holds.*

Together the results of Theorems 9.1 and 9.3 yield the following theorem.

THEOREM 9.4. *Suppose that for* $\alpha = 1,2$ *the coefficient functions of* (1.1^α) *satisfy either* $(\mathscr{A}_C^+)$ *or* $(\mathscr{A}_L^+)$, *and* $D'^1(I_0) = D'^2(I_0)$ *for arbitrary subintervals* I_0 *of* I, *while* $[a,b] \subset I$ *and* $J_a^{1,2}[\eta;a,b]$ *is positive definite on* $D_{*0}'^1[a,b] = D_{*0}'^2[a,b]$. *If relative to* $J_a^1[\eta;a,b]$ *there are* m *right-hand focal points* t_j^1 *to* $t = a$ *with* $a < t_1^1 < t_2^1 < \dots < t_m^1 \leq b$, *then relative to* $J_a^2[\eta;a,b]$ *there are right-hand focal points* t_k^2, $(k = 1,\dots,r)$ *on* $(a,b]$ *with* $r \geq m$ *and* $t_j^2 < t_j^1$, $(j = 1,\dots,m)$.

An elementary, but basic, fact for the deduction of this result from those of Theorems 9.1 and 9.3 is that if $[a_0,b_0] \subset [a,b]$ and $\eta_0 \in D_0'^1[a_0,b_0] = D_0'^2[a_0,b_0]$ then $\eta(t) = \eta_0(t)$ on $[a_0,b_0]$, $\eta(t) \equiv 0$ on $[a,a_0] \cup [b_0,b]$ is such that $\eta \in D_{*0}'^1[a,b] = D_{*0}'^2[a,b]$, and consequently the positive definiteness of $J_a^{1,2}[\eta;a,b]$ on $D_{*0}'^1[a,b] = D_{*0}'^2[a,b]$ implies the positive definiteness of $J^{1,2}[\eta;a_0,b_0]$ on $D_0'^1[a_0,b_0] = D_0'^2[a_0,b_0]$.

Analogous results for left-hand focal points hold under similar hypotheses for two functionals

$$J_b[\eta;a,b] = \gamma_b^\alpha \eta^2(b) + J^\alpha[\eta;a,b], \quad \alpha = 1,2, \qquad (9.9^\alpha)$$

where the $J^\alpha[\eta;a,b]$ are as in (9.2), the γ_b^α are real constants, and corresponding in (8.17b) we consider sets of boundary conditions

$$u(a) = 0, \quad \gamma_b^\alpha u(b) + v(b) = 0, \quad \alpha = 1,2. \qquad (9.10)$$

10. Morse Fundamental Quadratic Forms for Conjugate and Focal Points

In his extension of the classical Sturmian theory to self-adjoint differential systems, Morse [1,2] introduced certain algebraic quadratic forms whose (negative) index provided a count of the number of conjugate points or focal points to a given value $t = a$ on an interval with an end-point. In order to highlight the basic significance of this method, it seems desirable to introduce it for the simpler instance of the scalar equation of the classical Sturmian theory. That is, we continue to consider an equation (1.1) whose coefficients satisfy either $(\mathscr{A}_C^+)$ or $(\mathscr{A}_L^+)$ on an interval

I. If I_o is a given compact interval in I, then there exists a $\delta > 0$ such that (1.1) is disconjugate on any subinterval of I_o of length not exceeding δ. Indeed, if for $s \in I$ the solution $(u;v)$ of (1.2) satisfying $u(s) = 1$, $v(s) = 0$ is denoted by $u(t;s)$, $v(t;s)$, then for I_o a compact subinterval of I the function $u(t,s)$ is uniformly continuous on $I_o \times I_o$. Consequently, if $\delta > 0$ is such that $|u(t,s) - u(s,s)| < 1$ for $(t,s) \in I_o \times I_o$ with $|t - s| < \delta/2$, then $u(t,s) > 0$ for $t \in [s - \delta/2, s + \delta/2] \cap I_o$, and from Theorem 8.6 it follows that (1.1) is disconjugate on any such subinterval of I_o.

If $[a,b]$ is a compact subinterval of I, then a partition

$$\Pi : a = t_o < t_1 < \dots < t_k < t_{k+1} = b \qquad (10.1)$$

is called fundamental partition of $[a,b]$, {relative to (1.1) or (1.2)}, in case (1.1) is disconjugate on each of the component subintervals $[t_{j-1},t_j]$, $(j = 1,\dots,m+1)$. The comments of the preceding paragraph assures the existence of fundamental partitions of $[a,b]$, and provides an estimate of the number of division points appearing in a fundamental partition. It is to be emphasized, however, that there does not exist a finite upper bound to the number of division points in a fundamental partition. In particular, if Π is a fundamental partition of $[a,b]$ then any refinement of Π is also a fundamental partition of this interval.

For a fixed positive integer k let T_k denote the set of $(k+2)$-tuples $T = \{t_o,t_1,\dots,t_{k+1}\}$ belonging to fundamental partitions Π of compact subintervals $[a,b]$ of I as in (10.1), and denote by $X(\Pi)$ the totality of real sequences $x = (x^0, x^1, \ldots, x^{k+1})$. Since (1.1) is disconjugate on each component subinterval $[t_{j-1}, t_j]$ of a fundamental partition there exists a unique solution $(u^{(j)}(t); v^{(j)}(t)) = (u^{(j)}(t;T,x); v^{(j)}(t;T,x))$ of (1.2) which satisfies the end-conditions

$$u^{(j)}(t_{j-1}) = x^{j-1}, \quad u^{(j)}(t_j) = x^j, \quad (j = 1, \ldots, k+1). \quad (10.2)$$

From the specific formulas for these functions in terms of a given pair of linearly independent solutions of (1.2) on I, it follows immediately that $u^{(j)}(t;T,x)$, $v^{(j)}(t;T,x)$ are linear functions in the vector x with coefficients that are continuous functions of $t, t_0, t_1, \ldots, t_{k+1}$ on $I \times T_k$. Also, for $x \in X(\Pi)$ the function $u_x(t)$ defined by

$$u_x(t) = u^{(j)}(t) \text{ for } t \in [t_{j-1}, t_j], \quad (j = 1, \ldots, k+1), \quad (10.3)$$

is such that $u_x \in D'[a,b]:v_x$, where $v_x(t)$ is a piecewise continuous function on $[a,b]$ satisfying

$$v_x(t) = v^{(j)}(t) \text{ for } t \in (t_{j-1}, t_j), \quad (j = 1, \ldots, k+1). \quad (10.4)$$

The subclass of elements $x = (x^0, x^1, \ldots, x^{k+1})$ of $X(\Pi)$ for which $x^{k+1} = 0$ will be denoted by $X_{*0}(\Pi)$, and the subclass of elements x with $x^0 = 0$ will be designated by $X_{*0}(\Pi)$. Also, we set $X_0(\Pi) = X_{0*}(\Pi) \cap X_{*0}(\Pi)$, the subclass of elements x for which $x^0 = 0 = x^{k+1}$. That is, the function u_x defined by (10.3) belongs to $D'_0[a,b]$, $D'_{*0}[a,b]$, or $D'_{0*}[a,b]$, according as $x = (x^0, x^1, \ldots, x^{k+1})$ is in the respective set $X_0(\Pi)$, $X_{*0}(\Pi)$, or $X_{0*}(\Pi)$.

For Π a fundamental partition of $[a,b]$, consider the functional

$$Q^0\{x,y:\Pi\} = J[u_x, u_y:a,b] \text{ for } x \in X_0(\Pi), \ y \in X_0(\Pi). \quad (10.5)$$

As u_x and u_y belong to $D_0'[a,b]$, and $J[u_x,u_y;a,b] = J[u_y,u_x;a,b]$, it follows that

$$Q^o\{x,y:\Pi\} = \sum_{\alpha,\beta=1}^{k} Q^o_{\alpha\beta}\{\Pi\}y^\alpha x^\beta \tag{10.6}$$

and

$$Q^o_{\alpha\beta}\{\Pi\} = Q^o_{\beta\alpha}\{\Pi\} \quad (\alpha,\beta = 1,\ldots,k). \tag{10.7}$$

In accord with the notation employed for other quadratic functions, we write $Q^o\{x:\Pi\}$ for $Q^o\{x,x:\Pi\}$.

Since for $x \in X_0(\Pi)$ the functions u_x, v_x are a solution of (1.2) on each subinterval $[t_{j-1},t_j]$, with the aid of formula (8.5) for the individual subintervals of Π it follows that

$$\begin{gathered} Q^o\{x,y:\Pi\} = \sum_{\alpha=1}^{k} y^\alpha[v_x(t_\alpha^-) - v_x(t_\alpha^+)], \\ \text{for } x \in X_0(\Pi),\ y \in X_o(\Pi). \end{gathered} \tag{10.8}$$

In particular, in view of the continuity properties of the $u^{(j)}(t)$, $v^{(j)}(t)$ stated above, it follows that the $k \times k$ matrix $[Q^o_{\alpha\beta}\{\Pi\}]$, $(\alpha,\beta = 1,\ldots,k)$, is a real symmetric matrix whose elements are continuous functions of $(t_0,t_1,\ldots,t_{k+1})$ on T_k.

In general, a real quadratic form $Q\{\zeta\} = \sum_{i,j=1}^{N} Q_{ij}\zeta_i\zeta_j$, with $Q_{ij} = Q_{ji}$, $(i,j = 1,\ldots,N)$, is called *singular* if the matrix $[Q_{ij}]$ is singular, or, equivalently, if $\lambda = 0$ is an eigenvalue of $Q\{\zeta\}$; that is, $\lambda = 0$ is a root of the polynomial equation

$$\mathrm{Det}_Q(\lambda) \equiv \det[\lambda E - Q] = 0, \tag{10.9}$$

or there exists a non-zero N-tuple $\zeta^o = (\zeta_i^o)$ such that $Q\{\zeta;\zeta^o\} \equiv \sum_{i,j=1}^{N} Q_{ij}\zeta_i^o\zeta_j = 0$ for arbitrary real n-tuples

$\zeta = (\zeta_i)$. The order of λ as a root of (10.9), or equivalently, the dimension of the null space

$$\{\zeta : \zeta = (\zeta_i),\ \sum_{j=1}^{N} Q_{ij}\zeta_j = 0,\ (i = 1,\dots,N) \tag{10.10}$$

is called the *nullity* of the quadratic form $Q\{\zeta\}$. Also, such a quadratic form is said to have (*negative*) *index* equal to h, if h is the largest non-negative integer such that there is a subspace of N-dimensional real Euclidean space of dimension h on which $Q\{\zeta\}$ is negative definite. In particular, the index of $Q\{\zeta\}$ is equal to zero if and only if $Q\{\zeta\} \geq 0$ for all $\zeta = (\zeta_i)$. From the elementary theory of quadratic forms it follows that $Q\{\zeta\}$ has index h if and only if $Q\{\zeta\}$ has exactly h negative eigenvalues, where each eigenvalue is counted a number of times equal to its multiplicity. Also, the sum of the nullity and index of $Q\{\zeta\}$ is equal to ℓ if and only if ℓ is the largest integer such that there is a subspace of N-dimensional real Euclidean space of dimension ℓ on which $Q\{\zeta\}$ is non-positive; equivalently, ℓ is the number of non-positive roots of the equation (10.9), where each root is counted a number of times equal to its multiplicity. Moreover, if the coefficients Q_{ij} of the quadratic form $Q\{\zeta\}$ are continuous functions of a parameter $\mu = (\mu_1,\dots,\mu_p)$ on a set of D in p-dimensional real Euclidean space, then the roots of the equation (10.9) are continuous functions of μ on D.

A basic result for the Morse theorem is the following theorem, which will be stated without proof as it follows from an application of the above stated results for general quadratic forms to the particular form $Q^0\{x:\Pi\}$.

THEOREM 10.1. *If* $\Pi : a = t_0 < t_1 < \ldots < t_k < t_{k+1} = b$ *is a fundamental partition of* $[a,b]$, *then* $Q^0\{x:\Pi\}$ *is singular if and only if* $t = b$ *is conjugate to* $t = a$, *with respect to the differential equation* (1.1). *Moreover, if* $Q^0\{x:\Pi\}$ *is singular then its nullity is equal to one.*

Now suppose that $\Pi : a = t_0 < t_1 < \ldots < t_k < t_{k+1} = b$ is a fundamental partition of $[a,b]$, and $\hat{\Pi}$ is a partition of this interval obtained by inserting an additional division point $\hat{t}$ which belongs to the open interval (t_{m-1}, t_m). Clearly $\hat{\Pi}$ is also a fundamental partition of $[a,b]$. Also, if $\hat{x} \in X_0(\hat{\Pi})$, and $x \in X_0(\Pi)$ with $x^j = \hat{x}^j$ for $j = 0,1,\ldots,m-1$ and $x^j = \hat{x}^{j+1}$ for $j = m,m+1,\ldots,k+1$, then $J[u_{\hat{x}};t_{j-1},t_j] = J[u_x;t_{j-1},t_j]$ for $j = 1,\ldots,k+1$ and $j \neq m$, while in view of Theorem 3.5 and Corollary 2 to Theorem 3.1 we have that $J[u_{\hat{x}};t_{m-1},t_m] \geq J[u_x;t_{m-1},t_m]$, with the equality sign holding if and only if $\hat{x}^m = u_x(\hat{t})$. Therefore, for x thus determined by a given $\hat{x} \in X_0(\Pi)$, it follows that $Q^0\{x:\Pi\} \leq Q^0\{\hat{x}:\hat{\Pi}\}$ and hence the index of $Q^0\{\hat{x}:\hat{\Pi}\}$ does not exceed the index of $Q^0\{x:\Pi\}$. On the other hand, if $x \in X_0(\Pi)$, and $\hat{x} \in X_0(\hat{\Pi})$ is defined as $\hat{x}^j = x^j$ for $j = 0,1,\ldots,m-1$, $\hat{x}_m = u_x(\hat{t})$, $\hat{x}^j = x^{j-1}$ for $j = m+1,\ldots,k+2$, then $Q^0\{\hat{x}:\hat{\Pi}\} = Q^0\{x:\Pi\}$, and therefore, the index of $Q^0\{x:\Pi\}$ does not exceed the index of $Q^0\{\hat{x}:\hat{\Pi}\}$. Thus if Π is a fundamental partition of $[a,b]$, and $\hat{\Pi}$ is a refinement of Π obtained by the insertion of a single additional division point, the quadratic form $Q^0\{\hat{x}:\hat{\Pi}\}$ has the same index as $Q^0\{x:\Pi\}$. Now any partition Π' that is a refinement of a fundamental partition Π is the result of a finite number of successive refinements, each of which involves the insertion of a single additional division point, and consequently, the index of the corresponding quadratic form $Q^o\{x:\Pi'\}$ is equal to the index of $Q^o\{x:\Pi\}$. Finally, if Π^1 and Π^2 are two fundamental partitions of $[a,b]$, and Π^3 denotes the partition whose division points consist of those points that are division points of either Π^1 or Π^2, then Π^3 is a refinement of each of the partitions Π^1 and Π^2. This fact, together with Theorem 10.1, imply the following result.

THEOREM 10.2. *If* Π^1 *and* Π^2 *are two fundamental partitions of the compact interval* $[a,b]$, *then the two associated quadratic forms* $Q^o\{x:\Pi^1\}$ *and* $Q^o\{x:\Pi^2\}$ *have the same index and the same nullity.*

The following property is also basic for the Morse theory.

THEOREM 10.3. *Suppose that* $[a,b_1]$ *and* $[a,b_2]$ *are compact non-degenerate subintervals of* I *with* $[a,b_1]\subset[a,b_2]$, *and let* Π^1 *and* Π^2 *be fundamental partitions of* $[a,b_1]$ *and* $[a,b_2]$, *respectively. If* i_1 *and* n_1 *denote the index and nullity of* $Q^o\{x:\Pi^1\}$, *while* i_2 *and* n_2 *denote the index and nullity of* $Q^o\{x:\Pi^2\}$, *then* $i_1 \le i_2$ *and* $i_1 + n_1 \le i_2 + n_2$.

In view of Theorem 10.2, we may assume that $\Pi^1: a = t_0 < t_1 < \ldots < t_{k+1} = b_1$ and $\Pi^2: a = t_0 < t_1 < \ldots < t_{k+1} < \ldots < t_{h+1} = b_2$. The result of the lemma is then an immediate consequence of the fact that if $x \in X_o\{\Pi^1\}$ then $y^\alpha = x^\alpha$ for $\alpha = 0,1,\ldots,k+1$, $y^\beta = 0$ for $\beta = k+2,.=.,h+1$ defines an element $y \in X_o(\Pi^2)$ for which $Q^o\{y:\Pi^2\} = Q^o\{x:\Pi^1\}$.

The fundamental result on the relationship between the quadratic forms $Q^o\{x:\Pi\}$ and the existence of conjugate points is the following theorem.

THEOREM 10.4. *If* Π *is a fundamental partition of a compact non-degenerate subinterval* $[a,b]$ *of* I, *then the index of* $Q^o\{x:\Pi\}$ *is equal to the number of points on the open interval* (a,b) *which are conjugate to* $t = a$.

If Π is a fundamental partition (10.1) of $[a,b]$, and $\eta \in D_o'[a,b]$, then for $x^j[\eta] = \eta(t_j)$, $(j = 0,1,2,\ldots,k+1)$, it follows with the aid of Theorem 8.5 and Corollary 2 to Theorem 8.1 that $J[\eta;a,b] \geq Q^o\{x[\eta]:\Pi\}$, with the equality sign holding if and only if $\eta(t) \equiv u_x[\eta](t)$ on $[a,b]$. Consequently, $J[\eta;a,b]$ is non-negative or positive definite on $D_o'[a,b]$ if and only if $Q^o\{x:\Pi\}$ is correspondingly non-negative or positive definite on $X_o(\Pi)$. In view of Theorem 8.6 it then follows that the index of $Q^o\{x:\Pi\}$ is zero if and only if there are no points on the open interval (a,b) which are conjugate to $t = a$, or equally well if and only if there are no points on (a,b) which are conjugate to $t = b$.

For $[a,b]$ a given non-degenerate subinterval of I, denote by $\tau_1,\ldots,\tau_m$ the points on (a,b) which are conjugate to $t = a$, with $a < \tau_1 < \ldots < \tau_m < b$. Also, let k be a positive integer such that $(b-a)/(k+1) < \delta$, where δ is a positive constant such that (1.1) is disconjugate on any subinterval of $[a,b]$ with length not exceeding δ. For $0 < \mu \leq 1$ and $t_j = a + j(b-a)/(k+1)$, let Π^μ denote the partition $\Pi^\mu: a = t_o < t_1 < \ldots < t_{k+1}$ of $[a,a+\mu(b-a)]$. For μ sufficiently small and on the interval $(0,1]$ the value t_{k+1} lies in the interval (a,τ_1), and for such small values of μ the quadratic form $Q^o\{x:\Pi^\mu\}$ is positive definite. Moreover, if i_μ and n_μ denote the respective index and nullity of $Q^o\{x:\Pi^\mu\}$, then it follows from Theorem 10.3 that i_μ and $i_\mu + n_\mu$ are monotone non-decreasing functions of μ on $(0,1]$. The values i_μ and $i_\mu + n_\mu$ are, respectively, the number of negative and non-positive eigenvalues of $Q^0\{x:\Pi^\mu\}$. Moreover, if $\lambda_1(\mu) \leq \lambda_2(\mu) \leq \ldots \leq \lambda_k(\mu)$ denote the eigenvalues of $Q^0\{x:\Pi^\mu\}$ in non-decreasing order, then it follows from the continuity of the elements $Q^0_{\alpha\beta}\{\Pi\}$ as functions of $t_0,t_1,\ldots,t_{k+1}$ that each $\lambda_\sigma(\mu)$ is a continuous function of μ on $(0,1]$. In particular, if $\lambda_\alpha(\mu_0) = 0$ for a value $\mu_0 \in (0,1)$, from the non-decreasing nature of $i_\mu + n_\mu$ we have that $\lambda_\sigma(\mu) \leq 0$ for $\mu \in (\mu_0,1]$. Indeed, since the points conjugate to $t = a$ are isolated, and as the condition $\lambda_\sigma(\mu) = 0$ implies that t_{k+1} is conjugate to $t = a$, it follows that if $\lambda_\sigma(\mu_0) = 0$ then $\lambda_\sigma(\mu) < 0$ for $\mu \in (\mu_0,1]$. Consequently, the index of $Q^0\{x:\Pi^\mu\}$ remains constant as μ ranges over the interval defined by the condition that $a + \mu(b-a) \in (\tau_{j-1},\tau_j)$, where $\tau_0 = a$, and the index of this form increases by one as μ passes through the value specified by $a + \mu(b-a) = \tau_j$, $(j = 1,\ldots,m)$. Hence for $\mu = 1$ the index of $Q^0\{x:\Pi^1\}$ is equal to m, the number of points on the open interval (a,b) which are conjugate to $t = a$.

Turning to the problem of focal points, for $[a,b]$ a compact subinterval of I and γ_a a real constant, consider the quadratic functional

$$J_a[\eta;a,b] = \gamma_a\eta^2(a) + J[\eta;a,b], \tag{10.11}$$

and for Π a fundamental partition of $[a,b]$ define $Q^{*0}\{x,y:\Pi\}$ as

$$Q^{*0}\{x,y:\Pi\} = J[u_x,u_y:a,b], \text{ for } x \in X_{*0}\{\Pi\}, \; y \in X_{*0}\{\Pi\}; \tag{10.12}$$

also, as usual, we write $Q^{*0}\{x:\Pi\}$ for $Q^{*0}\{x,x:\Pi\}$. Corresponding to (10.6), (10.7), and (10.8) we now have the relations

$$Q^{*0}\{x,y:\Pi\} = \sum_{\alpha,\beta=0}^{k} Q^{*0}_{\alpha\beta}\{\Pi\}y^{\alpha}x^{\beta}, \tag{10.13}$$

$$Q^{*0}_{\alpha\beta}\{\Pi\} = Q^{*0}_{\alpha\beta}\{\Pi\}, \quad (\alpha,\beta = 0,1,\dots,k), \tag{10.14}$$

$$Q^{*0}\{x,y:\Pi\} = y^{0}[\gamma_a x^{0} - v_x(a)] + \sum_{\alpha=1}^{k} y^{\alpha}[v_x(t_\alpha^-) - v_x(t_\alpha^+)], \tag{10.15}$$
$$\text{for } x \in X_0\{\Pi\}, \quad y \in X_0\{\Pi\}.$$

By an argument similar to that used in the proof of Theorem 10.1, and with the aid of Theorems 8.7 and 8.8, there is established the following result relating the quadratic form $Q^{*0}\{x:\Pi\}$ to the existence of right-hand focal points to $t = a$ relative to the functional (10.11) or relative to the differential equation (1.1) with initial condition $\gamma_a u(a) - v(a) = 0$.

THEOREM 10.5. *If* Π *is a fundamental partition of* $[a,b]$, *then* $Q^{*0}\{x:\Pi\}$ *is singular if and only if* $t = b$ *is a right-hand focal point to* $t = a$ *relative to the functional* $J_a[\eta;a,b]$ *of* (10.11). *Moreover, if* $Q^{*0}\{x:\Pi\}$ *is singular then its nullity is equal to one.*

Results corresponding to those of Theorems 10.2, 10.3 and 10.4 hold for the quadratic form $Q^{*0}\{x:\Pi\}$. In particular, the analogue of Theorem 10.4 is the following theorem.

THEOREM 10.6. *If* Π *is a fundamental partition of* $[a,b]$, *then the index of the quadratic form* $Q^{*0}\{x:\Pi\}$ *is equal to the number of points on the open interval* (a,b) *which are right-hand focal points to* $t = a$, *relative to the functional* $J[\eta;a,b]$ *of* (10.11).

For corresponding results for left-hand focal points to $t = b$, relative to the functional

$$J_b[\eta;a,b] = \gamma_b \eta^2(b) + J[\eta;a,b], \tag{10.16}$$

or relative to the differential equation (1.1) with initial condition

$$\gamma_b u(b) + v(b) = 0,$$

one considers the quadratic form

$$Q^{0*}\{x:\Pi\} = J[u_x, u_y:a,b], \text{ for } x \in X_{0*}\{\Pi\},\ y \in X_{0*}\{\Pi\}. \tag{10.17}$$

Corresponding to Theorem 10.6, one now has the following result.

THEOREM 10.7. *If* Π *is a fundamental partition of* $[a,b]$, *then the index of the quadratic form* $Q^{0*}\{x:\Pi\}$ *is equal to the number of points on the open interval* (a,b) *which are left-hand focal points to* $t = b$, *relative to the functional* $J_b[\eta;a,b]$ *of* (10.16).

In regard to the chosen forms for $J_a[\eta;a,b]$ and $J_b[\eta;a,b]$ as in (10.11) and (10.16), respectively, it is to be noted that if $u(t)$ is any non-identically vanishing solution of (1.1) on I, and $t = a$ is a point on I where $u(a) \neq 0$, and $b > a$ is such that $[a,b] \subset I$, then the zeros of $u(t)$ on $(a,b]$ are the right-hand focal points to $t = a$ on this interval, relative to the functional $J_a[\eta;a,b]$ of (10.11) with $\gamma_a = v(a)/u(a)$. Corresponding, if $u(b) \neq 0$ and $a < b$ is such that $[a,b] \subset I$, then the zeros of $u(t)$ on $[a,b)$ are the left-hand focal points to $t = b$ on this interval, relative to the functional $J_b[\eta;a,b]$ of (10.16) with $\gamma_b = -v(b)/u(b)$.

The introduction of the algebraic quadratic forms Q^{o}, Q^{*o}, Q^{o*}, and their systematic use, was due to Marston Morse [1,2], who employed them as basic tools in extending the results of the Sturmian theory for scalar real second-order linear differential equations to self-adjoint systems of the second order. A more detailed discussion of results using this method will be reserved for Chapter V. A basic ingredient of the Morse treatment of conjugate points is the continuity of the coefficients of the quadratic form $Q^{o}\{x:\Pi\}$ as functions of $t_o,t_1,\ldots,t_{k+1}$ on $I \times T_k$. This property, together with the results of Theorems 10.2, 10.3 and 10.4, lead readily to the result that for a given $a \in I$ the j-th right-hand, or left-hand, conjugate point to $t = a$ is a continuous functional of the coefficients of the differential equation (1.1) in an appropriate sense, and advances or regresses continuously with $t = a$ as long as it remains on I. Corresponding results also hold for the quadratic forms $Q^{*o}\{x:\Pi\}$, $Q^{o*}\{x:\Pi\}$ and the above described focal points with respect to the functionals (10.11) and (10.16). In addition, separation and comparison theorems of the sort presented in Sections 8 and 9 may be established by the use of the quadratic forms introduced above. Basically, this method has intimate connections with the variational methods of the preceding sections, for the quadratic forms are the values of the functionals $J[\eta;a,b]$, $J_a[\eta;a,b]$, $J_b[\eta\ a,b]$ on "broken solutions" of the differential equation (1.1), but with the important ingredient added that the problems of conjugate and focal points are reduced to the determination of the (negative) index of a finite-dimensional quadratic form. In the following chapter another approach will be encountered, wherein the finite dimensionality occurs as the dimension of a maximal subspace of an infinite dimensional function space on which the functions $J[\eta;a,b]$, $J_a[\eta;a,b]$ and $J_b[\eta;a,b]$ are negative definite, and are shown to be equal to the number of eigenvalues of an associated eigenvalue problem associated with the differential operator $\ell[u]$ of (1.1).

11. Survey of Recent Literature

The preceding sections have been presented as an overview of material which to date has proved to be basic for the expanded Sturmian theory for a real linear homogeneous second order ordinary differential equation on a compact interval. The present section is devoted to a discussion of certain aspects that have received particular attention within relatively recent times. It is to be emphasized that in many cases an individual reference concerns concepts other than those introduced in the earlier sections on this Chapter, notably in the area of boundary problems and/or in the consideration of like problems on a non-compact interval. In such instances the basic reference to the work might have been included in the corresponding section of one of the two following chapters, and its listing here can be attributed to the subjective feeling of the author.

(i) New proofs of known general criteria for conjugacy and/or disconjugacy, with modifications and extensions of such criteria. Since any one differential equation of known behavior on a given interval may be used as a Sturmian comparison equation, theoretically this area is limitless. In general, such treatments have been based upon use of the Riccati differential equation, the study of the functionals J, J_a, J_b

as introduced in Section 8 above on special classes of functions, and the frequent use of auxiliary substitutions. In this category belong the following papers listed in the Bibliography: Nicolenco [1], Oakley [1], Gagliardo [1], Leighton [1, 8, 14].

(ii) Estimates of distance between consecutive conjugate points. A great deal of work has been done, and continues to be done, in connection with bounds, both upper and lower, on the length of interval between consecutive conjugate points for differential equations of the form (1.1). Most of the specific estimates have been presented for an equation of the form (1.1^{o}) or ($1.1^{\#}$), and of course lead to criteria for equations of the form (1.1) in view of the transformations presented in Section 1 above.

(a) Criteria involving point bounds of coefficient functions. Picard [1] derived results on the distance between consecutive zeros of a solution of a non-linear second order differential equation $y''(t) = f(t,y(t),y'(t))$, wherein the function f satisfies a Lipschitz condition in the y, y' arguments. When interpreted for a linear homogeneous equation

$$u''(t) + p_1(t)u'(t) + p_0(t)u(t) = 0, \tag{11.1}$$

where p_0 and p_1 are continuous real-valued functions on an interval I, one has the result that if

$$|p_0(t)| \le L_0, \quad |p_1(t)| \le L_1 \tag{11.2}$$

on a subinterval I_0, then if $u(t)$ is a non-identically vanishing solution of (11.1) with successive zeros at $t = a$ and $t = b$ of I_0, then $h = b - a$ satisfies the inequality

$$\frac{1}{2} L_o h^2 + L_1 h - 1 \geq 0. \tag{11.3}$$

Many years later, de la Vallée Poussin [1] presented a like criterion on the length of an interval on which a non-identically vanishing solution of an n-th order linear homogeneous differential equation might possess a set of n zeros, and for linear equations criteria of this form are frequently called "of Vallée Poussin type". Opial [6] obtained the improved result that the distance $h = b - a$ between consecutive conjugate points for (11.1) satisfies the inequality

$$\frac{1}{\pi^2} L_o h^2 + \frac{2}{\pi^2} L_1 h - 1 \geq 0, \tag{11.3'}$$

and in a sense this inequality is the best of its kind. Recently, Bailey and Waltman [1] investigated estimates of the minimum distance between a zero of a non-identically vanishing solution of (11.1) and a neighboring zero of its derivative, and with the aid of such estimates obtained a result that is superior to (11.3'), and which is sharp in the case of constant coefficients. Specifically, if $u(t)$ is a solution of (11.1) with consecutive zeros at $t = a$ and $t = b$ and for $\alpha = 1,2$ the respective infimum and supremum of $p_\alpha(t)$ on $[a,b]$ are denoted by M^I and M^S, then the result of Bailey and Waltman is expressible in the form

$$\alpha(M_1^I, M_0^I) + \beta(M_1^S, M_o^1) \geq b - a \geq \beta(M_1^I, M_0^S) + \alpha(M_1^S, M_0^S) \tag{11.4}$$

where α and β are complicated functions of their arguments that will not be given here.

(b) <u>Criteria involving integral bounds of coefficient functions</u>. An interpretation of a result of Liapunov [1] yields the result that if $t = a$ and $t = b$ are consecutive zeros of a non-identically vanishing solution of $(1.1^{\#})$, then

$$(b - a)\int_a^b q^+(t)dt > 4, \tag{11.5}$$

where $q^+(t) = \frac{1}{2}[q(t) + |q(t)|]$. Various proofs of this result have been given, (see, in particular, Borg [1] and Reid [30] for two quite different proofs of this result). This inequality was generalized to the condition

$$\int_a^b (t - a)(b - t)q^+(t)dt > b - a \tag{11.6}$$

by Hartman and Wintner [5]; an alternate proof of this inequality, due to Nehari [1], is given in Hartman [13, Th. 5.1 of Ch. XI]. Inequality (11.6) clearly implies

$$\int_a^b (t - a)q^+(t)dt > 1, \tag{11.7}$$

an inequality which was established independently by Bargman [1]. For the equation (11.1), Hartman and Wintner [10] established the inequality

$$b-a < \int_a^b (t-a)(b-t)p_0^+(t)dt + \mathrm{Max}\left\{\int_a^b (t-a)|p_1(t)|dt, \int_a^b (b-t)|p_1(t)|dt\right\}, \tag{11.8}$$

which reduces to (11.6) in case $p_1(t) \equiv 0$. In particular, (11.8) implies the "de la Vallée Poussin inequality" $\frac{1}{6} L_0 h^2 + \frac{1}{2} L_1 h - 1 > 0$, which, as Hartman [13, p. 346] points out, can readily be improved to obtain the inequality $\frac{1}{\pi^2} L_0 h^2 + \frac{1}{\pi} L_1 h - 1 \geq 0$, which Opial [6] improved to (11.3'). In a paper concerned with estimating the length of an interval on which a non-identically vanishing solution of an n-th order linear homogeneous differential equation has n zeros, Nehari [3] announced a result, later proved valid by Hartman

[18], which in the case of a second order equation (11.1) yields the inequality

$$(b - a)\int_a^b |p_0(t)|dt + 2\int_a^b |p_1(t)|dt > 4. \quad (11.9)$$

Fink and St. Mary [2] showed that if a and b are consecutive zeros of a solution of (11.1) then

$$(b - a)\int_a^b p_0^+ dt - 4 \exp\{-\tfrac{1}{2}\int_a^b |p_1(t)|dt\} > 0, \quad (11.10)$$

which in turn implies an inequality of the form (11.9) with the integrand $|p_0(t)|$ of the first integral replaced by $p_0^+(t)$. Hochstadt [1] has noted that from the result of Fink and St. Mary [2] one may extract the corresponding inequality

$$[(b - a)\int_a^b |p_0(t)|dt]^{\frac{1}{2}} + \tfrac{1}{2}\int_a^b |p_1(t)|dt \geq 2, \quad (11.11)$$

and generalized this inequality in the corresponding oscillation problem for the n-th order differential equation $y^{[n]}(t) - p(t)y^{[n-1]}(t) - q(t)y(t) = 0$, $n \geq 2$. Cohn [2] has presented an alternate proof of (11.6), based upon the preliminary result that if $t = a$ and $t = b$ are consecutive zeros of a non-identically vanishing real solution of $(1.1^{\#})$, and $c \in (a,b)$ is a value such that $y'(c) = 0$, then

$$\int_a^c (t - a)q^+(t)dt > 1, \quad \int_c^b (b - t)q^+(t)dt > 1.$$

Also, Cohn [3] has shown that

$$\int_a^b \sqrt{q^+(t)}dt > \pi/2, \text{ if } q^+(t) \text{ is monotone on } [a,b]. \quad (11.12)$$

If $f:[a,b] \to R$ is a positive function which has continuous derivatives of the first two orders, then under the substitution $u(t) = z(s)f(t)$, $s = s(t) = \int_a^t [f(\xi)]^{-2}d\xi$, $a \leq t \leq b$, equation $(1.1^{\#})$ is transformed into a like equation

$$D^2z(s) + Q(s)z(s) = 0, \quad s(a) = 0 \leq s \leq s(b),$$

and application of inequalities (11.6), (11.12) to this latter equation yield for the original equation $(1.1^{\#})$ respective criteria.

For an equation of the form

$$[r(t)u'(t)]' + q(t)u(t) = 0, \tag{11.13}$$

with $r > 0$, $q \geq 0$, and q and r are respectively continuous and continuously differentiable on an interval I, while $t = a$ and $t = b$ are adjacent zeros of a non-identically vanishing solution of (11.13), Willett [2,3] has formulated sequences (b_j^+) and (b_j^-) tending monotonically to b in respective decreasing and increasing fashion. Specifically, for $u(t)$ a solution of (11.13) determining a and b as conjugate points, and $u(t) > 0$ on (a,b), let $w(t)$ be defined on (a,b) by

$$w = [1-M(t)z^{-1}(t)]M(t), \quad z(t) = [1-r(t)R(t)u'(t)/u(t)]R(t) \tag{11.14}$$

where

$$R(t) = \int_a^t r^{-1}(s)ds, \quad M(s) = \int_a^t R^2(s)q(s)ds. \tag{11.15}$$

Then $w(t)$ satisfies on (a,b) the Riccati differential equation

$$w'(t) = \frac{M^2(t)}{r(t)R^2(t)} + \frac{R^2(t)}{M^2(t)}q(t)w^2(t). \tag{11.16}$$

In particular, $0 < w(t) < M(t)$ for $t \in (a,b)$ so that $w(a^+) = 0$ and the above mentioned sequences (b_j^+), (b_j^-) are obtained in terms of the iterates appearing in the solution of (11.16) by the successive approximations

$$w_j(t) = \int_a^t M^2(s)r^{-1}(s)R^{-2}(s)ds + \int_a^t R^2(s)M^{-2}(s)q(s)w_{j-1}^2(s)ds, \quad (j = 1,2,\ldots), \tag{11.17}$$

starting with the respective initial functions $w_0(t) = \int_a^t M^2(s)r^{-1}(s)R^{-2}(s)ds$ and $w_0(t) = M(t)\int_a^t M(s)r^{-1}(s)R^{-2}(s)ds$. In the special case of $(1.1^{\#})$ with $q(t) \geq 0$ the results of Willett [3] are as follows: Let

$$H_0(t) = \left[\int_a^t (s - a)(t - s)q(s)ds\right] / [b - a],$$

$$H_1(t) = \left[\int_a^t (s - a)q(s)H_0(s)ds\right] / \left[\int_a^t (s - a)^2 q(s)ds\right],$$

$$H_2(t) = \left[\int_a^t q(s)H_0^2(s)ds\right] / \left[\int_a^t (s - a)^2 q(s)ds\right].$$

Then $(1.1^{\#})$ is disconjugate on $[a,b]$ if either $H_0(b) \leq 1$ or $H_0(b) - (b - a)H_1(b) + (b - a)^2 H_2(b) \leq 1$, and $(1.1^{\#})$ is conjugate on $[a,b]$ if either $H_0(b) - (b - a)H_1(b) \geq 1$ or $H_0(b) - (b - a)H_1(b) + (b - a)^2 H_2(b) - (b - a)^2 H_1^2(b) \geq 1$. Willett [3] also obtained analogous results for estimates of the distance between a zero of a non-identically vanishing solution $u(t)$ of (11.13) and the nearest zero of the derivative $u'(t)$; that is, in the terminology of Picone [2], the distance between a value a and its first right or left pseudoconjugate.

(c) Criteria for $(1.1^{\#})$, wherein $q(t)$ is dominated by certain types of known functions. Fink [2,3] has investigated extensively the behavior of the function appearing as the left-hand member of (11.5), which for simplicity of notation may by a translation be normalized to the form

$$F(T|q) = T \int_0^T q(s)ds, \tag{11.18}$$

where it is to be understood that q is a real-valued non-negative function which is locally integrable on an interval I containing $t = 0$, and that T is a positive value belonging to I and such that $t = T$ is the first right-hand conjugate point to $t = 0$ relative to $(1.1^{\#})$. Under these conditions, Fink has termed $q(t)$ "admissible". Fink [2] is devoted to the case in which the function $q(t)$ is required to satisfy the inequalities $\ell(t) \leq q(t) \leq m(t)$, where $\ell(t) \geq 0$ and $\ell(t)$, $m(t)$ are locally bounded integrable functions on I. Using basic comparison lemmas due to Banks [1], Fink showed that in the class of such functions for fixed T there exists a function $q = q_*(t)$ which minimizes $F(T|q)$, and that for certain values c_*, d_* satisfying $0 \leq c_* \leq d_* \leq T$ we have $q_*(t) = \ell(t)$ on $(0,c_*) \cup (d_*,T)$, $q^*(t) = m(t)$ for $t \in (c_*,d_*)$. There exists a corresponding function $q = q^*(t)$ which maximizes $F(T|q)$, and its description is similar with the roles of $\ell(t)$ and $m(t)$ interchanged, and where in general the associated intermediate values c^*, d^* differ from the corresponding c_*, d_*. In particular, if $q(t)$ is restricted by $0 \leq h^2 \leq q(t) \leq H^2$, and T is a value satisfying $\pi/H \leq T \leq \pi/h$, then Fink [2; Th. 3] established best possible bounds $T[TH^2 - 2c(H^2 - h^2)] \leq F(t|q) \leq T[Th^2 - 2c'(h^2 - H^2)]$, where c and c' are values on $[0,\pi/2]$ given by $H \tan hc = h \operatorname{ctn} H(T/2 - c)$, $h \tan Hc' = H \operatorname{ctn} h(T/2 - c')$; in case $H = +\infty$ and $T < \pi/h$ best bounds are given by $T[Th^2 + 2h \operatorname{ctn}(hT/2)] \leq F(t|q) < +\infty$.

Considerable attention has been devoted to the problem of distance between consecutive conjugate points of an equation $(1.1^{\#})$, wherein $q(t)$ is either linear, convex or concave on its interval of definition. In the case of $q(t)$ concave or convex there are available the Sturm comparison theorems for $(1.1^{\#})$ and a similar equation with $q(t)$ a linear function. Consequently, it is important to note that when k and d are constants with $k > 0$ then under the substitution $s = kt + d$ the equation

$$u''(t) + [kt + d]u(t) = 0, \tag{11.19}$$

becomes

$$D^2\hat{u}(s) + (s/k^2)\hat{u}(s) = 0, \tag{11.20}$$

the general solution of which is expressible in terms of Bessel functions $J_{1/3}$ and $Y_{1/3}$ as

$$\hat{u}(s) = s^{1/2}[c_1 J_{1/3}(\tfrac{2}{3k} s^{3/2}) + c_2 Y_{1/3}(\tfrac{2}{3k} s^{3/2})].$$

If $q(t)$ is linear on $[a,b]$, while $t = a$ and $t = b$ are consecutive conjugate points with respect to $(1.1^{\#})$, then

$$F(b - a|q) \leq \pi^2; \tag{11.21}$$

indeed, Galbraith [1] established this inequality for equations $(1.1^{\#})$ with $q(t)$ a continuous function satisfying

$$\int_a^b q(s) \cos\left[\frac{2\pi(s - a)}{b - a}\right]dt \leq 0. \tag{11.22}$$

By a result of Makai [1], inequality (11.22) is satisfied by any continuous function $q:[a,b] \to \mathbf{R}$ that is concave, and Galbraith proceeded to show that if $q(t)$ is a non-negative, monotone and concave which satisfies the inequality

$$F(b - a|q) \geq (9/8)n^2\pi^2, \tag{11.23}$$

where n is a positive integer, then every real-valued solution of $(1.1^{\#})$ has at least n zeros on $[a,b]$. Moreover, in view of the asymptotic form of the positive zeros of $J_{1/3}$, it follows that the constant $9/8$ in (11.23) cannot be replaced by a smaller one. A result similar to (11.22), but under more restrictive hypotheses, was established earlier by Makai [2].

Leighton [7] presented various results on estimates on the distance between adjacent conjugate points $t = a$ and $t = b$, $a < b$, of $(1.1^{\#})$ under the hypothesis that the positive function $q(t)$ is convex or concave and of class $\mathscr{C}'$ on $[a,b]$. In particular, if $q(t)$ is positive convex and $t = b$ is conjugate to $t = a$ relative to $(1.1^{\#})$, then Theorem 1 of Leighton [7] presents the result that $kq(b) \geq k\{[q(a)]^{3/2} + 3k\lambda_0/2\}^{2/3}$, where $k = [q(b)-q(a)]/[b-a]$ and λ_0 is the smallest positive zero of $J_{1/3}(\lambda)$, which is approximately 2.9. Also, if $q(t)$ is a positive concave function then $(1.1^{\#})$ is oscillatory on (a,b) if $kq(b) \geq k\{[q(a)]^{3/2} + 3k\pi/2\}^{2/3}$ and either $q(a) > q(b)$, or $q(a) < q(b)$ and $[q(a)]^{3/2} + 3k\pi > 0$. {Leighton [7, Th. 2, corrected in line with author's corrections on p. 10 of Vol. 46 of same journal]}.

Fink [3; Th. 1 and Corollaries] showed that if q is a linear function on $[a,b]$ and $t = b$ is conjugate to $t = a$ with respect to $(1.1^{\#})$ then for fixed T the function $F(T|q)$ is an even function of the slope of q, that is decreasing for positive slopes, and the maximum is obtained for $q(t) \equiv \pi^2$. Moreover, if $q(t)$ is non-negative and linear on

$[0,T]$ then $(9/8)\lambda_0^2 \le F(t|q) \le \pi^2$, where λ_0 is the smallest positive zero of $J_{1/3}(\lambda)$; also, if for $T > 0$ the symbol $N(T)$ denotes the number of values on $(0,T]$ conjugate to $t = 0$ with respect to $(1.1^\#)$ then

$$\left[\frac{2\sqrt{2}}{3\pi}\{F(T|q)\}^{1/2}\right] \le N(T) \le \left[\frac{2\sqrt{2}}{3\lambda_0}\{F(T|q)\}^{1/2}\right]$$

where $[x]$ signifies the largest integer not exceeding x. Furthermore, if q is non-negative, increasing {decreasing} and linear, and for each $a \in I$ the distance to the first right-hand conjugate point to $t = a$ is denoted by $T(a)$, then $T(a)\int_a^{a+T(a)} q(t)dt$ is an increasing {decreasing} function of a. Finally, if $q(t)$ is non-negative and linear on $[a,b]$, where $t = a$ and $t = b$ are consecutive conjugate points for $(1.1^\#)$, then

$$\frac{3}{2}\lambda_0/\sqrt{q(a) + q(b)} \le (b-a) \le \pi\sqrt{2}/\sqrt{q(a) + q(b)}.$$

A direct application of a Sturmian comparison theorem yields the result that relative to $(1.1^\#)$ there is a point conjugate to $t = a$ on $(a,b]$ whenever $q(t)$ is a non-negative concave function on $[a,b]$ satisfying $b-a > \pi\sqrt{2}/\sqrt{q(a) + q(b)}$, {Fink [3, Th. 3]}. Also, by comparing $(1.1^\#)$ with a similar equation wherein the coefficient of $u(t)$ is the linear function joining $(a,q(a))$ and $(c,q(c))$, $a < c \le b$, Fink [3, Th. 2] proved that $(1.1^\#)$ is disconjugate on $[a,b]$ whenever $q(t)$ is a non-negative convex function on this interval with $b - a < (3/2)\lambda_0/\sqrt{q(a) + q(b)}$.

Leighton [8, Th. 6], (see also Leighton and Oo Kian Ke [2, Lemma 1.4]), showed that if $q(t)$ is a continuous positive and monotone non-decreasing function on a compact interval $[-a,a]$, and $(1.1^\#)$ has consevutive conjugate points at

$t = -a$ and $t = a$, then for any $q_1(t)$ that is continuous monotone non-decreasing function on $[-a,a]$ for which the function $d(t) = q_1(t) - q(t)$ is such that $d(0) = 0$, $d(t) \geq 0$ and $d(t) \geq -d(-t)$ for $t \in [0,a]$, there is a value $t_0 \in (-a,a)$ which is conjugate to $t = -a$ relative to the equation $u''(t) + q_1(t)u(t) = 0$. In particular, if $q(t)$ is a continuous positive function which is non-decreasing and convex on $[0,b]$, and b^* is a positive number determined by the equation $b^{*2}q(b^*/2) = \pi^2$, then with the aid of the result stated above it follows that b^* is an upper bound for the first right-hand conjugate point to $t = 0$, with respect to the equation $(1.1^\#)$, {Leighton and Oo Kian Ke [2, Th. 2.1]}. If $q(t)$ is a continuous positive function of class $\mathscr{C}'$ and concave on $[0,b]$, and if b^* is an upper bound for the first right-hand conjugate point to $t = 0$ with respect to $(1.1^\#)$, let $q_1(t) = q'(b^*/2)[t - b^*/2] + q(b^*/2)$, so that $y = q_1(t)$ is the equation of the tangent line to the curve $y = q(t)$ at $b = b^*/2$. If b_* is a lower bound for the first right-hand conjugate point to $t = 0$ with respect to the equation $u''(t) + q_1(t)u(t) = 0$, then b_* is also a lower bound for the first right-hand conjugate point to $t = 0$, relative to $(1.1^\#)$.

Eliason [1] has continued earlier work of Leighton, Galbraith, Banks and Fink, establishing more upper bounds for $q(t)$ of the form $\alpha|t - c|^r + d_c$, where α and r are fixed constants satisfying $\alpha > 0$, $1 \leq r \leq 2$, while c and d_c vary so that $q(t)$ is still admissible for $(1.1^\#)$ on $[-T/2,T/2]$, in the sense introduced by Fink. In general, a continuous function $q:[-T/2,T/2] \to \mathbf{R}$ is called *left-balanced* on $[-T/2,T/2]$ if $q(t) \leq q(-t)$ for $t \in [0,T/2]$, and also if there exists a $c \in (-T/2,T/2)$ such that $q(c) < 0$ then $q(t) \leq 0$ for $t \in [c,T/2]$; a continuous function $q:[-T/2,T/2] \to R$ is called *right-balanced* on $[-T/2,T/2]$ if $q_0(t) = q(-t)$ is left-balanced. If $q(t)$ is admissible and left-balanced on $[-T/2,T/2]$, and $u(t)$ is a solution of $(1.1^{\#})$ determining $T/2$ as conjugate to $-T/2$, then $u(t) \leq u(-t)$ for $t \in [0,T/2]$, $\int_{-T/2}^{T/2} s(t)u^2(t)dt \leq 0$ for arbitrary odd, integrable functions $s(t)$ satisfying $s(t) \geq 0$ a.e. on $[0,T/2]$, and $u(t)$ is monotone decreasing on $[0,T/2]$. The principal result of Eliason [1] involves two functions $q_1(t)$ and $q_2(t)$, each of which is admissible on $[-T/2,T/2]$, while $q_1(t)$ is left-balanced on this interval, and for $p(t) = q_1(t) - q_2(t)$, $s(t) = p(t)$ for $t < 0$, $s(0) = 0$, and $s(t) = -p(t)$ for $t > 0$ we have: (i) there exists a value $a \in [0,T/2]$ such that $p(t) \leq 0$ on $[-T/2,a]$ and $p(t) \geq 0$ on $[a,T/2]$, with strict inequality holding for some $t_0 \in [-T/2,T/2]$, and (ii) if $p(b) > s(b)$ for any $b \in (0,T/2)$, then $p(t) > s(t)$ on $[b,T/2]$. Under these conditions there exists a value $c \in [0,T/2)$ such that $p(t) \leq s(t)$ on $[0,c]$, $p(t) > s(t)$ on $(c,T/2]$, and $I\{q_1|T\} > I\{q_2|T\}$, where for brevity we set

$$I\{q|T\} = \int_{-T/2}^{T/2} q(t)dt. \tag{11.24}$$

In particular, let $q(t)$ be an even continuous function such that

$$q'(t) > 0,\ q''(t) \geq 0,\ q'''(t) \leq 0 \quad \text{for } t > 0, \tag{11.25}$$

and for each constant c let d_c be the value such that the function $q_c(t) = q(t-c) + d_c$ is admissible on $[-T/2,T/2]$.

Eliason [1, Th. 2 and its Corollaries] established the following results:

(i) If $0 \le c_1 < c_2$ and $p_{c_1}(t)$ is left-balanced, then $I\{p_{c_1}|T\} \le I\{p_{c_2}|T\}$, where equality holds if and only if $T/2 \le c_1$ and $p_{c_1}(t) \equiv p_{c_2}(t)$ on $[-T/2,T/2]$.

(ii) If $c_2 < c_1 \le 0$ and $p_{c_1}(t)$ is right-balanced, then $I\{p_{c_1}|T\} \ge I\{p_{c_2}|T\}$, where equality holds if and only if $c_1 \le -T/2$ and $p_{c_1}(t) \equiv p_{c_2}(t)$ on $[-T/2,T/2]$.

In particular, if $\alpha_k > 0$ and $1 \le r_k \le 2$ for $1 \le k \le n$, then $q(t) = \sum_{k=1}^{n} \alpha_k|t|^{r_k}$ is a possible function for (i) and (ii) above.

In general, for a given value a let $T(a)$ be such that $t = a + T(a)$ is the first right-hand conjugate point to $t = a$ with respect to $(1.1^{\#})$, if such a value exists. Eliason [1, Th. 3] showed that if $q(t)$ is an even, non-negative continuous function which is admissible on $[-T/2,T/2]$ and satisfies (11.25), then

$$T(a)\int_a^{a+T(a)} q(t)dt < T\int_{-T/2}^{T/2} q(t)dt, \text{ when } a \ne -T/2.$$

The proof of this result uses Lemma 3 of Beesack and Schwarz [1, p. 512], and Theorem 1 of Fink [3]. In particular, for $q(t) = \alpha|t|^r + d$ admissible on $[-T/2,T/2]$, where $\alpha > 0$, $d > 0$ and $1 \le r \le 2$, Eliason [1, Th. 4] established the inequality $T(a)\int_a^{a+T(a)} q(t)dt \le B_r$, where $B_r = (r + 2)^2 z_r^2/(r + 1)$, and z_r is the smallest positive zero of the Bessel function $J_{-\nu}(t)$ with $\nu = 1/(r + 2)$; also, if c and d_c are constants such that $q_c(t) = q(t - c) + d_c$ is admissible on $[-T/2,T/2]$, then

$I\{p_c|T\} \leq B_r$.

Stevens [1] has also obtained for $(1.1^{\#})$ criteria for conjugacy and disconjugacy involving the iterates of the operator $T_{a,b}$ defined on the class of continuous functions on $[a,b]$ by the condition that $z = T_{ab}f$ is the solution of the differential system

$$z''(t) + q(t)f(t) = 0,\ z(a) = f(a),\ z(b) = f(b).$$

In addition, Stevens presents like conditions for the existence of a right pseudo-conjugate and of a right hemi-conjugate to $t = a$ with respect to $(1.1^{\#})$.

12. Topics and Exercises

1. If $r(t)$ is a real-valued continuous function satisfying $r(t) > 0$ for $t \in [a,b]$, and $\eta(t)$ is a real a.c. function with $\eta'(t) \in \mathscr{L}^2[a,b]$, and $\eta(a) = 0 = \eta(b)$, then for $R = \int_a^b [1/r(t)]dt$ we have

$$\int_a^b r(t)[\eta'(t)]^2 dt \geq [4/R]\eta^2(s),\ \text{for}\ \ s \in (a,b):$$

moreover, if $\eta(t) \not\equiv 0$ on $[a,b]$, then equality holds only if s is the unique point on $[a,b]$ such that $\int_a^s [1/r(t)]dt = \int_s^b [1/r(t)]dt$, and $u(t) = [2/R]\eta(s)\int_a^t [1/r(\xi)]d\xi$ for $t \in [a,s]$, $u(t) = [2/R]\eta(s)\int_t^b [1/r(\xi)]d\xi$ for $t \in [a,b]$. {Picone [3, p. 533]. A ready proof of this inequality is provided by application of the Corollary 2 of Theorem 8.1 to the solution of the differential equation $[r(t)u'(t)]' = 0$ joining the points $(a,0)$, $(s,\eta(s))$, and the solution of this equation joining the points $(s,\eta(s))$ and $(b,0)$}.

2. If $a = t_0 < t_1 < \ldots < t_{m+1} = b$ is a partition of $[a,b]$, and $\eta(t)$ is an a.c. real-valued function on $[a,b]$ with $\eta'(t) \in \mathscr{L}^2[a,b]$, then

$$\sum_{\beta=0}^{m} \frac{[\eta(t_{\beta+1}) - \eta(t_\beta)]^2}{t_{\beta+1} - t_\beta} \le \int_a^b [\eta'(t)]^2 dt,$$

and the equality sign holds if and only if $\eta(t)$ is linear on each subinterval $(t_\beta, t_{\beta+1})$. {Picone [3, p. 528]}.

3. Suppose that $u(t)$ is a real-valued solution of the differential system

(i) $[r(t)u'(t)]' - p(t)u(t) = f(t)$, $t \in [a,b]$,

(ii) $u(a) = 0$, $u(b) = 0$,

where $r(t)$ and $p(t)$ are real-valued continuous functions on $[a,b]$, and $r(t) \ge \rho > 0$ on this interval.

(a) If $p(t) \ge 0$ for $t \in [a,b]$, then

$$|u(t)| \le \frac{(b-a)^2}{4\rho} \operatorname{Max}\{|f(s)| : s \in [a,b]\}, \text{ for } t \in [a,b].$$

(b) If M is a non-negative number such that $-p(t) \le M$ for $t \in [a,b]$, and $4\rho - (b-a)^2 M > 0$, then

$$|u(t)| \le \frac{(b-a)^2}{4\rho - (b-a)^2 M} \operatorname{Max}\{|f(s)| : s \in [a,b]\}, \text{ for } t \in [a,b];$$

(c) If $R = \int_a^b [1/r(t)]dt$, M is as in (b), and $4 - (b-a)RM > 0$, then

$$|u(t)| \le \frac{(b-a)R}{4-(b-a)RM} \operatorname{Max}\{|f(s)| : s \in [a,b]\}, \text{ for } t \in [a,b].$$

{Picone [3, pp. 529-533]}.

4. Let $g(t)$, $q(t)$, $q_1(t)$ be real-valued continuous functions on $[a,b]$ with $q_1(t) \ge q(t)$ on this interval. If $u(t)$ is a non-identically vanishing real-valued solution of

$$\ell[u](t) \equiv u''(t) + g(t)u'(t) + q(t)u(t) = 0, \quad t \in [a,b]$$

satisfying $u(a) = 0 = u(b)$, prove that there cannot exist a real-valued solution $u_1(t)$ of

$$\ell_1[u_1](t) = u_1''(t) + g(t)u_1'(t) + q_1(t)u_1(t) = 0, \; t \in [a,b]$$

with $u_1(t) > 0$ on $[a,b]$ by the following type of argument. If there did exist a real-valued solution of $\ell_1[u_1](t) = 0$ different from zero on $[a,b]$, then without loss of generality one might assume $u_1(t) > 0$ on this interval, and there would exist a value c such that $z(t) = u_1(t) - cu(t)$ is non-negative on $[a,b]$ and a value $t_0 \in (a,b)$ such that $z(t_0) = 0$. Show first that this is impossible if $q_1(t) > q(t)$ throughout $[a,b]$, and establish the general case by a limiting argument employing a corresponding equation $\ell_{1,\varepsilon}[u_1](t) = 0$ involving $q_{1,\varepsilon}(t) = q_1(t) + \varepsilon$, $\varepsilon > 0$. {Giuliano [1]}.

5. If $\phi_1, \phi_2, \psi_1, \psi_2$ are continuously differentiable functions on $[a,b]$ such that $\delta(\phi_1,\phi_2,\psi_1,\psi_2) \equiv \phi_1\psi_2 - \phi_2\psi_1$ is constant on $[a,b]$, then $\{\phi_1,\phi_2;\psi_1,\psi_2\} \equiv \{\psi_1,\psi_2;\phi_1,\phi_2\}$ and the functions $\omega_1(t)$, $\omega_2(t)$ of (4.5) are such that $\omega_1(t) + \omega_2(t) \equiv 0$.

6. If r_α, p_α, q_α, $(\alpha = 1,2)$, are real-valued piecewise continuous functions with $r_2(t) > 0$ for $t \in I$, show that the results of Theorem 9.1 for the differential equations $\ell_\alpha[u] = 0$, $(\alpha = 1,2)$, remain valid when the functional $J^{1,2}[\eta;a,b]$ of (9.3) is replaced by $c_1J_1[\eta;a,b] - c_2J_2[\eta;a,b]$, where c_1, c_2 are given positive constants. {Bôcher, [5; Comment on p. 173]; Leighton [8]}.

7. Suppose that $r_\alpha(t)$, $p_\alpha(t)$, $(\alpha = 1,2)$, satisfy hypothesis $(\mathscr{A}_C^+)$ and that $u_1(t)$ is a non-identically vanishing solution of (1.1_1^0) satisfying $u_1(a) = 0$, $u_1'(c) = 0$, where $a < c \leq b$ and $u_1'(t) \neq 0$ for $t \in [a,c)$. If

$$\int_a^c \{[r_1 - r_2]u_1'^2 + [p_1 - p_2]u_1^2\}dt \geq 0$$

then for any non-identically vanishing solution $u_2(t)$ of (1.1_2^0) satisfying $u_2(a) = 0$ there is a value $\xi \in (a,c]$ such that $u_2'(\xi) = 0$; moreover, $\xi = c$ if and only if $u = u_1(t)$ is also a solution of (1.1_2^0). {Leighton [8, Th. 1]. The last phrase is a correction of the erroneous statement of Leighton that $\xi = c$ implies that the two differential equations are identical}.

8. Suppose that $q(t)$ is a positive continuous function on $[-a,a]$, $a > 0$, and that $t = -a$ and $t = a$ are successive zeros of a real solution $u(t)$ of

$$\ell[u|q](t) \equiv u''(t) + q(t)u(t) = 0, \qquad (12.1|q)$$

and that $t = c$ is the unique value on $(-a,a)$ at which $u'(c) = 0$. If $q(t)$ is a monotone non-decreasing (non-increasing) function on $[-a,a]$ and $q(-a) \neq q(a)$, then $c > 0$, $\{c < 0\}$. {Leighton [14, Lemma 1.2]; the proof given for the cited lemma corrects an earlier faulty proof of an equivalent result stated as Th. 4 in Leighton [8]}.

9. Suppose that $r(t)$ and $p(t)$ are continuous and $r(t) > 0$ on an interval I. If $[a,b] \subset I$ and there exists a function $\eta(t)$ such that $\eta(t)$ and $r(t)\eta'(t)$ are of class $\mathscr{L}'[a,b]$, while $\eta(a) = 0 = \eta(b)$ and

$$\int_a^b \eta(t)\ell[\eta](t)dt > 0,$$

then relative to the differential equation (1.1^o) there exists a value $c \in (a,b)$ which is conjugate to $t = a$. (Leighton 13, Theorem 1).

10. If $u(t)$ is a non-identically vanishing solution of $(12.1|q)$ such that $u(0) = 0 = u(c)$, then for $b > 1$ and sufficiently close to 1 the function $\eta_b(t) = u(t/b)$ is such that

$$\int_0^{bc} \{r(t)[\eta_b'(t)]^2 + p(t)\eta_b^2(t)\}dt < 0.$$

(Leighton [13, Th. 4]).

11. If the functions $r(t)$ and $p(t)$ are continuous on an interval $[a,d)$, $(a > 0)$, and if there exists a $c > 0$ such that

$$r(bt) \geq r(t), \quad p(t) \leq b^2 p(bt) \text{ on } [a,d) \text{ for } b \in (1,1+c),$$

with strict inequality holding in one of these conditions, then a solution $u(t)$ of (1.1^o) satisfying $u(a) = 0$, $u'(a) \neq 0$ cannot vanish more than once on (a,d). (Leighton [13, Th. 5]).

12. If $q(t)$ is a real-valued piecewise continuous function on an interval I, then each of the following conditions is sufficient for $(12.1|q)$ to be disconjugate on I;

(i) there exists a value $s \in I$ such that

$$4\left[\int_t^s q(\xi)d\xi\right]^2 \leq q(t) \quad \text{for } t \in I;$$

(ii) $I \in (0,\infty)$ and there exists a value $s \in I$ such that

$$-3 \leq 4t \int_t^s q(\xi)d\xi \leq 1 \quad \text{for } t \in I.$$

If there exists a value t_0 such that $(t_0,\infty) \subset I$ and for $t_1 \in (t_0,\infty)$ the integral $\int_{t_1}^{\infty} q(\xi)d\xi = \lim_{t\to\infty} \int_{t_1}^{t} q(\xi)d\xi$ exists and is finite, then either (i) or (ii) with $s = \infty$ is a sufficient condition for disconjugacy on I. {Wintner [5]: Compute $k[w]$, for $w(t) = 2\int_t^s q(\xi)d\xi$ in (i), and $w(t) = \int_t^s q(\xi)d\xi + 1/(4t)$ in (ii)}.

13. Consider two differential equations (1.1_α), $(\alpha = 1,2)$, for which the coefficient functions $r_\alpha, p_\alpha, q_\alpha$ satisfy hypothesis $(\mathscr{A})$, and let

$$2\omega_\alpha(t,\eta,\zeta) = r_\alpha(t)\zeta^2 + 2q_\alpha(t)\zeta\eta + p_\alpha(t)\eta^2, \ (\alpha = 1,2).$$

If I_o is a non-degenerate subinterval of I, and $u_1 \in D''(I_o;\ell_1):v_1$ then (8.11) for $(\eta,\zeta) = (u_1;v_1)$ is equivalent to the identity

$$2\omega_1(t,u_1,u_1') = [u_1(t)v_1(t)]' - u_1(t)\ell_1[u_1](t). \quad (12.2)$$

Also, if $u_2 \in D''[I_o;\ell_2]$ and $u_2(t) \neq 0$ for $t \in I_o$, then the identity (8.11) for $u = u_2$ and $\eta = u_1$ becomes

$$2\omega_2(t,u_1,u_1') = r_2\left\{u_2\left(\frac{u_1}{u_2}\right)'\right\}^2 + \left\{u_1^2\,\frac{v_2}{u_2}\right\}' - \left(\frac{u_1}{u_2}\right)^2 u_2\ell_2[u_2]. \quad (12.3)$$

Combining these relations we have for $t \in I_o$ the identity

$$\begin{aligned}2\omega_1(t,u_1,u_1') - 2\omega_2(t,u_1,u_1') + \left\{\frac{u_1}{u_2}[v_2u_1 - u_2v_1]\right\}' \\ + r_2\left[u_1' - u_2'\,\frac{u_1}{u_2}\right]^2 + u_1\ell_1[u_1] + \left(\frac{u_1}{u_2}\right)^2 u_2\ell_2[u_2] \equiv 0.\end{aligned} \quad (12.4)$$

In particular, if $q_\alpha(t) \equiv 0$ for $(\alpha = 1,2)$, this relation reduces to the Picone identity (3.2). In the derivation of (12.4) a major portion of the argument is embodied in the identity (8.11) of the Legendre or Clebsch transformation.

14. If $r(t)$, $q(t)$ are real-valued continuous functions on $[a,b]$ with $r(t) > 0$ on this interval, then the differential equation

$$[r(t)u'(t)]' + q(t)u(t) = 0$$

is disconjugate on $[a,b]$ if there exists a monotone differentiable function $f:[a,b] \to \mathbf{R}$ with $f(a) = 0$, $f(b) = \pm\infty$, and a positive function $w(t)$ of class $\mathscr{C}'$ on $[a,b]$ for which

$$\left|\frac{w}{r} + \frac{q}{w}\right| + \left|\frac{w}{r} - \frac{q}{w}\right| + \left|\frac{w'}{w}\right| \leq 2f' \text{ for } t \in [a,b],$$

with strict inequality for some $t \in [a,b]$. A particular admissible f is $f(t) = \pi(t - a)/(b - a)$, in which case the right-hand member of the above inequality is the constant $2\pi/(b - a)$.

{McCarthy [1]. The presented proof involves the modified polar coordinate transformation $u(t) = \rho(t)\sin\theta(t)$, $r(t)u'(t) = w(t)\rho(t)\cos\theta(t)$ introduced by Barrett [1]}.

15. Suppose that $q_0(t)$, $q_1(t)$, $q_2(t)$ are real-valued continuous functions on $[a,b]$ with $q_2(t) > 0$ on this interval. Then every real-valued solution of the differential equation

$$\ell[u](t) \equiv q_2(t)u''(t) + 2q_1(t)u'(t) + q_0(t)u(t) = 0 \quad (12.5)$$

has a zero on the open interval (a,b) if there exists a continuous function $g:[a,b] \to \mathbf{R}$ such that $g(t)q_2(t) - q_1^2(t) > 0$ on this interval, and there is an a.c. $\eta_1:[a,b] \to \mathbf{R}$ with $\eta_1' \in \mathscr{L}^2[a,b]$, $\eta_1(a) = 0 = \eta_1(b)$, and

$$\int_a^b \{q_2(t)\eta_1'^2 - 2q_1(t)\eta_1(t)\eta_1'(t) + [g(t)-q_0(t)]\eta_1^2(t)\}dt < 0. \quad (12.6)$$

Swanson [1], Lemma 1.11, with his equations (1.19) and (1.20) corrected by inserting a factor 2 on the "middle terms". This result is a ready consequence of Theorem 8.5 in view of the following comments: If $\mu(t) = \exp\{2\int_\tau^t [q_1(s)/q_2(s)]ds\}$, where $\tau \in [a,b]$, then $\mu(t)\ell[u](t)$ is of the form (1.1) with $r(t) = \mu(t)q_2(t)$, $q(t) \equiv 0$, $p(t) = -\mu(t)q_0(t)$, and the corresponding functional $J[\eta;a,b]$ is

$$J[\eta] = \int_a^b \mu(t)\{[\eta'(t)]^2 + \eta^2(t)\}dt;$$

moreover, if $\eta_1(t)$ is a function as described above satisfying (12.6), then $\eta(t) = [1/\sqrt{\mu(t)}]\eta_1(t) \in D_0[a,b]$ and

$$J[\eta] = \int_a^b \{q_2\eta_1'^2 - 2q_1\eta_1\eta_1' + [(q_1^2/q_2) - q_0]\eta_1^2\}dt.$$

Indeed, the direct proof of the result given by Swanson is essentially the argument used to establish the Jacobi transformation (8.13), which is the crucial part of the proof of Theorem 8.5.}.

16. Consider the differential equations

$$\ell_\alpha^o[u](t) \equiv [r_\alpha(t)u'(t)] + q_\alpha(t)u(t) = 0, \quad t \in I,$$

where for $\alpha = 1,2$ the functions r_α, q_α are real-valued and continuous on the interval I, and each r_α is positive on this interval. If on I we have

$$q_1 \geq 0,\ q_2/r_2 \leq q_1/r_1,\ r_2/r_1 \text{ is monotone non-decreasing,} \tag{12.7}$$

then for $t_0 \in I$ the existence of the first right-hand focal point $\tau_{1,2}^+$ to t_0 with respect to $\ell_2^o[u] = 0$ implies the existence of the first right-hand focal point $\tau_{1,1}^+$ to t_0 with respect to $\ell_1^o[u] = 0$ and $\tau_{1,1}^+ \leq \tau_{1,2}^+$; moreover, if there exists a value $t_1 \in (t_0,\tau_{1,2}^+]$ such that $q_2(t_0)/r_2(t_0) < q_1(t_0)/r_1(t_0)$ then $\tau_{1,1}^+ < \tau_{1,2}^+$. Correspondingly, if on I we have

$$q_2 \geq 0,\ q_1/r_1 \leq q_2/r_2,\ r_2/r_1 \text{ is monotone non-decreasing,} \tag{12.8}$$

then for $t_o \in I$ the existence of the first left-hand focal point $\tau_{1,1}^-$ to t_o with respect to $\ell_1^o[u] = 0$ implies the existence of the first left-hand focal point $\tau_{1,2}^-$ to t_o with respect to $\ell_2^o[u] = 0$ and $\tau_{1,2}^- \geq \tau_{1,1}^-$; moreover, $\tau_{1,1}^- < \tau_{1,2}^-$ in case there is a value $t_2 \in [\tau_{1,1}^-, t_o]$ for which $q_1(t_2)/r_1(t_2) < q_2(t_2)/r_2(t_2)$.

{Under the additional hypotheses that the $r_\alpha(t)$ are continuously differentiable, Reid [13] established the conclusions involving the inequalities $\tau_{1,1}^+ \leq \tau_{1,2}^+$ and $\tau_{1,2}^- \geq \tau_{1,1}^-$. The stronger result stated above is proved by Morse [9, Sec. 22]}.

17. Let $u = u(t;a)$ be the solution of $(12.1|q)$ satisfying $u(a) = 0$, $u'(a) = 1$. If $t_j^+(a)$ denotes the j-th right-nand conjugate point to $t = a$, then $(d/da)t_j^+(a) = 1/[u'(t_j^+(a);a)]^2$. (Leighton and Oo Kian Ke [1]; Lemma).

18. Consider the differential equations $(12.1|q_\alpha)$, $(\alpha = 1,2)$, where $q_1(t)$ and $q_2(t)$ are positive continuous functions on an interval $[a,b+\delta]$, and $u = u_\alpha(t)$ is a non-trivial solution of $(12.1|q_\alpha)$ with consecutive zeros at $t = a$ and $t = b$. If $q_1(a) > q_2(a)$ and $q_2(b) > q_1(b)$, and the curves $y = q_\alpha(t)$ on $[a,b]$ either intersect in a single point or in a single closed subinterval of (a,b), then for $\varepsilon > 0$ and sufficiently small the first right-hand conjugate point to $t = a+\varepsilon$ with respect to the equation $(12.1|q_2)$ precedes the first right-hand conjugate point of $t = a+\varepsilon$ with respect to the equation $(12.1|q_1)$. (Leighton and Oo Kian Ke [1, Theorem 1]).

19. Let $f(t)$ be continuous on (a,b) with $f(t) = O[(t-a)^{-2}]$ as $t \to a^+$, $f(t) = O[(b-t)^{-2}]$ as $t \to b^-$, while the Riccati differential equation $g'(t) = f(t) + g^2(t)$ has a solution g of class $\mathscr{L}'$ on (a,b) with $g(t) = O[(t-a)^{-1}]$ as $t \to a^+$, $g(t) = O[(b-t)^{-1}]$ as $t \to b^-$. If $\eta(t)$ is piecewise smooth on $[a,b]$ and $\eta(a) = 0 = \eta(b)$, then

$$\int_a^b (\eta'^2 - f\eta^2)dt = \int_a^b (\eta' + g\eta)^2 dt$$

{Beesack [1], Lemma 1.1; special cases appear in Ths. 257 and 262 of Hardy, Littlewood and Polya [1]}.

20. Let $q(t)$, $q_1(t)$ be continuous, non-negative, symmetric functions on $(-\alpha,\alpha)$ and suppose that the differential equation $(12.1|q)$ has a solution $u(t)$ which is symmetric with respect to $t = 0$, and is positive on $(-\alpha,\alpha)$. If

$$\int_0^t q_1(s)u(s)ds \le \int_0^t q(s)u(s)ds, \quad 0 \le s < \alpha, \tag{12.9}$$

then the differential equation $(12.1|q_1)$ has a solution which does not vanish on $(-\alpha,\alpha)$. {Beesack [1], Lemma 5.1]}.

21. Let $q(t)$ be continuous, and non-negative on $[-\alpha,\alpha]$, while $q(t)$ is symmetric with respect to $t = 0$, and non-increasing on $[0,\alpha]$. If $(12.1|q)$ has a solution which does not vanish on $(-\alpha,\alpha)$ and $q_1(t) = q(\alpha-t)$ on $[0,\alpha]$, $q_1(t) = q_1(-t)$ on $[-\alpha,\alpha]$, then equation $(12.1|q_1)$ has a solution with the same property. {Beesack [1], Lemma 5.2; proved as an application of preceding lemma}.

22. Let $q(t)$ be continuous and non-negative on $[-\alpha,\alpha]$, symmetric with respect to $t = 0$, and suppose $(12.1|q)$ has a solution which does not vanish on $(-\alpha,\alpha)$. If

$0 < \beta < \alpha$, and $q_1(t) = q(t+\beta)$, $0 \le t \le \alpha-\beta$, $q_1(t) = q(-t)$, $\beta - \alpha \le t \le 0$, then the equation $(12.1|q_1)$ has a solution which does not vanish on $\beta - \alpha < t < \alpha - \beta$. {Beesack [1], Lemma 5.3.}.

23. Let $q(t)$ be continuous on $(-\alpha,\alpha)$, with $q(-t) = q(t) \ge 0$, and suppose $q(t)$ is non-decreasing on $[0,\alpha)$.

A. If $(12.1|q)$ has a solution $u(t)$ with $u(0) = 0$ and $u'(t) \ge 0$, $0 \le t \le \alpha$, then this equation is disconjugate on $(-\alpha,\alpha)$.

B. If $(12.1|q)$ is disconjugate on $(0,\alpha)$ and $u(t)$ is a solution of this equation with $u(0) = 0$, $u'(0) > 0$, then $u'(t) > 0$ on $[0,\alpha/2)$. {Beesack [1, Lemma 5.4]}.

24. Suppose that $(12.1|q)$ has a solution $u_1(t)$ with $u_1(0) = 0$, $u_1'(a) \ge 0$, and $u_1(t) > 0$ on $(0,a]$.

(i) If η is a.c. on $[0,a]$ with $\eta' \in \mathscr{L}^2[0,a]$ and $\eta(0) = 0$, then

$$\int_0^a [\eta'(t)]^2 dt \ge \int_0^a q(t)\eta^2(t)dt, \tag{12.10}$$

with equality only if $\eta(t)$ is a constant multiple of $u_1(t)$, and $\eta(t) \equiv 0$ if $u_1'(a) > 0$.

(ii) If, in addition, $\int_0^a q(t)dt \ge 0$ and η is a.c. on $[0,a]$ with $\eta' \in \mathscr{L}^2[0,a]$ with $\eta(0)\int_0^a q(t)\eta(t)dt \le 0$, then (12.10) holds. {Beesack [2], Ths. 1.1 and 1.1* for (i) and (ii), respectively}.

25. Suppose that $(12.1|q)$ has a solution $u_2(t)$ with $u_2(0) > 0$, $u_2(a) < 0$ and $u_2'(t) < 0$ for $t \in (0,a)$, while $\int_0^a q(t)dt \ge 0$. If η is a.c. on $[0,a]$ with $\eta' \in \mathscr{L}^2[0,a]$ and $\int_0^a q(t)\eta(t)dt = 0$, then (12.10) holds. If either

$u_2'(0) \neq 0$ or $u_2'(a) \neq 0$, equality in (12.10) holds only if η is constant on $[0,a]$, and only if $\eta(t) \equiv 0$ in case $\int_0^a q(t)dt > 0$. If $u_2'(0) = u_2'(a) = 0$, then equality in (12.10) holds only if $\eta(t) = k_o + k_1u_2(t)$, and $k_o = 0$ if $\int_0^a q(t)dt > 0$. {Beesack [2], Th. 1.2}.

26. Suppose that $q(t)$ is continuous on $(-a,a)$ and $q(t) = O[(t+a)^{-2}]$ as $t \to -a^+$, and $q(t) = O[(t-a)^{-2}]$ as $t \to a^-$. If $(12.1|q)$ has a solution $u_1(t) > 0$ for $t \in (-a,a)$, then for η a.c. on $[-a,a]$ with $\eta' \in \mathscr{L}^2[-a,a]$ and $\eta(a) = 0 = \eta(-a)$ we have

$$\int_{-a}^a [\eta'(t)]^2 dt \geq \int_{-a}^a q(t)\eta^2(t)dt, \tag{12.11}$$

with equality holding only if $\eta(t)$ is a constant multiple of $u_1(t)$, and $\eta(t) = 0$ if either $u_1(a) \neq 0$ or $u_1(-a) \neq 0$.

{Beesack [2], Th. 1.3. As examples of the result Beesack lists the following, the first of which is due to Nehari [1], and the second is given by Hardy, Littlewood and Polya [1, Ex. 262, p. 193]:

(i) $\displaystyle\int_{-1}^1 [\eta'(t)]^2 dt > \int_{-1}^1 \frac{\eta^2(t)}{(1-t^2)^2}\, dt,$

unless $\eta(t) = k(1-t^2)^{1/2}$,

(ii) $\displaystyle\int_{-1}^1 [\eta'(t)]^2 dt > \int_{-1}^1 \frac{2\eta^2(t)}{1-t^2}\, dt,$

unless $\eta(t) = c(1-t^2)$,

where in each case η is supposed to be a.c. on $[-1,1]$, with $\eta' \in \mathscr{L}^2[-1,1]$, and $\eta(-1) = 0 = \eta(1)$}.

27. Suppose that the continuous function $q:[-a,a] \to \mathbf{R}$ is positive and symmetric on $[-a,a]$, and non-increasing on $[0,a]$, and that $(12.1|q)$ has an even solution $u_1(t)$ with

consecutive zeros at $t = -a$ and $t = a$. If η is a.c. on $[-a,a]$, with $\eta' \in \mathscr{L}^2[-a,a]$, $\eta(-a) = \eta(a)$, and there exists a t_0 such that $\eta(t_0) = 0$ for some $t_0 \in [-a,a]$ then inequality (12.11) is valid, with equality holding only if $\eta(t)$ is a constant multiple of $u_1(t)$. {Beesack [2, Th. 1.4]}.

28. For $\alpha = 1,2$ let $q_\alpha : [a,b] \to \mathbf{R}$ be continuous functions, and consider solutions $u = u_\alpha(t)$ of $\ell[u|q_\alpha](t) = 0$, $(\alpha = 1,2)$, with $u_1(t) \neq 0$ for $t \in [a,b]$.

(i) If $u_2(a) \neq 0$, and $-u_1'(a)/u_1(a) + \int_a^t q_1(s)ds > |-u_2'(a)/u_2(a) + \int_a^t q_2(s)ds|$ for $t \in [a,b]$, then $u_2(t) \neq 0$ on $[a,b]$ and $-u_1'(t)/u_1(t) > |u_2'(t)/u_2(t)|$ for $t \in [a,b]$;

(ii) If $u_2(b) \neq 0$, and $u_1'(b)/u_1(b) + \int_t^b q_1(s)ds > |u_2'(b)/u_2(b) + \int_t^b q_2(s)ds|$ for $t \in [a,b]$, then $u_2(t) \neq 0$ on $[a,b]$ and $u_1'(t)/u_1(t) > |u_2'(t)/u_2(t)|$ for $t \in [a,b]$;

(iii) If there exists a non-identically vanishing solution $u_2(t)$ of $\ell[u|q_2](t) = 0$ satisfying $u_2(a) = 0 = u_2(b)$ and there exists a value $c \in (a,b)$ at which $u_2'(c) = 0$ and

$$\int_t^c q_1(s)ds \geq \left|\int_t^c q_2(s)ds\right|, \quad \text{for } t \in [a,c],$$

$$\int_c^t q_1(s)ds \geq \left|\int_c^t q_2(s)ds\right|, \quad \text{for } t \in [c,b],$$

then every real solution of $\ell[u|q_1](t) = 0$ has at least one zero on $[a,b]$.

{A. Ju. Levin [1]. Proofs of these results are to be found in Sec. 7 of Ch. 1 of Swanson [1]}.

29. Suppose that $q:[a,b] \to R$ is a positive function of class $\mathscr{C}'$ and has a piecewise continuous second derivative satisfying $qq'' - (q')^2 < 0$ on $[a,b]$. If $t = a$ and $t = b$ are successive conjugate points which respect to $(12.1|q)$, then $\int_a^b \sqrt{q(t)}dt < \pi$. {Makai [1]}.

30. Suppose that $q(t):(-\infty,\infty) \to R$ is a continuous function and let $u_1(t)$, $u_2(t)$ be two solutions of $(12.1|q)$ with Wronskian $W(t;u_2,u_1) \equiv u_2(t)u_1'(t) - u_1(t)u_2'(t) \equiv 1$. The curve $x = u_2(t)$, $y = u_1(t)$ in the (x,y)-plane is called the *indicatrix* for $(12.1|q)$. Geometrically, as t increases the radius vector moves counter-clockwise and twice the area swept out in moving from $P_1 = (u_2(t_1),u_1(t_1))$ to $P_2(u_2(t_2),u_1(t_2))$ is $t_2 - t_1$. Let $F(t_1,t_2) = u_1(t_2)u_2(t_1) - u_1(t_1)u_2(t_2)$, which is twice the signed area of the triangle with vertices at the origin z and P_1, P_2, with the sign interpreted in the usual fashion. Also, set $G(t_1,t_2) = u_1'(t_1)u_2(t_2) - u_1(t_2)u_2'(t_1)$. Geometrically, if the line through z and P_2 intersects the tangent to the indicatrix at P_1 in the point P_2^*, then $G(t_1,t_2) = zP_2/zP_2^*$ if z does not separate P_2 and P_2^*, $G(t_1,t_2) = -zP_2/zP_2^*$ if z separates P_2 and P_2^*, while $G(t_1,t_2) = 0$ if the line zP_2 is parallel to the tangent to the indicatrix at P_1. The functions $F(t_1,t_2)$ and $G(t_1,t_2)$ are independent of which pair of solutions with $W(t;u_2,u_1) \equiv 1$ is chosen, in particular, for any t_o we may choose $u_1(t)$, $u_2(t)$ with initial conditions $u_1(t_o) = 0$, $u_1'(t_o) = 1$, $u_2(t_o) = 1$, $u_2'(t_o) = 0$. For these particular solutions, a zero of $u_1(t)$ is a conjugate point to t_o, and a zero of $u_2(t)$ a focal point of t_o. If every point t_o has a focal point which follows it,

then let t_0^* denote the first such focal point and set $m(t_0) = t_0^* - t_0$, $\alpha(t_0) = \mu_1(t_0^*)$. Then $G(t,t + m(t)) \equiv 0$, and $\alpha(t) \equiv F(t,t + m(t))$. The tangent to the indicatrix at $P(t)$ is parallel to the line through z and $P(t + m(t))$, and $m(t)$ is twice the area swept out by the radius vector as the parameter increases from t to $t + m(t)$. Also, $\alpha(t)$ is equal to twice the area of the triangle determined by z, $P(t)$ and $P(t + m(t))$. If $(12.1|q)$ is oscillatory for large t, define $\lambda(t)$ by $\lambda(t_0) = t_0^* - t_0$, where t_0^* is the smallest conjugate point to t_0 exceeding t_0. Then $F(t,t + \lambda(t)) = 0$ and the points on the indicatrix corresponding to t and $t + \lambda(t)$ lie on a line through z and are separated by z; moreover, the quantity $\lambda(t)$ is twice the area swept out by the radius vector as it moves counterclockwise through the corresponding straight angle.

The corresponding curve defined parametrically by $x = u_2'(t)$, $y = u_1'(t)$, which Petty and Barry call the *hodograph*, may be constructed from the indicatrix by a polar reciprocation with respect to z followed by a counterclockwise rotation of $\pi/2$. If P is the point on the indicatrix corresponding to $t + m(t)$, and Q is the point on the hodograph corresponding to t, then these two points lie on a common ray from the origin z and $\alpha^{-1}(t) = zQ/zP$. Geometrically, this latter equation follows from the triangle area interpretation of $\alpha(t)$ and the parallelogram area interpretation of the Wronskian. As $u_1''(t)u_2'(t) - u_1'(t)u_2''(t) = q(t)$, with increasing t the radius vector of the hodograph moves counterclockwise if $q(t) > 0$, clockwise if $q(t) < 0$, and is stationary if $q(t) = 0$. The integral

$\int_{t_1}^{t_2} q(s)ds$ is twice the signed area swept out by the radius vector of the hodograph as t runs from t_1 to t_2. An easy way to compute examples is to use the following result, given as Theorem 3.9 of Petty and Barry [1]. If $g(t)$ is any solution of the differential equation

$$g''(t) - [g'(t)]^2 + e^{4g(t)} = q(t), \tag{12.12}$$

and $\theta(t) = \int_{t_o}^{t} e^{2g(\tau)}d\tau$, then $u_1(t) = e^{-g(t)}\sin\theta(t)$ and $u_2(t) = e^{-g(t)}\cos\theta(t)$ are independent solutions of $(12.1|q)$ with $W(t|u_2,u_1) \equiv 1$. Also, if u_1, u_2 are any two solutions of $(12.1|q)$ with $W(t\ u_2,u_1) \equiv 1$, then $g = \ell n[u_1^2(t) + u_2^2(t)]^{-\frac{1}{2}}$ is a solution of (12.12). Let λ be any positive constant, and $g(t)$ any periodic function of class $\mathscr{C}^2$ with period λ and such that $\int_0^{\lambda} e^{2g(\tau)}d\tau = \pi$. Then $q(t)$ is such that: (i) $\lambda(t)$ is constant; (ii) for any solution $u(t)$, $|u''(t)|$ has the same value at the zeros of $u(t)$; (iii) for any non-trivial solution $u(t)$, there exists a non-trivial solution $u^*(t)$ such that $u^{*\prime}(t) = 0$ whenever $u(t) = 0$. If $q(t)$ is non-negative, and characterized by the above property, the differential equation is called Minkowskian. The curvature of the indicatrix is non-negative, and it is a closed convex curve with center z. {See Petty and Barry [1], where this geometric interpretation of oscillation phenomena for $(12.1|q)$ is discussed in considerable detail. Related discussion is also to be found in Guggenheimer [1,2] and Borůvka [4]}.

31. Consider an equation $(12.1|q)$ with $q(t)$ locally of class $\mathscr{L}$ on an open interval I of the real line, and for $\tau \in I$ let $u_1(t) = u_1(t;\tau)$ and $u_2(t) = u_2(t;\tau)$ be linearly

independent solutions of this equation satisfying $u_1(\tau) = 0$, $u_2'(\tau) = 0$. In the terminology of Borůvka, a value $x \in I$ distinct from τ is a *conjugate* to τ, {with respect to (12.1|q)}, of the *first kind* if $u_1(x) = 0$, of the *second kind* if $u_2'(x) = 0$, of the *third kind* if $u_1'(x) = 0$, and of the *fourth kind* if $u_2(x) = 0$. As remarked in Section I.3, these kinds of conjugates correspond, respectively, to *conjugate, deconjugate, pseudoconjugate,* and *hemiconjugate* in the terminology of Picone [1].

We shall restrict attention to the case of $q(t)$ a continuous, positive function on I, so that if x is a conjugate to τ of either of the four kinds then in the class of this kind of conjugates x is an isolated value. Moreover, for simplicity of discussion it will be assumed that an arbitrary non-identically vanishing real solution of (12.1|q) has infinitely many zeros in arbitrary neighborhoods of each of the end-points of I. For $\tau \in I$, let $\phi_o(\tau) = \tau$ and denote by $\phi_n(\tau)$, $\{\phi_{-n}(\tau)\}$, $(n = 1,2,\ldots)$, the n-th right-hand, {left-hand} conjugate to τ on I of the first kind. Also, let $\chi_n(\tau)$, $\{\chi_{-n}(\tau)\}$, $(n = 1,2,\ldots)$, denote the n-th right-hand, {left-hand} conjugate to τ on I of the third kind. In view of the condition of unlimited oscillation in the neighborhood of each end-point of I, each of the sequences $\{\phi_n(\tau)\}, \{\phi_{-n}(\tau)\}$ is infinite, and consequently each of the sequences $\{\chi_n(\tau)\}$, $\{\chi_{-n}(\tau)\}$ is infinite and

$$\ldots < \chi_{-2}(\tau) < \phi_{-1}(\tau) < \chi_{-1}(\tau) < \phi_o(\tau)$$
$$= \tau < \chi_1(\tau) < \phi_1(\tau) < \chi_2(\tau) < \ldots$$

Correspondingly, let $\psi_o(\tau) = \tau$ and denote by $\psi_n(\tau)$, $\{\psi_{-n}(\tau)\}$, $(n = 1,2,\ldots)$, the sequences of right- and left-hand conjugates to τ of the second kind. Moreover, if $\omega_n(\tau)$, $\{\omega_{-n}(\tau)\}$, $(n = 1,2,\ldots)$, denote the sequences of right- and left-hand conjugates to τ of the fourth kind, then

$$\ldots < \omega_{-2}(\tau) < \psi_{-1}(\tau) < \omega_{-1}(\tau) < \psi_0(\tau)$$
$$= \tau < \omega_1(\tau) < \psi_1(\tau) < \omega_2(\tau) \quad \ldots$$

For $\nu = 0,\pm 1,\pm 2,\ldots$, the functions $\phi_\nu : I \to I$, $\psi_\nu : I \to I$ have been called the ν-th *central dispersions of the first and second kind*, respectively; similarly for $\nu = \pm 1,\pm 2,\ldots$, the functions $\chi_\nu : I \to I$, $\omega_\nu : I \to I$ have been called the ν-th *central dispersions of the third and fourth kinds*, respectively. Under the above stated conditions, each central dispersion is a continuous, monotone increasing function whose range is I.

Let Φ, Ψ, χ, Ω denote the totality of central dispersions of the same kind, whose individual functions have been represented above by the corresponding lower case Greek letter. With the usual operation of functional composition, the defined sets Φ and Ψ are infinite cyclic groups with respective unit elements ϕ_0 and ψ_0. In general, $\phi_\mu\phi_\nu = \phi_{\mu+\nu}$ and $\psi_\mu\psi_\nu = \psi_{\mu+\nu}$; in particular, $\phi_\nu\phi_{-\nu} = \phi_0$ and $\psi_\nu\psi_{-\nu} = \psi_0$. For other combinations of central dispersions the composition formulas are more complicated to describe, (see, for instance, Borůvka [4, p. 108]). For example, $\chi_\rho\omega_\sigma = \psi_{\rho+\sigma}$ and $\omega_\rho\chi_\sigma = \phi_{\rho+\sigma}$ for $\rho > 0$, $\alpha < 0$ and for $\rho < 0$, $\sigma > 0$; $\chi_\rho\omega_\sigma = \psi_{\rho+\sigma-1}$ and $\omega_\rho\chi_\sigma = \phi_{\rho+\sigma-1}$ for $\rho > 0$, $\sigma > 0$; $\chi_\rho\omega_\sigma = \psi_{\rho+\sigma+1}$, $\omega_\rho\chi_\sigma = \phi_{\rho+\sigma+1}$ for $\rho < 0$, $\sigma < 0$; in particular, $\chi_\rho\omega_{-\rho} = \psi_0$ and $\omega_\rho\chi_{-\rho} = \phi_0$.

If $u(t)$ and $v(t)$ are two linearly independent solutions of $(12.1|q)$, let $W_\alpha(t,\tau)$, $(\alpha = 1,2,3,4)$, $r(t)$ and

$s(t)$ be defined as follows: $W_1(t,\tau) = u(t)v(\tau) - u(\tau)v(t)$, $W_2(t,\tau) = u'(t)v'(\tau) - u'(\tau)v'(t)$, $W_3(t,\tau) = u(t)v'(\tau) - u'(\tau)v(t)$, $W_4(t,\tau) = u'(t)v(\tau) - u(\tau)v'(\tau) = -W_3(\tau,t)$, $r(t) = \sqrt{u^2(t) + v^2(t)}$, $s(t) = \sqrt{[u'(t)]^2 + [v'(t)]^2}$. Then the central dispersion functions satisfy the following functional equations on I:

$$W_1(t,\phi(t)) = 0,\ W_2(t,\psi(t)) = 0,\ W_3(t,\chi(t)) = 0,$$
$$W_4(t,\omega(t)) = 0.$$

Moreover, each central dispersion function is of class $\mathscr{C}'$ on I, with derivatives given by:

$$\phi'(t) = -\frac{W_4(t,\phi(t))}{W_3(t,\phi(t))} = \frac{r^2(\phi(t))}{r^2(t)};$$

$$\psi'(t) = -\frac{q(t)}{q(\psi(t))}\frac{W_3(t,\psi(t))}{W_4(t,\psi(t))} = \frac{q(t)}{q(\psi(t))}\frac{s^2(\psi(t))}{s^2(t)};$$

$$\chi'(t) = \frac{1}{q(\chi(t))}\frac{W_2(t,\chi(t))}{W_1(t,\chi(t))} = \frac{1}{q(\chi(t))}\frac{s^2(\chi(t))}{r^2(t)};$$

$$\omega'(t) = q(t)\frac{W_1(t,\omega(t))}{W_2(t,\omega(t))} = q(t)\frac{r^2(\omega(t))}{r^2(t)}.$$

Also, in terms of a single non-identically vanishing solution $u(t)$ of $(12.1|q)$ and its derivative, we have the following formulas:

$$\phi'(t) = u^2(\phi(t))/u^2(t) \quad \text{if} \quad u(t) \neq 0,$$
$$= [u'(t)]^2/[u'(\phi(t))]^2 \quad \text{if} \quad u(t) = 0;$$

$$\frac{q[\psi(t)]}{q(t)}\psi'(t) = [u'(\psi(t))]^2/[u'(t)]^2 \quad \text{if} \quad u'(t) \neq 0,$$
$$= u^2(t)/u^2(\psi(t)) \quad \text{if} \quad u'(t) = 0;$$

$$q(\chi(t))\chi'(t) = [u'(\chi(t))]^2/u^2(t) \quad \text{if} \quad u(t) \neq 0,$$
$$= [u'(t)]^2/u^2(\chi(t)) \quad \text{if} \quad u(t) = 0;$$

$$q(t)\omega'(t) = u^2(\omega(t))/[u'(t)]^2 \quad \text{if} \quad u'(t) \neq 0,$$
$$= u^2(t)/[u'(\omega(t))]^2 \quad \text{if} \quad u'(t) = 0.$$

{See Borůvka [1-4]; expecially [4-Part II-Chps. 12,13], for these and many other properties of the central dispersion functions. It is to be remarked that the $q(t)$ in equation $(12.1|q)$ is the negative of the function $q(t)$ in Borůvka's equation}.

32. Consider an equation (11.13) on (a,∞), where $r(t) > 0$, $q(t) > 0$, and there exists an $n \geq 2$ such that $q' \in M_n(a,\infty)$, $1/r \in M_{n-1}(a,\infty)$, and $[q/r]' \in M_{n-2}(a,\infty)$, where for k a positive integer $\mathcal{M}_k(a,\infty)$ denotes the class of functions satisfying $(-1)^j\phi^{[j]}(t) \geq 0$, $(j = 0,1,\ldots,k;$ $t \in (a,\infty)$; moreover, suppose that at least one of the two functions q' and $[q/r]'$ is positive on (a,∞). Let $u(t)$ be a solution of this equation such that $u(t)$ and $u'(t)$ have consecutive zeros at values belonging to the respective sequences $t_1 < t_2 < \ldots$ and $t_1' < t_2' < \ldots$, while $t_1' > a$. Then the sequence $\mu_\alpha = u^2(t_\alpha')$, $(\alpha = 1,2,\ldots)$, is $(n-1)$-times completely monotonic, (i.e., $(-1)^j\Delta^j\mu_\alpha \geq 0$, $(j = 0,1,\ldots,n-1;\ \alpha = 0,1,\ldots)$ where $\Delta\mu_\alpha = \mu_{\alpha+1} - \mu_\alpha$, $\Delta^{h+1}\mu_\alpha = \Delta(\Delta^h\mu_\alpha)$, $h = 1,2,\ldots)$. {Lorch, Szego, and Muldoon I-IV, Th. 4.3}.

CHAPTER III.
SELF-ADJOINT BOUNDARY PROBLEMS ASSOCIATED WITH SECOND ORDER LINEAR DIFFERENTIAL EQUATIONS

1. A Canonical Form for Boundary Conditions

Attention will now be directed to differential systems of the form

$$(a)\quad \ell[u](t) \equiv [r(t)u'(t) + q(t)u(t)]' - [q(t)u'(t) + p(t)u(t)] = 0, \qquad (\mathscr{B})$$

$$(b)\quad s_\alpha[u,v] \equiv M_{\alpha 1}u(a) + M_{\alpha 2}v(a) + M_{\alpha 3}u(b) + M_{\alpha 4}v(b) = 0,$$

where the compact interval $[a,b]$ is a subinterval of I and the coefficient functions r, p, q of ($\mathscr{B}$-a) satisfy either hypothesis ($\mathscr{A}_C^+$) of Section II.1 or the following strengthened form of hypothesis ($\mathscr{A}_L^+$):

($\mathscr{A}_L^{++}$) r, p, q *are real-valued, Lebesgue measurable on* I *with* $r(t) > 0$ *on this interval, while the functions* r, $1/r$, q *are locally of class* $\mathscr{L}^\infty$ *and* p *is locally of class* $\mathscr{L}$ *on* I.

Clearly hypothesis ($\mathscr{A}_C^+$) implies ($\mathscr{A}_L^{++}$); moreover, under hypothesis ($\mathscr{A}_L^{++}$) the functions $a = -q/r$ and $b = 1/r$ are locally of class $\mathscr{L}^\infty$, and $c = p - q^2/r$ is locally of class

$\mathscr{L}$ on I. In ($\mathscr{B}$) the symbol $v(t)$ denotes the "canonical variable" $v(t) = r(t)u'(t) + q(t)u(t)$ associated with a $u(t)$, and $M_{\alpha j}$, $(\alpha = 1,2;\ j = 1,2,3,4)$ are real constants such that the 2×4 matrix $M = [M_{\alpha j}]$ is of rank 2. For a given non-degenerate compact subinterval $[a,b]$ of I, the class of functions $u \in D''[a,b]:v$ as defined in Section II.8, and satisfying the boundary conditions ($\mathscr{B}$-b), will be denoted by $D[\mathscr{B}]$.

Now if $u_\alpha \in D''[a,b]:v_\alpha$, $(\alpha = 1,2)$, then from equation (II.8.6) we have

$$\int_a^b u_2 \ell[u_1]dt - \int_a^b \ell[u_2]u_1 dt = u_2 v_1 - v_2 u_1 \Big|_a^b . \tag{1.1}$$

If the value of the left-hand member of (1.1) is zero for all $u_\alpha \in D$, $(\alpha = 1,2)$, then the (formal) differential operator $\ell[u]$ with domain D is said to be *symmetric* or *self-adjoint*, and correspondingly the differential system ($\mathscr{B}$) is called symmetric or self-adjoint. Clearly this condition holds if and only if the algebraic condition

$$\begin{aligned} [u_2 v_1 - v_2 u_1]\Big|_a^b = 0 \text{ for arbitrary } u_\alpha(a),\ v_\alpha(a), \\ u_\alpha(b),\ v_\alpha(b) \text{ satisfying } (\mathscr{B}\text{-b}) \end{aligned} \tag{1.2}$$

holds. In turn, this algebraic condition is readily seen to be equivalent to the 2×2 matrix equation

$$M[\operatorname{diag}\{-J,J\}]M^* = 0 \tag{1.3}$$

where J is the real skew matrix

$$J = \begin{bmatrix} 0 & -1 \\ 1 & 0 \end{bmatrix}$$

introduced in (II.1.5). Moreover, if for brevity we set

$$S^{\rho\sigma} = \begin{vmatrix} M_{1\rho} & M_{1\sigma} \\ M_{2\rho} & M_{2\sigma} \end{vmatrix}, \quad D^{\rho\sigma} = \det S^{\rho\sigma}, \ (\rho,\sigma = 1,2,3,4), \tag{1.4}$$

one may verify directly that the left-hand member of (1.3) is equal to $(D^{34} - D^{12})J$, and consequently, the differential system ($\mathscr{B}$) is self-adjoint if and only if $D^{12} = D^{34}$.

Now the restraints on the end-values of u,v at $t = a$ and $t = b$ are unchanged whenever the 2×4 matrix of coefficients M of rank two in ($\mathscr{B}$-b) is replaced by KM, where K is a non-singular 2×2 matrix. One may proceed to establish that whenever condition (1.2) holds there exists a value of K such that the boundary conditions reduce to one of the following "canonical" forms, where the involved coefficients $\gamma_{\alpha\beta}$, $(\alpha,\beta = 1,2)$, ψ_1 and ψ_2 are real constants, and the classification is based upon the rank of the 2×2 matrix S^{24}, (see, for example, Reid [35, Sec. VI.1]).

$$\begin{aligned}
&\text{I.} && u(a) = 0, \quad u(b) = 0; \\
&\text{II}_1. && u(a) = 0, \quad \gamma_{22}u(b) + v(b) = 0; \\
&\text{II}_2. && u(b) = 0, \quad \gamma_{11}u(a) - v(a) = 0; \\
&\text{II}_3. && \psi_1 u(a) + \psi_2 u(b) = 0 \\
& && -\psi_2[\gamma_{11}u(a) - v(a)] + \psi_1[\gamma_{22}u(b) + v(b)] = 0, \\
& && |\psi_1| + |\psi_2| > 0; \\
&\text{III.} && \gamma_{11}u(a) + \gamma_{12}u(b) - v(a) = 0; \\
& && \gamma_{12}u(a) + \gamma_{22}u(b) + v(b) = 0.
\end{aligned} \tag{1.5}$$

It is to be noted that in all three cases II_1, II_2, II_3, the boundary conditions may be written in the alternate parametric form

$$\begin{aligned} \psi_1 u(a) + \psi_2 u(b) &= 0, \\ \gamma_{11} u(a) - v(a) + \psi_1 \nu &= 0, \\ \gamma_{22} u(b) + v(b) + \psi_2 \nu &= 0. \end{aligned} \tag{1.5-II'}$$

where ψ_1, ψ_2 are real constants with $|\psi_1| + |\psi_2| > 0$, and ν is a real parameter. In this form the forms II_1, II_2, II_3 correspond to the respective conditions $\psi_2 = 0$, $\psi_1 = 0$, $\psi_1\psi_2 \neq 0$.

For each of the canonical forms (1.5) of the boundary conditions ($\mathscr{B}$-b), let $S[u:\mathscr{B}] = (s_\beta[u:\mathscr{B}])$, $(\beta = 1,\ldots,k_{\mathscr{B}})$, denote the boundary conditions independent of $v(a)$, $v(b)$. The boundary conditions $s[u] = 0$ are known as the "essential boundary conditions", and the remaining boundary conditions are called the associated "natural boundary conditions" or "transversality conditions". Also, for each set of boundary conditions in (1.5), let

$$Q[\eta_1,\eta_2] = \eta_2(a)[\gamma_{11}\eta_1(a)+\gamma_{12}\eta_1(b)] + \eta_2(b)[\gamma_{12}\eta_1(a)+\gamma_{22}\eta_1(b)],$$

and write $Q[\eta] = Q[\eta,\eta]$, where γ_{11}, γ_{12}, γ_{22} are the constants in the boundary conditions (1.5) if they appear explicitly, and zero otherwise. The individual cases are presented in the following table.

CASE	k_B	$s[\eta;B] = 0$	$Q[\eta]$
I	2	$\eta(a) = 0,\ \eta(b) = 0$	-------------
II_1	1	$\eta(a) = 0$	$\gamma_{22}\eta^2(b)$
II_2	1	$\eta(b) = 0$	$\gamma_{11}\eta^2(a)$
II_3	1	$\psi_1\eta(a) + \psi_2\eta(b) = 0$ $\lvert\psi_1\rvert + \lvert\psi_2\rvert > 0$	$\gamma_{11}\eta^2(a) + \gamma_{22}\eta^2(b)$
III	0	---------------	$\gamma_{11}\eta^2(a)+2\gamma_{12}\eta(a)\eta(b)$ $+ \gamma_{22}\eta^2(b).$

For a general real quadratic form

$$Q[\eta] = \gamma_{11}\eta^2(a) + 2\gamma_{12}\eta(a)\eta(b) + \gamma_{22}\eta^2(b) \tag{1.6}$$

in the end-values $\eta(a)$, $\eta(b)$, the notations $Q_a[\eta]$ and $Q_b[\eta]$ are introduced by the "partial derivative" linear form

$$Q_a[\eta] = \gamma_{11}\eta(a) + \gamma_{12}\eta(b), \quad Q_b[\eta] = \gamma_{12}\eta(a) + \gamma_{22}\eta(b).$$

In particular, we have the identities

$$Q[\eta_1,\eta_2] = \eta_2(a)Q_a[\eta_1] + \eta_2(b)Q_b[\eta_1] = \eta_1(a)Q_a[\eta_2] + \eta_1(b)Q_b[\eta_2].$$

In matrix and vector notation let Γ denote the 2×2 matrix $\Gamma = [\gamma_{\alpha\beta}]$, $(\alpha,\beta = 1,2)$, with $\gamma_{21} = \gamma_{12}$, and for a function $f:[a,b] \to R$ let $\hat{f}$ denote the two-dimensional vector $(\hat{f}_\alpha)$, $(\alpha = 1,2)$, with $\hat{f}_1 = f(a)$ and $\hat{f}_2 = f(b)$. Then $Q[\eta] = \hat{\eta}^*\Gamma\hat{\eta}$ and the two-dimensional vector with components $Q_a[\eta]$, $Q_b[\eta]$ may be written as $\Gamma\hat{\eta}$. Moreover, let D be the 2×2 matrix $\begin{bmatrix} -1 & 0 \\ 1 & 1 \end{bmatrix}$. For self-adjoint systems ($\mathscr{B}$) we then have the following characterization.

THEOREM 1.1. *A differential system* $(\mathscr{B})$ *is self-adjoint if and only if there exists a real quadratic form* $Q = Q[\eta:\mathscr{B}]$ *of the form* (1.6), *and a linear subspace* $S = S[\mathscr{B}]$ *of real two-dimensional Euclidean space* R_2, *such that end-values* $u(a)$, $v(a)$, $u(b)$, $v(b)$ *satisfy the boundary conditions* $(\mathscr{B}\text{-b})$ *if and only if*

$$\hat{u} \in S[\mathscr{B}], \quad \Gamma\hat{u} + D\hat{v} \in S^{\perp}[\mathscr{B}], \tag{1.7}$$

where $S^{\perp}[\mathscr{B}]$ *denotes the orthogonal complement of* $S[\mathscr{B}]$ *in* R_2.

If $(\mathscr{B})$ is self-adjoint, then the boundary conditions (1.5) are seen to be of the form (1.7), where $S[\mathscr{B}]$ is of dimensional 0,1 and 2 in the respective cases I, II and III. Conversely, if the end-values $u(a)$, $v(a)$, $u(b)$, $v(b)$ are required to satisfy conditions (1.7), then whenever $S[\mathscr{B}]$ is of dimension 0 or 2 the requirement is readily seen to be equivalent to conditions of the respective forms (1.5-I) and (1.5-III). If $S[\mathscr{B}]$ is of dimension 1, then there exist real constants ψ_1, ψ_2 such that $|\psi_1| + |\psi_2| > 0$, and conditions (1.7) are equivalent to (1.5-II'), or

$$\begin{aligned} \psi_1 u(a) + \psi_2 u(b) &= 0, \\ -\psi_2\{Q_a[u] - v(a)\} + \psi_1\{Q_b[u] + v(b)\} &= 0. \end{aligned} \tag{1.8}$$

In this case, (1.8) is of the form $(\mathscr{B}\text{-b})$, with the coefficient matrix M of the form

$$M = \begin{bmatrix} \psi_1 & 0 & \psi_2 & 0 \\ M_{21} & \psi_2 & M_{23} & \psi_1 \end{bmatrix},$$

so that $D^{12} = \psi_1\psi_2 = D^{24}$.

For a given self-adjoint boundary problem $(\mathscr{B})$, let $D_e[\mathscr{B}]$ denote the set of functions $\eta \in D'[a,b]$, with end-values $\eta(a)$, $\eta(b)$ satisfying the essential boundary conditions, or equivalently, $\hat{\eta} \in S[\mathscr{B}]$. In particular, in the cases I, II_1, II_2, II_3 and III the set $D_e[\mathscr{B}]$ is respectively equal to $D_0'[a,b]$, $D_{0*}'[a,b]$, $D_{*0}'[a,b]$, $\{\eta : \eta \in D'[a,b], \psi_1\eta(a) + \psi_2\eta(b) = 0\}$, and $D'[a,b]$, with the involved classes of functions defined as in Chapter II. The following simple consequence of the definition of $S[\mathscr{B}]$ is of frequent use in the consideration of self-adjoint systems $(\mathscr{B})$.

COROLLARY. *If the differential system* $(\mathscr{B})$ *is self-adjoint, and* $S[\mathscr{B}]$ *is determined as in Theorem* 1.1, *then a set of end-values* $u(a)$, $v(a)$, $u(b)$, $v(b)$ *satisfies the bound-boundary conditions* $(\mathscr{B}\text{-b})$ *if and only if* $u \in S[\mathscr{B}]$, *and*

$$Q[u,\eta] + \eta(t)v(t)\Big|_a^b = 0, \text{ for arbitrary } \eta \in D_e[\mathscr{B}]. \quad (1.9)$$

If $\dim S[\mathscr{B}] = d_B = 2 - k_B$ is not zero, and N is a $2 \times d_B$ real matrix whose column vectors form a basis for $S[\mathscr{B}]$, then $u \in S[\mathscr{B}]$ if and only if there is a real d_B-dimensional vector σ such that $u = N\sigma$. Then $Q[\eta] = \sigma^*\Theta\sigma$, where Θ is the $d_B \times d_B$ real symmetric matrix $\Theta = N^*\Gamma N$, and the second condition of (1.7) may be written as $\Theta\sigma + N^*D\hat{v} = 0$. That is, the general boundary conditions of a self-adjoint differential system are either those of the null-end point problem

$$u(a) = 0, \quad u(b) = 0 \quad (1.10a)$$

or has the parametric representation

$$u = N\sigma, \quad \Theta\sigma + N^*D\hat{v} = 0, \quad (1.10b)$$

where N is a real $2 \times d_B$ matrix, and Θ is a real symmetric 2×2 matrix. This form for the self-adjoint boundary conditions has been preferred by Morse [1,2].

In accord with the notation of Chapter II, for a problem $(\mathscr{B})$ with associated end-form $Q = Q[\eta:\mathscr{B}]$ and $\eta_\alpha \in D'[a,b]:\zeta_\alpha$, $(\alpha = 1,2)$, let

$$\begin{aligned} J[\eta_1,\eta_2:\mathscr{B}] &= Q[\eta_1,\eta_2:\mathscr{B}] + \int_a^b \{\eta_2'[r\eta_1'+q\eta_1] + \eta_2[q\eta_1'+p\eta_1]\}dt \\ &= Q[\eta_1,\eta_2:\mathscr{B}] + J[\eta_1,\eta_2;a,b], \end{aligned} \tag{1.11}$$

and as usual we write $J[\eta_1:\mathscr{B}] = J[\eta_1,\eta_1:\mathscr{B}]$. Corresponding to the results of Theorems II.8.1 and II.8.7 for the above described cases I and II_2, for the general self-adjoint system $(\mathscr{B})$ we have that u is a solution of this system if and only if $u \in D_e[\mathscr{B}]$ and $J[u,\eta:\mathscr{B}] = 0$ for arbitrary $\eta \in D_e[\mathscr{B}]$. Also, as an extension of results established in Chapter II for the special cases I and II_2 of the above-formulated problem, we have that if $(\mathscr{B})$ is self-adjoint and $u(t)$ is a solution of the differential system

$$\ell[u](t) + f(t) = 0, \quad s_\alpha[u,v] = 0, \quad (\alpha = 1,2), \tag{1.12}$$

where f is an integrable function on $[a,b]$, then

$$J[u,\eta:\mathscr{B}] = \int_a^b \eta(t)f(t)dt, \text{ for arbitrary } \eta \in D_e[\mathscr{B}]; \tag{1.13}$$

in particular,

$$J[u:\mathscr{B}] = \int_a^b u(t)f(t)dt. \tag{1.14}$$

2. Extremum Properties for Self-Adjoint Systems

For a self-adjoint system $(\mathscr{B})$ with $D[\mathscr{B}]$, $D_e[\mathscr{B}]$, $Q[\eta:\mathscr{B}]$ and $J[\eta:\mathscr{B}]$ specified as in the preceding section, let $D_N[\mathscr{B}]$ denote the set

$$D_N[\mathscr{B}] = \{\eta:\eta \in D_e[\mathscr{B}], \int_a^b \eta^2(t)dt = 1\}. \tag{2.1}$$

One of the most basic properties for such a system is presented in the following lemma.

LEMMA 2.1. *Suppose that* $(\mathscr{B})$ *is self-adjoint, and that* $J[\eta:\mathscr{B}]$ *is non-negative on* $D_e[\mathscr{B}]$. *Then the infimum of* $J[\eta:\mathscr{B}]$ *on* $D_N[\mathscr{B}]$ *is zero if and only if* $(\mathscr{B})$ *has a non-identically vanishing solution; moreover, if* $u \in D_e[\mathscr{B}]$ *and* $J[u:\mathscr{B}] = 0$, *then* u *is a solution of* $(\mathscr{B})$.

Now suppose that $\mathscr{F} = [f_1,\ldots,f_r]$ is a set of real-valued continuous functions on a given interval $[a,b] \subset I$ which is linearly independent, so that, in particular, the $r \times r$ matrix $\left[\int_a^b f_i(t)f_j(t)dt\right]$, $(i,j = 1,\ldots,r)$, is non-singular. Then one has the following generalization of the above lemma to the case of a system involving isoperimetric orthogonality conditions with respect to the functions of $\mathscr{F}$. For brevity, the class of functions in $D_e[\mathscr{B}]$ and $D_N[\mathscr{B}]$, respectively, and which satisfy the boundary conditions

$$\begin{aligned} &(a)\quad s_\alpha[u,v] = 0, && (\alpha = 1,2), \\ &(b)\quad \int_a^b f_i(t)u(t)dt, && (i = 1,\ldots,r), \end{aligned} \tag{2.2}$$

will be denoted by $D_e[\mathscr{B}|\mathscr{F}]$ and $D_N[\mathscr{B}|\mathscr{F}]$, respectively.

LEMMA 2.2. *Syppose that* $(\mathscr{B})$ *is self-adjoint, and that* $J[\eta:\mathscr{B}]$ *is non-negative on* $D_e[\mathscr{B}|\mathscr{F}]$. *Then the infimum of*

$J[\eta:\mathscr{B}]$ *on* $D_N[\mathscr{B}|\mathscr{F}]$ *is zero if and only if there exist real constants* $k_1,\ldots,k_r$ *such that there is a real-valued non-identically vanishing solution of the differential equation*

$$\ell[u](t) + \sum_{j=1}^{r} k_j f_j(t) = 0, \qquad t \in [a,b], \tag{2.3}$$

which satisfies the boundary conditions (2.2-a,b); *moreover, if* $u_o \in D_e[\mathscr{B}|\mathscr{F}]$ *and* $J[u_o:\mathscr{B}] = 0$, *then* u_o *is a solution of the system* (2.3), (2.2) *for suitable real constants* $k_1,\ldots,k_r$. *In particular, if the functions* f_i *of* $\mathscr{F}$ *are such that there exist constants* γ_i *and functions* u_i *satisfying for* $i = 1,\ldots,r$ *the differential system*

$$\begin{aligned} \ell[u_i](t) + \gamma_i f_i(t) &= 0, \qquad t \in [a,b], \\ s_\alpha[u_i,v_i] &= 0, \qquad (\alpha = 1,2), \end{aligned} \tag{2.4}$$

and the $r \times r$ *matrix* $\left[\int_a^b u_i(t) f_j(t)\,dt\right]$, $(i,j = 1,\ldots,r)$, *is non-singular, then whenever* u *is a solution of* (2.3), (2.2) *with constants* $k_1,\ldots,k_r$ *each* k_j, $(j = 1,\ldots,r)$, *is equal to zero.*

Each of the above two lemmas is concerned with the solution of a variational problem involving the minimization of the quadratic functional in a certain class of functions, and both in special instances and in the general form stated above various types of proofs have been given. Some are indirect, wherein under the assumption that the stated conclusion does not hold the solvability theorems for associated differential systems are employed to obtain a member of the considered class of functions which provides a negative value to $J[\eta:\mathscr{B}]$, and thus contradicting the assumption that this functional is non-negative on the considered class, (see, for example, Reid [35, Sec. VII.2]). Others are by direct methods, involving the proof that a minimizing sequence of functions for the functional $J[\eta:\mathscr{B}]$ in the considered class $D_N[\mathscr{B}]$ or $D_N[\mathscr{B}|\mathscr{F}]$ possesses a type of "compactness" that permits the extraction of a subsequence of functions which converges in a suitable manner to a function that subsequently is shown to satisfy the differential equation and set of boundary conditions of the given system; for simple examples of this method, see Tonelli [1, Secs. 86, 138]. Moreover, in following either of these general procedures use may be made of auxiliary disciplines, such as the method of integral equations in the first instance, (see, for example, Mason [1]), an approximating system of algebraic difference equations (see, for example, Courant [2]), or a finite dimension problem as occurs in applying the method of Ritz, (see, for example, Sansone, [Part II, Ch. XI, §5]).

In particular, the results of the above two lemmas may be used to establish readily the existence and properties of eigenvalues and eigenfunctions for a self-adjoint boundary problem of the form

$$\begin{aligned} &(a) \quad \ell[u;\lambda](t) \equiv \ell[u](t) + \lambda k(t)u(t) = 0, \quad t \in [a,b], \\ &(b) \quad s_\alpha[u,v] \equiv M_{\alpha 1}u(a) + M_{\alpha 2}v(a) + M_{\alpha 3}u(b) + M_{\alpha 4}v(b) = 0, \\ &\qquad\qquad (\alpha = 1,2) \end{aligned} \tag{2.5}$$

where $\ell[u]$ and $s_\alpha[u,v]$ are as in $(\mathscr{B})$, and the following hypothesis is satisfied:

$(\mathscr{A})$ *The functions* r, p, q *satisfy hypothesis* $(\mathscr{A}_L^{++})$, *and* k(t) *is a positive (Lebesgue) measurable function which is such that* k *and* 1/k *are locally of class* $\mathscr{L}^\infty$

on I, *while the real coefficients* $M_{\alpha j}$, $(\alpha = 1,2;$ $j = 1,2,3,4)$ *in* (2.5-b) *are such that the* 2×4 *matrix* $M = [M_{\alpha j}]$ *is of rank two and the self-adjointness condition* (1.3) *holds.*

In particular, in view of the assumption that the functions r and k are positive and such that r, $1/r$, k and $1/k$ are locally of class $\mathscr{L}^\infty$ on I, for $[a,b]$ a compact subinterval of I there exist positive constants $\kappa = \kappa[a,b]$ and $\kappa_1 = \kappa_1[a,b]$ such that

$$\kappa \le r(t) \le 1/\kappa, \quad \kappa_1 \le k(t) \le 1/\kappa_1 \text{ for } t \text{ a.e. on } [a,b]. \tag{2.6}$$

In particular, when hypothesis $(\mathscr{U})$ is satisfied, as in Section 1 there is an associated quadratic form

$$\begin{aligned} Q[\eta:\mathscr{B}] = \gamma_{11}[\mathscr{B}]\eta^2(a) &+ 2\gamma_{12}[\mathscr{B}]\eta(a)\eta(b) \\ &+ \gamma_{22}[\mathscr{B}]\eta^2(b), \end{aligned} \tag{2.7}$$

the linear space $S[\mathscr{B}]$ of real end-values $\hat{\eta} = (\hat{\eta}_\alpha)$, $(\alpha = 1,2)$, with $\hat{\eta}_1 = \eta(a)$, $\hat{\eta}_2 = \eta(b)$, the associated function spaces

$$\begin{aligned} &\text{(a)} \quad D_e[\mathscr{B}] = \{\eta : \eta \in D'[a,b],\ \hat{\eta} \in S[\mathscr{B}]\} \\ &\text{(b)} \quad D[\mathscr{B}] = \{u : u \in D''[a,b] : v,\ s_\alpha[u,v] = 0,\ \alpha = 1,2\}, \end{aligned} \tag{2.8}$$

and the quadratic functional

$$\begin{aligned} J[\eta:\mathscr{B}] &= Q[\eta:\mathscr{B}] + \int_a^b \{r\eta'^2 + 2q\eta\eta' + p\eta^2\}dt, \\ &= Q[\eta:\mathscr{B}] + J[\eta:a,b]. \end{aligned} \tag{2.9}$$

Also, for brevity we set

$$K[\eta_1,\eta_2] = \int_a^b \eta_2(t)k(t)\eta_1(t)dt, \quad K[\eta] = K[\eta,\eta]. \tag{2.10}$$

With the aid of elementary algebraic inequalities one may establish the existence of a λ_0 such that

$$J[\eta;\lambda:B] = J[\eta:B] - \lambda K[\eta] \tag{2.11}$$

is positive definite on $D'[a,b]$ if $\lambda < \lambda_0$. Indeed, one may establish the following result.

LEMMA 2.3. *When hypothesis* ($\mathscr{A}$) *holds there exist constants* λ_o, $\mu > 0$ *such that for* $\lambda < \lambda_0$ *we have*

$$\begin{aligned} J[\eta;\lambda:B] &\geq \mu\left[\eta^2(a) + \eta^2(b) + \int_a^b \eta'^2 dt\right] \\ &\quad + (\lambda_0 - \lambda)\int_a^b \eta^2 dt, \quad \text{for } \eta \in D'[a,b]. \end{aligned} \tag{2.12}$$

To prove the result of this lemma, let $p_1(t) = \int_\tau^t p(s)ds$, where τ is some fixed value on $[a,b]$, and set $q_1(t) = q - p_1(t)$, $Q_1[\eta:B] = Q[\eta:\mathscr{B}] + [p_1(b) - 1]\eta^2(b) - [p_1(a) + 1]\eta^2(a)$. Then an integration by parts yields the relation

$$\begin{aligned} J[\eta:\mathscr{B}] &= \eta^2(a) + \eta^2(b) + Q_1[\eta:\mathscr{B}] \\ &\quad + \int_a^b \{r\eta'^2 + 2q_1\eta\eta'\}dt. \end{aligned}$$

Now if c is a constant such that $Q_1[\eta:\mathscr{B}] \geq -c[\eta^2(a) + \eta^2(b)]$ and $\phi(t) = [a + b - 2t]/[b - a]$, we have

$$\begin{aligned} Q_1[\eta:\mathscr{B}] &\geq -c[\eta^2(a) + \eta^2(b)] \\ &= \int_a^b c\{2\phi\eta\eta' + \phi'\eta^2\}dt, \quad \text{for } \eta \in D'[a,b]. \end{aligned}$$

Also, since q is locally of class $\mathscr{L}^\infty$ by hypothesis ($\mathscr{A}_L^{++}$), there exists a positive constant c_1 such that

$$\frac{2c}{b-a} + \frac{2}{\kappa}[q_1(t) + c\phi(t)]^2 \leq c_1, \text{ for } t \in [a,b], \tag{2.13}$$

where κ is as in (2.6). Consequently, since $\phi'(t) = -2/[b-a]$, with the aid of elementary algebraic inequalities it follows that for c_1 as in (2.13) and $c_0 = \kappa/2$ we have

$$J[\eta:\mathscr{B}] \geq \eta^2(a) + \eta^2(b) + c_0\int_a^b \eta'^2 dt - c_1\int_a^b \eta^2 dt, \quad \text{for } \eta \in D'[a,b]. \tag{2.14}$$

Moreover, since

$$K[\eta] \geq \kappa_1 \int_a^b \eta^2 dt \tag{2.15}$$

where κ_1 is as in (2.6), we have that

$$J[\eta:\mathscr{B};\lambda] \geq \eta^2(a)+\eta^2(b) + c_0\int_a^b \eta'^2 dt + [-c_1/\kappa_1-\lambda]\int_a^b \eta^2 dt,$$

so that inequality (2.12) holds with μ the smaller of the values 1, κ_1 and c_0, and $\lambda_0 = -c_1/\kappa_1$. It is to be emphasized that the thus determined value λ_0 is independent of the particular boundary conditions (2.5-b) of this problem.

It is also to be noted that by an even more elementary algebraic argument one may establish the existence of a positive constant μ_1 such that

$$J[\eta:\mathscr{B}] \leq \mu_1\left[\eta^2(a) + \eta^2(b) + \int_a^b \{\eta'^2 + \eta^2\}dt\right], \quad \text{for } \eta \in D'[a,b]. \tag{2.16}$$

If $\lambda = \lambda_0$ is an eigenvalue of (2.5), and $u_0 = u_1+iu_2$ is a corresponding eigenfunction with canonical variable $v_0 = v_1 + iv_2$, where u_1,u_2,v_1,v_2 are real-valued, then a suitable integration by parts and use of the fact that $s_\alpha[u_\beta,v_\beta] = 0$, $(\alpha,\beta = 1,2)$, yields the result that

$$\lambda\{K[u_1] + K[u_2]\} = J[u_1:\mathscr{B}] + J[u_2:\mathscr{B}],$$

and consequently, under hypotheses ($\mathscr{A}$) all eigenvalues of (2.5) are real and greater than $\check{\lambda}$, where $\check{\lambda}$ is a real value such that $J[\eta;\check{\lambda}:\mathscr{B}]$ is positive definite on $D'[a,b]$; moreover, the eigenfunctions corresponding to an eigenvalue may be chosen to be real-valued.

The basic existence theorem for the boundary problem (2.5) is presented in the following theorem.

THEOREM 2.1. *Whenever hypothesis* ($\mathscr{A}$) *is satisfied there exists for the boundary problem* (2.5) *an infinite sequence of real eigenvalues* $\lambda_1 \leq \lambda_2 \leq \ldots$, *with corresponding real eigenfunctions* $u = u_j(t)$ *for* $\lambda = \lambda_j$ *such that:*

(a) $K[u_i,u_j] = \delta_{ij}, \qquad (i,j = 1,2,\ldots);$

(b) $\lambda_1 = J[u_1:\mathscr{B}]$ *is the minimum of* $J[\eta:\mathscr{B}]$ *on the class* $D_N[\mathscr{B}|K] = \{\eta : \eta \in D_e[\mathscr{B}],\ K[\eta] = 1\}$;

(c) *for* $j = 2,3,\ldots,$ *the class* (2.17)

$$D_{Nj}[\mathscr{B}|K] = \{\eta : \eta \in D_N[\mathscr{B}|K],\ K[\eta,u_i] = 0,\quad i = 1,\ldots,j-1\}, \tag{2.18}$$

is non-empty, and $\lambda_j = J[u_j:\mathscr{B}]$ *is the minimum of* $J[\eta:\mathscr{B}]$ *on* $D_{Nj}[\mathscr{B}|K]$.

(d) $\{\lambda_j\} \to +\infty$ *as* $j \to \infty$.

As ready consequences of the conclusions of this theorem, one may easily establish the following results.

COROLLARY. *Suppose that hypothesis* ($\mathscr{A}$) *holds, and* $\{\lambda_j, u_j\}$ *is a sequence of eigenvalues and corresponding eigenfunctions as specified in Theorem* 2.1.

(a) *if* k *is a positive integer and* $c_1,\dots,c_k$ *are real constants such that* $c_1^2 + \dots + c_k^2 = 1$, *then* $\eta(t) = c_1 u_1(t) + \dots + c_k u_k(t)$ *belongs to* $D_N[\mathscr{B}]$, *and* $J[\eta:\mathscr{B}] \leq \lambda_k$.

(b) MAXIMUM-MINIMUM PROPERTY. *If* $\mathscr{F} = \{f_1,\dots,f_r\}$ *is a set of real-valued integrable functions on* $[a,b]$, *and* $\lambda\{\mathscr{F}\}$ *denotes the minimum of* $J[\eta:\mathscr{B}]$ *on the set*

$$\{\eta : \eta \in D_N[\mathscr{B}|K],\ \int_a^b f_i \eta\, dt = 0,\quad (i = 1,\dots,r)\},$$

then λ_{k+1} *is the maximum of* $\lambda\{\mathscr{F}\}$.

Now if λ is not an eigenvalue of (2.5) there exists a Green's function $g(t,s;\lambda)$, with the definitive property that for arbitrary integrable functions f on $[a,b]$ the unique solution of the nonhomogeneous differential system

$$\ell[u;\lambda](t) = f(t),\quad s_\alpha[u,v] = 0,\quad (\alpha = 1,2) \tag{2.19}$$

is given by

$$u(t) = \int_a^b g(t,s;\lambda) f(s)\, ds. \tag{2.20}$$

Indeed, in view of the self-adjointness property of (2.5) and the above stated reality of eigenvalues of such a system, it follows readily that

$$g(t,s;\bar{\lambda}) \equiv \overline{g(s,t;\lambda)}, \tag{2.21}$$

and for λ_0 a real number not an eigenvalue the Green's function $g(t,s;\lambda_0)$ is real-symmetric. In particular, it then follows that the theory of a self-adjoint boundary problem (2.5) is equivalent to the theory of the integral equation

$$u(t) = \mu \int_a^b g(t,s;\lambda_0) k(s) u(s)\, ds. \tag{2.22}$$

Indeed, λ is an eigenvalue of (2.5) with corresponding eigenfunction $u(t)$ if and only if $\mu = \lambda_o - \lambda$ is an eigenvalue of (2.22) for which $u(t)$ is a corresponding eigenfunction.

If $h(t)$ is a real-valued measurable function which is of integrable square on $[a,b]$, and we denote by $c_j[h]$ the generalized Fourier coefficient $K[h,u_j]$ of h with respect to the K-orthonormal sequence $\{u_j\}$ of eigenfunctions of (2.5), then in view of the non-negativeness of the functional $K[\eta]$ we have the (Bessel) equality

$$K\left[h - \sum_{j=1}^{k} c_j[h]u_j\right] = K[h] - \sum_{j=1}^{k} c_j^2[h], \; k = 1,2,\ldots \qquad (2.23)$$

In particular, from this equality it follows that the infinite series $\sum_{j=1}^{\infty} c_j^2[h]$ converges, and

$$\sum_{j=1}^{\infty} c_j^2[h] \leq K[h]. \qquad (2.24)$$

As in the above discussion, for a self-adjoint problem $(\mathscr{B})$ satisfying hypothesis $(\mathscr{A})$ let $\{\lambda_j, u_j\}$, $(j = 1,2,\ldots)$, denote a sequence of eigenvalues and eigenfunctions as specified in Theorem 2.1, and for a function h as described a above let $c_j[h] = K[h,u_j]$, $(j = 1,2,\ldots)$. If $\eta \in D_e[\mathscr{B}]$, and $\eta_k(t) = \eta(t) - \sum_{j=1}^{k} c_j[\eta]u_j(t)$, we have $K[\eta_k,u_i] = 0$ for $i = 1,\ldots,k$, and $K[\eta_k] = K[\eta] - \sum_{j=1}^{k} c_j^2[\eta]$. If λ_o is a real number less than the smallest eigenvalue λ_1, then the minimizing property of λ_1 implies that $J[\eta;\lambda_o:\mathscr{B}]$ is positive definite on $D_e[\mathscr{B}]$, and $(\lambda_{k+1} - \lambda_o)K[\eta_k] \leq J[\eta_k;\lambda_o:\mathscr{B}]$. Moreover, since the orthonormal character of the eigenfunctions u_i implies that $J[u_i,u_j;\lambda_o:\mathscr{B}] = (\lambda_j - \lambda_o)\delta_{ij}$, $(i,j = 1,2,\ldots)$, we also have

$$J[\eta_k;\lambda_o:\mathscr{B}] = J[\eta;\lambda_o:\mathscr{B}] - \sum_{j=1}^{k}(\lambda_j - \lambda_o)c_j^2[\eta] \le J[\eta;\lambda_o:\mathscr{B}].$$

As conclusion (d) of Theorem 2.1 implies that $\lambda_{k+1} - \lambda_o \to \infty$ as $k \to \infty$, it then follows that $K[\eta] - \sum_{j=1}^{k} c_j^2[\eta] = K[\eta_k] \to 0$ as $k \to \infty$, so that

$$K[\eta] = \sum_{j=1}^{\infty} c_j^2[\eta], \text{ for arbitrary } \eta \in D_e[\mathscr{B}]. \tag{2.25}$$

Indeed, if h is merely measurable and square integrable on $[a,b]$, the fact that for a given positive integer k the minimum of $K[h - \sum_{j=1}^{k} d_j u_j]$ is provided by $d_j = c_j[h]$, together with the ability to determine for arbitrary $\varepsilon > 0$ a function $\eta_\varepsilon \in D_e[\mathscr{B}]$ such that $K[h - \eta_\varepsilon] < \varepsilon$, yields the result that

$$K[h] = \sum_{j=1}^{\infty} c_j^2[h] \tag{2.26}$$

for arbitrary measurable and square integrable h on $[a,b]$. This result is the so-called *completeness property* of the sequence of K-orthogonal eigenfunctions $\{u_j\}$, $(j = 1,2,\ldots)$. For the reader not using the theory of the Lebesgue integral, and supposing that hypothesis $(\mathscr{A}_C^+)$ holds, the function h appearing in (2.23), (2.24), (2.26) and the above statements may be supposed to be piecewise continuous. Also, as a ready consequence of the above result, if h is continuous and the infinite series $\sum_{j=1}^{\infty} c_j[h]u_j(t)$ converges uniformly on $[a,b]$, then the sum of this series is equal to $h(t)$ for $t \in [a,b]$; clearly such a continuous function h must satisfy the essential boundary conditions of $\mathscr{B}$.

Inequality (2.24), together with the definitive property of the Green's function, yields a ready proof of the following theorem.

THEOREM 2.2. *Suppose that hypothesis* ($\mathscr{A}$) *holds, and* $\{\lambda_j, u_j\}$, $(j = 1,2,\ldots)$, *is a sequence of eigenvalues and corresponding eigenfunctions as specified in Theorem* 2.1. *If* λ_0 *is a real value satisfying* $\lambda_0 < \lambda_1$, *then the infinite series*

$$\sum_{j=1}^{\infty}(\lambda_j - \lambda_0)^{-2}u_j^2(t), \quad \sum_{j=1}^{\infty}(\lambda_j - \lambda_0)^{-2}v_j^2(t), \quad t \in [a,b] \tag{2.27}$$

converge and the sums of these series do not exceed the respective values

$$\int_a^b k(s)g^2(t,s;\lambda_0)dt, \quad \int_a^b k(s)g_1^2(t,s;\lambda_0)dt \tag{2.28}$$

where $g(t,s:\lambda_0)$ *is the Green's function for the incompatible system*

$$\ell[u:\lambda_0](t) = 0, \quad s_\alpha[u,v] = 0, \quad \alpha = 1,2 \tag{2.29}$$

and $g_1(t,s;\lambda_0) = r(t)g_t(t,s;\lambda_0) + q(t)g(t,s;\lambda_0)$. *Moreover, the infinite series* $\sum_{j=1}^{\infty}(\lambda_j - \lambda_0)^{-2}$ *converges, and*

$$\sum_{j=1}^{\infty}(\lambda_j - \lambda_0)^{-2} \leq \int_a^b \int_a^b k(t)k(s)g^2(t,s;\lambda_0)dtds. \tag{2.30}$$

In particular, the convergence of the series $\sum_{j=1}^{\infty}(\lambda_j-\lambda_0)^{-2}$ yields an independent proof of conclusion (d) of Theorem 2.1, as well as provide some elementary results on the order of growth of the eigenvalues λ_j.

One may proceed to establish the following additional expansion theorems.

THEOREM 2.3. *If* $\eta \in D_e[\mathscr{B}]$, *then the infinite series* $\sum_{j=1}^{\infty} c_j[\eta]u_j(t)$ *converges to* $\eta(t)$ *uniformly on* $[a,b]$; *moreover,*

$$\int_a^b \{\eta'(t) - \sum_{j=1}^{k} c_j[\eta]u_j(t)\}^2 dt \to 0 \quad \text{as} \quad k \to \infty; \tag{2.31}$$

$$J[\eta:\mathscr{B}] = \sum_{j=1}^{\infty} \lambda_j c_j^2[\eta]. \tag{2.32}$$

THEOREM 2.4. *If* λ *is not an eigenvalue of* $(\mathscr{B})$, *then the infinite series* $\sum_{j=1}^{\infty}(\lambda_j - \lambda)^{-1}u_j(t)u_j(s)$ *converges absolutely and uniformly for* $(t,s) \in [a,b] \times [a,b]$, *and*

$$\sum_{j=1}^{\infty}(\lambda_j - \lambda)^{-1}u_j(t)u_j(s) = -g(t,s;\lambda), \tag{2.33}$$

where $g(t,s;\lambda)$ *is the Green's function for the incompatible differential system*

$$\ell[u;\lambda](t) = 0, \quad s_\alpha[u,v] = 0, \quad (\alpha = 1,2). \tag{2.34}$$

Also,

$$\sum_{j=1}^{\infty}(\lambda_j - \lambda_0)^{-1} = -\int_a^b k(t)g(t,t;\lambda_0)dt. \tag{2.35}$$

Under the hypotheses of Theorem 2.4, for $m = 2,3,\ldots,$ the infinite series $\sum_{j=1}^{\infty}(\lambda_j - \lambda)^{-m}u_j(t)u_j(s)$ converges absolutely and uniformly for $(t,s) \in [a,b] \times [a,b]$, and the series $\sum_{j=1}^{\infty}(\lambda_j - \lambda)^{-m}$ converges. Moreover, the sums of these series are equal to certain integrals involving the Green's function and the coefficient function $k(t)$. In particular, for $m = 2$ we have

$$\sum_{j=1}^{\infty}(\lambda_j - \lambda)^{-2}u_j(t)u_j(s) = \int_a^b g(t,r;\lambda)k(r)g(r,s;\lambda)dr, \tag{2.36}$$

$$\sum_{j=1}^{\infty}(\lambda_j - \lambda)^{-2} = \int_a^b \int_a^b k(t)g(t,r;\lambda)k(r)g(r,t;\lambda)drdt. \tag{2.37}$$

THEOREM 2.5. *If* $h(t)$ *is a function of class* $\mathscr{L}^2$ *on* $[a,b]$, *and* $u(t)$ *is a solution of the differential system*

$$\ell[u](t) + k(t)h(t) = 0, \quad s_\alpha[u,v] = 0, \quad (\alpha = 1,2), \tag{2.38}$$

then the infinite series

$$\sum_{j=1}^{\infty} c_j[u]u_j(t), \quad \sum_{j=1}^{\infty} c_j[u]v_j(t) \tag{2.39}$$

converge absolutely and uniformly for $t \in [a,b]$, *and have sums equal to* $u(t)$ *and* $v(t) = r(t)u'(t) + q(t)u(t)$, *respectively.*

For the case of a system (2.5) whose coefficients satisfy $(\mathscr{A}_C^{++})$ these results are established in Section 5 of Chapter VI of Reid [35], while in case hypothesis $(\mathscr{A}_L^{++})$ holds they are consequences of the results of Section 11 in Chapter VII of the same reference, when applied to the first order system equivalent to (2.5).

In view of the positiveness of the function $k(t)$, the completeness property (2.25) may equally well be phrased as "a real-valued function $\eta(t)$ of class $\mathscr{L}^2[a,b]$ is equal to zero a.e. if and only if $c_j[\eta] = 0$, $(j = 1,2,\ldots)$". There are varied proofs of the completeness of the set of eigenfunctions of a system $(\mathscr{B})$. Many depend upon establishing an expansion theorem of the form (2.25), (2.32), or of the type in Theorem 2.5, utilizing preliminary derivation of the Green's function and results of the theory of integral equations with real symmetric kernel, which directly or indirectly involves the fact that for a real λ_0 not an eigenvalue of $(\mathscr{B})$ the function Tf defined by $(Tf)(t) = \int_a^b \sqrt{k(t)}g(t,s;\lambda_0)\sqrt{k(s)}f(s)ds$ provides a compact, (completely continuous) symmetric operator on $\mathscr{L}^2[a,b]$ into $\mathscr{L}^2[a,b]$, (see, for example, Coddington and Levinson [1, Ch. 7, Sec. 4], Hille [2, Ch. 8, Sec. 5], Hartman [13, Ch. XI, Sec. 4]). The approximation method of Schwarz produces a completeness proof, (for Sturm-Liouville systems see Ince [1, Ch. 11, Sec. 5]) that has been a central feature of the Schmidt [1] theory of integral equations and subsequent theories of definite boundary problems, (Bliss [4], Reid [35, Ch. IV, Sec. 6]). Another method, dating from the time of Liouville, involves the asymptotic nature of the eigenfunctions, (see, for example, G. D. Birkhoff [1,2,3], Titchmarsh [1, Ch. I]). For Sturm-Liouville problems a direct proof of the completeness of the set of eigenfunctions is to be found in Birkhoff and Rota [2; 1, Ch. XI, Secs. 9,10,11], using the asymptotic form of eigenfunctions and the fact that if $\{\phi_k\}$, $(k = 1,2,\ldots)$ is an orthonormal basis in a Hilbert space, and $\{\psi_k\}$, $(k = 1,2,\ldots)$ is an orthonormal sequence in Hilbert space such that $\sum_{k=1}^{\infty} \|\psi_k - \psi_k\|^2 < \infty$, then $\{\psi_k\}$ is also an orthonormal basis.

3. Comparison Theorems

A boundary problem $\mathscr{B}$ of the form (2.5) involves the real quadratic integrand form

$$2\omega(t,\eta,\zeta:\mathscr{B}) = r(t:\mathscr{B})\zeta^2 + 2q(t:\mathscr{B})\zeta\eta + p(t:\mathscr{B})\eta^2, \quad (3.1)$$

the real quadratic end-form

$$Q[\eta:\mathscr{B}] = \gamma_{11}^{\mathscr{B}}\eta^2(a) + 2\gamma_{12}^{\mathscr{B}}\eta(a)\eta(b) + \gamma_{22}^{\mathscr{B}}\eta^2(b), \quad (3.2)$$

the quadratic functional

$$J[\eta:\mathscr{B}] = Q[\eta:\mathscr{B}] + \int_a^b 2\omega(t,\eta(t),\eta'(t):\mathscr{B})dt, \quad (3.3)$$

the linear subspace $S[\mathscr{B}]$ of the real plane, the set

$$D_e[\mathscr{B}] = \{\eta:\eta \in D'[a,b],\ \hat{\eta} \in S[\mathscr{B}]\}, \quad (3.4)$$

and also the coefficient function $k(t) = k(t:\mathscr{B})$. We shall now consider some comparison theorems for a problem $\mathscr{B}$ and a second problem $\tilde{\mathscr{B}}$ involving corresponding $\omega(t,\eta,\pi:\tilde{\mathscr{B}})$, $Q[\eta:\tilde{\mathscr{B}}]$, $S[\tilde{\mathscr{B}}]$ and $k(t:\tilde{\mathscr{B}})$. In particular, if $S[\mathscr{B}] = S[\tilde{\mathscr{B}}]$, then $D_e[\mathscr{B}] = D_e[\tilde{\mathscr{B}}]$. *In all cases it will be assumed that each of the considered problems* $\mathscr{B}$ *and* $\tilde{\mathscr{B}}$ *satisfies hypotheses* ($\mathscr{A}$) *of Section* 2, *unless specifically stated otherwise.* Moreover, for these respective problems a set of eigenvalues and eigenfunctions satisfying the conditions of Theorem 2.1 will be denoted by $\{\lambda_j,u_j\}$ and $\{\tilde{\lambda}_j,\tilde{u}_j\}$. One of the most readily established results is the monotoneity property of eigenvalues of the following theorem, which is a ready consequence of the minimizing property of λ_j and conclusion (a) of the Corollary to Theorem 2.1.

THEOREM 3.1. *Suppose that* $\mathscr{B}$ *and* $\tilde{\mathscr{B}}$ *are such that* $S[\mathscr{B}] = S[\tilde{\mathscr{B}}]$ *and* $k(t:\mathscr{B}) \equiv k(t:\tilde{\mathscr{B}})$. *If* $\Delta J[\eta:\mathscr{B},\tilde{\mathscr{B}}] = J[\eta:\tilde{\mathscr{B}}] - J[\eta:\mathscr{B}]$ *is non-negative on* $D_e[\mathscr{B}] = D_e[\tilde{\mathscr{B}}]$, *then* $\tilde{\lambda}_j \geq \lambda_j$, $(j = 1,2,\ldots)$; *moreover, if* $\Delta J[\eta:\mathscr{B},\tilde{\mathscr{B}}]$ *is positive definite on* $D_e[\mathscr{B}] = D_e[\tilde{\mathscr{B}}]$, *then* $\tilde{\lambda}_j > \lambda_j$, $(j = 1,2,\ldots)$.

If two problems $\mathscr{B}$ and $\tilde{\mathscr{B}}$ are such that $S[\mathscr{B}] = S[\tilde{\mathscr{B}}] = S$, and $k(t:\mathscr{B}) = k(t:\tilde{\mathscr{B}}) = k(t)$, then the *difference problem* involving S, $k(t)$ and

$$\begin{aligned} J[\eta:\mathscr{B},\tilde{\mathscr{B}}] &= J[\eta:\tilde{\mathscr{B}}] - J[\eta:\tilde{\mathscr{B}}] \\ &= \gamma^{\Delta}_{11}\eta^2(a) + 2\gamma^{\Delta}_{12}\eta(a)\eta(b) + \gamma^{\Delta}_{22}\eta^2(b) \\ &\quad + \int_a^b \{(\hat{r}-r)\eta'^2 + 2(\hat{q}-q)\eta'\eta + (\hat{p}-p)\eta^2\}dt, \end{aligned}$$

where r,p,q and $\tilde{r},\tilde{p},\tilde{q}$ are written in place of $r(t:\mathscr{B})$, $r(t:\tilde{\mathscr{B}})$, etc., and $\gamma^{\Delta}_{\alpha\beta} = \gamma^{\tilde{\mathscr{B}}}_{\alpha\beta} - \gamma^{\mathscr{B}}_{\alpha\beta}$, $(\alpha,\beta = 1,2;\ \alpha \leq \beta)$, is denoted by $\Delta(\mathscr{B},\tilde{\mathscr{B}})$. The conditions of hypothesis ($\mathscr{A}$) are satisfied by $(\mathscr{B},\tilde{\mathscr{B}})$, with the possible exception of the non-zero nature of $\tilde{r}(t) - r(t)$, and the conditions that $\tilde{r}(t) - r(t)$ and $1/[\tilde{r}(t) - r(t)]$ are locally of class $\mathscr{L}^\infty$. If these further conditions are satisfied, however, the corresponding difference boundary problem

$$[(\tilde{r}-r)u' + (\tilde{q}-q)u]' - [(\tilde{q}-q)u' + (\tilde{p}-p)u] + \lambda ku = 0,$$

$$u \in S, \quad \Gamma^\Delta u + Dw \in S^\perp,$$

where Γ^Δ is the 2×2 matrix $[\gamma^\Delta_{\alpha\beta}]$, and $w = (\tilde{r}-r)u' + (\tilde{q}-q)u$, has an infinite sequence of eigenvalues and corresponding eigenfunctions, determined as in Theorem 2.1. Again, in view of the minimizing properties of eigenvalues and Conclusion (a) of the Corollary to Theorem 2.1, we have the following result.

THEOREM 3.2. *Suppose that* $S[\mathscr{B}] = S[\tilde{\mathscr{B}}] = S$ *and* $k(t:\mathscr{B}) = k(t:\tilde{\mathscr{B}}) = k(t)$ *for problems* $\mathscr{B}$ *and* $\tilde{\mathscr{B}}$*, while hypothesis* ($\mathscr{A}$) *is satisfied by each of the problems* $\mathscr{B}$, $\tilde{\mathscr{B}}$ *and* $\Delta(\mathscr{B},\tilde{\mathscr{B}})$. *If* $\{\lambda^\Delta_j, u_j\}$ *denotes a sequence of eigenvalues and corresponding eigenfunctions for* $\Delta(\mathscr{B},\tilde{\mathscr{B}})$ *determined as in Theorem* 2.1, *then* $\tilde{\lambda}_{j+i-1} \geq \lambda^\Delta_j + \lambda_i$, $(i,j = 1,2,\ldots)$.

Now consider two problems $\mathscr{B}$ and $\tilde{\mathscr{B}}$ that satisfy hypothesis ($\mathscr{A}$), and which differ only in the spaces $S[\mathscr{B}]$ and $S[\tilde{\mathscr{B}}]$. Problem $\tilde{\mathscr{B}}$ is said to be a *subproblem* of $\mathscr{B}$ if $S[\tilde{\mathscr{B}}] \subset S[\mathscr{B}]$; if $d = \dim S[\mathscr{B}]$ and $\tilde{d} = \dim S[\tilde{\mathscr{B}}]$, then $d - \tilde{d} \geq 0$, and $\tilde{\mathscr{B}}$ is said to be a *subproblem of of dimension* $d - \tilde{d}$. If $d > \tilde{d}$, then there exist $d - \tilde{d}$ independent linear forms $x^\alpha[\eta] \equiv x^\alpha_1\eta(a) + x^\alpha_2\eta(b)$, $(\alpha = 1,\ldots,d-\tilde{d})$ such that

$$S[\tilde{\mathscr{B}}] = \{\eta : \hat{\eta} \in S[\mathscr{B}],\ x^\alpha[\eta] = 0,\ \alpha = 1,\ldots,d-\tilde{d}\}.$$

In particular, for a boundary problem $\mathscr{B}$ of the form (2.5) satisfying hypothesis ($\mathscr{U}$), the problem $\tilde{\mathscr{B}}$ involving the same differential equation and the *null end-conditions* $\eta(a) = 0 = \eta(b)$ is a subproblem of $\mathscr{B}$ of dimension equal to $\dim S[\mathscr{B}]$.

If I is an interval on the real line, *the symbol* $\sigma(I:\mathscr{B})$ *will denote the number of eigenvalues of the problem* $\mathscr{B}$ *on* I, *each counted a number of times equal to its multiplicity*. Also, corresponding to a real number x, we shall denote by $V_x(\mathscr{B})$, $\{W_x(\mathscr{B})\}$, the *number of eigenvalues of* $\mathscr{B}$ *which are less, {not greater}, than* x; *that is*, $V_x(\mathscr{B}) = \sigma((-\infty,x):\mathscr{B})$ and $W_x(\mathscr{B}) = \sigma((-\infty,x]:\mathscr{B})$.

In view of the above remarks, the following results are ready consequences of the minimizing properties of the eigenvalues of $\mathscr{B}$ and $\tilde{\mathscr{B}}$, together with Conclusion (a) of the Corollary to Theorem 2.1.

THEOREM 3.3. *If each of the problems* $\mathscr{B}$ *and* $\tilde{\mathscr{B}}$ *satisfies hypothesis* ($\mathscr{U}$), *and* $\tilde{\mathscr{B}}$ *is a subproblem of* $\mathscr{B}$ *of dimension* $\delta = d - \tilde{d}$, *then* $\lambda_{j+\delta} \geq \tilde{\lambda}_j \geq \lambda_j$, $(j = 1,2,\ldots)$. *For each real number* x *we have*

$$V_x(\mathscr{B}) - \delta \leq V_x(\tilde{\mathscr{B}}) \leq V_x(\mathscr{B}),\quad W_x(\mathscr{B}) - \delta \leq W_x(\tilde{\mathscr{B}}) \leq W_x(\mathscr{B});$$

moreover, $|\sigma(I:\mathscr{B}) - \sigma(I:\tilde{\mathscr{B}})| \leq \delta$ *for every bounded subinterval* I *of the real line.*

One may also establish more sophisticated comparison theorems, involving in particular two problems that have different end-forms. A more detailed discussion of such comparison theorems will be left until Chapter VI, however, wherein the associated algebraic problem will have more content.

Consequently, further discussion of self-adjoint two-point boundary problems will be limited to a comparison theorem for systems involving different functions $k(t)$, and to a result on the continuity of eigenvalues as functionals of the coefficients and involved end-forms.

The following result is also a ready consequence of the extremizing properties of eigenvalues.

THEOREM 3.4. *Suppose that* $\mathscr{B}$ *and* $\tilde{\mathscr{B}}$ *are boundary problems of the form* (2.5) *that satisfy hypothesis* ($\mathscr{A}$), *with* $\omega(t,\eta,\zeta:\tilde{\mathscr{B}}) = \omega(t,\eta,\zeta:\mathscr{B})$, $Q[\eta:\tilde{\mathscr{B}}] = Q[\eta:\mathscr{B}]$, $S[\tilde{\mathscr{B}}] = S[\mathscr{B}]$, *and* $k(t:\tilde{\mathscr{B}}) \geq k(t:\mathscr{B}) > 0$ *on* $[a,b]$. *Let* $\{\lambda_j, u_j\}$ *and* $\{\tilde{\lambda}_j, \tilde{u}_j\}$ *denote sets of eigenvalues and corresponding eigenfunctions for the respective problems* $\mathscr{B}$, $\tilde{\mathscr{B}}$ *which individually satisfy the conditions of Theorem* 2.1, *while* p *and* q *are integers such that* $\lambda_p < 0 < \lambda_q$. *Then* $\lambda_j \leq \tilde{\lambda}_j < 0$ *for* $j = 1,\dots,p$ *and* $0 < \tilde{\lambda}_j \leq \lambda_j$ *and* $j = q,q+1,\dots,$; *moreover, if* $q > p+1$ *and* $\lambda_j = 0$ *for* $p < j < q$, *then also* $\tilde{\lambda}_j = 0$ *for* $p < j < q$.

Now the argument used to establish the result of Lemma 2.3 may also be used to establish the following result.

LEMMA 3.1. *If* $\lambda_o < \lambda_1$, *the smallest eigenvalue of a boundary problem* ($\mathscr{B}$), *then there exists a* $\mu_o > 0$ *such that*

$$J[\eta;\lambda_o:\mathscr{B}] \geq \mu_o[\eta^2(a) + \eta^2(b) + \int_a^b \{\eta'^2 + \eta^2\}dt] \qquad (3.5)$$

$$\text{for } \eta \in D_e(\mathscr{B}).$$

A very simple, but useful, comparison result is that of the following theorem.

THEOREM 3.5. *Suppose that* $\mathscr{B}$ *and* $\tilde{\mathscr{B}}$ *are such that* $S[\mathscr{B}] = S[\tilde{\mathscr{B}}] = S$, *and* $k(t:\mathscr{B}) = k(t:\tilde{\mathscr{B}}) = k(t)$. *If* κ *is such that*

$$|J[\eta:\mathscr{B}] - J[\eta:\mathscr{B}]| \leq \kappa\Big[\eta^2(a)+\eta^2(b) + \int_a^b \{\eta'^2+\eta^2\}dt\Big], \qquad (3.6)$$
$$\textit{for} \quad \eta \in D_e[\mathscr{B}] = D_e[\mathscr{B}],$$

and for some λ_0 *less than the smallest eigenvalue* λ_1 *of* $\mathscr{B}$, μ_0 *is a constant satisfying inequality* (3.5), *then*

$$\tilde{\lambda}_j \leq (1 + \kappa/\mu)\lambda_j - (\kappa/\mu)\lambda_0, \ (j = 1,2,\ldots); \qquad (3.7)$$

moreover, if $\kappa < \mu$, *then also*

$$\tilde{\lambda}_j \geq (1 - \kappa/\mu)\lambda_j + (\kappa/\mu)\lambda_0, \ (j = 1,2,\ldots). \qquad (3.8)$$

Indeed, condition (3.6) implies the inequalities

$$(1 - \kappa/\mu)J[\eta;\lambda_0:\mathscr{B}] \leq J[\eta;\lambda_0:\tilde{\mathscr{B}}] \leq (1+ \kappa/\mu)J[\eta;\lambda_0:\mathscr{B}]$$

for arbitrary $\eta \in D_e[\mathscr{B}] = D_e[\tilde{\mathscr{B}}]$. Also, the two boundary problems with respective functionals $J[\eta;\lambda_0:\tilde{\mathscr{B}}]$ and $(1 + \kappa/\mu)J[\eta;\lambda_0:\mathscr{B}]$, together with the same end-space S and norming function $k(t)$, have sequences of eigenvalues and eigenfunctions $\{\tilde{\lambda}_j - \lambda_0,\tilde{u}_j\}$ and $\{(1 + \kappa/\mu)(\lambda_j - \lambda_0),u_j\}$, respectively. Consequently, Theorem 3.1 applied to these functionals yields the inequalities

$$\tilde{\lambda}_j - \lambda_0 \leq (1 + \kappa/\mu)(\lambda_j - \lambda_0), \quad (j = 1,2,\ldots),$$

which is equivalent to (3.7). Whenever $\kappa < \mu$, inequality (3.8) follows from a similar argument applied to the boundary problems with respective functionals $(1 - \kappa/\mu)J[\eta;\lambda_0:\mathscr{B}]$ and $J[\eta;\lambda_0:\tilde{\mathscr{B}}]$, together with the same end-space S and norming function $k(t)$.

It is to be noted that condition (3.6) holds whenever κ is a constant which satisfies the following algebraic inequalities

$$|2\omega(t,\lambda,\zeta:\tilde{\mathscr{B}}) - 2\omega(t,\eta,\zeta:\mathscr{B})| \le \kappa\{\zeta^2 + \eta^2\}, \quad \text{for arbitrary } (t,\eta,\zeta) \text{ with } t \in [a,b]; \tag{3.9}$$

$$|Q[\eta:\tilde{\mathscr{B}}] - Q[\eta:\mathscr{B}]| \le \kappa\{\eta^2(a) + \eta^2(b)\}, \quad \text{for arbitrary real } (\eta(a),\eta(b)). \tag{3.10}$$

With this remark, the following result is an immediate corollary to the above theorem.

COROLLARY 1. *Let* $B(\nu)$, $\nu \in \mathscr{N}$, *be a boundary problem in which the real quadratic integrand form*

$$2\omega(t,\eta,\zeta) = 2\omega(t,\eta,\zeta:\nu) = r(t,\nu)\zeta^2 + 2q(t,\nu)\zeta\eta + p(t,\nu)\eta^2$$

and the real quadratic end-form

$$Q[\eta:\nu] = \gamma_{11}(\nu)\eta^2(a) + 2\gamma_{12}(\nu)\eta(a)\eta(b) + \gamma_{22}(\nu)\eta^2(b) = \hat{\eta}^*\Gamma(\nu)\hat{\eta}$$

depend upon the parameter ν, *while the end-space* S *and the norming function* $k(t)$ *are independent of* ν. *Moreover, suppose that:*

(a) $\gamma_{\alpha\beta}(\nu)$, $(\alpha,\beta = 1,2)$, *are continuous in* ν *on* $\mathscr{N}$;

(b) $r(t,\nu)$, $q(t,\nu)$ *and* $p(t,\nu)$ *are continuous on* S; *uniformly with respect to* t *on* $[a,b]$;

(c) *for each* $\nu \in \mathscr{N}$, *the boundary problem*

$$\begin{aligned} \ell[u;\lambda](t) \equiv {} & [r(t,\nu)u'(t) + q(t,\nu)u(t)]' \\ & - [q(t,\nu)u'(t) + p(t,\nu)u(t)] + \lambda k(t)u(t) = 0, \end{aligned}$$

$$u \in S, \qquad \Gamma(\nu)u + Dv \in S^{\perp} \tag{3.11}$$

satisfies hypothesis ($\mathscr{A}$).

If $\{\lambda_j(\nu), u_j(t:\nu)\}$ *denotes a set of eigenvalues and corresponding eigenfunctions of* $\mathscr{B}(\nu)$, *satisfying the conditions of Theorem* 2.1, *then each eigenvalue* $\lambda_j(\nu)$ *is a continuous function of* ν *on* $\mathscr{N}$.

It is to be remarked that in the above corollary the domain $\mathscr{N}$ of the parameter ν has not been specified as a subset of the real line. The result is equally valid for ν a vector parameter $(\nu_1,\ldots,\nu_k)$ with domain a subset of k-dimensional Euclidean space, and, indeed, for much more general cases.

Particular attention is directed to the paper of Gottlieb [1], which is devoted to a detailed presentation of Morse's theory to the Sturm-Liouville boundary problem involving a differential equation of the form

$$[r(t,\lambda)u(t)]' - p(t,\lambda)u(t) = 0, \quad t \in [a,b]$$

and self-adjoint boundary conditions at $t = a$ and $t = b$.

4. Comments on Recent Literature

In the recent literature dealing specifically with self-adjoint two-point boundary problems involving a real scalar linear homogeneous second order differential equation one may isolate the following areas, many of which are illustrated in the following section on Topics and Exercises.

(a) Treatment based upon trigonometric substitutions, and direct generalizations of such substitutions. Included in this category are the papers of Prüfer [1], Whyburn [2], Sturdivant [1], Kamke [3,4,5], Barrett[4,5], and Atkinson [1].

(b) The study of integral inequalities, either for their own sake or in conjunction with allied consideration of boundary problems. In general, for problems in this area the extremizing properties of eigenvalues and eigenfunctions plays a central role. Papers dealing with this topic include Beesack [1,2,3,4], Banks [4], Bradley and Everitt [1,2], Coles [2,4], and Horgan [1]. In particular, Beesack [4] is a report on integral inequalities involving a function and its derivative, and no attempt is made to include here the extensive bibliography on this topic which he provides.

(c) Estimates of eigenvalues and eigenfunctions of boundary problems wherein coefficients satisfy certain inequality conditions. Considerable interest in this area has been stimulated by the results of Krein [1], illustrated by the fact that if $q:[a,b] \to \mathbf{R}$ is a non-negative integrable function, and $\mu_1[q] \leq \mu_2[q] \leq \ldots$ denotes the sequence of eigenvalues of the boundary problem

$$u''(t) + \mu q(t)u(t) = 0, \quad u(a) = 0 = u(b),$$

then whenever q is required to satisfy the additional restraints $0 \leq q(t) \leq H$, $\int_a^b q(t)dt = M$, where H and M are given positive constants, we have

$$\frac{4Hn^2}{M^2} X\left(\frac{M}{H[b-a]}\right) \leq \mu_n[q] \leq \frac{Hn^2\pi^2}{M^2},$$

where $X(t)$ is the least positive root of the equation $\sqrt{X} \tan X = t/(1-t)$; moreover, in this class of functions q there exists an extremizing q_1 for which $\mu_1[q_1]$ is the infimum of $\mu_1[q]$ on this class. Papers by Banks [1,2,3] and Breuer and Gottlieb [1,3] are in this area.

(d) Estimates of eigenvalues and eigenfunctions of boundary problems wherein coefficients satisfy functional conditions such as monotoneity, convexity, or concavity. The papers of Banks [1,2,3], Makai [1,2] and Abramovich [1] are concerned with such questions.

5. Topics and Exercises

1. If the boundary problem (2.5) satisfies hypothesis $(\mathscr{A})$, then there exist constants c_1, d_1, c_2, d_2 such that for $j = 1,2,\ldots$, we have

$$c_1 j^2 + d_1 \leq \lambda_j \leq c_2 j^2 + d_2.$$

In particular, the series $\sum_j^{!}(1/|\lambda_j|^p)$ converges for $p > 1/2$, but diverges for $p \leq 1/2$, where the symbol $\sum_j^{!}$ denotes summation over those values of j for which $\lambda_j \neq 0$.

2. Suppose that the coefficient functions r,p,q,k of (2.5a) satisfy the conditions of hypothesis $(\mathscr{A})$, and consider the boundary problem defined by this equation and the boundary conditions $u(a) = 0$, $u(b) = 0$. If λ_0 is a real number that is not an eigenvalue of this problem, then $g(t,s;\lambda_0)$ is of constant sign on $[a,b] \times [a,b]$ if and only if λ_0 is less than the smallest eigenvalue of the boundary problem.

3. Suppose that a boundary problem (2.5) satisfies hypothesis ($\mathscr{A}$), and that $\{\lambda_j, u_j(t)\}$ is a sequence of eigenvalues and eigenfunctions satisfying the conditions of Theorem 2.1. If $f_\alpha(t)$, $(\alpha = 1,\ldots,k)$, are given functions of $D_e[\mathscr{B}]$ which are linearly independent, let

$$J_{\alpha\beta} = J[f_\alpha, f_\beta : \mathscr{B}], \quad K_{\alpha\beta} = K[f_\alpha, f_\beta : \mathscr{B}], \ \alpha,\beta = 1,\ldots,k.$$

Then $J = [J_{\alpha\beta}]$ and $K = [K_{\alpha\beta}]$ are real, symmetric $k \times k$ matrices, with K positive definite, and if $\sigma_1 \le \sigma_2 \le \ldots \le \sigma_k$ denote the zeros of the characteristic equation

$$\det [J - \sigma K] = 0,$$

each repeated a number of times equal to its multiplicity then $\lambda_j \le \sigma_j$, $(j = 1,\ldots,k)$.

4. The Sturm-Liouville system

$$\begin{aligned} u''(t) + \lambda k(t)u(t) &= 0, \\ \beta_a u(a) - u'(a) &= 0, \\ \beta_b u(b) + u'(b) &= 0, \end{aligned} \tag{5.1}$$

where $k(t)$ is a real-valued non-identically vanishing function of class $\mathscr{L}^\infty$ on $[a,b]$, and $\beta_a > 0$, $\beta_b > 0$, is equivalent to the system

$$\begin{aligned} \tilde{u}''(t) + \lambda \tilde{k}(t)\tilde{u}(t) &= 0, \\ \tilde{u}(\tilde{a}) = 0, \quad \tilde{u}(\tilde{b}) &= 0, \end{aligned} \tag{5.2}$$

where $\tilde{a} = a - (1/\beta_a)$, $\tilde{b} = b + (1/\beta_b)$, $\tilde{k}(t) = k(t)$ on $[a,b]$, $\tilde{k}(t) = 0$ on $[\tilde{a},a) \cup (b,\tilde{b}]$ in the sense that $u(t)$ is an eigenfunction of (5.1) for an eigenvalue λ if and only if for this value of λ there is a solution $u(t)$ of (5.2) which is equal to $\tilde{u}(t)$ on $[a,b]$. The totality of eigenfunctions of (5.2) is not complete, however, as is easily seen since all such functions are linear on the subintervals $[\tilde{a},a]$ and $[b,\tilde{b}]$. {Lichtenstein [5]}.

5. In the following suppose that on the indicated interval of integration the real-valued function η is a.c. and $\eta' \in \mathscr{L}^2$.

(a) If $\eta(0) = 0$, then $\int_0^{\pi/2} [\eta'(t)\ ^2dt] > \int_0^{\pi/2} \eta^2(t)dt$, unless $\eta(t)$ is a constant multiple of $\sin t$.

(b) If $\eta(-\pi/2) = 0 = \eta(\pi/2)$, then $\int_{-\pi/2}^{\pi/2} [\eta'(t)]^2dt >$ $\int_{-\pi/2}^{\pi/2} \eta^2(t)dt$ unless $\eta(t)$ is a constant multiple of $\cos t$.

(c) If $\eta(-\pi) = \eta(\pi)$, and $\int_{-\pi}^{\pi} \eta(t)dt = 0$, then $\int_{-\pi}^{\pi} [\eta'(t)]^2dt > \int_{-\pi}^{\pi} \eta^2(t)dt$ unless $\eta(t)$ is of the form $\eta(t) = c_1 \cos t + c_2 \sin t$. {Results (a) and (b) express the extremizing property of the smallest eigenvalue of associated boundary problems. The result of (c), which is known as "Wirtinger's inequality" expresses the minimizing property of the second eigenvalue of the boundary problem $u''(t) + u(t) = 0$, $u(-\pi) - u(\pi) = 0$, $u'(-\pi) - u'(\pi) = 0$. Various proofs of these inequalities are to be found in many places, one of which is Hardy, Littlewood and Pólya [1, Chapter VII]}.

6. In the following suppose that on the indicated interval of integration the real-valued function η is continuously differentiable, with $\eta'(t)$ a.c. and $\eta'' \in \mathscr{L}^2$.

(a) If $L > 0$ and $\eta(0) = \eta'(0) = \eta(L) = \eta'(L) = 0$, then

$$\int_0^L [\eta''(t)]^2dt \geq \pi^4 KL^{-4} \int_0^L \eta^2(t)dt,$$

where K is the smallest positive root of the equation

$\tanh(\frac{1}{2}\pi K^{\frac{1}{4}}) + \tan(\frac{1}{2}\pi K^{\frac{1}{4}}) = 0.$

(b) If $\eta(0) = 0 = \eta(\pi)$, then $\int_0^\pi [\eta''(t)]^2 dt >$ $\int_0^\pi \eta^2(t)dt$ unless $\eta(t)$ is a constant multiple of $\sin t$.

(c) If $\eta'(0) = 0 = \eta'(\pi)$ and $\int_0^\pi \eta(t)dt = 0$, then $\int_0^\pi [\eta''(t)]^2 dt > \int_0^\pi \eta^2(t)dt$ unless $\eta(t)$ is a constant multiple of $\cos t$.

{Again, results (a) and (b) express the extremizing property of the smallest eigenvalue of an associated boundary problem, and (c) expresses the minimizing property of the second eigenvalue of an associated boundary problem. Result (a) is to be found in the paper of Anderson, Arthurs and Hall [1], and results (b), (c) in the paper by Fan, Taussky and Todd [1]}.

7. Let $\mathscr{B}$ and $\tilde{\mathscr{B}}$ be boundary problems satisfying hypothesis ($\mathscr{A}$) with $\omega(t,\eta,\pi:\mathscr{B}) = \omega(t,\eta,\pi:\tilde{\mathscr{B}})$, $Q[\eta:\mathscr{B}] = Q[\eta:\tilde{\mathscr{B}}]$, $S[\mathscr{B}] = S[\tilde{\mathscr{B}}]$, and suppose that $J[\eta:\mathscr{B}] \equiv J[\eta:\tilde{\mathscr{B}}]$ is positive definite on $S_e[\mathscr{B}] = S_e[\tilde{\mathscr{B}}]$. Let $\tilde{\tilde{\mathscr{B}}}$ denote the boundary problem with $\omega(t,\eta,\pi:\tilde{\tilde{\mathscr{B}}})$, $Q[\eta:\tilde{\tilde{\mathscr{B}}}]$ and $S[\tilde{\tilde{\mathscr{B}}}]$ equal to the respective common elements of $\tilde{\mathscr{B}}$ and $\mathscr{B}$, while $k(t:\tilde{\tilde{\mathscr{B}}}) = k(t:\mathscr{B}) + k(t:\hat{\mathscr{B}})$. If $\{\lambda_j,u_j\}$, $\{\tilde{\lambda}_j,\tilde{u}_j\}$, $\{\tilde{\tilde{\lambda}}_j,\tilde{\tilde{u}}_j\}$ are sequences of eigenvalues and eigenfunctions for the respective boundary problems $\mathscr{B}$, $\tilde{\mathscr{B}}$, and $\tilde{\tilde{\mathscr{B}}}$ which individually satisfy the conditions of Theorem 2.1, then $\lambda_j \geq \tilde{\tilde{\lambda}}_j$ and $\tilde{\lambda}_j \geq \tilde{\tilde{\lambda}}_j$, $(j = 1,2,\ldots)$, and

$$\lambda_j^{-1} + \tilde{\lambda}_k^{-1} \geq \tilde{\tilde{\lambda}}_{j+k-1}^{-1}, \quad (j,k = 1,2,\ldots).$$

8. For a given compact interval $[a,b]$, let $\mathscr{B}_{oo}[a,b]$, $\mathscr{B}_{o*}[a,b]$, $\mathscr{B}_{*o}[a,b]$, $\mathscr{B}_{**}[a,b]$ denote the four boundary problems involving the differential equation $\ell[u:\lambda](t) = 0$

of (2.5) and the respective boundary conditions:

$$u(a) = 0, \quad u(b) = 0;$$

$$u(a) = 0, \quad v(b) = 0;$$

$$v(a) = 0, \quad u(b) = 0;$$

$$v(a) = 0, \quad v(b) = 0.$$

Then $\mathscr{B}_{oo}[a,b]$ is a subproblem of dimension one of each of the problems $\mathscr{B}_{o*}[a,b]$ and $\mathscr{B}_{*o}[a,b]$, while $\mathscr{B}_{o*}[a,b]$ and $\mathscr{B}_{*o}[a,b]$ are individually subproblems of $\mathscr{B}_{**}[a,b]$ of dimension one. For (σ,τ) any one of the sets $(0,0)$, $(0,*)$, $(*,0)$, and $(*,*)$, let $\{\lambda_j^{\sigma\tau}, u_j^{\sigma\tau}\}$ be a sequence of eigenvalues and eigenfunctions of $\mathscr{B}_{\sigma\tau}[a,b]$ satisfying the conditions of Theorem 2.1. Then:

(a) for $j = 1,2,\ldots$, we have

$$\lambda_j^{o*}[a,b] \leq \lambda_j^{oo}[a,b] \leq \lambda_{j+1}^{o*}[a,b],$$

$$\lambda_j^{*o}[a,b] \leq \lambda_j^{oo}[a,b] \leq \lambda_{j+1}^{*o}[a,b],$$

$$\lambda_j^{**}[a,b] \leq \lambda_j^{o*}[a,b] \leq \lambda_{j+1}^{**}[a,b],$$

$$\lambda_j^{**}[a,b] \leq \lambda_j^{*o}[a,b] \leq \lambda_{j+1}^{**}[a,b];$$

(b) if a,b,c are points of I satisfying $a < c < b$ then

$$\lambda_{j+k-1}^{oo}[a,b] \leq \operatorname{Max}\{\lambda_j^{o*}[a,c], \lambda_k^{*o}[c,b]\}.$$

{The inequalities of (a) are some of the most classic of Sturmian theory. Conclusion (b) has been noted specifically in Reid [19, Th. 2.2]}.

9. Suppose that $r(t)$, $p_1(t)$, $p_2(t)$ are real-valued continuous on $[0,b)$, while $r(t) > 0$ and $p_1(t) \geq p_2(t)$ on this interval. If $u_\alpha(t)$, $(\alpha = 1,2)$ is the solution of

the differential system

$$\ell_\alpha[u](t) \equiv [r(t)u'(t)]' - p_\alpha(t)u(t) = 0, \ u(0) = A, \quad u'(0) = B, \tag{5.3$_\alpha$}$$

where A and B are real constants not both zero, and $u_1(t)$, $u_2(t)$ are both different from zero on $[a,b)$, except possibly at $t = 0$, and $u_1(t)/u_2(t)$ is continuously differentiable on $[0,b)$, then throughout this interval $(u_1/u_2)' \geq 0$, $1 \leq u_1(t)/u_2(t) \leq \exp G_{21}(t)$, and $u_2(t)/u_1(t) \leq \exp G_{12}(t) \leq 1$, where

$$G_{ij}(t) = \int_0^t \frac{1}{r(s)u_i^2(s)}\left\{\int_0^s [f_j(t)-f_i(t)]u_i^2(t)\,dt\right\}ds, \quad i,j = 1,2.$$

{Breuer and Gottlieb [1, Lemma 3]}.

10. Let $r(t)$, $p(t)$ be continuous real-valued functions with $r(t) > 0$ on $[a,b]$, and set $k^2 = \operatorname{Min}\{r(t)p(t): t \in [a,b]\}$ and $K^2 = \operatorname{Max}\{r(t)p(t): t \in [a,b]\}$, where either of the numbers k^2, K^2 may be negative, positive, or zero.

Let $u(t)$ be the solution of the initial value problem

$$\ell_o[u](t) \equiv [r(t)u'(t)]' - p(t)u(t) = 0, \quad t \in [0,b], \quad u(0) = A, \qquad u'(0) = B, \tag{5.4}$$

where A and B are real constants not both zero, and define for arbitrary values λ associated functions

$$z(t;\lambda) = A \cosh\left[\lambda\int_0^t \frac{ds}{r(s)}\right] + \frac{Br(0)}{\lambda} \sinh\left[\lambda\int_0^t \frac{ds}{r(s)}\right], \tag{5.5}$$

$$G(t;\lambda) = \int_0^t \frac{1}{r(s)z^2(s;\lambda)}\left\{\int_0^s\left[p(\xi) - \frac{\lambda^2}{r(\xi)}\right]z^2(\xi;\lambda)\,d\xi\right\}ds. \tag{5.6}$$

(i) If $k^2 > 0$, $AB \geq 0$ with either $A > 0$ or $A = 0$,

$B > 0$, then

$$0 \le z(t;k) \le u(t) \le z(t;k) \exp G(t;k), \quad t \in [a,b]; \quad (5.7)$$

if $k^2 > 0$, $AB \ge 0$, with either $A < 0$ or $A = 0$, $B < 0$ the inequalities in (5.7) are reversed.

{Breuer and Gottlieb [1, Th. 3.1]. Proof involves the fact that $z(t;k)$ defined by (5.5) is the solution of the particular initial value problem

$$(r(t)z')' - (k^2/r(t))z = 0, \; z(0) = A, \; A'(0) = B, \quad (5.8)$$

and application of result of preceding Exercise 9 to the functions $u_1(t) = u(t)$, $u_2(t) = z(t;k)$}.

(ii) If $k^2 > 0$, $AB \ge 0$ with either $A > 0$ or $A = 0$, $B > 0$, then

$$u(t) \le z(t;k) \exp G(t;k), \quad t \in [a,b]; \quad (5.9)$$

if $k^2 > 0$, $AB \ge 0$ with either $A < 0$ or $A = 0$, $B < 0$ the inequality is reversed. {Breuer and Gottlieb [1, Th. 4.1]}.

(iii) if $k^2 > 0$, $AB < 0$ and $A > 0$, then

$$0 \le z(t;k) \le u(t) \le z(t;k) \exp G(t;k), \; t \in I^+(k), \quad (5.10)$$

where $I^+(k) = [0,b_k)$ if there exists a value $b_k \in (0,b]$ such that

$$\tanh\left[k \int_0^{b_k} \frac{ds}{r(s)}\right] = \left|\frac{Ak}{Br(0)}\right|, \quad 0 < k = \sqrt{k^2},$$

and $I^+(k) = [0,b]$ otherwise; inequalities in (5.10) are reversed if $A < 0$. {Breuer and Gottlieb [1, Th. 3.2]}.

(iv) if $k^2 > 0$, $AB < 0$ and $A > 0$, then

$$u(t) \le z(t;k) \exp G(t;k), \quad t \in I^+(k), \quad (5.11)$$

the inequality being reversed if $A < 0$. {Breuer and Gottlieb [1, Th. 4.2]}.

Corresponding bounds in cases of $k^2 = 0$, $AB < 0$, $k^2 = -c^2 < 0$, as well as corresponding results in terms of K^2 are obtained in Breuer and Gottlieb [1].

11. Suppose that $r(t)$, $q(t)$ and $\tilde{q}(t)$ are continuous and real-valued on $[a,b]$, and

$$\int_t^b q(s)ds \geq \left|\int_t^b \tilde{q}(s)ds\right|, \quad t \in [a,b]. \tag{5.12}$$

If λ_1 and $\tilde{\lambda}_1$ denote, respectively, the smallest positive eigenvalues of the boundary problems

(i) $[r(t)u'(t)]' + \lambda q(t)u(t) = 0, \quad u(a) = 0 = u'(b),$

(ii) $[r(t)u'(t)]' + \lambda\tilde{q}(t)u(t) = 0, \quad u(a) = 0 = u'(b),$

then $\tilde{\lambda}_1 \geq \lambda_1$. {St. Mary [1; Th. 4]}.

12. Suppose that $r(t)$, $p(t)$, $k(t)$ are real-valued continuous functions on $[a,b]$ with $r(t) > 0$, $k(t) > 0$ on this interval and denote by $\{\lambda_n\}$ the sequence of eigenvalues of the Sturm-Liouville system

$$\begin{gathered} \ell[u;\lambda](t) \equiv [r(t)u'(t)]' + [\lambda k(t) - p(t)]u(t) = 0, \\ t \in [a,b], \qquad u(a) = 0 = u(b). \end{gathered} \tag{5.13}$$

If $m = \text{Min}\{p(t)/k(t); t \in [a,b]\}$, while $f:[a,b] \to \mathbb{R}$ is any continuous function satisfying

$$0 < f(t) \leq r(t), \quad t \in [a,b], \tag{5.14}$$

and c is a positive constant satisfying

$$c^2 \geq k(t)f(t), \quad t \in [a,b], \tag{5.15}$$

then

$$\lambda_n < m + n^2\pi^2/\left[c \int_o^b \frac{dt}{f(t)}\right]^2, \quad n = 1,2,\ldots . \tag{5.16}$$

Correspondingly, if $M = \operatorname{Max}\{p(t)/k(t): t \in [a,b]\}$, while $F:[a,b] \to \mathbb{R}$ is any continuous function satisfying $F(t) \geq r(t)$ on $[a,b]$, and C is a positive constant satisfying

$$C^2 \leq k(t)F(t), \quad t \in [a,b], \tag{5.17}$$

then

$$\lambda_n < M + n^2\pi^2/\left[C \int_o^b \frac{dt}{F(t)}\right]^2, \quad n = 1,2,\ldots . \tag{5.18}$$

{Breuer and Gottlieb [2, Ths. 1.2]. The proof of (5.16) is by considering the associated differential system

$$[f(t)z'(t)]' + \mu[c^2/f(t)]z(t) = 0, \quad z(a) = 0 = z(b)$$

which possesses eigenvalues and corresponding eigenfunctions

$$\mu_n = n^2\pi^2/\left[c \int_a^b \frac{dt}{f(t)}\right]^2, \quad z_n(t) = \sin\left[\sqrt{\mu_n}\, c \int_a^t \frac{ds}{f(s)}\right],$$
$$n = 1,2,\ldots,$$

and the establishment of the existence of at least one value on (a,b) at which $\mu_n c^2/f(t) < \lambda_n k(t) - p(t)$. Inequality (5.18) is established by a corresponding proof involving the comparison system

$$[F(t)z'(t)]' + \nu[C^2/F(t)]z(t) = 0, \quad z(a) = 0 = z(b)\}.$$

13. Let $r(t)$, $k(t)$ be continuous positive functions on $[a,b]$ and c, C positive constants such that $C^2 \leq r(t)k(t) \leq c^2$ on $[a,b]$. Consider the boundary problems

$$\begin{aligned} &\text{(i)} \quad [r(t)u'(t)]' + \lambda k(t)u(t) = 0, \\ &\qquad B_1[u] \equiv \beta_a u(a) - \alpha_a r(a)u'(a) = 0, \\ &\qquad B_2[u] \equiv \beta_b u(b) + \alpha_b r(b)u'(b) = 0, \end{aligned} \tag{5.19}$$

and the associated problems

$$[r(t)u'(t)]' + \mu[c^2/r(t)]u(t) = 0, \quad B_\alpha[u] = 0, \ (\alpha=1,2) \quad (5.20)$$

$$[r(t)u'(t)]' + \nu[C^2/r(t)]u(t) = 0, \quad B_\alpha[u] = 0, \ (\alpha=1,2). \quad (5.21)$$

If λ_1, μ_1, ν_1 are the smallest eigenvalues of (5.19), (5.20), (5.21), respectively, then $\mu_1 < \lambda_1 < \nu_1$. {Breuer and Gottlieb [2, Th. 3]}.

14. If $r(t)$, $k(t)$ are continuous positive functions on $[a,b]$, and λ_1 is the smallest eigenvalue of

$$[r(t)u'(t)]' + \lambda k(t)u(t) = 0, \quad u(a) = 0 = u'(b), \quad (5.22)$$

then

$$\frac{(\pi/2)^2}{K^2\left[\int_a^b \frac{dt}{r(t)}\right]^2} < \lambda_1 < \frac{(\pi/2)^2}{k^2\left[\int_a^b \frac{dt}{r(t)}\right]^2}$$

where $k^2 = \mathrm{Min}\{r(t)k(t) : t \in [a,b]\}$ and $K^2 = \mathrm{Max}\{r(t)k(t) : t \in [a,b]\}$. {Breuer and Gottlieb [2, Th. 4]}.

15. If $r(t)$, $k(t)$ are continuous functions on $[a,b]$ with $r(t) > 0$ on this interval, and λ_1 is the smallest eigenvalue of the boundary problem (5.22), then

$$\lambda_1 \geq (\pi/2)^2 \Big/ \left\{D\left[\int_a^b \frac{ds}{r(s)}\right]\right\}^2, \quad (5.23)$$

where $D^2 = \mathrm{Max}\left\{\left|\int_t^b k(s)ds\right| \Big/ \int_t^b \frac{ds}{r(s)} : t \in [a,b]\right\}$. Also, if

$d^2 = \mathrm{Min}\left\{\int_t^b k(s)ds \Big/ \int_t^b \frac{ds}{r(s)} : t \in [a,b]\right\} > 0$, where it is now to be emphasized that hypothesis involves the positiveness of this value, then

$$\lambda_1 \leq (\pi/2)^2 \Big/ \left\{d\left[\int_a^b \frac{ds}{r(s)}\right]\right\}^2. \quad (5.24)$$

{Breuer and Gottlieb [2; Ths. 5,6]. The presented proof of (5.23) involves the comparison equation $[r(t)u'(t)]' + \mu[D^2/r(t)]u(t) = 0$, and the result of St. Mary [1], given above as Exercise 11. Inequality (5.24) is established by similar argument involving the comparison equation $[r(t)u'(t)]' + \mu[d^2/r(t)]u(t) = 0$}.

16. Let $r_\alpha(t)$, $(\alpha = 1,2)$ be positive functions of class $\mathscr{C}''[a,b]$, and $q(t)$ a positive continuous function on $[a,b]$. If $\psi(t)$ is a positive function of class $\mathscr{C}'[a,b]$ such that $r_1^2(t)\psi(t)$ is non-decreasing on this interval, and

$$\int_a^t r_2(s)\psi(s)ds \leq \int_a^t r_1(s)\psi(s)ds, \quad t \in (a,b),$$

and λ_1^α, $(\alpha = 1,2)$ denotes the smallest eigenvalue of the boundary problem

$$\begin{aligned} \ell_\alpha[u](t) &\equiv [r_\alpha(t)u(t)]' + \lambda q(t)u(t) = 0 \\ u(a) &= 0 = u'(b), \end{aligned} \qquad (5.25_\alpha)$$

then $\lambda_1^1 \geq \lambda_1^2$. {Howard [1, Th. 4.10]}.

17. Let λ_1 be the smallest eigenvalue of the boundary problem (5.22), where $r(t)$ and $k(t)$ are positive functions of class $\mathscr{C}''[a,b]$. Then the smallest eigenvalue of this problem satisfies

$$\frac{(\pi/2)^2 k^2}{\left[\int_a^b k(s)ds\right]^2} \leq \lambda_1 \leq \frac{(\pi/2)^2 K^2}{\left[\int_a^b k(s)ds\right]^2}, \qquad (5.26)$$

where k^2 and K^2 are defined as in Exercise 10 above.

{Breuer and Gottlieb [2, Th. 7]. To prove the left-hand inequality of (5.26), apply the result of Exercise 12 to the system (5.25_α), $(\alpha = 1,2)$, wherein $r_1(t) = r(t)$, $r_2(t) = k^2/k(t)$, $q(t) = k(t)$, $\psi(t) = k^2/r^2(t)$, where k^2 is the minimum of $r(t)k(t)$ on $[a,b]$. The right-hand inequality of (5.26) may be established in a similar fashion using systems (5.25_α) wherein $r_1(t) = K^2/k(t)$, $r_2(t) = r(t)$, $q(t) = k(t)$ and $\psi(t) = K^2/r^2(t)$, where K^2 is the maximum of $r(t)k(t)$ on $[a,b]$.}

18. For a Sturm-Liouville system

$$[r(t)u'(t)]' - [p(t) - \lambda k(t)]u(t) = 0,$$

$$u(a) = 0 = u(b)$$

with $r(t)$, $p(t)$, $k(t)$ real-valued, continuous and $r(t) > 0$, $k(t) > 0$ for $t \in [a,b]$, let $\{\lambda_n, u_n(t)\}$ denote the sequence of eigenvalues in increasing order and corresponding eigenfunctions. If for a closed subinterval $[\alpha,\beta]$ of $[a,b]$ the symbol $G_n[\alpha,\beta]$ denotes the number of zeros of $u_n(t)$ on $[\alpha,\beta]$, then

$$\lim_{n\to\infty} \frac{G_n[\alpha,\beta]}{n+1} = \frac{\int_\alpha^\beta \sqrt{k(t)/r(t)}\,dt}{\int_a^b \sqrt{k(t)/r(t)}\,dt}.$$

{Opial [4]}.

19. If $q:[0,2T] \to \mathbb{R}$ is continuous, then in order that the Sturm-Liouville boundary problem

$$u''(t) + [\lambda + q(t)]u(t) = 0, \quad u(0) = 0 = u(2T) \tag{5.27}$$

have a non-negative eigenvalue it is necessary that

$$T\int_0^T [q^+(t) + q^+(2T - t)]dt > 2. \tag{5.28}$$

A sufficient condition that (5.27) have a negative eigenvalue is that

$$M(q) \equiv T^{-1}\int_0^T t^2[q(t) + q(2T - t)]dt > 2. \tag{5.29}$$

Furthermore, the constant 2 occurring as the right-hand member of (5.29) is the best possible in the sense that for every $\varepsilon > 0$ there is a continuous function $q(t)$ satisfying $M(q) > 2 - \varepsilon$, and such that all eigenvalues of (5.27) are positive. {Putnam [3]}.

20. Suppose that $r(t)$ and $p(t)$ are continuous real-valued functions on $[a,b]$ with $r(t) > 0$, and for $\Pi: a = t_0 < t_1 < \ldots < t_k = b$, let $r_M(t)$, $p_M(t)$, $r_m(t)$, $p_m(t)$ be step functions on $[a,b]$ such that on each sub-interval (t_{j-1},t_j) the functions r_M and p_M are the maxima of $r(t)$ and $p(t)$ on $[t_{j-1},t_j]$, while r_m and p_m are the respective minima of these functions. Also, let $\check{r}(t)$ and $\check{p}(t)$ be defined as

$$\check{r}(t) = \frac{1}{t_j - t_{j-1}} \int_{t_{j-1}}^{t_j} r(s)ds, \quad \check{p}(t) = \frac{1}{t_j - t_{j-1}} \int_{t_{j-1}}^{t_j} p(s)ds,$$

$$\text{for } t \in (t_{j-1},t_j).$$

For given real values γ_{11}, γ_{22} let γ_{11}^M, γ_{22}^M and γ_{11}^m, γ_{22}^m, $\check{\gamma}_{11}$, $\check{\gamma}_{22}$ be such that $\gamma_{\alpha\alpha}^M \geq \gamma_{\alpha\alpha} \geq \gamma_{\alpha\alpha}^m$, and $\gamma_{\alpha\alpha}^M \geq \check{\gamma}_{\alpha\alpha} \geq \gamma_{\alpha\alpha}^m$, $(\alpha = 1,2)$, and denote by $\mathscr{B}$, $\mathscr{B}^M$, $\mathscr{B}^m$, $\check{\mathscr{B}}$ the respective boundary problems associated with the functionals

$$J[\eta;a,b] = \gamma_{11}\eta^2(a) + \gamma_{22}\eta^2(b) + \int_a^b \{r\eta'^2 + p\eta^2\}dt,$$

$$J^M[\eta;a,b] = \gamma_{11}^M\eta^2(a) + \gamma_{22}^M\eta^2(b) + \int_a^b \{r_M\eta'^2 + p_M\eta^2\}dt,$$

$$J^m[\eta;a,b] = \gamma_{11}^m\eta^2(a) + \gamma_{22}^m\eta^2(b) + \int_a^b \{r_m\eta'^2 + p_m\eta^2\}dt,$$

$$\check{J}[\eta;a,b] = \check{\gamma}_{11}\eta^2(a) + \check{\gamma}_{22}\eta^2(b) + \int_a^b \{\check{r}\eta'^2 + \check{p}\eta^2\}dt,$$

a fixed set S in the plane, and a fixed norming function $k(t)$. If $\{\lambda_j, u_j(t)\}$, $\{\lambda_j^M, u_j^M(t)\}$, $\{\lambda_j^m, u_j^m(t)\}$ and $\{\check{\lambda}_j, u_j(t)\}$, $(j = 1,2,\ldots)$, denote sequences of eigenvalues and associated eigenfunctions of these respective problems satisfying the conditions of Theorem 2.1, then

(a) $\lambda_j^M \geq \lambda_j \geq \lambda_j^m$, and $\lambda_j^M \geq \check{\lambda}_j \geq \lambda_j^m$, $j = 1,2,\ldots$

(b) if $|\Pi|$ denotes the norm of the partition Π, that is, $|\Pi| = \mathrm{Max}\{t_j - t_{j-1}: j = 1,\ldots,k\}$, then

$$\lim_{|\Pi|\to 0} \lambda_j^M = \lambda_j, \quad \lim_{|\Pi|\to 0} \lambda_j^m = \lambda_j, \quad \lim_{|\Pi|\to 0} \check{\lambda}_j = \lambda_j, \quad (j = 1,2,\ldots).$$

It is to be noted that a solution of $(r_M u')' - p_M u = 0$ is a continuous function on $[a,b]$ which on the subinterval (t_{j-1}, t_j) satisfies the differential equation $u'' - (p_M/r_M)u = 0$ with constant coefficients, and for which $r_M(t_\alpha^-)u'(t_\alpha^-) = r_M(t_\alpha^+)u'(t_\alpha^+)$, $(\alpha = 1,\ldots,k-1)$, with similar comments for the differential equations of the boundary problems $\mathscr{B}^m$ and $\check{\mathscr{B}}$.

{The general concepts of this topic have been basic for the approximation of eigenvalues and eigenfunctions, both theoretical and practical. Historically, the rigorous passage to the limit from difference equations to differential equations was first achieved by Porter [1]. Works dealing with similar problems are Whyburn [3,4] and Fort [3, Ch. X], and a survey of interrelations between differential equations and difference equations is to be found in the papers of Carmichael [1,2,3]. For the general theory of approximation the monograph of Kryloff [1] is noteworthy. Recently Colautti [1] has considered the specific problems of the approximation of the eigenvalues of Sturm-Liouville systems by methods in this general area}.

21. For $N = \{\nu : N' < \nu < N''\}$, and $-\infty \leq N' < N'' \leq \infty$, let $B(\nu)$ be a boundary problem as in Corollary 1 to Theorem 3.5 involving $\omega(t,\eta,\zeta:\nu)$, $Q[\eta:\nu]$, and

$$J[\eta:\nu] = Q[\eta:\nu] + \int_a^b 2\omega(t,\eta(t),\eta'(t):\nu)dt, \tag{5.30}$$

a fixed end-space S and norming function $k(t)$. Moreover, suppose that conditions (a), (b), (c) of this Corollary are satisfied, and denote by $\{\lambda_j(\nu), u_j(t:\nu)\}$, $(j = 1,2,\ldots)$, a set of eigenvalues and eigenfunctions satisfying the conditions of Theorem 2.1. From the cited Corollary, each of the eigenvalues $\lambda_j(\nu)$ is a continuous function of ν on N. Now suppose that the following conditions hold:

1°. there exists a ν' such that for $\nu \in (N',\nu')$ the functional $J[\eta:\nu]$ is positive definite on D_e;

2°. for an arbitrary positive integer m there exists a value ν^m such that if $\nu \in (\nu^m,N'')$ then there is a subspace of D_e of dimension m on which $J[\eta:\nu]$ is negative definite.

Then for $j = 1,2,\ldots,$ there is a sequence of disjoint compact subsets N_j of N such that $\lambda_j(\nu) = 0$ for $\nu \in N_j$. That is, for $\nu \in N_j$ there is a non-identically vanishing solution of the boundary problem

$$\begin{gathered}[r(t,\nu)u'(t)+q(t,\nu)u(t)]' - [q(t,\nu)u'(t)+p(t,\nu)u(t)] = 0 \\ (u(a);u(b)) \in S, \quad (Q_1[u,v]-v(a:\nu);Q_2[u,v]+v(b:\nu)) \in S^{\perp}.\end{gathered} \tag{5.31}$$

Also, any oscillation property possessed by an eigenfunction belonging to the j-th eigenvalue is possessed by a non-identically vanishing solution of (5.31) for $\nu \in N_j$. For example, if S has dimension zero then the boundary conditions of the problem are $u(a) = 0$, $u(b) = 0$, and for each $\nu \in N_j$ a non-identically vanishing solution of (5.31) has exactly $j - 1$ zeros on (a,b).

If for arbitrary $\eta \in D_e$ we have

$$J[\eta:\nu''] - J[\eta:\nu'] \leq 0, \text{ for } N' < \nu' < \nu'' < N'', \quad (5.32)$$

then for each j we have that $\lambda_j(\nu)$ is a monotone non-increasing function of ν on N, and consequently, if $\nu \in N_j$ and $\nu' \in N_{j+1}$ we have $\nu \leq \nu'$. In particular, the sets N_j reduce to singleton sets of one element if the strict inequality in (5.32) holds for arbitrary non-identically vanishing $\eta \in D_e$, or more generally, if in addition to the above conditions 1^0 and 2^0 we have the condition

3^0. there exists a function $g(\nu',\nu)$ defined for ν', ν elements of N satisfying $\nu' < \nu$, and such that if $N_1 < \nu' < \nu < N_2$ then $J[\eta:\nu] < g(\nu',\nu)J[\eta:\nu']$, for η an arbitrary non-identically vanishing element of D_e.

{Various special problems of the general nature of this Exercise have been considered in Richardson [4], Whyburn [1,2], Reid [7,40], Hartman [18] and Eisenfeld [1]}.

22. Suppose that $k:[a,b] \to \mathbf{R}$ is a non-identically vanishing and bounded positive function, which is of class $\mathscr{L}''$ on (a,b). If $f(t) = 5[k'(t)]^2 - 4k(t)k''(t)$ is non-negative on (a,b), and positive on some subinterval (a',b'), then the n-th eigenvalue λ_n of the boundary problem

$$u''(t) + \lambda k(t)u(t) = 0, \quad u(a) = 0 = u(b)$$

satisfies the inequality $\lambda_n \leq \left[\pi n / \int_a^b \sqrt{k(t)}dt\right]^2$. When the function $f(t)$ is non-positive on (a,b), and negative on a finite number of subintervals, then $\lambda_n > \left[\pi n / \int_a^b \sqrt{k(t)}dt\right]^2$. {Makai [2; Th. 1]}.

23. Suppose $q(t)$ is positive and continuous on $[-1,1]$, non-increasing on $[-1,\ell]$, non-decreasing on $[\ell,1]$, and by the process of "continuous symmetrization" of Polya and Szegö [1, p. 200] define the associated family $q(t,\alpha)$, $0 \leq \alpha \leq 1$, of equi-measurable functions. Specifically, for $t \in [-1,\ell]$ denote the function inverse to $q(t)$ by $t_1(s)$, and for $t \in [\ell,1]$ the corresponding inverse function by $t_2(s)$. For $t \in [-1,\ell(1-\alpha)]$ the function inverse to $q(t,\alpha)$ is denoted by $t_{1\alpha}(s)$, and for $t \in [\ell(1-\alpha),1]$ the function inverse to $q(t,\alpha)$ by $t_{2\alpha}(s)$, where $t_{1\alpha}(s) = (1-\alpha/2)t_1(s) - (\alpha/2)t_2(s)$, $t_{2\alpha}(s) = (1-\alpha/2)t_2(s) - (\alpha/2)t_1(s)$; if $q(-1) > q(1)$, then $t_2(s)$ is extended as the constant value 1 on $[q(1),q(-1)]$. Also, to complete the definition of $q(t,\alpha)$, if $q(t)$ attains the same constant value k in two subintervals $[a,b]$ and $[c,d]$, $a \leq b \leq c \leq d$, and if $t_1(k) = (1-m)a + mb$, $0 \leq m \leq 1$, then $t_2(k)$ is chosen to be $mc + (1-m)d$. In particular, $q(t,0) = q(t)$ and $q(t,1)$ is the symmetrically increasing rearrangement of $q(t)$. For the family of boundary problems

$$u''(t) + \lambda q(t,\alpha)u(t) = 0, \quad u(-1) = 0 = u(1), \tag{5.33}$$

$0 \leq \alpha \leq 1$, with eigenvalues $\lambda_1(\alpha) \leq \lambda_2(\alpha) \leq \cdots$ we have the following results:

(i) if $q(t)$ is also left-balanced then $\lambda_1(\alpha_1) \leq \lambda_1(\alpha_2)$ for $0 \leq \alpha_1 \leq \alpha_2 \leq 1$, and equality holds only if

$q(t)$ is symmetrically increasing;

(ii) $\sum_{n=1}^{\infty} [\lambda_n(\alpha_2)]^{-1} \leq \sum_{n=1}^{\infty} [\lambda_n(\alpha_1)]^{-1}$ for $0 \leq \alpha_1 \leq \alpha_2 \leq 1$, and equality holds only if $q(t)$ is symmetrically decreasing; moreover, if $q(t)$ is left-balanced then $\sum_{n=1}^{\infty} [\lambda_n(\alpha)]^{-1}$ is monotone also for $-1 \leq \alpha \leq 0$. {Abramovich [1]; the proof of (i) utilizes a result of Eliason [1] that $u_1(t)$, the eigenfunction of (5.33) for $\lambda = \lambda_1(\alpha)$, may be chosen to be left-balanced}.

24. Consider the boundary problem

$$u''(t) + \lambda q(t)u(t) = 0, \quad u(0) = 0 = u(1), \tag{5.34}$$

where $q:[0,1] \to \mathbf{R}$ is a non-negative integrable function. The sequence of eigenvalues of (5.34) will be denoted by $\lambda_1[q] \leq \lambda_2[q] \leq \ldots$

(a) If $q(t)$ can be expressed as

$$q(t) = \int_0^1 K(t,s)g(s)df(s), \tag{5.35}$$

where f is a monotone increasing bounded function, g is non-negative and continuous, while $K(t,s)$ is non-negative and $\int_0^1 K(t,s)dt = 1$ for $s \in [0,1]$, then

$$\lambda_1[q] \geq \left[\int_0^1 q(s)ds\right]^{-1} \inf_{s \in [0,1]} \lambda_1[K(\ ,s)].$$

{Banks [1, Lemma 1.1]}.

(b) Suppose that $q(t)$ is an increasing function on $[0,1]$, and for $\tau \in [0,1]$ define $H(t,\tau)$ as $H(t,\tau) = 0$ for $0 \leq t \leq \tau$, $H(t,\tau) = 1/(1-\tau)$ for $\tau < t \leq 1$. Then

(i) $\lambda_1[q] \int_0^1 q(t)dt \geq \lambda_0$, where $\lambda_0 = 7.88$ approximately, and equality is attained for $q(t) = H(t,t_0)$, where t_0 is approximately 0.357.

(ii) For $n = 1,2,\ldots$, there exists a value $t_n \in [0,1]$ such that

$$\left[\int_0^1 q(t)dt\right]^{-1} \sum_{k=1}^{n} 1/\lambda_k[q] \leq \sum_{k=1}^{n} 1/\lambda_k[H(\ ,t_n)].$$

(iii) For $n = 1,2,\ldots$, there exists a monotone non-decreasing step function q_n with at most n jumps such that $\int_0^1 q_n(t)dt = \int_0^1 q(t)dt$ and $\lambda_n[q] \geq \lambda_n[q_n]$.

{Banks [1, Ths. 2.1, 2.2, 2.3]. The proof of conclusion (iii) utilizes the following auxiliary result which has proved useful in other considerations as well}.

(c) Suppose that $p(t)$ and $q(t)$ are non-negative integrable functions on $[a,b]$, and $f(t)$ is non-negative, continuous and monotone increasing on this interval. If there exists a $c \in (a,b)$ such that $p(t) \geq q(t)$ on $[a,c)$ and $p(t) \leq q(t)$ on $(c,b]$ and also $\int_a^b p(t)dt = \int_a^b q(t)dt$, then $\int_a^b p(t)f(t)dt \leq \int_a^b q(t)f(t)dt$. If $f(t)$ is monotone decreasing, then the inequality is reversed. {Banks [1, Lemma 1.2]}.

(d) Suppose that $q(t)$ is a continuous convex function on $[0,1]$, and for $\tau \in [0,1]$ define $G(t,\tau)$ as $G(t,\tau) = 0$ for $0 \leq t \leq \tau$, $G(t,\tau) = 2(t-\tau)/(1-\tau)^2$ for $\tau < t \leq 1$. Then

(i) $\lambda_1[q] \int_0^1 q(t)dt \geq \lambda_0'$, where $\lambda_0' = 0.297$ approximately, and equality is attained for $q(t) = G(t,t_0')$, where t_0' is approximately 0.104.

(ii) For $n = 1,2,\ldots$, there is a value $t_n' \in [0,1]$ such that

$$\left[\int_0^1 q(t)dt\right]^{-1} \sum_{k=1}^{n} 1/\lambda_k[q] \leq \sum_{k=1}^{n} 1/\lambda_k[G(\ ,t_n')].$$

(iii) For $n = 1,2,\ldots$, there exists a piecewise linear convex function q_n with at most $n + 1$ distinct linear segments such that $\int_0^1 q_n(t)dt = \int_0^1 q(t)dt$ and $\lambda_n[q] \geq \lambda_n[q_n]$. {Banks [1, Ths. 3.1, 3.2, 3.4]}.

(e) Suppose that $q(t)$ is a continuous concave function on $[0,1]$. Then

(i) $\lambda_1[q] \int_0^1 q(t)dt \geq \lambda_0''$, where $\lambda_0'' = 6.952$ approximately, and equality is attained for the function $q(t) = T(t)$ defined as $T(t) = 4t$ for $t \in [0,1/2]$, $T(t) = 4(1-t)$ for $t \in [1/2,1]$.

(ii) If $G_0(t,s)$ is defined on $[0,1] \times [0,1]$ as $G_0(t,s) = 2t/s$ for $0 \leq t \leq s$, $G_0(t,s) = 2(1-t)/(1-s)$ for $s < t \leq 1$, then for $n = 1,2,\ldots,$ there exists a value $t_n'' \in [0,1]$ such that

$$\left[\int_0^1 q(t)dt\right]^{-1} \sum_{k=1}^{n} 1/\lambda_k[q] \leq \sum_{k=1}^{n} 1/\lambda_k[G_0(\ ,t_n'')].$$

(iii) For $n = 1,2,\ldots,$ there exists a piecewise linear concave function q_n whose graph has at most $n + 1$ linear segments such that $\int_0^1 q_n(t)dt = \int_0^1 q(t)dt$, $q_n(0) = 0 = q_n(1)$, and $\lambda_n[q] \geq \lambda_n[q_n]$. {Banks [1, Ths. 4.1, 4.2, 4.3]}.

25. If $p:[0,1] \to \mathsf{R}$ is concave, and $k:[0,1] \to \mathsf{R}$ is non-negative and integrable, then the smallest eigenvalue $\lambda_1[k,p]$ of the Sturm-Liouville system

$$u''(t) + [\lambda k(t) - p(t)]u(t) = 0,\ u(0) = 0 = u(1), \tag{5.36}$$

satisfies the inequality

$$\lambda_1[k,p] \geq \mathrm{Min}\{\lambda_1[k,PG(\ ,s)]:s \in [0,1]\},$$

where $\lambda_1[k,PG(\ ,s)]$ denotes the smallest eigenvalue of (5.36) with $p(t)$ replaced by $PG(t,s)$, where $P = \int_0^1 p(t)dt$ and $G(t,s) = 2t/s$ for $t \in [0,s]$, $G(t,s) = 2(1-t)/(1-s)$

for $t \in [s,1]$. {Banks [1; Th. 5.2]}.

26. For given positive M, H and a, let $E_1(M,H,a)$ denote the set of monotone increasing functions $q(t)$ on $[0,a]$ satisfying

$$\text{(i)} \quad M = \int_0^a q(t)dt, \qquad \text{(ii)} \quad 0 \leq q(t) \leq H, \tag{5.37}$$

$E_2(M,H,a)$ denote the set of continuous convex functions on $[0,a]$ satisfying conditions (5.37), and $E_3(M,a)$ denote the set of continuous concave functions on $[0,a]$ satisfying (i) of (5.37). Also, let $0 < \lambda_1(q) < \lambda_2(q) < \dots$ denote the sequence of eigenvalues of the boundary problem

$$u''(t) + \lambda q(t)u(t) = 0, \quad u(0) = 0 = u(a).$$

(a) For each positive integer n there exists a function $\rho_n \in E_1(M,H,a)$ that is a step function with at least one and at most n discontinuities on the open interval $(0,a)$ such that $\lambda_n(q) \leq \lambda_n(\rho_n)$ for all $q \in E_1(M,H,a)$. {Banks [3, Th. 1]}.

(b) For each positive integer n there exists a function $\rho_n \in E_2(M,H,a)$ that is a piecewise linear convex function with at most $(n + 2)$ pieces, and such that $\lambda_n(q) \leq \lambda_n(\rho_n)$ for all $q \in E_2(M,H,a)$. {Banks [3, Th. 3]}.

(c) For each positive integer n there exists a $\hat{\rho}_n \in E_3(M,a)$ that is a piecewise linear concave function with at most n pieces, and such that $\lambda_n(q) \leq \lambda_n(\hat{\rho}_n)$ for all $q \in E_3(M,a)$. {Banks [3, Th. 4]}.

27. Suppose that all eigenvalues of (2.5) are positive, so that $J[\eta:\mathscr{B}] > 0$ for arbitrary non-identically vanishing $\eta \in D_e[\mathscr{B}]$. For $\eta \in D[\mathscr{B}]$ and ζ such that $\eta \in D'[a,b]:\zeta$, then in view of (II.8.5) we have $J[\eta:\mathscr{B}] = -\int_a^b \eta \ell[\eta]dt$ and

by the Schwarz inequality we have $\left(-\int_a^b \eta\ell[\eta]dt\right)^2 \le \left(\int_a^b k\eta^2 dt\right)\left(\int_a^b (1/k)\{\ell[\eta]\}^2 dt\right)$. Consequently, if for a non-identically vanishing $\eta \in D$ we set

$$R_1[\eta] = \left(-\int_a^b \eta\ell[\eta]dt\right)\Big/\int_a^b k\eta^2 dt,$$

$$R_2[\eta] = \int_a^b (1/k)\{\ell[\eta]\}^2 dt\Big/\left(\int_a^b -\eta\ell[\eta]dt\right)$$

then for $j = 1,2,\ldots$, we have $R_2[\eta] \ge R_1[\eta] \ge \lambda_j$ for $\eta \in D \cap D_{N_j}[\mathscr{B}]$, where $D_{N_j}[\mathscr{B}]$ is defined by (2.18), and $u_j \in D \cap D_{N_j}[\mathscr{B}]$ with $R_2[u_j] = R_1[u_j] = \lambda_j$.

28. Suppose that: (i) the coefficients r, p, q of (2.5a) and the coefficients $M_{\sigma j}$ of the boundary conditions (2.5b) satisfy the conditions of hypothesis $(\mathscr{A})$; (ii) $k:[a,b] \to \mathbf{R}$ is a (Lebesgue) measurable function such that k and $1/k$ are locally of class $\mathscr{L}^\infty$ on I, while $[a,b]$ is a compact subinterval of I such that each of the sets $\{t:t \in [a,b], k(t) > 0\}$ and $\{t:t \in [a,b], k(t) < 0\}$ has positive measure; (iii) the functional $J[\eta:\mathscr{B}]$ defined by (2.9) is positive definite on $D_e[\mathscr{B}]$. Then there exists a sequence of positive eigenvalues $0 < \lambda_1 \le \lambda_2 \le \ldots$ with corresponding real eigenfunctions $u = u_\alpha(t)$ for $\lambda = \lambda_\alpha$, $(\alpha = 1,2,\ldots)$, and a sequence of negative eigenvaiues $0 > \lambda_{-1} \ge \lambda_{-2} \ge \ldots$ with corresponding real eigenfunctions $u = u_\beta(t)$ for $\lambda = \lambda_\beta$, $(\beta = -1,-2,\ldots)$, such that

(a) $K[u_i,u_j] = (\lambda_j/|\lambda_j|)\delta_{ij}$, $(i,j = \pm 1,\pm 2,\ldots)$;

(b) λ_1 is the minimum of $J[\eta:\mathscr{B}]$ on the class $D_N[\mathscr{B}|K^+] = \{\eta:\eta \in D_e[\mathscr{B}], K[\eta] = 1\}$, and $-\lambda_{-1}$ is the minimum of $J[\eta:\mathscr{B}]$ on the class $D_N[\mathscr{B}|K^-] = \{\eta:\eta \in D_e[\mathscr{B}], K[\eta] = -1\}$.

(c) for $j = 2,3,\ldots$ the classes

$$D_{Nj}[\mathscr{B}|K^+] = \{\eta : \eta \in D_N[\mathscr{B}|K^+],\ K[\eta,u_\alpha] = 0,\ \alpha = 1,\ldots,j-1\},$$

$$D_{Nj}[\mathscr{B}|K^-] = \{\eta : \eta \in D_N[\mathscr{B}|K^-],\ K[\eta,u_\beta] = 0,\ \beta = -1,\ldots,-j+1\},$$

are non-empty, with λ_j the minimum of $J[\eta:\mathscr{B}]$ on $D_{Nj}[\mathscr{B}|K^+]$, and $-\lambda_{-j}$ the minimum of $J[\eta|\mathscr{B}]$ on $D_{Nj}[\mathscr{B}|K^-]$.

(d) $\lambda_j \to \infty$ as $j \to \infty$, and $\lambda_j \to -\infty$ as $j \to -\infty$.

(e) if $\eta \in D_e[\mathscr{B}]$, and $d_j[\eta] = (|\lambda_j|/\lambda_j)K[\eta,u_j]$, $(j = \pm 1,\pm 2,\ldots)$, then the series $\sum_{j=-\infty}^{\infty}(|\lambda_j|/\lambda_j)d_j^2[\eta]$, $\sum_{j=-\infty}^{\infty}|\lambda_j|d_j^2[\eta]$ converge and

$$K[\eta] = \sum_{j=-\infty}^{\infty}(|\lambda_j|/\lambda_j)d_j^2[\eta], \quad J[\eta:\mathscr{B}] \geq \sum_{j=-\infty}^{\infty}|\lambda_j|d_j^2[\eta];$$

moreover, if $h \in D_e[\mathscr{B}]$ and $u(t)$ is the solution of (2.38), then $J[u:\mathscr{B}] = \sum_{j=-\infty}^{\infty}|\lambda_j|d_j^2[u]$ and $K[u] = \sum_{j=-\infty}^{\infty}(|\lambda_j|/\lambda_j)d_j^2[u]$.

(f) in particular, the functional $J[\eta:\mathscr{B}] - K[\eta]$ is positive definite, {non-negative definite} on $D_e[\eta]$ if and only if (2.5) has no eigenvalues satisfying $0 < \lambda \leq 1$, $\{0 < \lambda < 1\}$.

It is to be commented that for variational problems various authors, (see, for example, Lichtenstein [1,3,4]), have formulated the positive and non-negative definiteness of the second variation in terms of a boundary problem of the above sort. Specifically, for the fixed end-point problem discussed in Section 2 of Chapter I, under suitable differentiability conditions for the supposed minimizing arc the second variation functional of (I.2.6) may be written as

$$J_2[\eta] = \int_{t_1}^{t_2} \{f^o_{rr}(t)[\eta'(t)]^2 + (f^o_{yy}(t) - [f^o_{yr}(t)]')\eta^2(t)\}dt$$

and consequently if $f^o_{rr}(t) > 0$ for $t \in [t_1,t_2]$ the positive, {non-negative}, definiteness of $J_2[\eta]$ on V_o may be formulated as the condition that the eigenvalue problem

$$[f^o_{rr}(t)u'(t)]' + \lambda([f^o_{yr}(t)]' - f^o_{yy}(t))u(t) = 0$$

$$u(t_1) = 0 = u(t_2)$$

have no eigenvalue λ satisfying $0 < \lambda \leq 1$, $\{0 < \lambda < 1\}$.

{Boundary problems of the above form, sometimes termed "of polar form" have been considered in varying degrees of generality by many authors. See, for example, Courant and Hilbert [1-I, p. 161; 2-I, p. 136], Reid [2; 35, Ch. IV, Sec 6], Kamke [6]}.

29. If $q:[0,T] \to \mathbf{R}$ is continuous and the smallest eigenvalue λ_1 of the boundary problem

$$u''(t) + \lambda q(t)u(t) = 0, \quad u(0) = 0 = u(T)$$

is positive then the smallest eigenvalue μ_1 of the boundary problem

$$u''(t) + \mu q^2(t)u(t) = 0, \quad u(0) = 0 = u(T)$$

satisfies $0 < \mu_1 \leq (T\lambda_1)/\pi^2$. {Fink [5]}.

30. Suppose that $r(t)$ and $q_\alpha(t)$, $(\alpha = 1,2)$ are real-valued continuous functions on $[-a,a]$ with $r(t) > 0$ on this interval, and denote by $\lambda_1[q_\alpha]$ the smallest positive eigenvalues of the respective boundary problems

$$\ell^o[u:q_\alpha](t) \equiv [r(t)u'(t)]' + \lambda q_\alpha(t)u(t) = 0, \; u(-a) = 0 = u(a).$$

(i) If $b \in (-a,a)$ is a zero of the derivative of an

eigenfunction of $\ell^0[u:q_1](t) = 0$ for the eigenvalue $\lambda_1[q_1]$, and $0 \leq (b-t)\int_t^b q_1(s)ds \leq (b-t)\int_t^b q_2(s)ds$ for all $t \in (-a,a)$ then $\lambda_1[q_1] \leq \lambda_1[q_2]$, and strict inequality holds unless $\int_t^b q_1(s)ds = \int_t^b q_2(s)ds$ for all $t \in (-a,a)$.

(ii) If $r(t)$ and $q(t)$ are even functions on $(-a,a)$ and $0 \leq \int_{-t}^t q_1(s)ds \leq \int_{-t}^t q_2(s)ds$ for $t \in (0,a)$, then $\lambda_1[q_1] \leq \lambda_1[q_2]$ and strict inequality holds if $\int_{-\tau}^{\tau} q_1(s)ds <$ $\int_{-\tau}^{\tau} q_2(s)ds$ for some τ.

(iii) Suppose that $r(t)$ is even on $(-a,a)$, while $\int_{-t}^t q_1(s)ds > 0$ for $t \in (0,a)$. If there exists a $b \in (0,a)$ such that $q_1(t)$ is even on $(-b,b)$, and $q_2(t)$ is such that $q_1(t) \leq q_2(t)$ on $(-b,b)$, $q_1(t) \geq q_2(t)$ on $(-a,-b) \cup (b,a)$, and $\int_{-a}^a q_1(s)ds \leq \int_{-a}^a q_2(s)ds$, then $\lambda_1[q_1] \leq \lambda_1[q_2]$.

(iv) If $q(t)$ is a continuous, even, convex function on $(-a,a)$ with $\int_{-a}^a q(s)ds > 0$, then the smallest positive eigenvalue $\lambda_1[q]$ of the boundary problem

$$u''(t) + q(t)u(t) = 0, \quad u(-a) = u(a) = 0$$

satisfies the inequality $\lambda_1[q] \geq \pi^2/[2a\int_{-a}^a q(s)ds]$.

{Fink [6]. Conclusions (i) and (ii) are the respective Theorems 1 and 2 of this paper, while (iii) and (iv) are the Corollaries 1,2 of Theorem 2. In particular, (iii) generalizes a result of Beesack [1]}.

31. Suppose that $q(t)$ is continuous and not identically zero on $(-a,a)$, where $0 < a < \infty$, and let $q^{\wedge}(t)$ and $q_{\wedge}(t)$ denote the rearrangement of $q(t)$ in symmetrically

increasing and symmetrically decreasing order, respectively, in the sense of Hardy, Littlewood and Pólya [1, Ch. X] and of Pólya and Szegö [1, Ch. VII]. Moreover, let λ_1, $\lambda_1^{\wedge}$ and $\lambda_{1\wedge}$ denote the least positive eigenvalues of the respective differential systems

$$u''(t) + \lambda q(t)u(t) = 0, \qquad u(\pm a) = 0, \tag{5.38}$$

$$u''(t) + \lambda q^{\wedge}(t)u(t) = 0, \qquad u(\pm a) = 0, \tag{5.38$^\wedge$}$$

$$u''(t) + \lambda q_{\wedge}(t)u(t) = 0, \qquad u(\pm a) = 0. \tag{5.38$_\wedge$}$$

Then $\lambda_{1\wedge} \leq \lambda_1$ even if $q(t)$ changes sign finitely often, while $\lambda_1 \leq \lambda_1^{\wedge}$ whenever $q(t) \geq 0$ on $[-a,a]$.

{Beesack and Schwarz [1; Th. 2]. In addition to the extremizing property of the smallest eigenvalues of the given systems, the proof uses the following result from the cited books of Hardy, Littlewood and Pólya [1, Theorem 378] and Pólya and Szegö [1, p. 153]. If f, f_1, f_2, g, g_1 and g_2 are continuous functions on $(-a,a)$, $0 < a < \infty$, with f_1 and g_1 similarly ordered, f_2 and g_2 oppositely ordered, f, f_1, f_2 equimeasurable, and also g, g_1, g_2 equimeasurable, then

$$\int_{-a}^{a} f_2 g_2 dt \leq \int_{-a}^{a} fg dt \leq \int_{-a}^{a} f_1 g_1 dt.$$

Beesack and Schwarz also note that the above results include the following result announced by Pokornyi [DOKL, 79(1951), 743-746] and proved by Beesack [TAMS 81(1956), 211-242];

"If $q(t)$ is continuous non-negative even function on $(-a,a)$ which is non-increasing on $[0,a]$, and the equation $u''(t) + q(t)u(t) = 0$ has a solution which does not vanish on $(-a,a)$, then for $q_1(t)$ the even function on $[-a,a]$

defined as $q_1(t) = q(a-t)$ on $[0,a]$ the equation $u''(t) + q_1(t)u(t) = 0$ also has a solution which does not vanish on $(-a,a)$."}.

32. Suppose that $q(t)$ is continuous on $[a,b]$ and $q^{\sim}(t)$, $q_{\sim}(t)$ are determined from $q(t)$ by the condition that $q(t)$, $q^{\sim}(t)$ and $q(t)q_{\sim}(t)$ are equimeasurable, with $q^{\sim}(t)$ non-increasing and $q_{\sim}(t)$ non-decreasing, see Hardy, Littlewood and Pólya [1; Ch. 10]. Moreover, let λ_1, $\lambda_1^{\sim}$, $\lambda_{1\sim}$ denote the least positive eigenvalues of the respective boundary problems

$$u''(t) + \lambda q(t)u(t) = 0, \quad u(a) = 0, \ u'(b) = 0, \tag{5.39}$$

$$u''(t) + \lambda q^{\sim}(t)u(t) = 0, \quad u(a) = 0, \ u'(b) = 0, \tag{5.39$^\sim$}$$

$$u''(t) + \lambda q_{\sim}(t)u(t) = 0, \quad u(a) = 0, \ u'(b) = 0. \tag{5.39$_\sim$}$$

Then $\lambda_{1\sim} \le \lambda_1$ even if $q(t)$ changes sign at most a finite number of times, while $\lambda_1 \le \lambda_1^{\sim}$ whenever $q(t) \ge 0$ on $[a,b]$.

{St. Mary [1; Th. 5]. This result is obtained by a direct application of the result of Beesack and Schwarz [1; Th. 2], listed as the preceding Exercise, to the null-end point boundary problems obtained by reflecting about the line $t = b$ the graphs of the functions $q(t)$, $q^{\sim}(t)$, $q_{\sim}(t)$ and the corresponding solutions of the above boundary problems}.

33. Suppose that $q(t)$ is a real valued, continuous, even function on $(-\infty,\infty)$ which is non-increasing on $(0,\infty)$ and that $u(t)$ is an even solution of the differential equation

$$u''(t) + q(t)u(t) = 0, \quad t \in (-\infty,\infty) \tag{5.40}$$

which has a least positive zero at $t = a$. If a_1, a_2, where

$-\infty < a_1 < a_2 < +\infty$, is any pair of consecutive zeros of a real solution $u(t)$ of (5.40) then $a_2 - a_1 \geq 2a$; moreover, if $q(t)$ is strictly decreasing on $(0,a)$ then $a_2 - a_1 > 2a$ unless $a_1 = -a$, $a_2 = a$. Also, if $q(t)$ is a non-negative, continuous, even function that is non-decreasing on $(0,\infty)$, then $a_2 - a_1 \leq 2a$, and if $q(t)$ is strictly increasing on $(0,a)$ then $a_2 - a_1 < 2a$ unless $a_1 = -a$, $a_2 = a$.

{Beesack and Schwarz [1, Lemma 3]}.

34. If λ_α, $\phi_\alpha(t)$, $(\alpha = 0,1,\ldots)$, are the eigenvalues and eigenfunctions of a Sturm-Liouville problem, then λ_β, $\psi_\beta(t) = \phi_\beta'(t) - \phi_0'(t)\phi_\beta(t)/\phi_0(t)$, $(\beta = 1,2,\ldots)$ are the eigenvalues and eigenfunctions of another Sturm-Liouville system, termed the "first associated system" by Crum [1]. Repeating this process, the n-th associated system is determined, for which Crum gives an explicit determinatal form.

CHAPTER IV.
OSCILLATION THEORY ON A NON-COMPACT INTERVAL

1. Introduction

We shall consider in this chapter the behavior of non-identically vanishing real solutions of a linear second order differential equation of the form (II.1.1), (II.1.1^{0}) or (II.1.1$^{\#}$) on a non-compact interval which for the major portion of the discussion will be taken to be of the form $I = [a,\infty)$, where a is a finite value. Such an equation is said to be *oscillatory* in case one non-identically vanishing real solution, and hence all such solutions, have infinitely many zeros on I; clearly an equivalent statement is that the equation is not disconjugate on any non-degenerate subinterval $I_0 = [a_0,\infty)$ of I. It is to be remarked that alternate terminologies for this concept are "oscillatory in a neighborhood of ∞", or "oscillatory for large t". If an equation fails to be oscillatory it is said to be "non-oscillatory", with the corresponding qualifications "in a neighborhood of ∞" or "for large t".

In the study of the qualitative nature of solutions on an interval $[a,\infty)$ two types of questions naturally arise and have been studied in great detail. One is the asymptotic nature of solutions, and the other is the question of oscillation and/or non-oscillation of solutions. Although there are important interrelations between these concepts, the first will not be considered here except in a most cursory fashion. In particular, it is intimately related to the stability of solutions and thus very appropriately would be considered in an article on this latter topic for differential equations. In regard to the second type of question, the comprehensive article by Willett [2] provides an excellent resume of literature and unification of certain aspects of the theory to the date of its publication, and there is no intention to present excessive duplication here. There are certain aspects of the theory which have been the foundation for recent studies of oscillation phenomena for differential systems, however, and the presented discussion of specific results will be limited largely to such topics. The same general comment applies to the theory of singular, self-adjoint boundary problems emanating from the 1909 memoir of Hermann Weyl. Many accounts of this theory are available, as in Naimark [1], Titchmarsh [1], and especially in the extensive Chapter XIII of Dunford-Schwartz [1-II] which includes an extensive list of exercises and historical comments on various spectral problems connected with linear differential operators of the second order, with special attention to Sturm-Liouville operators and results concerned with the essential spectrum.

2. Integral Criteria for Oscillation and Non-Oscillation

Most of our attention in this section will be to an equation of the form (II.1.1$^{\#}$), which for the purpose of relating stated results to referenced papers will be written in the form

$$\ell^{\#}[u](t) \equiv u''(t) + q(t)u(t), \quad t \in [a,\infty), \qquad (2.1|q)$$

where $q:[a,\infty) \to \mathbf{R}$ is supposed to satisfy $(\mathscr{A}_C)$ or $(\mathscr{A}_L)$ on $I = [a,\infty)$. One of the earlier papers on the oscillatory and non-oscillatory character of solutions of such an equation, and one that has exerted a tremendous influence on subsequent work in this area, is that of Kneser [1] in 1892. Noting that the Euler type differential equation

$$u''(t) + \mu t^{-2}u(t) = 0, \quad t \in [1,\infty) \qquad (2.2)$$

is non-oscillatory for $\mu \leq 1/4$ and oscillatory for $\mu > 1/4$, Kneser employed the Sturm comparison theorem to conclude that if $q:[1,\infty) \to \mathbf{R}$ is continuous and $t^2q(t) \to L$ as $t \to \infty$, then (2.1) is non-oscillatory if $L < 1/4$ and oscillatory if $L \geq 1/4$. Over the years, equation (2.2) has remained as a most useful one for comparison, and has motivated many criteria for oscillation and non-oscillation.

Kneser [1] also considered oscillation properties of solutions of a higher order linear differential equation

$$u^{[n]}(t) + q(t)u(t) = 0, \quad t \in [a,\infty), \qquad (2.3)$$

where $q(t)$ is a continuous real-valued function on $[a,\infty)$. In particular, for n even he showed that if there exist positive constants m, k such that $2m \leq n$ and $t^m q(t) \geq k$ for large t then every real solution of (2.3) has infinitely many zeros on $[a,\infty)$. The problem of oscillation of solutions of an equation (2.3) was also treated by Fite [1], who showed that if $q(t) > 0$ and the integral $\int_a^\infty q(s)ds = \lim_{t\to\infty} \int_a^t q(s)ds = +\infty$, then every non-identically vanishing real-valued solution of (2.3) must change sign infinitely many times on $[a,\infty)$. For $n = 2$ a relatively slight extension of this result of Fite was re-discovered by Wintner [4] in 1947, who showed that if $q(t)$ is non-negative and equation (2.1) is non-oscillatory on $[a,\infty)$, then $q \in \mathscr{L}[a,\infty)$: that is, $\int_0^\infty q(t)dt$ is finite. This paper of Wintner served as impetus for the paper [1] of Hille, who considered for an equation (2.1) both types of problems mentioned above. In particular, Hille showed that if $q(t) \geq 0$ and (2.1) is non-oscillatory on $[a,\infty)$, then $t^\sigma q(t) \in \mathscr{L}[\tau,\infty)$ for each $\sigma < 1$ and $\tau > 0$, and that the function

$$\gamma(t|q) = t\int_t^\infty q(s)ds \tag{2.4}$$

remains bounded as $t \to \infty$.

Clearly, (2.1) is non-oscillatory on $[a,\infty)$ if and only if there exists a neighborhood $[a_0,\infty)$ such that there is a real solution $w(t)$ of the Riccati differential equation

$$w'(t) + w^2(t) + q(t) = 0, \quad t \in [a_0,\infty), \tag{2.5}$$

or, indeed, in view of Theorem II.9.2, if and only if there exists a real-valued function $w(t)$ on $[a,\infty)$ which is locally a.c. on this interval, and

$$w'(t) + w^2(t) + q(t) \leq 0, \quad \text{for } t \text{ a.e. on } [a_0,\infty). \tag{2.5'}$$

Continuing with the assumption that $q(t) \geq 0$ on I,

it then follows by elementary argument that (2.1) is non-oscillatory on $I = [a,\infty)$ if and only if there exists an $[a_0,\infty) \subset [a,\infty)$ such that there is a real solution of the integral equation

$$w(t) = \int_t^\infty w^2(s)ds + \int_t^\infty q(s)ds, \quad t \in [a_0,\infty). \tag{2.6}$$

An integral equation equivalent to (2.6), and which Hille found to be more appropriate for the study of existence and qualitative nature of solutions, is

$$z(t) = t\int_t^\infty s^{-2}z^2(s)ds + \gamma(t|q), \tag{2.7|q}$$

in the associated function $z(t) = tw(t)$. In particular, for $q(t) \geq 0$ Hille showed that if (2.1|q) is non-oscillatory on $[a,\infty)$ then $\liminf_{t\to\infty} \gamma(t|q) \leq 1/4$ and $\limsup_{t\to\infty} \gamma(t|q) \leq 1$; moreover, if $\limsup_{t\to\infty} \gamma(t|q) < 1/4$ then (2.1|q) is non-oscillatory. Hille proceeded to show that if $q_\alpha(t) \geq 0$, $(\alpha = 1,2)$, and $\gamma(t|q_1) \geq \gamma(t|q_2)$ for $t \in [a,\infty)$, then non-oscillation of $(2.1|q_1)$ on $[a,\infty)$ implies non-oscillation of $(2.1|q_2)$ on $[a,\infty)$.

In addition, Hille derived a sequence of results elaborating the extension of Kneser's result formulated as: (2.1|q) is non-oscillatory for large t if $\limsup_{t\to\infty} t^2q(t) < 1/4$ and oscillatory for large t if $\liminf_{t\to\infty} t^2q(t) > 1/4$. The first test of this sequence is given by: If $F_1(t) = (t \ln t)^2[q(t) - (1/4t^2)]$ then (2.1|q) is non-oscillatory for large t if $\limsup_{t\to\infty} F_1(t) < 1/4$ and oscillatory for large t if $\liminf_{t\to\infty} F_1(t) > 1/4$. In particular, Hille [1, Th. 12] showed that (2.1|q) is non-oscillatory for large

t if there exists an integer p such that

$$\gamma(t|q) \leq \frac{1}{4}\int_t^\infty S_p(s)ds,$$

where $S_p(t) = \sum_{k=0}^{p} [L_k(t)]^{-2}$ with $L_0(t) = t$, $L_j(t) = L_{j-1}(t)\ell n_j t$, and $\ell n_1 t = \ell n\ t$, $\ell n_h t = \ell n(\ell n_{h-1} t)$, $h = 2,3,\ldots$ Subsequently, Wintner [6] showed without any assumption on the algebraic sign of q that if

$$\gamma_1(t|q) = \frac{1}{t}\int_0^t\left[\int_0^\sigma q(s)ds\right]d\sigma \tag{2.8}$$

is such that $\gamma_1(t|q) \to +\infty$ as $t \to +\infty$ then $(2.1|q)$ is oscillatory, and consequently if this equation is non-oscillatory then $\int_0^\sigma q(s)ds$ cannot diverge to $+\infty$ as $t \to +\infty$.

In view of the possible reduction of an equation $(II.1.1^0)$ to the form $(II.1.1^\#)$ by the change of independent variable (II.1.9), from this result of Wintner [6] it follows that if $r(t)$ and $q(t)$ are continuous on $[a,\infty)$ with $r(t) > 0$ on this interval, then the equation

$$\ell^0[u](t) \equiv [r(t)u'(t)]' + q(t)u(t) = 0,\ t \in [a,\infty) \tag{2.9}$$

is oscillatory on $[a,\infty)$ if

$$\int_a^\infty [1/r(t)]dt = +\infty, \quad \int_a^\infty q(t)dt = +\infty, \tag{2.10}$$

where the second integral is to be interpreted as the Cauchy-Lebesgue integral $\lim_{\tau\to\infty}\int_a^\tau q(t)dt$. This result was also proved independently by Leighton in his paper [2]. It is to be remarked that the paper [8] of Coles presents a very simple proof of this Wintner-Leighton result.

In 1957 Wintner [10] completed the Kneser-Hille comparison theorem and removed the condition on the algebraic sign on the q_α's to obtain the result that if $q_\alpha(t)$, $(\alpha = 1,2)$, are continuous real-valued functions on (a,∞) such that $\gamma_0(t|q_\alpha) = \int_t^\infty q_\alpha(s)ds = \lim_{\tau\to\infty} \int_t^\tau q_\alpha(s)ds$ exist and are finite, and $0 \le \gamma_0(t|q_2) \le \gamma_0(t|q_1)$ for t on a subinterval $[a_0,\infty)$ of I, then if $(2.1|q_1)$ is non-oscillatory the equation $(2.1|q_2)$ is also non-oscillatory and consequently if $(2.1|q_2)$ is oscillatory then $(2.1|q_1)$ is oscillatory.

Various extensions of the Wintner-Leighton comparison result have been obtained. In particular, largely through the use of judicious change of dependent variable in (2.9), Moore [1] showed that (2.9) is oscillatory provided one of the following conditions is satisfied:

$$\begin{aligned} &\text{(a)} \quad \int_a^\infty [1/r(t)]dt = +\infty \;\; \textit{and} \;\; \int_a^\infty q(t)g^n(t)dt = +\infty \;\; \textit{for} \\ &\qquad \textit{some} \;\; n < 1, \;\textit{where}\;\; g(t) = 1 + \int_a^t [1/r(s)]ds; \\ &\text{(b)} \quad \int_a^\infty [1/r(t)]dt < +\infty, \;\textit{and}\;\; \int_a^\infty q(t)h^m(t)dt = +\infty \;\; \textit{for} \\ &\qquad \textit{some} \;\; m, \;\textit{where}\;\; h(t) = \int_t^\infty [1/r(s)]ds. \end{aligned} \tag{2.11}$$

Also, Moore showed that (2.9) is non-oscillatory for large t if either of the following conditions is satisfied, where $g(t)$ and $h(t)$ are as in (a) and (b) above:

$$\begin{aligned} &\text{(a}'\text{)} \quad \textit{there exists a finite value } A \textit{ such that} \\ &\qquad 0 < A - \int_b^t q(t)g(t)dt < 1 \quad \textit{for} \quad a \le b \le t; \\ &\text{(b}'\text{)} \quad \textit{there exists a finite value } B \textit{ such that} \\ &\qquad 0 < B + \int_b^t q(t)h(t)dt < 1 \quad \textit{for} \quad a \le b \le t. \end{aligned} \tag{2.12}$$

In particular, as a corollary to a more general result, Moore shows that (2.9) is non-oscillatory provided

$$\int_a^\infty 1/r(t)\,dt = \infty, \quad \text{and} \quad \limsup_{t\to\infty}\left|\int_a^t q(s)ds\right| < +\infty. \tag{2.13}$$

Wintner [7] showed that if $q(t) > 0$ and

$$\gamma_0(t|q) = \int_t^\infty q(s)ds, \quad t \in [0,\infty) \tag{2.14}$$

exists, then $(2.1|q)$ is non-oscillatory on $[0,\infty)$ if the inequality

$$\gamma_0^2(t|q) \le (1/4)q(t)$$

holds for large t. That $1/4$ is the optimum constant for such an inequality was shown by Opial [2], who showed that if $\gamma_0(t|q)$ exists as an improper integral and $\gamma_0(t|q) > 0$ in a neighborhood of ∞, then $(2.1|q)$ is non-oscillatory if for large t we have

$$\gamma_0(t|\gamma_0^2(\ |q)) \le (1/4)\gamma_0(t|q),$$

while $(2.1|q)$ is oscillatory if for some $\varepsilon > 0$ we have

$$\gamma_0(t|\gamma_0^2(\ |q)) \ge ([1+\varepsilon]/4)\gamma_0(t|q)$$

for large t. Oscillation results which generalized those of Fite and Wintner mentioned above were also given in by Olech, Opial and Ważewski [1], who showed that $(2.1|q)$ is oscillatory for large t if $\lim\text{appr}_{t\to\infty} \int_0^t q(s)ds = \infty$, and also that the condition (see MR19-650 for definition)

$$\lim\text{appr inf} \int_0^t q(s)ds < \lim\text{appr sup} \int_0^t q(s)ds$$

is sufficient for oscillation.

Coles [7] obtained extensions of the mentioned results of Hartman and Wintner through the introduction of "mean value functionals" or "weighted averages" of the form

$$A(t) = A(t,\tau|q,f) = \frac{\int_t^\tau f(s)\left(\int_t^s q(\xi)d\xi\right)}{\int_t^\tau f(s)ds}, \qquad (2.15)$$

where $f(t)$ belongs to the class of non-negative locally integrable functions on $[0,\infty)$ for which there exists a value $a > 0$ and a constant $k \in [0,1)$ such that

$$f_k(t|f) = \int_a^t f(s)\left\{\left[\int_0^s f(\tau)d\tau\right]^k \Big/ \left[\int_0^s f^2(\tau)\right]d\tau\right\}ds \to \infty$$
$$\text{as } t \to \infty.$$

Coles [7] proved that if $\gamma_1(t|q)$ does not tend to a limit as $t \to \infty$, and if there exists a function f of the above class such that $\liminf_{t\to\infty} A(t,\tau|q,f) > -\infty$, then $(2.1|q)$ is oscillatory. Willett [1] extended the concept of weighted average introduced by Coles, employing the following classes of functions:

$$F = \left\{f{:}f \text{ locally integrable on } [a,\infty),\ f(t) \geq 0, \int_a^\infty f(t)dt = \infty\right\}$$

$$F_0 = \left\{f{:}f \in F,\ \lim_{t\to\infty}\left[\int_a^t f^2(s)ds\right]\Big/\left[\int_a^t f(s)ds\right]^2 = 0\right\}$$

$$F_1 = \left\{f{:}f \in F,\ \limsup_{t\to\infty}\left[\left(\int_a^t f(s)ds\right)\left(\int_t^\infty f(s)\left\{\int_a^s f^2(\xi)d\xi\right\}^{-1}ds\right)\right] > 0.\right.$$

A continuous function $q{:}[a,\infty) \to R$ is said to have an average integral with respect to F, F_0 or F_1 if there exists an f of the respective class such that in the extended real number system the limit of $A(t,\tau|q,f)$ as $\tau \to \infty$ exists, and the value of this average integral is denoted by $A(t|q,f)$. It follows readily that if this limit exists and is finite for one value t_1 on $[a,\infty)$ then it exists and is finite for all values of t and

$$A(t|q,f) = A(t_1 | q,f) - \int_{t_1}^{t} q(s)ds,$$

so that $A(t|q,f)$ has derivative equal to $-q(t)$; furthermore, if $\int_t^\infty q(s)ds = \lim_{\tau\to\infty} \int_t^\tau q(s)ds$ exists then $A(t|q,f) = \int_t^\infty q(s)ds$. Corresponding to the above mentioned result of Coles, Willett [1] proved that if there exists an $f \in F_1$ such that $A(a|q,f) = \infty$ then $(2.1|q)$ is oscillatory. Also, if there exists an $f \in F$ such that $\liminf_{t\to\infty} A(t,a|q;f) > -\infty$ then either $(2.1|q)$ is oscillatory, or for arbitrary $g \in F_0$ and $t \in [a,\infty)$ the average integral $A(t|q,g)$ exists and is finite. Moreover, Willett [1] showed that if q has a finite averaged integral $A(t|q,f)$ with respect to F_0, then $(2.1|q)$ is disconjugate on $[a,\infty)$ if and only if there exists on this interval a solution of the integral inequality

$$w(t) \geq A(t|q,f) + \int_t^\infty w^2(s)ds. \tag{2.16}$$

For an arbitrary continuous function $\Phi: [a,\infty) \to R$, let

$$K(t,s|\Phi) = \exp\left\{2 \int_t^s \Phi(\xi)d\xi\right\}, \tag{2.17}$$

$$M(t|\Phi) = \int_t^\infty K(t,s|\Phi)\Phi^2(s)ds. \tag{2.18}$$

In case q has a finite averaged integral $A(t|q,f)$ with $f \in F_0$, Willett [1] also showed that $(2.1|q)$ is disconjugate on $[a,\infty)$ if and only if there exists on this interval a solution of the Riccati integral inequality

$$w(t) \geq M(t|A(\ |q,f)) + \int_t^\infty K(t,s|A(\ |q,f))w^2(s)ds, \tag{2.19}$$

and proceeded to generalize various known criteria of oscillation and non-oscillation. In particular, if $Q(t) = Q(t;q)$

is any continuously differentiable function on $[a,\infty)$ such that $Q'(t) = -q(t)$ on this interval, then:

(i) $(2.1|q)$ is disconjugate on $[a,\infty)$ if $M(t|Q) < \infty$ and the function

$$M_1(t|Q) = \int_t^\infty K(t,s|Q)M^2(s|Q)ds \tag{2.20}$$

satisfies the inequality

$$M_1(t|Q) \le (1/4)M(t|Q), \quad \text{for} \quad t \in [a,\infty). \tag{2.21}$$

(ii) If q has a finite average integral $A(t|q,f)$ with $f \in F_o$, and either $M(t|Q) = \infty$ or there exists an $\varepsilon > 0$ such that

$$M_1(t|Q) \ge ([1+\varepsilon]/4)M(t|Q), \quad \text{for} \quad t \in [a,\infty) \tag{2.22}$$

then $(2.1|q)$ is oscillatory on $[a,\infty)$. Moreover, Willett [1] showed that $(2.1|q)$ is oscillatory in case there exists a finite averaged integral $A(t|q,f)$ with $f \in F_o$ for which there is a constant k, $0 < k \le 4$ satisfying $\int_a^\infty \exp\{-\gamma\int_a^t A(s|q,f)ds\}dt < \infty$ a result which generalized a result of Hartman [7]. The study of such criteria was continued in Willett [3], wherein an iterative procedure was presented for the derivation of a necessary and sufficient condition for the existence of a solution of (2.19).

This program was continued and generalized in Coles and Willett [1], wherein the ideas of the earlier papers were exploited to relate classical summability methods to the problem of oscillation of $(2.1|q)$. In particular, as a Corollary to a general result it is shown that if there exists a positive integer n such that the function $Q_0(t) = \int_0^t q(s)ds$

is either summable by Cesàro means $\{C,n\}$ to $+\infty$, or summable by Hölder means $\{H,n\}$ to $+\infty$, then $(2.1|q)$ is oscillatory.

The averaging technique introduced by Coles [1] was also extended in a fashion somewhat different than the above by Macki and Wong [1], using an "averaging pair" of functions (σ,α). Special cases of their Theorem 2 are the well-known criterion of Wintner [6], that of Coles [7] cited above, and a result of Howard [2] when specialized to linear equations. Also, for the averaging pair $(\sigma,\alpha) = (1,1)$ in the author's Theorem 1 there is obtained the non-oscillation criterion of Hartman [7; Th. 2].

The paper of Wong [1] is in the general format of the Coles-Willett studies, and concerned primarily with the consideration of equations $(2.1|q)$ for which $\gamma_0(t|q)$ does not remain non-negative in a neighborhood of ∞.

In particular, if $\int_a^\infty q(s)ds = \lim_{t\to\infty} \int_a^t q(s)ds$ exists and is finite, then in (2.20), (2.21) we may choose $Q(t) = \gamma_0(t|q)$, and for brevity we set

$$\hat{K}_1(t,s|q) = \exp\left\{2 \int_t^s \gamma_0(\xi|q)d\xi\right\} \tag{2.23}$$

$$\hat{M}_1(t|q) = M(t;\gamma_0(\ |q)) = \int_t^\infty K_1(t,s|q)\gamma_0^2(s|q)ds. \tag{2.24}$$

Wong [1] then obtained the following general results.

(i) The non-oscillation of $(2.1|q)$ for large t is equivalent to the existence of a solution of the non-linear Riccati integral equation

$$z(t) = \hat{M}_1(t|q) + \int_t^\infty \hat{K}_1(t,s|q)z^2(s)ds \tag{2.25}$$

for t on some neighborhood $[c,\infty)$ of ∞.

(ii) If $\lim_{t\to\infty} \int_0^t \gamma_0(s|q)ds$ exists and is finite and $\hat{M}_1(t|q)$ satisfies the inequality

$$\int_0^\infty \exp\left\{-4 \int_0^s \hat{M}_1(\tau|q)d\tau\right\}ds < \infty, \tag{2.26}$$

then $(2.1|q)$ is oscillatory.

(iii) If $\lim_{t\to\infty} \int_0^t \gamma_0(s|q)ds$ exists and is finite, and

$$\hat{K}_2(t,s|q) = \exp\left\{2 \int_t^s [\gamma_0(\tau|q) + \hat{M}_1(\tau|q)]d\tau\right\} \tag{2.27}$$

$$\hat{M}_2(t|q) = \int_t^\infty \hat{K}_2(t,s|q)\gamma_0^2(s|q)ds, \tag{2.28}$$

then $(2.1|q)$ is non-oscillatory on a neighborhood of ∞ if and only if there is a solution of the non-linear Riccati integral equation

$$z(t) = \hat{M}_2(t|q) + \int_t^\infty \hat{K}_2(t,s|q)z^2(s)ds \tag{2.29}$$

for t in some neighborhood of ∞. Using these general results, Wong [1] derived a number of oscillation and non-oscillation theorems, including as corollaries certain results of Opial [1] and Willett [1].

3. Principal Solutions

In 1936, Morse and Leighton [1] made a major contribution to the theory of singular quadratic functionals, wherein the coefficient functions of a differential equation (II.1.1) satisfy hypothesis ($\mathscr{A}$) on an open interval $I = (a_0,b_0)$, and one is concerned with the problem of minimizing an associated quadratic functional on an appropriate class of functions defined on the open interval I. In particular, the work of

Morse and Leighton [1] involved an extension of the conjugate point theory for an equation of the form (II.1.1).

In the following discussion it will be assumed that the coefficients of (II.1.1) satisfy hypothesis ($\mathscr{A}$) on an open interval $I = (a_0,b_0)$ where $-\infty \leq a_0 < b_0 \leq \infty$. Now if (II.1.1) is disconjugate on I, then for $s \in I$ and $u(t)$ a solution of this equation satisfying $u(s) = 0$, $u'(s) \neq 0$ it follows that $u(t) \neq 0$ for all t distinct form s in I. On the other hand, if (II.1.1) fails to be disconjugate on I, then for a value $s \in I$ in a suitable neighborhood of a_0 there exists a first right-hand conjugate point to s on I, which is denoted by $\tau_1^+(s)$. Moreover, in its interval of existence, $\tau_1^+(s)$ is a strictly monotone increasing function of s, in view of the separation of zeros of linearly independent solutions. Consequently, $\lim_{s \to a_0} \tau_1^+(s)$ exists, is denoted by $\tau_1^+(a_0)$, and is called the *first (right-hand) conjugate point to* a_0 *on* I. Correspondingly, for $s \in I$ in a suitable neighborhood of the end-point b_0 there exists a first left-hand conjugate point $\tau_1^-(s)$ to s on I, and $\tau_1^-(b_0) = \lim_{s \to b_0} \tau_1^-(s)$ is called the first (left-hand) conjugate point to b_0 on I. Clearly, either a_0 or b_0 may be its own first-hand conjugate point on I, as is exemplified by the equation $u'' + u = 0$ on $(-\infty,\infty)$.

If an end-point of I is not its own first conjugate point on I, however, then for $\{u_1(t),u_2(t)\}$ a set of linearly independent real-valued solutions of (II.1.1) on I there exist values a_1,b_1 such that both $u_1(t)$ and $u_2(t)$ are different from zero on $(a_0,a_1] \cup [b_1,b_0)$, so that both of the ratios $u_1(t)/u_2(t)$ and $u_2(t)/u_1(t)$ are well-defined and finite on a neighborhood of each end-point. Moreover, since on such neighborhoods we have $[u_1/u_2]' = -\{u_1,u_2\}/ru_2^2$, it follows that each of the ratios u_1/u_2 tends to a limit, finite or infinite, at each of the end-points a_o and b_o of I. Now if $\{u_1(t),u_2(t)\}$ is a set of linearly independent real-valued solutions for which $u_1(t)/u_2(t)$ tends to a finite limit k as $t \to a_o$, then it follows readily, (see, for example, Lemma 2.1 of Morse and Leighton [1]), that $u(t;a_o) = u_1(t) - ku_2(t)$ is a solution of (II.1.1) such that if $u(t)$ is any solution of this equation forming with $u(t;a_o)$ a linearly independent set then $u(t;a_o)/u(t) \to 0$ as $t \to a_o$, and $u(t;a_o)$ is uniquely determined except for a non-zero constant factor; moreover, the first conjugate point to $t = a_o$ on I, if its exists, is the smallest zero of $u(t;a_o)$ on I. Similarly, there exists a non-identically vanishing solution $u(t) = u(t;b_o)$ of (II.1.1) such that if $u(t)$ is any solution of this equation with $\{u(t;b_o),u(t)\}$ linearly independent then $u(t;b_o)/u(t) \to 0$ as $t \to b_o$. Also, as in the case of the end-point a_o, the solution $u(t;b_o)$ is unique except for a non-zero constant factor, and the first conjugate point to $t = b_o$ on I is the largest zero of $u(t;b_o)$ on I. The solutions $u(t;a_o)$ and $u(t;b_o)$ are called *principal solutions* of (II.1.1) at $t = a_o$ and $t = b_o$, respectively. It is to be emphasized that a principal solution is defined at an endpoint of I if and only if this equation is disconjugate on a subinterval I_o of I with this endpoint of I also an endpoint of I_o.

Now if $u(t;a_o)$ is a principal solution of (II.1.1) at a_o, and $u(t)$ is a real-valued solution of this equation such that the set $\{u(t;a_o), u(t)\}$ is linearly independent, let $a_1 \in I$ be such that both $u(t;a_o)$ and $u(t)$ are non-zero on $(a_o,a_1]$. From Theorem II.2.4 it then follows that there exist non-zero constants c_1', c_1'' such that for $t \in (a_o,a_1]$ we have

$$\begin{aligned} &(a) \quad \frac{u(t)}{u(t;a_o)} = \frac{u(a_1)}{u(a_1;a_o)} - c_1' \int_t^{a_1} \frac{ds}{r(s)u^2(s;a_o)}, \\ &(b) \quad \frac{u(t;a_o)}{u(t)} = \frac{u(a_1;a_o)}{u(a_1)} - c_1'' \int_t^{a_1} \frac{ds}{r(s)u^2(s)}. \end{aligned} \tag{3.1}$$

As $u(t;a_o)/u(t) \to 0$ and $|u(t)/u(t;a_o)| \to \infty$ as $t \to a_o$, it then follows that the improper integrals

$$\int_{a_o}^{a_1} \frac{ds}{r(s)u^2(s;a_o)}, \quad \int_{a_o}^{a_1} \frac{ds}{r(s)u^2(s)} \tag{3.2}$$

are, respectively, divergent and convergent. That is, a non-identically vanishing solution $u(t)$ of (II.1.1) is for the end-point a_o a principal solution or a non-principal solution according as the integral

$$\int_{a_o} \frac{ds}{r(s)u^2(s)} \tag{3.3}$$

diverges or converges. Analogously, a non-identically vanishing solution $u(t)$ of (II.1.1) is for the end-point b_o a principal solution or a non-principal solution according as the integral

$$\int^{b_o} \frac{ds}{r(s)u^2(s)} \tag{3.4}$$

diverges or converges. This discriminating property of solutions was discovered by Hartman and Wintner [5], and employed in the study of the asymptotic behavior of solutions.

Returning to the case of a principal solution $u(t;a_o)$, let $a_1 \in I$ be such that $u(t;a_o) \neq 0$ for $t \in (a_o,a_1]$, and for $a_o < \tau < c \leq a_1$ let $u = u(t;\tau,c)$ be a solution of (II.1.1) satisfying the conditions $u(c) = 1$, $u(\tau) = 0$. Again, employing the representation formula (2.1) of Theorem II.2.4, we have that (assume $u(c;a_o) = 1$)

$$u(t;\tau,c) = u(t;a_o)\left\{1 - \left[\int_t^c \frac{ds}{r(s)u^2(s;a_o)}\right]\Big/\left[\int_\tau^c \frac{ds}{r(s)u^2(s;a_o)}\right]\right\}. \tag{3.5}$$

In view of the divergence of the first integral of (3.2), it then follows that $u(t;\tau,c) \to u(t;a_o)$ as $\tau \to a_o$, and indeed the convergence is uniform on any compact subinterval of $(a_o,a_1]$.

In particular, for $a_o < c < a_1$, let $u = u_o(t;c,a_1)$ be the solution of (II.1.1) satisfying the conditions $u_o(c;c,a_1) = 1$, $u_o(a_1;c,a_1) = 0$, and for $a_o < \tau < c < a_1$ consider the function $\eta_\tau(t)$ defined as $\eta_\tau(t) \equiv 0$ on $[a_o,\tau]$, $\eta_\tau(t) = u(t;\tau,c)$ on $[\tau,c]$, and $\eta_\tau(t) = u_o(t;c,a_1)$ on $[c,a_1]$. As this equation is disconjugate on $(a_o,a_1]$, the functional $J[\eta:\tau,a_1]$ is positive definite on $D_o'[\tau,a_1]$, and therefore, $J[\eta_\tau:\tau,a_1] > 0$. Now

$$\begin{aligned} J[\eta_\tau:\tau,a_1] &= J[u(\ ;\tau,c):\tau,c] + J[u_o(\ ;c,a_1):c,a_1] \\ &= u(c;\tau,c)v(c|\tau,c) - u_o(c;c,a_1)v_o(c|c,a_1) \\ &= v(c;\tau,c) - v_o(c;c,a_1). \end{aligned}$$

Moreover, if $a_o < \tau_2 < \tau_1 < c < a_1$, then $u(\tau_2;\tau_2,c) = 0 = \eta_{\tau_1}(\tau_2)$, $u(c;\tau_2,c) = 1 = \eta_{\tau_1}(c)$, and since $u(t;\tau_2,c) \not\equiv 0$ on (τ_2,τ_1) it follows that

$$J[u(\ ;\tau_2,c):\tau_2,c] < J[\eta_{\tau_1}:\tau_2,c] = J[u(\ ;\tau_1,c):\tau_1,c].$$

As $J[u(\ ;\tau_\alpha,c):\tau_\alpha,c] = u(c;\tau_\alpha,c)v(c;\tau_\alpha,c) = v(c;\tau_\alpha,c)$ for

$\alpha = 1,2$, it then follows that for a given value $c \in (a_o,a_1)$ we have that $v(c;\tau,c)$ is a strictly monotone increasing function of τ which is bounded below by $v_o(c;c,a_1)$. Therefore, $v_{a_o} = \lim_{\tau \to a_o} v(c; ,c)$ exists and is not less than $v_o(c;c,a_1)$. In particular, from the above determination of $u(t;a_o)$ as the limit of $u(t;\tau,c)$ as $\tau \to a_o$, it follows that the principal solution of (II.1.1) at the end-point a_o is determined as the function $u(t)$ belonging to the solution $(u(t),v(t))$ of (II.1.2) satisfying the initial conditions $u(c) = 1$, $v(c) = v_{a_o}(c)$. Moreover, in view of the uniqueness of this principal solution to within a multiplicative constant, it follows that if $c \in (a_o,a_1)$ and (u,v) is a solution of (II.1.2) with $u(c) \neq 0$ and $v(c)/u(c) > v_{a_o}(c)$, then $u(t)$ vanishes for a value on (a_o,c), whereas if $v(c)/u(c) < v_{a_o}(c)$ the function $u(t)$ is non-zero on $(a_o,c]$.

Correspondingly, if $u(t;b_o)$ is a principal solution at b_o, and $b_1 \in (a_o,b_o)$ is such that (II.1.1) is disconjugate on $[b_1,b_o)$, then for $b_1 < c < \tau < b$ the solution $u(t;c,\tau)$ determined by the conditions $u(c;c,\tau) = 1$, $u(\tau;c,\tau) = 0$, and the solution $u_o(t;b_1,c)$ satisfying the conditions $u_o(b_1;b_1,c) = 0$, $u_o(c;b_1,c) = 1$ is such that $v_o(c;b_1,c) > v(c;c,\tau)$, and $v(c;c,\tau)$ is a strictly monotone increasing function of τ on (c,b_o). Consequently, $v_{b_o}(c) = \lim_{\tau \to b_o} v(c;c,\tau)$ does not exceed $v_o(c;b_1,c)$, and the solution (u,v) of (II.1.2) specified by the initial conditions $u(c) = 1$, $v(c) = v_{b_o}(c)$ is a principal solution $u(t;b_o)$ of (II.1.1) at b_o. Also, if (u,v) is a solution of (II.1.2) with $u(c) \neq 0$ and $v(c)/u(c) < v_{b_o}(c)$ then $u(t)$ vanishes for a value on (c,b_0), while if $v(c)/u(c) > v_{b_0}(c)$ the function $u(t)$ is non-zero on (c,b_0).

In particular, if (II.1.1) is disconjugate on the entire interval (a_0,b_0) then both principal solutions $u(t;a_0)$, $u(t;b_0)$ exist and are non-zero throughout this interval; also, in this case we have that

$$\frac{v(c;b_0)}{u(c;b_0)} \le \frac{v(c;a_0)}{u(c;a_0)}, \quad \text{for arbitrary } c \in (a_0,b_0). \tag{3.6}$$

Moreover, for $c \in (a_0,b_0)$ and $(u(t),v(t))$ a solution of (II.1.2) with $u(c) \neq 0$, we have the following possibilities:

(i) *if* $\dfrac{v(c)}{u(c)} > \dfrac{v(c;a_0)}{u(c;a_0)}$, *then* $u(t)$ *vanishes at a point on* (a_0,c), *and* $u(t) \neq 0$ *on* (c,b_0);

(ii) *if* $\dfrac{v(c)}{u(c)} < \dfrac{v(c;b_0)}{u(c;b_0)}$, *then* $u(t)$ *vanishes at a point on* (c,b_0), *and* $u(t) \neq 0$ *on* (a_0,c);

(iii) *if* $\dfrac{v(c;b_0)}{u(c;b_0)} < \dfrac{v(c)}{u(c)} < \dfrac{v(c;a_0)}{u(c;a_0)}$, *then* $u(t) \neq 0$ *on* (a_0,b_0).

Now, in general, if $(u(t),v(t))$ is a solution of (II.1.2) with $u(t) \neq 0$ on a subinterval I_0 of I, then $w(t) = v(t)/u(t)$ is a solution of the Riccati differential equation (II.2.5) on I. Consequently, whenever (II.1.1) possesses a principal solution $u(t;a_0)$ or $u(t;b_0)$ there exists a corresponding solution $w(t;a_0) = v(t;a_0)/u(t;a_0)$ or $w(t;b_0) = v(t;b_0)/u(t;b_0)$ in a neighborhood of the respective end-point. Such a solution of (II.2.5), which has been called a *distinguished solution* of this equation, is unique whenever it exists. Moreover, if (II.1.1) is disconjugate on the entire interval (a_0,b_0), then both $w(t;a_0)$ and $w(t;b_0)$ exist on (a_0,b_0), and the above results on the behavior of solutions $(u(t),v(t))$ translate into the following properties of solutions of the Riccati differential equation (II.2.5).

If (II.1.1) *is disconjugate on* (a_0,b_0), *then* $w(t;a_0)$ *and* $w(t;b_0)$ *exist on* (a_0,b_0); *and*

(a) $w(t;b_0) \leq w(t;a_0)$, *for* $t \in (a_0,b_0)$;

(b) *if* $c \in (a_0,b_0)$, *and* $w(t)$ *is a solution of* (II.2.5) *defined in a neighborhood of* $t = c$, *then:*

(i) *if* $w(c) > w(c;a_0)$, *the maximal interval of existence of* $w(t)$ *is of the form* (s,b_0), *where* $s \in [a_0,c)$;

(ii) *if* $w(c) < w(c;b_0)$, *the maximal interval of existence of* $w(t)$ *is of the form* (a_0,s), *where* $s \in (c,b_0]$;

(iii) *if* $w(c,b_0) < w(c) < w(c,a_0)$, *then the maximal interval of existence of* $w(t)$ *is* (a_0,b_0), *and* $w(t,c_0) < w(t) < w(t,a_0)$ *throughout this interval.*

The above determination of the principal solution $u(t;a_0)$ of (II.1.1) at the end-point a_0, and the allied discussion of the principal solution $w(t;a_0)$, follows the pattern used by Reid [15] in 1958 for the extension to these concepts to differential systems of the sort to be discussed in Chapter V.

4. Theory of Singular Quadratic Functionals

As in the preceding section, suppose that the coefficients of the differential equation (II.1.1) satisfy hypothesis ($\mathscr{A}$) on an open interval (a_0,b_0), where $-\infty \le a_0 < b_0 \le +\infty$. The theory of singular quadratic functionals as introduced by Morse and Leighton [1] involves the study of the behavior of the functional $J[\eta;s_1,s_2]$, $a_0 < s_1 < s_2 < b_0$, as $s_1 \to a_0$, $s_2 \to b_0$, and η is restricted to belong to certain classes of arcs $\eta:(a_0,b_0) \to \mathbf{R}$. Specifically, a basic problem is that of determining conditions which are necessary and/or sufficient for the relation

$$\liminf_{s_1\to a_0, s_2\to b_0} J[\eta;s_1,s_2] \ge 0 \tag{4.1}$$

to hold for arbitrary η in the prescribed class of "admissible arcs". Whenever (4.1) holds for a given class of arcs the function $\eta \equiv 0$, or the segment (a_0,b_0), is said to afford a *minimum limit* to J on the given class; whenever the minimum limit exists it is said to be *proper* if the equality sign in (4.1) holds only if $\eta \equiv 0$.

The major portion of Morse and Leighton [1] is devoted to the study of problems involving only a single singular end-point, and in the following discussion specific attention will also be limited to this case with the singular end-point chosen to be b_0. Corresponding to terminology of the initial paper of Morse and Leighton [1] and subsequent studies of Leighton [2,3], Leighton and Morse [1], Martin [1] and Morse [8,9], for the case of a singular end-point at b_0 and a given $c \in (a_0,b_0)$, the following classes of "admissible arcs" have received major attention.

$\mathscr{F}$-admissible arcs, $\eta:[c,b_0) \to R$ with $\eta(c) = 0$, while on arbitrary compact subintervals $[c,b]$ of $[c,b_0)$ the function η is a.c. and $\eta' \in \mathscr{L}^2[c,b]$.

$\mathscr{F}'$-admissible arcs, $\eta:[c,b_0) \to R$ with η an $\mathscr{F}$-admissible arc which is bounded on $[c,b_0)$.

$\mathscr{A}$-admissible arcs, $\eta:[c,b_0) \to R$ with η an $\mathscr{F}'$-admissible arc satisfying $\lim_{t \to b_0} \eta(t) = 0$; for the case of b_0 finite, equally well the class of $\mathscr{A}$-admissible arcs may be defined as the class of continuous functions $\eta:[c,b_0] \to R$ with $\eta(c) = 0 = \eta(b_0)$, and which are $\mathscr{F}$-admissible on $[c,b_0)$.

Following the notation of Morse [8,9], to show the dependence of each of the above classes upon the value c the superscript c is appended to the corresponding letters to yield the designations $\mathscr{F}^c$-, $\mathscr{F}'^c$-, and $\mathscr{A}^c$-admissible arcs. We shall proceed to discuss briefly the condition

$$\liminf_{s \to b_0} J[\eta;c,s] \geq 0 \quad \text{for} \quad \eta \in \mathscr{A}^c. \tag{4.1'}$$

Now if there exists a value $s_0 \in (c,b_0)$ which is conjugate to c relative to (II.1.1), then in view of Theorem II.8.4 it follows that for $b \in (s_0,b_0)$ there exists an arc $\eta_0 \in D_0'[c,b]$ such that $J[\eta_0;c,b] < 0$, and consequently $\eta:[c,b_0) \to R$ defined as $\eta(t) = \eta_0(t)$ for $t \in [c,b]$, $\eta(t) = 0$ for $t \in (b,b_0)$, is such that $\eta \in \mathscr{A}^c$ and $J[\eta;c,s] = J[\eta_0;c,b] < 0$ for $s \in [b,b_0)$ and hence condition (4.1) does not hold. That is, if $[c,b_0)$ affords a minimum limit to J on $\mathscr{A}^c$, then (II.1.1) is disconjugate on $[c,b_0)$.

On the other hand, if (II.1.1) is disconjugate on $[c,b_0)$, then the solution $(u_c;v_c)$ of (II.1.2) determined by the initial conditions $u_c(c) = 0$, $v_c(c) = 1$ is such that

$u_c(t) > 0$ for $t \in (c,b_0)$. For $\tau \in (c,b_0)$ and $\eta \in \mathscr{A}^c$, let $\eta_{c\tau}:[c,b_0) \to R$ be defined as $\eta_{c\tau}(t) = [\eta(\tau)/u_c(\tau)]u_c(t)$ for $t \in [c,\tau]$, $\eta_{c\tau}(t) = \eta(t)$ for $t \in (\tau,b_0)$. Then $\eta_{c\tau} \in \mathscr{A}^c$ and in view of Corollary 2 to Theorem II.8.1 we have that $J[\eta;c,\tau] \geq J[\eta_{c\tau};c,\tau]$, with equality sign holding only if $\eta(t)$ is a multiple of u_c on $[c,\tau]$. Therefore, for $s \in [\tau,b_0)$ we have $J[\eta;c,s] \geq J[\eta_{c\tau};c,s]$ and consequently $[c,b_0)$ affords a minimum value to J if and only if

$$\liminf_{s \to b_0} J[\eta_{c\tau};c,s] \geq 0, \text{ for } \eta \in \mathscr{A}^c \text{ and } \tau \in (c,b_0). \tag{4.2}$$

Proceeding with an analysis of the functional $J[\eta_{c\tau};c,s]$ of (4.2), Leighton and Morse [1] established the following basis result.

THEOREM 4.1. *If* (II.1.1) *is disconjugate on* $[c,b_0)$, *then* $[c,b_0)$ *affords a minimum value to* J *on* $\mathscr{A}^c$, *if and only if*

$$\liminf_{\tau \to b_0} J[\eta_{c\tau};c,\tau] \geq 0, \quad \text{for all } \eta \in \mathscr{A}^c \tag{4.3}$$

for which $\liminf_{s \to b_0} J[\eta;c,s] < +\infty$.

The condition (4.3) has been called by Morse and Leighton the "singularity condition" on J belonging to the segment $[c,b_0)$.

In terms of the solution $(u_c(t);v_c(t))$ of (II.1.2) as defined above, one has the evaluation

$$\begin{aligned} J[\eta_{c\tau};c,\tau] &= [v_c(\tau)/u_c(\tau)]\eta^2(\tau) \\ &= w_c(\tau)\eta^2(\tau), \end{aligned} \tag{4.4}$$

where $w_c(t) = v_c(t)/u_c(t)$ is the corresponding solution of the Riccati differential equation (II.2.5).

Now suppose (II.1.1) is disconjugate on $[c,b_o)$ and $b \in (c,b_o)$. If η is a non-identically vanishing element of $\mathscr{A}^b$ and $\eta^o:[c,b_o) \to R$ is defined as $\eta^o(t) = \eta(t)$ for $t \in [b,b_o)$ and $\eta^o(t) = 0$ for $t \in [c,b)$, then $\eta^o \in \mathscr{A}^c$. Moreover, if $\tau \in (b,b_o)$ is such that $\eta(\tau) \neq 0$, then $\eta^o_{c\tau}$ is the solution of (II.1.1) such that $\eta^o_{c\tau}(c) = 0$, $\eta^o_{c\tau}(\tau) = \eta(\tau)$ and $\eta^o_{c\tau}(t) \not\equiv \eta(t)$ on $[c,\tau]$ so that $J[\eta^o;c,\tau] - J[\eta^o_{c\tau};c,s] > 0$. Consequently, for $s \in [\tau,b_o)$ we have $J[\eta;b,s] = J[\eta^o;c,s]$ and

$$J[\eta;b,s] = J[\eta^o_{c\tau};c,s] + \{J[\eta^o;c,\tau] - J[\eta^o_{c\tau};c,s]\}.$$

Therefore, if (4.1) holds then for $b \in (c,b_o)$ we have

$$\liminf_{s \to b_o} J[\eta;b,s] \geq J[\eta^o;c,\tau] - J[\eta^o_{c\tau};c,s] > 0$$

for $\eta \in \mathscr{A}^b$. That is, if (II.1.1) is disconjugate on $[c,b_o)$ and the segment $[c,b_o)$ affords a minimum limit to J on $\mathscr{A}^c$, then for $b \in (c,b_o)$ the segment $[b,b_o)$ affords a proper minimum limit to J on $\mathscr{A}^b$,

If (II.1.1) is disconjugate on $[c,b_o)$ and $\eta \in \mathscr{A}^c$, then for $s \in (c,b_o)$ and $b \in [c,s)$ let $\eta_b(t) = 0$ for $t \in [c,b]$, $\eta_b(t) = \eta_{bs}(t)$ for $t \in [s,b_o)$; that is, on $[b,s]$ the function $\eta_{bs}(t)$ is the solution $u(t)$ satisfying $u(b) = 0$, $u(s) = \eta(s)$, and $\eta_b(t) = \eta(t)$ for $t \in [s,b_o)$. Then $\eta_b \in \mathscr{A}^b$, and $\eta_b(t)$ is equal to $\eta_{cs}(t)$ for $b = c$. It then follows that $J[\eta_b;b,s] \to J[\eta_{cs};c,s]$ as $b \to c$. In particular, if (II.1.1) is disconjugate on $[c,b_o)$ and for each $b \in (c,b_o)$ the segment $[b,b_o)$ affords a minimum limit to J on $\mathscr{A}^b$, then the segment $[c,b_o)$ affords a minimum limit on J on $\mathscr{A}^c$.

As noted by Morse and Leighton [1, Example 4.3 on p. 263], in Theorem 4.1 one may not omit the stipulation that $\liminf_{s \to b_0} J[\eta;c,s] < +\infty$, as is shown by the example $u'' + u = 0$ on the interval $(-\pi,\pi)$ with $c = 0$. This equation is disconjugate on $[0,\pi)$ and the associated solution $w_c(t)$ of (II.2.5) is $w_c(t)$ is ctn t, so that the corresponding value of (4.4) $= \eta^2(\tau) \operatorname{ctn} \tau$. However, $\eta(t) = t(\pi-t)^{1/2}$ belongs to $\mathscr{A}^c$ and $J[\eta_{c\tau};c,\tau] \to -\pi^2$.

Morse and Leighton [1] also consider problems involving two singular end-points, and give necessary and sufficient conditions for the existence of the minimum limit in a class of arcs termed "$(\mathscr{A},\mathscr{A}^*)$ admissible". Also, there is given a preliminary sufficient condition for $\eta \equiv 0$ to afford a minimum limit to J in the class of $\tilde{t}$-admissible arcs for a problem involving a single singular end-point. In particular, the authors give special attention to the case of integrals J with $r(t) = t^{\alpha}g(t)$, $q(t) \equiv 0$, $p(t) = -t^{\alpha-2}h(t)$ where α is any real number g and h are real analytic functions with $g(t) > 0$ on an interval $[0,c)$, so that the corresponding equation (II.1.1) has a regular singular point at $t = 0$.

The paper [3] of Leighton is devoted to a more detailed study of problems involving a single singular end-point, and wherein $q(t) \equiv 0$ and $p(t)$ is of one sign near the singular end-point. In particular, this paper marked the beginning of the author's study of necessary and sufficient conditions for the existence of minimum limits in classes of types $\mathscr{A}$ and $\mathscr{F}$, with these conditions expressed in terms of the coefficient functions $r(t)$, $p(t)$ and the solutions of the differential equation (II.1.1), and undoubtedly provided the impetus for his subsequent extensive research on criteria of oscillation and non-oscillation for real scalar second order linear differential equations. The work of Leighton [3] is continued in Leighton and Martin [1] with attention devoted to functionals J wherein $q(t) \equiv 0$, but with the condition that p be of a fixed sign in a neighborhood of the singular end-point now removed.

The papers [1] and [2] of Martin extend further the results of Leighton and Martin [1]. In particular, Martin [1] considers the case wherein the assumption that $r(t) > 0$ on (a_0,b_0) is removed, and indeed covers cases in which $r(t) \geq 0$ with $r(t) = 0$ on a set of positive measure. For $[a,b]$ a compact subinterval of I and s a real number, Martin considers the class $\mathscr{F}_s[a,b]$ of real-valued functions η on $[a,b]$ which are a.c., with $\eta' \in \mathscr{L}^2[a,b]$ and $\eta(a) = s$. If for $a \in I$ there exists a $d > a$ such that $J[\eta;a,c] \geq 0$ for all $\eta \in \mathscr{F}_0[a,d]$, then the least upper bound of such values d is called the first right-hand conjugate point to $t = a$ and denoted by $c(a)$. Also, if $b > a$ and $b \in I$, the function $L(t,b)$, $a < t < b$, defined as the infimum of $J[\eta;t,b]$ on $\mathscr{F}_1(t,b)$ is finite if and only if $c(a) \geq b$, in which case on (a,b) the function $L(t,b)$ is right-hand continuous and satisfies almost everywhere on (a,b) the generalized Riccati inequality

$$r(t)[L'(t,b) + p(t)] - [L(t,b) + q(t)]^2 \geq 0,$$

and $L(t,b)$ is a.c. on every closed subinterval of (a,b) on which $r(t)$ is continuous and positive.

The paper [2] of Martin amplifies the results of Leighton and Martin [1] in that the function $q(t)$ is not required to be identically zero, and, more importantly, for the case of a single singular end-point the consideration of the minimum limit of the functional J on classes of $\mathscr{F}_0$- and $\mathscr{A}_0$-admissible functions comprised of those functions belonging to the respective classes $\mathscr{F}$ and $\mathscr{A}$ and for which the singular end-point is a limit point of the set of zeros of η.

5. Interrelations Between Oscillation Criteria and Boundary Problems

For a differential equation of the form

$$\ell^{\#}[u](t) \equiv u''(t) + q(t)u(t) = 0, \quad t \in [a,\infty), \qquad (2.1|q)$$

where q is a positive continuous function on $I = [a,\infty)$, Nehari [1] derived various specific oscillation criteria from the following general result.

THEOREM 5.1. *If* $q:[a,\infty) \to \mathrm{R}$ *is positive and continuous on* $I = [a,\infty)$, *then* $(2.1|q)$ *is non-oscillatory on* I *if and only if the smallest eigenvalue of the boundary problem*

$$u''(t) + \lambda q(t)u(t) = 0, \quad u(a) = 0, \quad u'(b) = 0, \qquad (5.1)$$

is greater than 1 *for arbitrary* $b \in (a,\infty)$.

This result, in view of the definitive extremizing property of the smallest eigenvalue of (5.1), may be stated as follows: If $q:[a,\infty) \to \mathrm{R}$ is positive and continuous on $I = [a,\infty)$, then $(2.1|q)$ is non-oscillatory on this interval if and only if

$$\int_a^b \{[\eta'(t)]^2 - q(t)\eta^2(t)\}dt \ \textit{ is positive definite on } \ D^2_{0*}[a,b] \ \textit{ for arbitrary } \ b \in (a,\infty), \qquad (5.2)$$

where $D^2_{0*}[a,b]$ denotes the class of $\eta:[a,b] \to \mathbf{R}$ which are a.c. with $\eta' \in \mathscr{L}^2[a,b]$ and $\eta(a) = 0$.

For $c \in (a,\infty)$ let $\eta(t) = (t-a)^{\beta/2}(c-a)^{-\beta/2}$ for $t \in [a,c]$ and $\eta(t) = (t-a)^{\alpha/2}(c-a)^{-\alpha/2}$ for $t \in (c,\infty)$, where $\beta > 1$ and $0 \leq \alpha < 1$. Upon evaluating the integral of (5.2) for this function η, and letting $b \to \infty$, one obtains the following result, where relation (5.4) is obtained from (5.3) using an integration by parts.

THEOREM 5.2. *If* $q:[a,\infty) \to \mathbf{R}$ *is positive and continuous, and* $(2.1|q)$ *is non-oscillatory on* $[a,\infty)$, *then for* $\beta > 1$ *and* $0 \leq \alpha < 1$, *we have:*

$$\begin{aligned}(t-a)^{1-\beta}&\int_a^t (s-a)^{\beta}q(s)ds \\ &+ (t-a)^{1-\alpha}\int_t^{\infty}(s-a)^{\alpha}q(s)ds \leq \frac{\beta-\alpha}{4}\left[1 + \frac{1}{(\beta-1)(1-\alpha)}\right];\end{aligned} \tag{5.3}$$

$$\begin{aligned}&(t-a)^{1-\beta}\int_a^t (s-a)^{\beta-2}\sigma(s)ds \leq \frac{1}{4} + \frac{1}{4(\beta-1)(1-\alpha)}, \\ &\text{where}\quad \sigma(t) = (t-a)^{1-\alpha}\int_t^{\infty}(s-a)^{\alpha}q(s)ds.\end{aligned} \tag{5.4}$$

As a consequence of (5.4), Nehari derived the following result, which for $\alpha = 0$ reduces to a result of Hille mentioned in Section 2 above. The bound is sharp, as is shown by the particular equation (2.2).

COROLLARY 1. *If* $q:[a,\infty) \to \mathbf{R}$ *is positive and continuous, and* $(2.1|q)$ *is non-oscillatory on* $[a,\infty)$, *then*

$$\liminf_{t\to\infty} t^{1-\alpha}\int_t^{\infty} s^{\alpha}q(s)ds \leq \frac{1}{4(1-\alpha)},\ \textit{for}\ 0 \leq \alpha < 1. \tag{5.5}$$

Nehari [1] also established the following result, wherein the lower bound is obtained by considering the particular equation (2.2), and the upper bound is derived with the aid of criterion (5.2) for particular functions η of the form

$$\eta(t) = \left(\int_a^t q(s)ds\right)^\nu \text{ for } t \in [a,c] \text{ and } \eta(t) = \left(\int_a^c q(s)ds\right)^\nu$$

for $t \in [c,b]$, with $\nu > 1/2$.

THEOREM 5.3. *If* $q:[a,\infty) \to \mathbf{R}$ *is a continuous function satisfying* $0 \le q(t) \le m^2$ *on* $I = [a,\infty)$, *and* $(2.1|q)$ *is non-oscillatory on* I, *then there exists a universal constant* c_o *such that*

$$\int_a^\infty q(t)dt \le cm$$

holds for $c = c_o$ *but not, in general, for* $c < c_o$; *also,*

$$1/2 \le c_o < 3^{3/4}/2^{1/2} = 1.61...$$

It is to be noted that whenever $q:[a,\infty) \to \mathbf{R}$ is positive and continuous then for $b \in [a,\infty)$ there exists a value $d_1 \in (b,\infty)$ such that $t = d_1$ is a focal point to $t = b$ with respect to $(2.1|q)$ and the initial condition $u'(b) = 0$, and if $d \in (d_1,\infty)$ then there exists an η which is a.c. on $[b,d]$, has $\eta' \in \mathscr{L}^2[b,d]$ and $\eta(d) = 0$, and

$$\int_b^d \{[\eta'(t)]^2 - q(t)\eta^2(t)\}dt < 0.$$

The existence of a focal point d_1 is a ready consequence of the fact that if $u = u(t)$ is the solution of $(2.1|q)$ satisfying $u(b) = 1$, $u'(b) = 0$, then the positivity of $q(t)$ implies that $u(t)$ is concave downward as long as this function is positive. This result is the crucial step in the proof of Nehari's result as given in the above Theorem 5.1. For general equations of the form (II.1.1) one has the following result, which is the result of Theorem 3.1 of Reid [19] in the case of a scalar equation

THEOREM 5.4. *Suppose that hypothesis* ($\mathscr{A}$) *is satisfied by the coefficients of* (II.1.1) *on* $I = [a,\infty)$, *and that for arbitrary* $b \in I$ *there exists a* $d \in (b,\infty)$ *such that the smallest eigenvalue of the boundary problem*

$$\ell[u] + \lambda u = 0, \quad u'(b) = 0, \quad u(d) = 0 \tag{5.6}$$

is negative. Then (II.1.1) *is disconjugate on* I *if and only if for arbitrary* $b \in (a,\infty)$ *the smallest eigenvalue of the boundary problem*

$$\ell[u] + \lambda u = 0, \quad u(a) = 0, \quad u'(b) = 0, \quad a < b < \infty \tag{5.7}$$

is positive, or equivalently, $J[\eta;a,b]$ *is positive definite on* $D'_{0*}[a,b]$.

6. Strong and Conditional Oscillation

If $q\colon[a,\infty) \to \mathbf{R}$ is positive and continuous on $I = [a,\infty)$, then the differential equation $(2.1|q)$ is said to be *strongly oscillatory* if the differential equation

$$u''(t) + \lambda q(t)u(t) = 0, \quad t \in [a,\infty) \tag{6.1}$$

is oscillatory for all positive values of λ. Also, $(2.1|q)$ is said to be *strongly non-oscillatory* on I if (6.1) is non-oscillatory on I for all positive λ, and $(2.1|q)$ is termed *conditionally oscillatory* on I if (6.1) is oscillatory on I for some positive λ and non-oscillatory on I for other positive λ.

By the Sturm comparison theorem it follows that in case $(2.1|q)$ is conditionally oscillatory on I, there is a finite positive μ such that (6.1) is oscillatory on I for $\lambda > \mu$, and non-oscillatory on I for $0 < \lambda < \mu$. The value μ is called the *oscillation constant* of $(2.1|q)$, or the *oscillation constant of the function* q. The case of strong oscillation may be designated by $\mu = 0$, and the case of strong non-oscillation by $\mu = \infty$. For example, as noted in the opening paragraph of Section 2, the oscillation constant of the differential equation $u''(t) + t^{-2}u(t) = 0$ on $[1,\infty)$ is $\mu = 1/4$.

Moreover, the results of Hille given in Theorem 2.1 yield the following theorem in terms of the function $\gamma(t|q) = t\int_t^\infty q(s)ds$, as defined in (2.2).

THEOREM 6.1. *If* $q:[a,\infty) \to R$ *is positive and continuous then* $(2.1|q)$ *is strongly oscillatory on* I *if and only if* $\limsup_{t\to\infty} \gamma(t|q) = \infty$, *and* $(2.1|q)$ *is strongly non-oscillatory on* I *if and only if* $\lim_{t\to\infty} \gamma(t|q) = 0$.

Nehari [1, Theorem VI] established the following result, comparing the oscillatory behavior of an equation $(2.1|q)$ with that of a conditionally oscillatory equation $(2.1|q_0)$ whose oscillation constant is known.

THEOREM 6.2. *Suppose that* q *and* q_0 *are continuous non-negative functions on* $I = [a,\infty)$, *and* q_0 *has a finite, positive oscillation constant* μ_0. *Then* $(2.1|q)$ *is oscillatory on* I *if*

$$A = \liminf_{t\to\infty} \left(\int_t^\infty q(s)ds\right)\left(\int_t^\infty q_0(s)ds\right) > \mu_0 \tag{6.2}$$

and, $(2.1|q)$ *is non-oscillatory on* I *if*

$$B = \limsup_{t\to\infty} \left(\int_t^\infty q(s)ds\right)\left(\int_t^\infty q_0(s)ds\right) < \mu_0 . \tag{6.3}$$

As Nehari notes, the result of this theorem may be deduced from Theorem VII of Hille [1], which was established by considering the non-linear singular integral equation $(2.5|q)$. Nehari presents an independent direct proof using linear methods.

7. A Class of Sturmian Boundary Problems on a Non-Compact Interval

In this section attention will be limited to a class of boundary problems involving a linear homogeneous differential equation of the second order

$$[r(t,\lambda)u'(t)]' - p(t,\lambda)u(t) = 0, \tag{7.1}$$

where $r(t,\lambda)$, $p(t,\lambda)$ are real-valued functions of (t,λ) on $I \times \Delta$, where I is a non-compact interval of the real line, and Δ is an open interval $\Delta = \{\lambda : \Lambda_1 < \lambda < \Lambda_2\}$, and for which the boundary conditions are of such a nature that each eigenfunction has only a finite number of zeros on I. That is, for such a problem an eigenfunction for a value $\lambda = \lambda_0$ is for this value of λ a principal solution of the differential equation at each singular end-point of I. In particular, under various conditions the condition that an eigenfunction be principal at a singular endpoint is equivalent to the condition that this function be of class $\mathscr{L}^2$ on a neighborhood of this end-point, and it is in this regard that the problem has direct contact with the theory of singular eigenvalue problems initiated by Hermann Weyl.

For example, consider the boundary problem involving the differential equation

$$u''(t) + [\lambda - p(t)]u(t) = 0, \quad t \in [0,\infty) \tag{7.2}$$

where $p:[0,\infty) \to R$ is a continuous function, together with the real Sturmian boundary condition

$$\alpha u(0) + \beta u'(0) = 0 \tag{7.3}$$

at the end-point $t = 0$, and the condition

$$\int_0^\infty u^2(t)dt < \infty. \tag{7.4}$$

According to Weyl's classification, the equation (7.2) is in the *limit point* case if for some value of λ there is at least one solution of (7.2) for which condition (7.4) fails to hold, in which instance it is true that for each value of λ there is at least one solution for which (7.4) does not hold. He then termed a value λ an eigenvalue of (7.2), (7.3), (7.4) if there is a corresponding non-identically vanishing solution of (7.2) satisfying (7.3) and (7.4). In particular, Weyl [2, pp. 251-257] showed that if $\mu = \liminf_{t \to \infty} p(t)$ is not equal to $-\infty$ then (7.2) is in the limit point case, and that the set of values λ in the spectrum of this problem satisfying $\lambda < \mu$ is either empty or consists of a finite or infinite sequence of eigenvalues $\lambda_1 < \lambda_2 < \ldots < \mu$ such that an eigenfunction of this boundary problem corresponding to the eigenvalue $\lambda = \lambda_j$ has exactly $j - 1$ zeros on $(0,\infty)$. Various extensions of this result have been obtained, but we shall limit specific reference to one. Hartman [2] has shown that if the continuous function $f:[0,\infty) \to R$ is such that every non-identically vanishing solution of (7.2) for real values λ has only a finite number of zeros on $[0,\infty)$, then the spectrum of the boundary problem (7.2), (7.3), (7.4) consists of an infinite sequence of eigenvalues $\lambda_1 < \lambda_2 < \ldots$ tending to $+\infty$, with an eigenfunction for $\lambda = \lambda_j$ having exactly $j - 1$ zeros on $(0,\infty)$; moreover, the set of eigenfunctions is complete in the space $\mathscr{L}^2[0,\infty)$. Various aspects of similar boundary problems involving an equation of the form (7.2) have been treated by many authors. In particular, for the problem involving an equation (7.2) on the interval $(-\infty,\infty)$, and where the equation is in the limit point case at each individual end-point, Wolfson [1] determined interrelations between the given problem and the component problems on the subintervals $(-\infty,0]$ and $[0,\infty)$ with the adjunction at $t = 0$ of a Sturmian condition (7.3).

For further discussion we shall assume that $I = (-\infty,\infty)$, and that the following conditions hold.

(i) On $I \times \Delta$ the functions r and p are real-valued and $r > 0$.

(ii) For $\lambda_o \in \Delta$ the functions $r(t,\lambda_o)$, $1/r(t,\lambda_o)$, and $p(t,\lambda_o)$ are locally of class $\mathscr{L}^\infty$; moreover, for t a.e. on R these functions are continuous in λ at λ_o.

(iii) For $t \in \mathrm{R}$ the functions $r(t,\lambda)$, $p(t,\lambda)$ are monotone non-increasing functions of λ on Δ, and such that for $[a,b]$ an arbitrary compact subinterval of R the functional

$$J_o[\eta;\lambda|a,b] = \int_a^b \{r(t,\lambda)\eta'^2(t)+p(t,\lambda)\eta^2(t)\}dt$$

is a strictly decreasing function of λ for arbitrary $\eta(t) \not\equiv 0$ belonging to the class $D[a,b]$ of real-valued $\eta(t)$ which are a.c. on $[a,b]$ and $\eta' \in \mathscr{L}^2[a,b]$.

(iv) For t a.e. on R, $\lim_{\lambda\to\Lambda_1} p(t,\lambda) = +\infty$.

(v) There exists a compact interval I_0 of $\mathbb{R}$ such that $\lim_{t\to\Lambda_2} p(t,\lambda) = -\infty$ for t a.e. on I_0.

(vi) For $\lambda \in \Delta$, $\liminf_{t\to\pm\infty} p(t,\lambda) > 0$.

In particular, the above conditions (i)-(vi) hold for $\Delta = \mathbb{R}$ whenever the following hypothesis holds:

($\mathscr{A}^0$) $r(t,\lambda) \equiv 1$, while on $\mathbb{R}\times\mathbb{R}$ the real-valued function $p(t,\lambda)$ is continuous, has a negative partial derivative with respect to λ, and

$$\lim_{\lambda\to-\infty} p(t,\lambda) = +\infty, \qquad \lim_{\lambda\to\infty} p(t,\lambda) = -\infty,$$

$$\lim_{t\to\pm\infty} p(t,\lambda) = +\infty.$$

The conditions ($\mathscr{A}^0$) are clearly satisfied if $r(t,\lambda) \equiv 1$ and $p(t,\lambda) = p_0(t) - \lambda$, where $p_0(t)$ is a real-valued continuous function on $\mathbb{R}$ such that $\lim_{t\to\pm\infty} p_0(t) = +\infty$.

Under hypotheses ($\mathscr{A}^0$) the boundary problem consisting of (7.1) and the boundary condition

$$\int_{-\infty}^{\infty} u^2(t)dt < \infty \tag{7.5}$$

was considered by Milne [2], who determined the existence an infinite sequence of eigenvalues $\lambda_1 < \lambda_2 < \ldots$ with $\lambda_j \to \infty$ as $j \to \infty$, and such that an eigenfunction for $\lambda = \lambda_j$ has exactly $j - 1$ zeros on $(-\infty,\infty)$. His method of treatment involved for arbitrary compact subintervals $[a,b]$ of $\mathbb{R}$ the consideration of self-adjoint boundary problems involving (7.1) and two-point boundary conditions at $t = a$ and $t = b$, and considering limiting values as $a \to -\infty$ and $b \to +\infty$.

Now in view of Condition (vi) it follows that if hypotheses (i)-(vi) hold then the differential equation (7.1) is such that for each λ an arbitrary real solution of this equation has only a finite number of zeros on $(-\infty,\infty)$, and the condition that a solution $u(t)$ of this equation be principal at $+\infty$ or at $-\infty$ is equivalent to the condition that this solution be of class $\mathscr{L}^2$ in a neighborhood of this end-point. Using these facts, Reid [40] showed that under hypotheses (i-vi) the real eigenvalues of (7.1), (7.5) form an infinite sequence $\lambda_1 < \lambda_2 < \ldots$ such that $\lambda_j \to \Lambda_2$ as $j \to \infty$ and for $j = 1,2,\ldots$ the corresponding eigenfunction of this system has exactly $j - 1$ zeros on $(-\infty,\infty)$. His method of proof involved using the principal solutions at the two end-points, and showing that for a given positive integer k the existence of the first k eigenvalues and corresponding oscillation property of the eigenfunctions could be reduced to the analogous results for a related problem of classical Sturmian type on a corresponding compact subinterval $[a_k,b_k]$ of $\mathbf{R}$. He also considered a second type of boundary problem involving an equation of the form

$$u''(t) - g(t,\lambda)u(t) = 0, \quad t \in (-\infty,\infty)$$

and boundary condition (7.5), under hypotheses involving "turning points". In this instance comparison theorems for an allied problem on a compact subinterval were used to establish the existence of a sequence of sets of eigenvalues such that the eigenfunctions corresponding to parameter values in the j-th set have exactly $j - 1$ zeros on $(-\infty,\infty)$.

The paper [19] of Hartman presents a general discussion of boundary problems involving (7.1) on $I \times \Delta$, where I is an open interval (α,β) and the boundary condition requires that an eigenfunction be principal at each end-point $t = \alpha$ and $t = \beta$. His methods provide results for some non-linear problems as well as linear problems. In particular, for linear problems his results generalize those of Reid, and his methods make clear the roles played by the assumptions of monotony and semi-boundedness occurring in the above hypotheses ($\mathcal{A}$i-vi).

8. Topics and Exercises

For brevity in the listing of results presented in various references, and in accord with the notation of the present chapter, we use the following notations for special forms of the differential equations that appear in most of the problems.

$$\ell[u|q](t) \equiv u''(t) + q(t)u(t) = 0, \qquad (2.1|q)$$

$$\ell[u|r,q](t) \equiv [r(t)u'(t)]' + q(t)u(t) = 0. \qquad (2.9|r,q)$$

Also, since most of the cited reference assume reality and continuity of coefficients, *unless stated otherwise it is assumed that the coefficient functions* $r(t)$, $q(t)$ *in these equations are real-valued and continuous with* $r(t) > 0$, *and that considered solutions of these equations are real-valued.*

1. Consider the differential equation $(2.9|r,q)$ where $r(t)$, $q(t)$ are real-valued non-negative measurable functions on $[a,\infty)$ such that r, $1/r$, q, $1/q$ are locally integrable on this interval. The equation

$$\left[\frac{1}{q(t)} v'(t)\right]' + \frac{1}{r(t)} v(t) = 0 \qquad (2.9|,1/q,1/r)$$

has been called the "reciprocal equation" of $(2.9|r,q)$. If $\int_a^\infty r^{-1}(s)ds = \infty$, and $(2.9|r,q)$ is disconjugate on $[a,\infty)$, then $(2.9|1/q,1/r)$ is also disconjugate on this interval. If $\int_a^\infty r^{-1}(s)ds = \infty$ and $(2.9|1/q,1/r)$ is disconjugate on $[a,\infty)$ then a non-identically vanishing solution $u(t)$ of $(2.9|r,q)$ can have at most one zero on $[a,\infty)$. {Potter [1]; Barrett [5]. If $u(t)$ is a solution of $(2.9|r,q)$ then $v(t) = r(t)u'(t)$ is a solution of $(2.9|1,q/1/r)$}.

2. Consider equations $(2.9|r_\alpha,q_\alpha)$, $(\alpha = 1,2)$, on $[a,\infty)$, with the integrals $\int_t^\infty q_\alpha(s)ds$ convergent and $\int_t^\infty q_1(s)ds \geq \left|\int_t^\infty q_2(s)ds\right|$ for $t \in [a,\infty)$, while $r_2(t) \geq r_1(t) > 0$, and there exists a constant A such that $r_1(t) \leq A$ on $[a,\infty)$. If $(2.9|r,q_1)$ is non-oscillatory on $[a,\infty)$, then $(2.9|r,q_2)$ is also non-oscillatory on this interval {Taam [1]}.

3. A necessary and sufficient condition that $(2.1|q)$ be non-oscillatory for large t is that there exist some positive function $\lambda(t)$, for which the functions $\lambda^*(t) = \lambda(t)\int_t^\infty \lambda^{-2}(s)ds$, $\ell[\lambda|q](t)$ are continuous and $\int_a^\infty \lambda^*(t)\ell[\lambda|q](t)dt < \infty$. {Wintner [5]. The sufficiency of this condition can be deduced from the following fact, which is a corollary to a general theorem of this paper, to the effect that if $r(t)$, $q(t)$ are continuous functions for large t such that $r(t) > 0$ then the condition

$$\int^\infty |q(t)|\left(\int_t^\infty r^{-1}(s)ds\right)dt < \infty$$

is sufficient that some solution of $(2.1|q)$ should tend to a finite non-zero limit}.

4. If $f(t)$ is a positive continuous function which is monotone on $[a,\infty)$, and the differential equation

$$u''(t) + [f(t)]^{-4}u(t) = 0 \tag{8.1}$$

is oscillatory on $[a,\infty)$, then $\lim_{t\to\infty} \int_a^t [f(t)]^{-2}dt = \infty$.

{Leighton [5, Th. 3.4]. Leighton's proof utilized the fact that under the substitution $u(t) = t^{1/2}z(t)$ equation (8.1) is transformed into $[tz'(t)]' + t^{-1}(t^2[f(t)]^{-4} - [1/4])z(t) = 0$}.

5. If $r(t)$ and $q(t)$ are positive continuous functions on $[a,\infty)$ with $r(t)q(t)$ monotone, and $(2.9|r,q)$ is oscillatory, then $\lim_{t\to\infty} \int_a^t [q(s)/r(s)]^{1/2}ds = \infty$.

{Leighton [5, Th. 4.1]. Transform $(2.9|r,q)$ by the change of independent variable $s = \int_a^t [r(s)]^{-1}ds$, and apply the result of Exercise 4}.

6. If $f:[a,\infty) \to R$ is a positive function of class $\mathscr{C}''$, and $1 + f^3(t)f''(t)$ is positive and monotone non-decreasing, then each solution of (8.1) is such that there exists a constant M satisfying $|u(t)| \leq Mf(t)$ on $[a,\infty)$. {Leighton [5, Th. 3.2]. Used in this proof is the fact that if f is a positive function of class $\mathscr{C}''$ on $[a,\infty)$, then (8.1) is transformed into the equation $[f^2(t)z'(t)]' + [f^{-2}(t) + f(t)f''(t)]z(t) = 0$ under the substitution $u(t) = f(t)z(t)$}.

7. For $x \geq 0$ let $\Omega(x)$ denote the class of real-valued continuous functions $\eta(t)$ with piecewise continuous derivatives on arbitrary compact subintervals of $[0,\infty)$ which are identically zero on $[0,x]$, are normalized functions of class $\mathscr{L}^2[0,\infty)$ so that $\int_0^\infty \eta^2(t)dt = 1$, and the integral

$\int_0^\infty \{[\eta'(t)]^2 + |q(t)|\eta^2(t)\}^2(t)dt$ is finite. Then the differential equation $(2.1|q)$ is oscillatory on $[0,\infty)$ if and only if for each $x \in [0,\infty)$ the infimum of $\int_0^\infty \{[\eta'(t)]^2 - q(t)\}^2(t)dt$ on $\Omega(x)$ is negative. {Putnam [1]}.

8. Suppose that $\int^\infty [1/r(s)]ds = \infty$. If there exists a positive function $g:[a,\infty) \to \mathrm{R}$ of class $\mathscr{C}'$ such that $\int^\infty \{r(s)[g'(s)]^2/g(s)\}ds < \infty$ and $\int^\infty g(s)q(s)ds = +\infty$, then the equation $(2.9|r,q)$ is oscillatory for large t. {Zlamal [1]}.

9. Suppose that $q:[a,\infty) \to \mathrm{R}$ is continuous, and define $q_0(t) = q(t)$, $q_1(t) = t^2 q_0(t) - 1/4$, $q_2(t) = (\ell n\ t)^2 q_1(t) - 1/4$, $q_3(t) = (\ell n_2 t)^2 q_2(t) - 1/4$, etc. If $i > 0$ and $\lim_{t\to\infty} q_{i-1}(t) = 0$ then all solutions of $(2.1|q)$ are oscillatory if there exists a positive constant C such that $q_i(t) \geq C$ for large t, and $(2.1|q)$ is non-oscillatory for large t if $q_i(t) \leq 0$ for t in a neighborhood of ∞. {J. C. P. Miller [1]}.

10. If $(2.1|q)$ is oscillatory for large t, then $\int^\infty t(\ell n\ t)[q(t) - 1/4t^2]^+ dt = \infty$. {Nikolenko [1]. As F. V. Atkinson points out in his review of this paper in MR 17-263, the test presented by the author is the second in a chain of tests that may be derived by a change of variables starting from the test $\int^\infty tq^+(t)dt = \infty$, in analogy with the chain of tests given by J. C. P. Miller [1]}.

11. If $q:[a,\infty) \to \mathrm{R}$ is non-negative with $\int_a^\infty q(s)ds < \infty$, and $u(t)$ is a non-identically vanishing solution of $(2.1|q)$ which possesses infinitely many zeros $t_1 < t_2 < \dots$ on $[a,\infty)$, then $\sum_{j=1}^\infty (t_{j+1} - t_j)^{-1} < \infty$. {Gagliardo [1]}.

12. If $Q:[a,b] \to \mathbf{R}$ is any function such that $Q'(t) = -q(t)$ on $[a,b]$ and the equation

$$u''(t) + 4Q^2(t)u(t) = 0 \tag{8.2}$$

is disconjugate on $[a,b]$, then $(2.1|q)$ is also disconjugate on this interval. {Hartman [6]. Disconjugacy of (8.2) on $[a,b]$ implies the existence of a solution $z(t)$ of the Riccati differential inequality $z'(t) + z^2(t) + 4Q^2(t) \leq 0$ on $[a,b]$, and $w(t) = Q(t) + \frac{1}{2} z(t)$ satisfies $w'(t) + w^2(t) + q(t) \leq 0$ on this interval}.

13. If $p(t)$ and $q(t)$ are real-valued continuous functions on $[a,\infty)$ then the differential equation

$$u''(t) + p(t)u'(t) + q(t)u(t) = 0$$

is oscillatory for large t if and only if there exists a pair of functions $\beta(t)$ and $\theta(t)$, which are respectively of class $\mathscr{C}'$ and $\mathscr{C}''$ such that $-p(t)/2 + \int_a^t [q(s) - q^2(s)/4]\,ds = \beta(t) - \theta(t)$, where

$$\int_a^\infty \exp\left\{-2\int_a^t \theta(s)ds\right\}dt = \int_a^\infty \exp\left\{2\int_a^t \theta(s)ds\right\}\{\beta'(t)+\theta^2(t)\}dt = +\infty.$$

{El'sin [4]}.

14. If $h:[a,\infty) \to \mathbf{R}$ is a positive function of class $\mathscr{C}''$, and

$$H_1(t) = \frac{1}{h(t)} - \frac{[h'(t)]^2}{4h(t)} + \frac{h''(t)}{2},$$

$$H_2(t) = \frac{1}{h(t)} - \frac{[h'(t)]^2}{4h(t)} - \frac{h''(t)}{2},$$

then the four differential equations

$$u''(t) + h^{-2}(t)u(t) = 0, \tag{8.3}$$

$$[h^2(t)z'(t)]' + z(t) = 0, \tag{8.4}$$

$$[h(t)\eta'(t)]' + H_1(t)\eta(t) = 0, \tag{8.5}$$

$$[h(t)\zeta'(t)]' + H_2(t)\zeta(t) = 0, \tag{8.6}$$

are oscillatory of non-oscillatory simultaneously on $[a,\infty)$. Moreover,

(i) If $\lim_{t\to\infty} \int_a^t H_2(s)ds = +\infty$, then (8.3) is oscillatory on $[a,\infty)$;

(ii) If $\lim_{t\to\infty} \int_a^t H_1(s)ds = +\infty$, then (8.4) is oscillatory on $[a,\infty)$ if and only if $\lim_{t\to\infty} \int_a^t [h(s)]^{-1}ds = \infty$;

(iii) If $h(t)H_2(t)$ is a positive monotone function, and (8.4) is oscillatory on $[a,\infty)$, then $\lim_{t\to\infty} \int_a^t [h(s)]^{-1}ds = \infty$.

(iv) (8.4) is oscillatory on $[a,\infty)$ if $\lim_{t\to\infty} \int_a^t [h(s)]^{-1}ds = \infty$, and there is a constant k such that $[h'(t)]^2 \le k^2 < 4$ for large values of ∞.

{Potter [1, Ths. 1.1, 1.2, 1.4, 1.3]. If $u(t)$ is a solution of (7.3) then $z(t) = u'(t)$ is a solution of (8.4), while $\eta(t) = [h(t)]^{-\frac{1}{2}}u(t)$ and $\zeta(t) = [h(t)]^{\frac{1}{2}}u(t)$ are solutions of (8.5) and (8.6), respectively}.

15. If $r(t)$, $q(t)$ are positive continuous functions on $[a,\infty)$ with $r(t)q(t)$ of class $\mathscr{L}'$ and $[r(t)q(t)]' \le 0$, then $(2.9|r,q)$ is oscillatory on $[a,\infty)$ whenever $\lim_{t\to\infty} \left\{\int_a^t [q(t)/r(t)]^{\frac{1}{2}}dt + \frac{1}{4}\ln[q(t)r(t)]\right\} = \infty$. {Barrett [1, Cor. 2.1 to Th. 2]}.

16. If $q:[a,\infty) \to \mathbf{R}$ is locally of class $\mathscr{L}$, then each of the following is a sufficient condition for $\ell[u|q](t) = 0$ to be non-oscillatory on $[a,\infty)$:

(a) $-3/4 \leq t \int_t^\infty q(s)ds \leq 1/4$ for large t, {Wintner [7, p. 370]};

(b) there exists a constant $k > 0$ such that

$$-k - \sqrt{k} \leq t \int_t^\infty q(s)ds \leq -k + \sqrt{k} \quad \text{for large } t;$$

{Moore [1, p. 132]};

(c) there exist a constant A such that

$$0 < A - \int_c^t sq(s)ds < 1 \quad \text{for large } t,$$

{Moore [1, p. 129]}.

17. Let $g: [a,\infty) \to \mathbf{R}$ be locally a.c., and set $\mu(t) = \mu(t|g) = \exp\left\{2 \int_a^t g(s)ds\right\}$. (i) for c an arbitrary real constant the number of zeros in $[a,t]$ of a non-identically vanishing real-valued solution $u(t)$ of the differential equation

$$u''(t) + [g'(t) - g^2(t) + c^2\mu^2(t|g)]u(t) = 0 \tag{8.7}$$

is given by $[|c|/\pi] \int_a^t \mu(s|g)ds + O(1)$, and accordingly (8.7) is non-oscillatory if either $c = 0$ or $c \neq 0$ and $\int_a^\infty \mu(t|g)dt < \infty$, but is oscillatory if $c \neq 0$ and $\int_a^\infty \mu(t|g)dt = \infty$.

(ii) if $\int_a^\infty t\left(\int_t^\infty g(s)ds\right)dt = \infty$ and $M(t|g) = \int_a^t \mu(t|g)dt$, then the equation

$$u''(t) + \left[g'(t) - g^2(t) + \frac{c\mu^2(t|g)}{4M^2(t|g)}\right]u(t) = 0 \tag{8.8}$$

is non-oscillatory or oscillatory on $[a,\infty)$ according as $c \leq 1$ or $c > 1$; if $\int_a^\infty \mu(t|g)dt < \infty$ and $M(t|g) = \int_t^\infty g(s)ds$, then the same conclusions on non-oscillation and oscillation hold for the differential equation (8.8).

{Wray [1, Ths. 2,3]. Conclusion (i) is an immediate consequence of the fact that $u_0(t) = \exp\left\{-\int_a^t [g(s) + ic\mu(s|g)]ds\right\}$ is a solution of (8.7), and hence the real and pure imaginary parts of $u_0(t)$ are real solutions of this equation. Conclusion (ii) is derived with the aid of conclusion (i) applied to the equation resulting from the transformation $\mu(t) = \mu(t|g)/[2M(t|g)]$, $g(t) = \mu'(t)/[2\mu(t)]$, under which for arbitrary constants λ one has

$$g' - g^2 + \lambda^2\mu^2 = g' - g^2 + \frac{(\lambda^2+1)\mu^2}{4M^2},$$

and an application of Sturm's comparison theorem}.

18. Let $q_0\colon [a,\infty) \to \mathbf{R}$ be a continuous function such that $(2.1|q_0)$ is non-oscillatory, and let $u = u_0(t)$ be a solution of this equation different from zero on $[b,\infty)$ and such that $\int_b^\infty [u_0(s)]^{-2}ds < \infty$. Then $\eta(t) = u_0(t)\int_t^\infty [u_0(s)]^{-2}ds$ is another solution of $(2.1|q_0)$ on $[b,\infty)$. Now let β be either 1 or -1, γ any positive constant, and $\rho(t)$, $\sigma(t)$ any two locally a.c. functions which satisfy $\rho(t)\sigma(t) = \beta\eta(t)u_0(t)$, $\rho'(t)\sigma(t) = -\gamma$ for $t \geq b$, and for $q\colon [a,\infty) \to \mathbf{R}$ a continuous function set

$$I(t) = \rho(t)\int_{t_0}^t \sigma(s)[q(s) - q_0(s)]ds,$$

where in this definition the value t_0 may be equal to ∞, in which case the integral need only be a Cauchy-Lebesgue integral. If $\mu > 0$ and $\varepsilon > 0$, then $(2.1|q)$ is non-oscillatory if

$$-\mu - \frac{\mu^2}{\gamma} + \frac{1-\mu^2}{4\gamma} \leq I(t) \leq \mu - \frac{\mu^2}{\gamma} + \frac{1-\mu^2}{4\gamma}$$

is oscillatory if

$$-\mu + \frac{\mu^2}{\gamma} + \frac{1+\mu^2}{4\gamma} + \varepsilon \leq I(t) \leq \mu + \frac{\mu^2}{\gamma} + \frac{1+\mu^2}{4\gamma} - \varepsilon,$$

the inequalities to hold on some half-line. In addition, $(2.1|q)$ is non-oscillatory if there is a constant τ such that the relation

$$\frac{\beta - 1}{2} - \varepsilon \leq I(t) \leq \frac{\beta + 1}{2} - \tau$$

holds on some half-line. If $\gamma = 1$, the above conditions become $-\mu - \mu^2 \leq I(t) \leq \mu - \mu^2$ and $-\mu + \mu^2 + \frac{1}{2} + \varepsilon \leq I(t) \leq \mu + \mu^2 + \frac{1}{2} - \varepsilon$. If $\mu = 1/2$ in these, it follows that $(2.1|q)$ is non-oscillatory if $-3/4 \leq I(t) \leq 1/4$ and oscillatory if $(1/4) + \varepsilon \leq I(t) \leq (5/4) - \varepsilon$, $\varepsilon > 0$. {Wray [1, Th. 5]}.

19. Equation $(2.1|q)$ is non-oscillatory if

$$\frac{1}{\alpha}\{\frac{1}{4} - (\mu + \frac{\alpha}{2})^2\} \leq t^{\alpha}\int_t^{\infty} s^{1-\alpha}q(s)ds \leq \frac{1}{\alpha}\{\frac{1}{4} - (\mu - \frac{\alpha}{2})^2\},$$

and oscillatory if

$$\varepsilon + \frac{1}{\alpha}\{\frac{1}{4} + (\mu - \frac{\alpha}{2})^2\} \leq t^{\alpha}\int_t^{\infty} s^{1-\alpha}q(s)ds \leq \frac{1}{\alpha}\{\frac{1}{4} + (\mu + \frac{\alpha}{2})^2\} - \varepsilon$$

for some $\alpha \neq 0$, $\mu > 0$, $\varepsilon > 0$. {Wray [1, Theorem 7]. This result includes that of Wintner mentioned in 16(a). Also, the first of Moore's results mentioned in 16 holds by taking in the preceding Exercise $q_0(t) \equiv 0$, $u_0(t) = t$, $\eta = \rho \equiv 1$, $\sigma = t$. Then $\beta = 1$ and $\ell[u|q](t) = 0$ is non-oscillatory if there exists a constant τ such that $-\tau \leq \int_t^{\infty} sq(s)ds \leq 1 - \tau$, in which case, either $\int_t^{\infty} sq(s)ds$ converges as $t \to \infty$, or $\int_t^{\infty} sq(s)ds$ eventually oscillates over an interval of length not exceeding 1 as $t \to \infty$, insures that $\ell[u|q](t) = 0$ is non-oscillatory. The second of Moore's results mentioned in 16 is obtained, with strict inequality no longer necessary, by taking $q_0(t) \equiv 0$, $u_0(t) = t$, $\eta = \rho \equiv 1$, $\sigma = t$, $A = 1 - \tau$}.

20. Suppose that the coefficients $r(t)$ and $q(t)$ of the differential equation $(2.9|r,q)$ satisfy hypothesis ($\mathscr{A}$) on $[a,\infty)$, and this equation is in the limit circle case; that is, all solutions of $(2.9|r,q)$ belong to $\mathscr{L}^2[a,\infty)$. Then

(i) if $\int^\infty [1/\sqrt{r(t)}]dt = \infty$, then $(2.9|r,q)$ is oscillatory;

(ii) if $r(t)$ is essentially bounded on $[a,\infty)$, then $(2.9|r,q)$ is oscillatory and for any non-identically vanishing real solution of this equation the distance between consecutive zeros tends to zero as $t \to \infty$.

{Patula and Waltman [1; Ths. 1,2]}.

21. If $\int_0^\infty [r(t)]^{-1}dt = +\infty$, and there exists a positive function $\omega(t)$ of class $\mathscr{C}'$ such that $\int_0^t \omega(s)\{q(s) - \frac{1}{4} r(s)[\omega'(s)/\omega(s)]^2\}ds \to \infty$ as $t \to \infty$, then $(2.9|r,q)$ is oscillatory on $[0,\infty)$. {Opial [3]}.

22. Suppose that $r:[a,\infty) \to \mathbf{R}$ is positive and of class $\mathscr{C}'$, while $q:[a,\infty) \to \mathbf{R}$ is continuous. Then the equation $(2.9|r,q)$ is oscillatory for large t if and only if there exists a positive function g of class $\mathscr{C}'$ on $[a,\infty)$ such that there exists a constant c such that $\lim_{t\to\infty} \int_a^t \frac{1}{r(s)} \exp\{2\, G(s|g;a)\}ds = \infty$, where $G(s|g;a) = \int_a^s \left[\frac{1}{r(\xi)g^2(\xi)} \int_a^\xi \{q(\tau)g^2(\tau) - r(\tau)g'^2(\tau)\}d\tau - c\right]d\xi$. {Ráb [2]}.

23. If $1 < s \leq 2$ there exists a function $q:[0,\infty) \to \mathbf{R}$ which is positive, piecewise continuous on arbitrary compact

subintervals, satisfies $\int_0^\infty t^{(2/s)-1}[q(t)]^{1/s}dt < \infty$, and such that the differential equation $(2.1|q)$ is oscillatory on $[0,\infty)$. {Simons [2]}.

24. Suppose that $q_\alpha(t)$, $(\alpha = 1,2)$ satisfy $q_1(t) \geq q_2(t)$ on $[a,\infty)$ and $(2.1|q_1)$ is non-oscillatory for large t. Then $(2.1|q_2)$ is also non-oscillatory for large t, and for a given solution $u = u_1(t)$ of $(2.1|q_1)$ there exists a solution $u = u_2(t)$ of $(2.1|q_2)$ such that $u_2(t) = O(|u_1(t)|)$ as $t \to \infty$. {Hartman and Wintner [4]}.

25. Suppose that $(2.1|q)$ is non-oscillatory.

(i) The condition

$$\sup_{0<v<\infty}\left|\int_u^{u+v} q(s)ds\right|/(1+v) \to 0 \quad \text{as} \quad u \to \infty \tag{8.9}$$

is necessary and sufficient in order that $u'(t)/u(t) \to 0$ as $t \to \infty$ for one (and/or every) non-identically vanishing solution $u(t)$ of $(2.1|q)$.

(ii) A necessary and sufficient condition that $\int^\infty [u'(t)/u(t)]^2 dt < \infty$ hold for one (and/or every) solution $u(t)$ of $(2.1|q)$ is that

$$\lim_{T\to\infty} \frac{1}{T}\int_0^T\left\{\int_0^t q(s)ds\right\}dt = M \tag{8.10}$$

exists and is finite.

(iii) Either (8.10) holds, or $\liminf_{T\to\infty} \frac{1}{T}\int_0^T\left\{\int_0^t q(s)ds\right\}dt = -\infty$,

(iv) If there exists a constant C such that $\int_0^t q(s)ds > -C$ for all t on a subset of $[a,\infty)$ of infinite measure, then condition (8.10) holds. {Hartman [7]}.

26. Suppose $q:[a,\infty) \to \mathbf{R}$ is a continuous function such that $F(t|q) = \lim_{\tau\to\infty}\int_t^\tau f(s)ds$ exists and is finite, and $(2.1|q)$ is non-oscillatory for large t. If the condition

$$\int_a^\infty \exp\left\{B \int_a^t F(s|q)ds\right\}dt = \infty$$

holds for $B = 4/3$ then no solution $u(t)$ of $(2.1|q)$ is of class $\mathscr{L}^2[a,\infty)$, but this assertion becomes false if $B = 4/3$ is replaced by any $B < 4/3$. {Hartman [6]}.

27. Suppose that $q:[0,\infty) \to \mathbf{R}$ is a continuous positive monotone non-increasing function, and $(2.1|q)$ is non-oscillatory. Then for any real solution $u(t)$ of this equation the function $V(t|u) = [u'(t)]^2 + q(t)u^2(t)$ is a positive monotone non-increasing function, so that $V_\infty(u) = \lim_{t\to\infty} V(t|u)$ is a non-negative finite value. There exists some non-identically vanishing solutions $u(t)$ such that $V_\infty(u) = 0$; in fact this condition holds for a principal solution of $(2.1|q)$ at ∞. In order that $V_\infty(u) = 0$ for every solution $u(t)$ of $(2.1|q)$ it is necessary and sufficient that $\int^\infty sq(s)ds = \infty$. {Hartman and Wintner [7]}.

28. Suppose that $q:[0,\infty) \to \mathbf{R}$ is continuous, and for $T \geq 0$, $M > 0$, let $E(M,T)$ denote the set $\{t : t \in [T,\infty)\}$ for which the function $\gamma_1(t|q) = t^{-1}\int_0^t\left[\int_0^\sigma q(s)ds\right]d\sigma$ satisfies $\gamma_1(t|q) > M$. (i) If there exists a pair of sequences T_n, M_n such that $T_n \to \infty$, $M_n \to \infty$ as $n \to \infty$ and for which $\operatorname{meas} E(M_n,T_n) \exp\{M_nT_n\} \to \infty$ as $n \to \infty$, then $\ell[u|q](t) = 0$ is oscillatory on $[0,\infty)$.

(ii) If $\limsup_{t\to\infty} \gamma_1(t|q) = +\infty$, and there exists a positive constant C such that $\gamma_1(t|q) > -\exp Ct$, then $\ell[u|q](t) = 0$ is oscillatory on $[0,\infty)$.

{Putnam [2]. His proof of (i) is a refinement of the argument of Wintner [6] showing that $\ell[u|q](t) = 0$ is oscillatory whenever $\gamma_1(t|q) \to \infty$ as $t \to \infty$, and conclusion (ii) is established as a corollary to (i)}.

29. If $r(t)$ is a positive function such that r and $1/r$ are locally integrable on $[a,\infty)$, and $q:[a,\infty) \to \mathbf{R}$ is locally integrable on $[a,\infty)$, then

$$\int_a^\infty r^{-1}(s)ds < \infty, \quad \int_a^\infty |q(s)|ds < \infty, \tag{8.11}$$

one may prove that $(2.9|r,q)$ is non-oscillatory on $[a,\infty)$ by noting that under the substitution $s = \int_a^t r^{-1}(\xi)d\xi = s(t)$ this equation is transformed to

$$(D^2u)(s) + q(s)u(t) = 0, \; 0 \leq s \leq S = \int_a^\infty r^{-1}(s)ds, \tag{8.12}$$

where $q:[0,S] \to \mathbf{R}$ is Lebesgue integrable, and hence there exists a solution $u_0(s)$ of (8.12) satisfying $u_0(S) = 1$, $(Du)(S) = 0$, and if $u_0(s) > 0$ on $[s_1,S]$ then the corresponding solution $u(t) = u_0(s(t))$ is positive on $[t(s_1),\infty)]$ and $(2.9|r,q)$ is disconjugate on this interval. Although the hypothesis (8.11) is decidedly stronger than (2.13), which implies non-oscillation for large t, with the above argument it follows that for arbitrary values ξ_0, ζ_0 there is a unique solution of (8.12) satisfying $u(S) = \xi_0$, $(Du)(S) = \zeta_0$, which is equivalent to the statement that there exists a unique solution $u(t)$ of $(2.9|r,q)$ satisfying $\lim_{t\to\infty} u(t) = \xi_0$, $\lim_{t\to\infty} v(t) = \zeta_0$, where $v(t)$ is the canonical variable $v(t) = r(t)u'(t)$ corresponding to $u(t)$.

30. If $q:[0,\infty) \to \mathbf{R}$ is non-positive and continuius, then $(2.1|q)$ has a non-identically vanishing solution $u(t)$ which is non-negative and never increasing on $(0,\infty)$, and $u(\infty) = \lim_{t\to\infty} u(t)$ cannot satisfy $u(\infty) \neq 0$ unless the integral $\int^\infty q(s)ds$ is convergent. {Wintner [4]}.

31. Suppose that $q:[0,\infty) \to R$ is a positive non-increasing continuous function, and $q(t) \to 0$ as $t \to \infty$. Then: (a) at least one solution of $(2.1|q)$ is unbounded; (b) there exists a non-identically vanishing, non-oscillatory solution of $(2.1|q)$ if and only if $\int_0^\infty sq(s)ds < \infty$; (c) $(2.1|q)$ can be oscillatory, and possess a bounded non-identically vanishing solution; (d) all solutions of $(2.1|q)$ are oscillatory and unbounded when either (i) $q \in C''[0,\infty)$ with $[-q'(t)/q(t)]' \le 0$ and $\limsup_{t\to\infty} [-q'(t)/q^{3/2}(t)] < 4$, or (ii) $q''(t) \ge 0$ and $q'(t) = O(q^{3/2}(t))$ as $t \to \infty$. {Hartman and Wintner [3]}.

32. Suppose that $q:[a,\infty) \to R$ is such that $\int_a^\infty q(s)ds = \lim_{t\to\infty} \int_a^t q(s)ds$ is finite, and $F(t) = \int_t^\infty q(s)ds$ is such that $\int_a^\infty F(s)ds$ converges also. Then $(2.1|q)$ possesses a pair of solutions $u_1(t)$, $u_2(t)$ such that $u_1(t) \to 1$, $u_1'(t) \to 0$, $u_2(t) \sim t$, $u_2'(t) = O(t)$ as $t \to \infty$ if and only if $\int_a^\infty sF^2(s)ds < \infty$. {Hartman and Wintner [6]}.

33. If $q:[a,\infty) \to R$ is monotone for large t and $q(t) \to \infty$ as $t \to \infty$, then the differential equation $(2.1|q)$ possesses at least one non-identically vanishing solution $u(t)$ satisfying $u(t) \to 0$ as $t \to \infty$. {Milloux [1]. Hartman [1] derived this result as a consequence of a theorem on solutions of differential systems}.

34. If $q:[a,\infty) \to R$ is monotone for large t, and $q(t)$ tends to a finite positive limit as $t \to \infty$, then $(2.1|q)$ possesses a pair of (linearly independent) solutions $u_1(t)$, $u_2(t)$ of the form

$$u_1(t) = \cos\left\{\int^t \sqrt{|q(s)|}ds\right\} + 0(1),$$

$$u_2(t) = \sin\left\{\int^t \sqrt{|q(s)|}ds\right\} + 0(1)$$

as $t \to \infty$. {Wintner [3]}.

35. If $(2.1|q)$ is non-oscillatory for large t, then as $t \to \infty$: (a) some solution must fail to be $0(t^{1/2})$; (b) every solution can, but need not, be $0(t^{1/2} \ln t)$; (c) every solution $u(t)$ can, but need not, satisfy $t^{1/2} = 0(|u(t)|)$. {Hartman and Wintner [4]}.

36. Suppose that $q:[a,\infty) \to \mathbf{R}$ is a real-valued continuous function, and $Q(t)$ is a non-decreasing function such that $q(t) \leq Q(t)$ on $[a,\infty)$. (i) If $u(t)$ is a solution of $(2.1|q)$ and $m(t)$ denotes the maximum of $|u(\xi)|$ for $|t-\xi| \leq 2/Q^{1/2}(t + 2/Q^{1/2}(t))$, then $|u'(t)| \leq 2Q^{1/2}(t + 2/Q^{1/2}(t))m(t)$ for $t \geq 2/Q^{1/2}(0)$.

(ii) If, in addition, $\int^{\infty} [1/Q(s)]ds = \infty$, (for example, if $\liminf_{t\to\infty} Q(t)/t < \infty$), and $(2.1|q)$ possesses a solution $u_1(t)$ which is bounded as $t \to \infty$, then no solution of $(2.1|q)$ linearly independent of $u_1(t)$ is of class $\mathscr{L}_2(0,\infty)$. In particular, if all solutions of $(2.1|q)$ are bounded (for example, if $q(t)$ is positive and non-decreasing), then no non-identically vanishing solution of this equation is of class $\mathscr{L}_2(0,\infty)$. {Hartman [9]}.

37. Consider two equations $(2.1|q_\alpha)$, $(\alpha = 1,2)$, on $[a,\infty)$, and suppose that $(2.1|q_1)$ is non-oscillatory for large t. If $U_1(t)$ and $U_2(t)$ are linearly independent solutions of $(2.1|q_1)$ with $U_1(t)$ a principal solution of this equation, and $q_2(t)$ is such that $\int_a^{\infty} |q_1(t)-q_2(t)|U_1(t)U_2(t)dt < \infty$, then $(2.1|q_2)$ is also non-oscillatory for large t and

possesses a fundamental set of solutions $u_1(t)$, $u_2(t)$ with $u_\alpha \sim U_\alpha$, $(\alpha = 1,2)$, as $t \to \infty$ and $u_1(t)$ is a principal solution of $(2.1|q_2)$. {Marik and Ráb [1]}.

38. Suppose that $p(t)$ and $q(t)$ are real-valued continuous functions on $[a,\infty)$, and for $\gamma(t)$, $\delta(t)$ real-valued continuously differentiable functions with $\delta(t) > 0$ on this interval, let $P = p\delta - \delta' + 2\gamma$ and $Q = q\delta^2 + \gamma^2 + p\gamma\delta + \gamma'\delta - \gamma\delta'$. Then a function $u:[a,\infty) \to \mathbf{R}$ is a principal solution of the differential equation $u''(t) + p(t)u'(t) + q(t)u(t) = 0$ at ∞ if and only if $z(t) = \delta(t)u'(t)/u(t) - \gamma(t)$ is a solution of the Riccati differential equation $\delta(t)z'(t) + z^2(t) + P(t)z(t) + Q(t) = 0$ which is such that

$$I(z,b,t) = \int_b^t \delta^{-1}(s)\exp\left\{-\int_b^s \delta^{-1}(\xi)[2z(\xi) + P(\xi)]d\xi\right\}ds$$

satisfies $I(z,b,\infty) = \infty$. {Marik and Ráb [2]}.

39. For the differential equation

$$t^2u''(t) + tp_1(t)u'(t) + p_2(t)u(t) = 0, \quad (8.13)$$

where $p_1(t)$ and $p_2(t)$ are of class $\mathscr{C}'$ on an interval $[0,b]$, suppose that the zeros σ, τ, $(\sigma > \tau)$ of $\rho^2 - [1 - p_1(0)]\rho + p_2(0)$ are real, and $Q(t) = \sigma^2 - [1 - \rho_1(t)]\sigma + p_2(t)$ is of sign on a subinterval $(0,\varepsilon)$, $\varepsilon > 0$. Then (8.13) has a solution of the form $u(t) = t^\sigma z(t)$, where $z(t)$ is continuous on $[0,b]$ with $z(0) \neq 0$, and this solution is the principal solution of (8.13) at the regular singular point $t = 0$. {Leighton [9]}.

40. If $r(t)$ and $q(t)$ are positive continuous functions with $r(t)q(t)$ of class $\mathscr{C}'$ on an interval I, and I_0 is a subinterval of I, then for a solution $u = u(t)$ of

(2.9|r,q) the quadratic functional $V_1(t|u) \equiv r(t)q(t)u^2 + (r(t)u')^2$ is monotone increasing (decreasing) on I_0 if and only if the function $r(t)q(t)$ is monotone increasing (decreasing), and the quadratic functional $V_2(t|u) = u^2 + (r(t)u')^2/[r(t)q(t)]$ is monotone increasing (decreasing) on I_0 if and only if the function $r(t)q(t)$ is monotone decreasing (increasing) on this interval. {Stickler [1]. Specifically, along a solution $u(t)$ of $\ell^0[u](t) = 0$ one has $[V_1(t|u)]' = u^2(t)[r(t)q(t)]'$ and $[V_2(t|u)]' = -[r(t)u'(t)]^2[r(t)q(t)]'/[r(t)q(t)]^2$}.

41. Consider the Riccati differential equations

$$k_\alpha[w](t) \equiv w'(t) + w^2(t) + q_\alpha(t) = 0, \quad (\alpha = 1,2) \qquad (\mathscr{E}_\alpha)$$

where $q_1(t)$ and $q_2(t)$ are continuous functions on $[a,\infty)$, with $a > 0$.

(i) If there exists a b satisfying $a < b \leq \infty$ such that $q_1(t)$ and $q_2(t)$ are non-negative on $[a,b)$, and

$$\int_a^t s^2 q_2(s)ds \leq \int_a^t s^2 q_1(s)ds, \text{ for } t \in [a,b),$$

then whenever $(\mathscr{E}_1)$ has a solution $w_1(t)$ on $[a,b)$ with $aw_1(a) < 1$ the equation $(\mathscr{E}_2)$ also has a solution $w_2(t)$ on $[a,b)$.

(ii) Suppose that there exists a value b satisfying $a < b \leq \infty$ and a continuous function $\mu(t)$ on $[a,b)$ such that $\mu(t) > 0$ and

$$\int_a^t \mu^2(s)q_2(s)ds \leq \int_a^t \mu^2(s)q_1(s)ds, \quad \text{for} \quad t \in [a,b).$$

If $(\mathscr{E}_1)$ has a solution $w_1(t)$ on $[a,b)$ which satisfies $\mu(t)w_1(t) \leq w'(t)$ and

$$2\mu'' + 2(\mu' - \mu w_1)^2/\mu + \mu(q_1 + q_2) \geq 0 \quad \text{for} \quad t \in [a,b),$$

then $(\mathscr{L}_2)$ has a solution $w_2(t)$ on $[a,b)$.

{Stafford and Heidel [1]. The proofs involve the transformation $\hat{w}(t) = tw(t)$, under which $(\mathscr{L}_\alpha)$ becomes

$$\hat{k}_\alpha[\hat{w}](t) \equiv t\hat{w}'(t) - \hat{w}(t) + \hat{w}^2(t) + t^2 q_\alpha(t) = 0. \qquad (\hat{\mathscr{L}}_\alpha)$$

In particular, if $\hat{w}_1(a) < 1$, and $\hat{w}_2(a)$ satisfies $\hat{w}_1(a) < \hat{w}_2(a) < 1$, then using the integrated form of the equations $(\hat{\mathscr{L}}_\alpha)$ it is established that $\hat{w}_1(t) \leq \hat{w}_2(t) \leq 1$ on $[a,b)$}.

42. Consider two equations $(2.1|q_\alpha)$, $(\alpha = 1,2)$ on $[a,\infty)$, where $a > 0$. If $(2.1|q_1)$ is disconjugate on $[a,\infty)$, and there exists a value $\alpha \geq 2$ such that

$$\left|\int_a^t (s-a)^\alpha q_2(s)ds\right| \leq \int_a^t (s-a)^\alpha q_1(s)ds, \quad \text{for} \quad t \in [a,\infty),$$

then $(2.1|q_2)$ is also disconjugate on $[a,\infty)$.

{Travis [1]. The result was established with the aid of the following generalization of the results of Stafford and Heidel presented in the preceding Exercise 41, where the notation of that exercise is continued: Suppose that $(\mathscr{L}_1)$ has a solution $w_1(t)$ on $[a,b]$, $0 < a < b \leq \infty$, and there exists a constant w_0 and a positive function μ of class $\mathscr{C}''[a,b]$ such that for $t \in [a,\infty)$ we have

$$\left|\mu(a)[\mu'(a) - \mu(a)w_0] + \int_a^t \mu^2(s)q_2(s)ds\right| \leq \mu(a)[\mu'(a) - \mu(a)w_1(a)] + \int_a^t \mu^2(s)q_1(s)ds. \tag{8.14}$$

Then $(\mathscr{L}_2)$ has a solution $w_2(t)$ on $[a,b)$ such that $w_2(a) = w_0$ and $|\mu'(t) - \mu(t)w_2(t)| \leq \mu'(t) - \mu(t)w_1(t)$; if, in addition, $\mu(a)w_1(a) \le \mu'(a)$ then (8.14) may be replaced by $\left|\int_a^t \mu^2(s)q_2(s)ds\right| \le \int_a^t \mu^2(s)q_1(s)ds$ for $t \in [a,\infty)$}.

43. Suppose that $\int_a^\infty q(s)ds = \lim_{t\to\infty} \int_a^t q(s)ds$ is finite and $\int_a^\infty [1/r(s)]ds = +\infty$; moreover, $\int_t^\infty q(s)ds > 0$ for t in a neighborhood of ∞. Then $(2.1|q)$ is oscillatory for large t if and only if there is a sequence of intervals $[a_n,b_n]$ with $a_n \to \infty$ as $n \to \infty$, and such that the least positive eigenvalue μ_n of the system

$$[r(t)u'(t)]' + \mu q(t)u(t) = 0, \; u(a_n) = 0 = u'(b_n) \qquad (8.15)$$

satisfies $\mu_n \le 1$, $(n = 1,2,\ldots)$. {St. Mary [1; Th. 1]}.

44. Suppose that there exists a sequence of intervals $[a_n,b_n]$, $n = 1,2,\ldots$ with $a_n \to \infty$ as $n \to \infty$, and such that the least positive eigenvalue μ_n of (8.15) satisfies $\mu_n \le 1$, $(n = 1,2,\ldots)$. If $q_0(t)$ is a continuous function on $[a,\infty)$ such that $\int_t^\infty q_0(s)ds \ge 0$ for t in a neighborhood of ∞, and $\int_t^{b_n} q_0(s)ds \ge \left|\int_t^{b_n} q_n(s)ds\right|$ for $t \in [a_n,b_n]$, and $n = 1,2,\ldots$, then the differential equation $(2.1|q_0)$ is oscillatory for large t. {St. Mary [1, Th. 8]}.

45. Consider the equation

$$[r(t)u'(t)]' + \lambda q(t)u(t) = 0, \quad t \in [0,\infty), \qquad \mathscr{E}(\lambda|q)$$

where $r(t)$ is a positive continuous function on $[0,\infty)$, and $q(t)$ is a non-negative function that is piecewise continuous on arbitrary compact subintervals of $[a,\infty)$. Let $\mathcal{O}(q)$ denote the "oscillation set" of positive λ for which $\mathscr{E}(\lambda|q)$ has non-identically vanishing solutions with arbitrarily large zeros on $(0,\infty)$, and denote by $\mathscr{NO}(q)$ the "non-oscillation set" $[0,\infty) - O(q)$. Since $0 \in \mathcal{NO}(q)$ this set is non-empty; in general, $\mathcal{NO}(q)$ is an interval of the form $[0,a)$ or $[0,a]$ where $a \in (0,\infty]$, or $\mathcal{NO}(q) = \{0\}$. If $\lambda \in \mathcal{NO}(q)$ then $\mathcal{L}(\lambda|q)$ is disconjugate on some interval of the form (c,∞), and the left-hand endpoint of the maximal such interval is denoted by $a(\lambda)$; also the supremum of the set $\mathcal{NO}(q)$ is denoted by $b(q)$. Then we have the following results.

(i) A necessary and sufficient condition that $b(q) \in \mathcal{NO}(q)$ is that the supremum of the set $\{a(\lambda) : 0 \leq \lambda < b(q)\}$ be finite.

(ii) For $\lambda \in \mathcal{NO}(q)$, let $n(\lambda)$ denote the number of points $t \in (0,\infty)$ that are conjugate to $t = 0$; that is, $n(\lambda)$ is equal to the number of positive zeros of the solution $u(t)$ of $\mathcal{L}(\lambda|q)$ determined by the initial conditions $u(0) = 0$, $u'(0) = 1$. Then a necessary and sufficient condition that the supremum of $\{a(\lambda) : 0 \leq \lambda < b(q)\}$ be finite is that the supremum of $\{n(\lambda) : 0 \leq \lambda < b(q)\}$ be finite.

(iii) For $n = 1,2,\ldots$ suppose that $q_n(t)$ satisfies the conditions required above on $q(t)$, and that $\{q_n\} \to q$ uniformly on compact subsets of $[0,\infty)$, while for each n we have $b(q_n) \in \mathcal{NO}(q_n)$ and $b(q_n) < \infty$. If $b(q) < \infty$, $b(q) \in c\ell[\bigcup_{n=m}^{\infty} \mathcal{NO}(q_n)]$ for every m, and the supremum of $\{a(b(q_n)) : n \geq 1\}$ is finite, then $b(q) \in \mathcal{NO}(q)$.

(iv) If $\int_a^\infty q(t)dt = \int_a^\infty r^{-1}(t)dt = \infty$ then $b(q) = 0$, while if both integrals are finite then $b(q) = +\infty$.

(v) Suppose that $r(t) \equiv 1$, while $q(t) > 0$ and $\gamma_0(t|q) = \int_t^\infty q(s)ds$ is finite. If one of the following limits exists

(a) $\lim_{t\to\infty} t^2 q(t) = L$: (b) $\lim_{t\to\infty} \gamma_0^2(t|q)/q(t) = L$;

(c) $\lim_{t\to\infty} \gamma(t|q) = \lim_{t\to\infty} t\gamma_0(t|q) = L$,

then $b(q) = +\infty$ if $L = 0$, $b(q) = 0$ if $L = +\infty$, and $b(q) = 1/4$ if $L \neq 0, +\infty$.

{Fink and St. Mary [1]. Conclusion (iv) was proved by Barrett [5]. Conclusion (v) under individual condition (a), (b), or (c) is a result of Kneser [1], Wintner [7] and Opial [1], and Hille [1], respectively}.

46. Suppose that $\omega:[0,\infty) \to \mathbf{R}$ is a positive continuous function, and for $u(t)$ a non-identically vanishing solution of $u''(t) + \omega^2(t)u(t) = 0$ for $T \in (0,\infty)$ let $N(T)$ denote the number of zeros of $u(t)$ on $[0,T]$.

(i) The asymptotic formula

$$N(t) \sim \pi^{-1} \int_0^t \omega(s)ds \tag{8.16}$$

holds whenever $\omega(t)$ has a continuous derivative satisfying $\omega' = o(\omega^2)$ as $t \to \infty$, but can fail to hold if "o" is relaxed to "O"; in fact, one may have $\omega' = O(\omega^2)$, $\omega(t)$ monotone with $\omega(t) \to \infty$ as $t \to \infty$, and (8.16) not hold. {Hartman and Wintner [1]}.

(ii) If $\omega(t)$ is of bounded variation on arbitrary compact subintervals of $[0,\infty)$, then for $T \in (0,\infty)$ there exists a $\tau = \tau(T)$ satisfying $|\tau| \leq \pi$ and

$$\pi N(T) = \int_0^T \omega(s)ds + \tau\left\{\frac{1}{2}\int_0^T \omega^{-1}(s)\,|d\omega(s)| + 2\right\}.$$

{Hartman [5]}.

47. Suppose that $p:[0,\infty) \to \mathbf{R}$ is a real-valued, continuous, monotone function such that $p(t) \to \infty$ as $t \to \infty$, and let $\{\lambda_j, u_j(t)\}$, $(j = 0,1,\dots)$, denote the sequence of

eigenvalues and corresponding eigenfunctions for the Sturm-Liouville boundary problem

$$\begin{aligned} &\text{(a)} \quad u''(t) + [\lambda - p(t)]u(t) = 0, \\ &\text{(b)} \quad u(0)\sin\theta - u'(0)\cos\theta = 0, \quad \int_0^\infty u^2(t)\,dt < \infty \end{aligned} \tag{8.17}$$

with $\lambda_0 < \lambda_1 < \ldots$, and $u_j(t)$ has exactly j zeros on $(0,\infty)$. If r_n denotes the value such that $p(r_j) = \lambda_j$, the asymptotic formula

$$n\pi \sim \int_0^{r_n} [\lambda_n - p(t)]^{1/2}\,dt \tag{8.18}$$

was established by Milne [2] under the additional assumption that p is of class $\mathscr{C}'''[0,\infty)$ with $p'(t) > 0$, $p''(t) \geq 0$, $p''(t) = o\{[p'(t)]^{4/3}\}$ as $t \to \infty$. Titchmarsh, (see Titchmarsh [1, First Edition, Ch. VII]), simplified Milne's proof and eliminated the requirement of the existence of the third derivative. Hartman [5] showed that (8.18) is valid for $p(t)$ a continuous, increasing function satisfying

$$\underset{t \leq u < v < \infty}{\text{g.}\ell\text{.b}} \left\{ \frac{p(v) - p(u)}{\int_u^v s^{-3}\,ds} \right\} \to +\infty \quad \text{as} \quad t \to \infty.$$

In particular, this condition holds if $p(t)$ is a continuously differentiable function with $p'(t) > 0$ for large t and $t^3 p'(t) \to \infty$ as $t \to \infty$. An even more elegant result, also due to Hartman, is that of the following Exercise.

48. Let $p:[0,\infty) \to \mathbf{R}$ be a positive increasing continuous function satisfying $p(t) \to \infty$ as $t \to \infty$. Then for λ a fixed real value a real non-identically vanishing solution $u(t)$ of (8.17a) has on $[0,\infty)$ a finite number $N(\lambda)$ of zeros, and the number $N(\lambda)$, up to a correction of -1, 0,

or 1, is independent of the particular solution used in its determination. If in addition $p(t)$ is convex on $[0,\infty)$, then

$$\pi N(\lambda) = \int_0^{\phi(\lambda)} [\lambda - p(s)]^{1/2} ds + 0(1) \text{ as } \lambda \to \infty, \tag{8.19}$$

where ϕ is the inverse function of p.

CHAPTER V.
STURMIAN THEORY FOR DIFFERENTIAL SYSTEMS

1. Introduction

This chapter is devoted to the extension of various results of the preceding chapters to differential systems. As a major portion of such results are valid in a setting wherein the coefficients of the system are complex-valued, the coefficient functions are no longer assumed to be real. Also, the hypotheses on the coefficient functions are stated specifically only for the case of solutions in the Carathéodory sense; that is, solution functions are supposed to be merely a.c. (absolutely continuous), and thus in general differentiable only a.e. (almost everywhere) on the interval of definition, so that a given derivative relation will be expected to hold only a.e. In only a few instances are the proofs under such hypotheses more complicated than in the case of systems whose coefficient functions are continuous. Moreover, as stated in the Introduction of Chapter II for scalar second order equations, for the reader unfamiliar with Lebesgue integration there should be no difficulty in interpreting the results in the setting wherein coefficients are continuous or piecewise continuous, the concept of absolute continuity is replaced by continuously differentiable or piecewise continuously differentiable, and statements on conditions holding a.e. are replaced by conditions holding everywhere or except for at most a finite number of values.

The contents of this chapter are the outgrowth of the work of many individuals, notably Morse, Hu, Birkhoff, Hestenes, Leighton, Nehari, the author, and their students. The notion of conjugate, or conjoined, solutions of a real self-adjoint differential system dates from von Escherich [1]. Properties of solutions of the Riccati matrix differential equation, and its relation to oscillation phenomena, are based upon the works of Radon [1,2], J. J. Levin [1], Sandor [1], and Reid [10, 20; see also, 38; Chs. III, IV]. For the reader familiar with variational theory, Theorem 6.1 and its Corollary embody the *Legendre* or *Clebsch transformation* of the functional $J[\eta;a,b]$, and the equivalent conditions of Theorem 6.3 present alternate forms of the *Jacobi* condition for this functional. In this connection, the reader is referred to Morse [2-Ch. I, 9-Ch. 1] and Bliss [7-Ch. VIII: Sec. 81, Ch. IX: Secs. 89-91]. The results of Theorems 6.3, 6.4, 6.5, 7.1, 7.2, 7.3, 7.4 provide the most basic tools for the study of oscillation and comparison phenomena for self-adjoint vector differential systems, and under varying degrees of generality are to be found in Morse [1, 2-Ch. IV, 9-Parts III, IV], Birkhoff and Hestenes [1], Hestenes [1], and Reid [12, 18, 21, 35-Ch. VII]. Section 8 presents the essential properties of the Morse fundamental forms, (see Morse [1, 4-Ch. III, 7, 9], and the basic relation between the negative index of these forms and the number of focal points of a given conjoined family of solutions of the associated differential system. Section 9 is devoted to the fundamental properties of the generalized polar coordinate transformation for Hamiltonian matrix differential systems as based on the work of Barrett [3], and Reid [17, 32]; within recent years contributions in this area include those of Etgen [1, 2,3] and Kreith [8, 10, 11]. The basis of the matrix oscillation theory of Lidskii-Jakubovic-Atkinson is presented in Section 10, and Section 11 is devoted to the concept and fundamental properties of principal solutions of Hamiltonian systems in the form introduced by Hartman [10] and Reid [15]. Except for Section 5, in which the basic definitions of normality and abnormality are given, the discussion of Sections 2-10 is limited in almost its entirity to systems that are identically normal. Section 12 contains some comments on modifications in theory and results that ensue which this condition is not present. Section 13 surveys briefly the results for higher order differential equations, both in the self-adjoint case when oscillation is defined via a subsidiary Hamiltonian system, and for general equations with conjugacy defined in the manner introduced by Leighton and Nehari [1]. As in previous chapters, the final section is devoted to selected exercises and topics from the literature in this field.

Matrix notation is used throughout, and, unless stated otherwise, the matrix elements are allowed to be complex-valued. In particular, matrices of one column are called vectors, and for a vector $\eta = (\eta_\alpha)$, $(\alpha = 1,\ldots,m)$, the system $|\eta|$ is used for the Euclidean norm $[|\eta_1|^2 + \ldots + |\eta_m|^2]^{1/2}$. The linear vector space of ordered m-tuples of complex numbers, with complex scalars, is denoted by $\mathbf{C}_m$. The symbol E_m is used for the $m \times m$ identity matrix, and reduced to merely E when there is no ambiguity; 0 is used indiscriminately for the zero matrix of any dimensions. The transpose of a matrix M is denoted by M^{Δ}, the conjugate of M by $\overline{M}$, and the conjugate transpose of M by M^*. If M is an $m \times n$ matrix, the symbol $|M|$ is used for the supremum of $|M\zeta|$ for ζ in the closed unit ball $\{\zeta : |\zeta| \leq 1\}$ of $\mathbf{C}_n$ which is also the supremum of $|M^*\eta|$ for η in the closed unit ball of $\mathbf{C}_m$. The notation $M \geq N$, $(M > N)$, is used to signify that M and N are hermitian matrices of the same dimensions and $M - N$ is a non-negative (positive) defined hermitian matrix. For typographical simplicity, if $M = [M_{\alpha\beta}]$ and $N = [N_{\alpha\beta}]$, $(\alpha = 1,\ldots,n;$ $\beta = 1,\ldots,r)$ are $n \times r$ matrices, then the $2n \times r$ matrix $P = [P_{\alpha\beta}]$, $(\alpha = 1,\ldots,2n;\ \beta = 1,\ldots,r)$, with $P_{\alpha\beta} = M_{\alpha\beta}$, $P_{n+\alpha,\beta} = N_{\alpha\beta}$ is denoted by $(M;N)$.

A matrix function $M(t) = [M_{\alpha\beta}(t)]$ is called continuous, integrable, etc., when each element $M_{\alpha\beta}$ possesses the specified property. If a matrix function $M(t)$ is a.c. on an interval $[a,b]$, then $M'(t)$ signifies the matrix of derivatives at values where these derivatives exist, and zero elsewhere. Similarly, if $M(t)$ is (Lebesgue) integrable on $[a,b]$ then $\int_a^b M(t)dt$ denotes the matrix of integrals of respective elements of $M(t)$. In the totality of finite dimensional rectangular matrix functions defined on a given interval I, we denote by $\mathscr{C}(I)$ the set of all such matrix functions that are continuous on I, and by $\mathscr{C}^k(I)$ the set of matrix functions that are continuous and have continuous derivatives of the first k orders on I. Also, we denote by $\mathscr{L}(I)$ the set of matrix functions whose elements are (Lebesgue) integrable on I, by $\mathscr{L}^2(I)$ the set of matrix functions whose elements $M_{\alpha\beta}(t)$ are (Lebesgue) measurable and $|M_{\alpha\beta}(t)|^2 \in \mathscr{L}(I)$, and by $\mathscr{L}^\infty(I)$ the set of all matrix functions whose elements are measurable and essentially bounded on I. Also, a matrix function M(t) is said to be locally of class $\mathscr{L}$, $\mathscr{L}^2$ or $\mathscr{L}^\infty$ on an interval I if M(t) belongs to the corresponding class $\mathscr{L}[a,b]$, $\mathscr{L}^2[a,b]$ or $\mathscr{L}^\infty[a,b]$ on arbitrary compact subintervals [a,b] of I.

2. Special Examples

The most direct generalization of the scalar differential equation (II.1.1) to a vector equation is

$$[R(t)u'(t) + Q(t)u(t)]' - [Q^*(t)u'(t) + P(t)u(t)] = 0 \quad (2.1)$$

wherein R(t), P(t), Q(t) are $n \times n$ matrix functions, $u(t) \equiv (u_\alpha(t))$ is an n-dimensional vector function, and these matrix coefficient functions satisfy the following hypothesis.

($\mathscr{A}_\omega$) *On the given interval* I *on the real line, the* $n \times n$ *matrix functions* R(t), P(t) *are hermitian,* R(t) *is non-singular, and the matrix functions* $R^{-1}(t)$, $R^{-1}(t)Q(t)$, $P(t) - Q^*(t)R^{-1}(t)Q(t)$ *are locally of class* $\mathscr{L}$ *on* I.

In the terminology of the calculus of variations, the equation (2.1) is the "vector Euler differential equation" for the hermitian functional

$$J[\eta;a,b] = \int_a^b 2\omega(t,\eta(t),\eta'(t))dt, \tag{2.2}$$

where $2\omega(t,\eta,\zeta)$ denotes the hermitian form

$$2\omega(t,\eta,\zeta) = \zeta^*[R(t)\zeta + Q(t)\eta] + \eta^*[Q^*(t)\zeta + P(t)\eta]. \tag{2.3}$$

In terms of the "canonical variable" vector functions

$$u(t),\ v(t) = R(t)u'(t) + Q(t)u(t) \tag{2.4}$$

the vector differential equation (2.1) may be written as

$$\begin{aligned} L_1[u,v](t) &\equiv -v'(t) + C(t)u(t) - A^*(t)v(t) = 0, \\ L_2[u,v](t) &\equiv \ u'(t) - A(t)u(t) - B(t)v(t) \ = 0, \end{aligned} \tag{2.5}$$

where the matrix functions $A(t)$, $B(t)$, $C(t)$ are defined as

$$A = -R^{-1}Q, \quad B = R^{-1}, \quad C = P - Q^*R^{-1}Q \tag{2.6}$$

whenever $R(t)$, $P(t)$, $Q(t)$ satisfy hypothesis $(\mathscr{H}_\omega)$ the matrix functions $A(t)$, $B(t)$, $C(t)$ defined by (2.6) satisfy the following hypothesis.

$(\mathscr{H}_L)$ *On the given interval* I *on the real line the* $n \times n$ *matrix functions* $B(t)$, $C(t)$ *are hermitian, and the matrix functions* $A(t)$, $B(t)$, $C(t)$ *are locally of class* $\mathscr{L}$ *on* I.

Clearly a most important instance of the above example is that wherein the matrix functions R, P, Q are all continuous on I, in which case the matrix coefficient functions A, B, C of (2.6) are all continuous.

A still more general example of a system (2.5) is provided by a differential system which is of the form of the accessory differential system for a variational problem of

Lagrange or Bolza type, (see, for example, Bliss [7, Sec. 81]). In addition to the hermitian form (2.3) in (η,π), consider a vector linear form

$$\Phi(t,\eta,\zeta) = \phi(t)\zeta + \theta(t)\eta, \tag{2.7}$$

where $\phi(t)$, $\theta(t)$ are $m \times n$, $(m < n)$, matrix functions. For the problem to be described we assume the following hypothesis, which leads to a system (2.5) whose coefficients satisfy conditions corresponding somewhat to those of hypothesis $(\mathscr{A}_L^{++})$ in Chapter III.

$(\mathscr{A}_\Omega)$

(a) *The* $n \times n$ *matrix functions* $R(t)$, $P(t)$ *are hermitian for* $t \in I$.

(b) *The* $(n + m) \times (n + m)$ *hermitian matrix function*

$$\begin{bmatrix} R(t) & \phi^*(t) \\ \phi(t) & 0 \end{bmatrix} \tag{2.8}$$

is non-singular for $t \in I$. [It then follows that the inverse of (2.8) is of the form

$$\begin{bmatrix} T(t) & \tau^*(t) \\ \tau(t) & \chi(t) \end{bmatrix} \tag{2.9}$$

where $T(t)$ *and* $\chi(t)$ *are hermitian matrix functions of orders* n *and* m, *respectively, and* $\tau(t)$ *is an* $m \times n$ *matrix function.*]

(c) *Each of the matrix functions* $R(t)$, $Q(t)$, $T(t)$, $\phi(t)$, $\tau(t)$, $\theta(t)$, $\chi(t)$ *is locally of class* $\mathscr{L}^\infty$, *and* $P(t)$ *is locally of class* $\mathscr{L}$ *on* I.

For the variational problem involving the functional (2.2) subject to the auxiliary m-dimensional vector differential equation

$$\Phi(t,\eta(t),\eta'(t)) = 0, \quad t \in [a,b], \tag{2.10}$$

the corresponding "Euler-Lagrange differential system" may be written in vector form as

$$\begin{aligned}[R(t)u'(t) + P(t)u(t) + \phi^*(t)\mu(t)]' \\ - [Q^*(t)u'(t) + P(t)u(t) + \theta^*(t)\mu(t)] = 0, \\ \Phi(t,u(t),u'(t)) = 0,\end{aligned} \tag{2.11}$$

where $u = u(t)$ is an n-dimensional vector function and $\mu(t)$ is an m-dimensional "multiplier" vector function. Moreover, in terms of the "canonical variables"

$$u(t),\ v(t) = R(t)u'(t) + Q(t)u(t) + \phi^*(t)\mu(t), \tag{2.12}$$

the system (2.11) reduces to the form (2.5), where now

$$\begin{aligned} A &= -(TQ + \tau^*\theta), \quad B = T, \\ C &= P - Q^*TQ - Q^*\tau^*\theta - \theta^*\tau Q - \theta^*\chi\theta. \end{aligned} \tag{2.13}$$

It follows readily that hypothesis $(\mathscr{A}_\Omega)$ implies that the matrix functions A,B,C of (2.13) satisfy hypothesis $(\mathscr{A})$; indeed, the matrix functions A, B and $C - P$ are all locally of class $\mathscr{L}^\infty$ on I.

In particular, the generality of the system (2.5) with coefficient matrix functions of the above form is illustrated by the fact that it includes systems which are equivalent to real self-adjoint differential equations of even order. Suppose that $r_j(t)$, $(j = 0,1,\ldots,n)$, are real-valued with $r_n(t) \neq 0$ for $t \in I$, while $r_j(t)$, $(j = 0,1,\ldots,n-1)$ and $1/r_n(t)$ are locally of class $\mathscr{L}$ on I. If

$$2\omega(t,\eta,\zeta) = r_n(t)\zeta_n^2 + r_{n-1}(t)\eta_n^2 + \ldots + r_0(t)\eta_1^2, \tag{2.14}$$

and $m = n-1$, with

$$\Phi(t,\eta,\zeta) = (\zeta_\alpha - \eta_{\alpha+1}), \ (\alpha = 1,\dots,n-1), \tag{2.15}$$

then $(u;v)$ is a solution of the system (2.5) with coefficient matrices given by the corresponding relations (2.13) if and only if there exists a function $f(t) \in \mathscr{C}^{n-1}(I)$ with $f^{[n-1]}(t)$ locally a.c. on I, and

$$\begin{aligned} u_j(t) &= f^{[j-1]}(t), \quad (j = 1,\dots,n), \\ v_n(t) &= r_n(t)f^{[n]}(t), \\ v_1'(t) &= r_0(t)f(t), \\ v_{\beta+1}'(t) &= r_\beta(t)f^{[\beta]}(t) - v_\beta(t), \quad (\beta = 1,\dots,n-1). \end{aligned} \tag{2.16}$$

In particular, if for $j = 0,1,\dots,n$ we have that $r_j(t) \in \mathscr{C}^j[a,b]$, then (2.14) is satisfied by vector functions $u(t) = (u_\alpha(t))$, $v(t) = (v_\alpha(t))$ and $f(t)$ if and only if $f(t)$ is a solution of the 2n-th order (formally) self-adjoint differential equation

$$\sum_{j=0}^{n} (-1)^j \{r_j(t)f^{[j]}(t)\}^{[j]} = 0; \tag{2.17}$$

in this case $u_j(t) = f^{[j-1]}(t)$, $(j = 1,\dots,n)$, and

$$v_j(t) = \sum_{\beta=j}^{n} (-1)^{\beta-j}\{r_\beta(t)f^{[\beta]}(t)\}^{[\beta-j]}, \quad (j = 1,\dots,n). \tag{2.18}$$

3. Preliminary Properties of Solutions of (2.5)

A vector differential system (2.5) in n-dimensional vector functions $u(t)$, $v(t)$ may be written in terms of the 2n-dimensional vector function $y(t) = (u(t);v(t))$ as

$$L[y] = \mathscr{J}y'(t) + \mathscr{A}(t)y(t) = 0, \tag{3.1}$$

where $\mathcal{J}$ and $\mathcal{A}(t)$ are the $2n \times 2n$ matrix functions

$$\mathcal{J} = \begin{bmatrix} 0 & -E_n \\ E_n & 0 \end{bmatrix}, \quad \mathcal{A}(t) = \begin{bmatrix} C(t) & -A^*(t) \\ -A(t) & -B(t) \end{bmatrix}. \tag{3.2}$$

The matrix function $\mathcal{A}(t)$ is supposed to satisfy the following hypothesis.

($\mathfrak{A}$) *On a given interval* I *on the real line the* $2n \times 2n$ *matrix function* $\mathcal{A}(t)$ *is hermitian, and locally of class* $\mathcal{L}$.

Because of its occurrence as the canonical form of an accessory system derived from a variational problem, and equation (3.1) may be called a "Hamiltonian system", or an "hermitian system".

It is to be commented that in this chapter there will be frequent references to Chapter VII of Reid [35], wherein the hypothesis corresponding to ($\mathfrak{A}$) above requires the matrix function $\mathcal{A}(t)$ to be locally of class $\mathcal{L}^\infty$. There the hypothesis was thus formulated so that the Dirichlet functional $J[\eta]$ defined by (6.1) below might be considered in the Hilbert space setting wherein the associated canonical variable is of class $\mathcal{L}^2[a,b]$. As far as the solutions of (3.1) are concerned, however, and associated integral conditions involving the functional J, like results under the above hypothesis ($\mathfrak{A}$) are derivable by identical proofs, where now the canonical variable ζ associated with an a.c. η is assumed to be locally of class $\mathcal{L}^\infty$. For a discussion of differential systems (3.2) in the context of the present hypothesis; reference is made to paper [22] of the author; his paper [12] is also couched in terms intimately related to this hypothesis.

From well-known existence theorems, (see, for example, Coddington and Levinson [1, Ch. II], or Reid [35, Ch. II]), it follows that for arbitrary $t_o \in I$ and given n-dimensional vectors u^o, v^o there is a unique solution $y = (u;v)$ of (3.1) on I satisfying $u(t_o) = u^o$, $v(t_o) = v^o$. If $y_\alpha = (u_\alpha;v_\alpha)$, $(\alpha = 1,2)$, are solutions of (3.1) then the function

$$\{y_1,y_2\}(t) \equiv y_2^*(t)\mathcal{J}y_1(t) = v_2^*(t)u_1(t)-u_2^*(t)v_1(t) \qquad (3.3)$$

is constant on I. If the value of this constant is zero these solutions are said to be *(mutually) conjoined*; an alternate terminology for this concept is *isotropic*, (see, for example, Coppel [2, p. 34]).

Corresponding to (3.1), we have the general matrix equation

$$L[Y](t) = \mathcal{J}Y'(t) + \mathcal{A}(t)Y(t) = 0. \qquad (3.1_M)$$

Moreover, if $Y_1 = (U_1;V_1)$ is an $n \times r_1$ matrix solution of (3.1_M), and $Y_2 = (U_2;V_2)$ is an $n \times r_2$ matrix solution of this equation, then $\{Y_1,Y_2\} \equiv Y_2^*(t)\mathcal{J}Y_1(t) = V_2^*(t)U_1(t) - U_2^*(t)V_1(t)$ is a constant $r_2 \times r_1$ matrix function on I. If $Y(t) = (U(t);V(t))$ is a $2n \times r$ matrix whose column vectors are linearly independent solutions of (3.1), and $\{Y_1,Y_2\} = 0$, these solutions form a basis for a conjoined family of solutions of dimension r, consisting of the set of all solutions of (3.1) which are linear combinations of these column vectors. If $Y(t) = (U(t);V(t))$ is a $2n \times n$ solution of (3.1_M) whose column vectors form a basis for an n-dimensional conjoined family of solutions of (3.1), then for brevity $Y(t)$ is referred to as a *conjoined basis for*

(3.1). Alternate terminologies are *isotropic solutions* of (3.1_M), or *prepared solution* of (3.1_M); this latter terminology is due to Hartman [10]. In particular, if $c \in I$ and $Y(t;c) = (U(t;c);V(t,c))$, $Y_0(t;c) = (U_0(t;c),V_0(t;c))$ are the solutions of (3.1_M) determined by the initial conditions

$$\begin{aligned} &(a)\quad U(c;c) = 0, \quad V(c;c) = E_n; \\ &(b)\quad U_0(c;c) = E_n, \quad V_0(c;c) = 0, \end{aligned} \tag{3.4}$$

then each of the matrix solutions $Y(t;c)$ and $Y_0(t;c)$ is a conjoined basis for (3.1). A basic property of solutions of (3.1) is presented in the following lemma.

LEMMA 3.1. *The maximal dimension of a conjoined family of solutions of* (3.1) *is* n; *moreover, a given conjoined family of solutions of dimension* $r < n$ *is contained in a conjoined basis for* (3.1).

If the elements of the matrix $\mathscr{A}(t)$ of (3.1) are real-valued, then two real-valued solutions y_1 and y_2 of this equation are conjoined if and only if they are conjugate in the sense introduced originally by von Escherich [1]. Now under the usual isomorphism between the algebra of $n \times n$ complex matrices and the algebra of an associated class of $2n \times 2n$ real matrices, the differential equation in $y(t) = (u(t);v(t))$ is equivalent to a real 4n-dimensional vector differential equation in $\mathsf{y}(t) = (\mathsf{u}(t);\mathsf{v}(t))$, where $\mathsf{u}(t) = (\mathrm{Re}\, u(t);\ \mathrm{Im}\, u(t))$ and $\mathsf{v}(t) = (\mathrm{Re}\, v(t);\ \mathrm{Im}\, v(t))$. If $y_\alpha(t) = (u_\alpha(t);v_\alpha(t))$, $(\alpha = 1,2)$, are conjoined solutions of an hermitian system (3.1), then the corresponding solutions $\mathsf{y}_\alpha(t)$ of the associated real system are conjugate. On the other hand, if $\mathsf{y}_\alpha(t) = (\mathsf{u}_\alpha(t):\mathsf{v}_\alpha(t))$, $\alpha = 1,2$, are real solutions of the related real system which are conjugate, and $y_\alpha(t) = (u_\alpha(t);v_\alpha(t))$ are the solutions of (3.1) such that $\mathrm{u}_\alpha(t) = (\mathrm{Re}\, u_\alpha(t);\mathrm{Im}\, v_\alpha(t))$, $\mathrm{v}_\alpha(t) = (\mathrm{Re}\, v_\alpha(t):\mathrm{Im}\, v_\alpha(t))$, it does not follow that $y_1(t)$ and $y_2(t)$ are solutions of (3.1) which are conjoined, but merely that $\mathrm{Re}\{y_1,y_2\} \equiv 0$ on I.

Correspondingly to the result of Theorem II.2.4, we have the following property of solutions of (3.1_M).

THEOREM 3.1. *Suppose that* $Y_0(t) = (U_0(t);V_0(t))$ *is a conjoined basis for* (3.1), *and* $U_0(t)$ *is non-singular on a subinterval* I_0 *of* I. *If* $c \in I_0$ *then* $Y(t) = (U(t):V(t))$ *is a solution of* (3.1_M) *on* I_0 *if and only if on this interval*

$$U(t) = U_0(t)H(t),\ V(t) = V_0(t)H(t) + U_0^{*-1}(t)K_1, \tag{3.5}$$

where

$$H(t) = K_0 + \left[\int_c^t U_0^{-1}(s)B(s)U_0^{*-1}(s)ds\right]K_1, \tag{3.6}$$

and K_0, K_1 *are constant matrices; moreover, in* (3.6),

$$K_0 = U_0^{-1}(c)U(c),\quad K_1 = -\{Y,Y_0\}. \tag{3.6'}$$

The result of this theorem is greatly simplified by the restriction of $Y_0(t)$ to be a conjoined basis for (3.1). As shown by Reid [15, Sec. 3; see also 35, Prob. VII.2.4], if $Y_0(t) = (U_0(t);V_0(t))$ is a solution of (3.1_M) with $U_0(t)$ non-singular on a subinterval I_0 of I, and $K = -\{Y_0,Y_0\}$, then $Y(t) = (U(t);V(t))$ is a solution of (3.1_M) on I_0 if and only if on I_0 we have

$$\begin{aligned} U(t) &= U_0(t)H(t), \\ V(t) &= V_0(t)H(t) + U_0^{*-1}(t)[K_1 - KH(t)], \end{aligned} \tag{3.5'}$$

where K_1 is a constant matrix and $H(t)$ is a solution of the differential equation

$$H'(t) = U_0^{-1}(t)B(t)U_0^{*-1}(t)[K_1 - KH(t)], \quad t \in I_0;$$

moreover, if $c \in I_0$ and $T = T(t,c;U_0)$ is the solution of the matrix differential system

$$T'(t) = -U_0^{-1}(t)B(t)U_0^{*-1}(t)KT(t), \quad T(c) = E, \tag{3.7}$$

then $H(t)$ is of the form

$$H(t) = T(t,s;U_0)\{K_0 + S(t,c;U_0)K_1\},$$

where

$$S(t,c;U_0) = \int_c^t T^{-1}(s,c,;U_0)U_0^{-1}(s)B(s)U_0^{*-1}(s)ds, \quad t\in I_0, \tag{3.8}$$

where K_0, K_1 are given by (3.6'). It is to be emphasized that in Theorem 3.1 and the following statements the solution $Y(t) = (U(t);V(t))$ of (3.1_M) may be a $2n \times r$ matrix function, in which case the matrix K is $n \times n$, the matrices K_0 and K_1 are $n \times r$, and the matrix function $H(t)$ is $n \times r$ on I_0.

Of particular significance is the following result for systems (3.1) with $B(t) \geq 0$ a.e. on I.

THEOREM 3.2. *Suppose that hypothesis* ($\mathscr{A}$) *holds and* $B(t) \geq 0$ a.e. *on* I. *If* $Y_0(t) = (U_0(t);V_0(t))$ *is a conjoined basis for* (3.1) *with* $U_0(t)$ *non-singular on a subinterval* $I_0 = (c_0,d_0)$ *of* I, *and for a given* $c \in I_0$ *the matrix function* $Y(t) = (U(t);V(t))$ *is defined on* I_0 *by*

$$\begin{aligned} U(t) &= U_0(t)\left[E + \int_c^t U_0^{-1}(s)B(s)U_0^{*-1}(s)ds\right] \\ V(t) &= V_0(t)\left[E + \int_c^t U_0^{-1}(s)B(s)U_0^{*-1}(s)ds\right] + U_0^{*-1}(t) \end{aligned} \tag{3.9}$$

then $Y(t)$ *is a conjoined basis for* (3.1) *satisfying* $\{Y,Y_0\} = E$. *Moreover, for* $t \in [c,d_0)$ *we have that* $U(t)$ *is non-singular and*

$$\begin{aligned} U_0(t) &= U(t)\left[E - \int_c^t U^{-1}(s)B(s)U^{*-1}(s)ds\right], \\ V_0(t) &= V(t)\left[E - \int_c^t U^{-1}(s)B(s)U^{*-1}(s)ds\right] - U^{*-1}(t). \end{aligned} \tag{3.9'}$$

Moreover, $\int_c^{d_0} U^{-1}(s)B(s)U^{*-1}(s)ds = \lim_{t\to d_0} \int_c^t U^{-1}(s)B(s)U^{*-1}(s)ds$ *is a non-negative hermitian matrix dominated by* E, *and* $Y_1(t) = (U_1(t);V_1(t))$ *defined on* $[c,d_0)$ *by*

$$\begin{aligned} U_1(t) &= U(t) \int_t^{d_0} U^{-1}(s)B(s)U^{*-1}(s)ds, \\ V_1(t) &= V(t) \int_t^{d_0} U^{-1}(s)B(s)U^{*-1}(s)ds + U^{*-1}(t) \end{aligned} \tag{3.10}$$

is a conjoined basis for (3.1) *satisfying* $\{Y_1,Y\} = E$ *and* $U^{-1}(t)U_1(t) \to 0$ *as* $t \to d_0^-$.

Relations (3.9), (3.9') are direct consequences of Theorem 3.1. The final statements of the theorem follow from the fact that (3.9), 3.9') imply that the matrix function $M_1(t) = E + \int_c^t U_0^{-1}(s)B(s)U_0^{*-1}(s)ds$ and $M_2(t) = E - \int_c^t U^{-1}(s)B(s)U^{*-1}(s)ds$ are inverses on $[c,d_0)$, and since the condition $B(t) \geq 0$ implies that $M_1(t) \geq E$ for $t \in [c,d_0)$ we have $0 < M_2(t) \leq E$ on this interval. In particular, if d_0 is a point of I, we have that $Y_1(t)$ is the solution of (3.1_M) satisfying $U_1(d_0) = 0$, $V_1(d_0) = U^{*-1}(d_0)$.

Of particular interest is a matrix differential system of the form

$$\begin{aligned} L_1[\Phi,\Psi](t) &\equiv -\Psi'(t) - Q(t)\Phi(t) = 0, \\ L_2[\Phi,\Psi](t) &\equiv \quad \Phi(t) - Q(t)\Psi(t) = 0, \end{aligned} \tag{3.11}$$

where $Q(t)$ is an $n \times n$ hermitian matrix which is locally of class $\mathscr{L}$ on I. Clearly (3.11) is of the form (3.1_M) with $A(t) = 0$ and $B(t) = -C(t) = Q(t)$. If $\Phi = \Phi_0(t)$, $\Psi = \Psi_0(t)$ is a solution of (3.11), then $\Phi = -\Psi_0(t)$, $\Psi = \Phi_0(t)$ is also a solution of this system. Therefore, in view of the above discussion of properties of (3.1_M), it follows that if $(\Phi(t);\Psi(t))$ is a solution of (3.11) then the matrices $\Psi^*(t)\Phi(t) - \Phi^*(t)\Psi(t)$ and $\Phi^*(t)\Phi(t) + \Psi^*(t)\Psi(t)$ are constant. In particular, if $(\Phi(t);\Psi(t))$ is a solution of (3.11) for which $\Psi^*(t)\Phi(t) - \Phi^*(t)\Psi(t) \equiv 0$ and $\Phi^*(t)\Phi(t) + \Psi^*(t)\Psi(t) \equiv E$, then the $2n \times 2n$ matrix function

$$\begin{bmatrix} \Phi(t) & -\Psi(t) \\ \Psi(t) & \Phi(t) \end{bmatrix} \tag{3.12}$$

is unitary throughout on I, and hence also $\Phi(t)\Phi^*(t) + \Psi(t)\Psi^*(t) \equiv E$ and $\Phi(t)\Psi^*(t) - \Psi(t)\Phi^*(t) \equiv 0$. In particular, if for $\tau \in I$ the solution $(\Phi;\Psi)$ of (3.11) satisfying $\Phi(\tau) = 0$, $\Psi(\tau) = E$ is denoted by $\Phi = S(t;\tau)$, $\Psi = C(t;\tau)$, it follows that for $(t,\tau) \in I \times I$ we have the following identities satisfied by the matrix functions $S = S(t;\tau)$ and $C = C(t;\tau)$:

$$\begin{aligned} S^*S + C^*C &\equiv E, & S^*C - C^*S &\equiv 0, \\ SS^* + CC^* &\equiv E, & SC^* - CS^* &\equiv 0. \end{aligned} \tag{3.13}$$

Also, for $(t,\tau,\sigma) \in I \times I \times I$, we have the identies:

$$
\begin{aligned}
C(t;\sigma) &\equiv C(t;\tau)C^*(\sigma;\tau) + S(t;\tau)S^*(\sigma;\tau),\\
S(t;\sigma) &\equiv S(t;\tau)C^*(\sigma;\tau) - C(t;\tau)S^*(\sigma;\tau).
\end{aligned}
\tag{3.14}
$$

In respect to both the differential equations (3.11) and the identities (3.13), (3.14), clearly the matrix functions $S(t;\tau)$ and $C(t;\tau)$ are generalizations of the trigonometric functions $\sin(t-\tau)$ and $\cos(t-\tau)$. This generalization is further emphasized by the following additional results, which for systems (3.11) involving real symmetric $Q(t)$ was first established by Etgen [2]. For $(t,\tau) \in I \times I$ the matrix functions $Y(t;\tau) = 2S(t;\tau)C^*(t;\tau)$ and $Z(t;\tau) = C(t;\tau)C^*(t;\tau) - S(t;\tau)S^*(t;\tau)$ satisfy the matrix differential system

$$
\begin{aligned}
Y'(t) &= Q(t)Z(t) + Z(t)Q(t),\\
Z'(t) &= -Q(t)Y(t) - Y(t)Q(t);
\end{aligned}
\tag{3.15}
$$

moreover,

$$
\begin{aligned}
Y^2(t;\tau) + Z^2(t;\tau) &\equiv E,\\
Y(t;\tau)Z(t;\tau) &= Z(t;\tau)Y(t;\tau).
\end{aligned}
\tag{3.16}
$$

For given positive integers n and m, let $M(n,m)$ denote the class of $n \times m$ matrices with complex elements, and suppose that $Q(t;\Phi,\Psi)$ satisfies the following condition.

$(\mathscr{C})$ *If* $(t;\Phi,\Psi) \in I \times M(n,n) \times M(n,n)$, *then* $Q(t;\Phi,\Psi)$ *is an* $n \times n$ *hermitian matrix function which in* t *is locally of class* $\mathscr{L}$ *on* I *for fixed* $\Phi,\Psi \in M(n,n) \times M(n,n)$, *and continuous in* (Φ,Ψ) *on* $M(n,n) \times M(n,n)$ *for fixed* $t \in I$. *Moreover, the solutions of the matrix differential system*

$$
-\Psi'(t) + Q(t;\Phi,\Psi)\Phi(t) = 0,\quad \Phi'(t) - \Phi(t;\Phi,\Psi)\Psi(t) = 0, \tag{3.17}
$$

are locally unique.

In particular, condition ($\mathscr{C}$) holds for $Q(t;\Phi,\Psi)$ of the form $Q_0(t) + \Psi B(t)\Psi^* + \Psi A(t)\Phi^* + \Phi A^*(t)\Psi^* - \Phi C(t)\Phi^*$, where $A(t)$, $B(t)$, $C(t)$ and $Q_0(t)$ are $n \times n$ matrix functions that are locally of class $\mathscr{L}$ on I, while $Q_0(t)$, $B(t)$ and $C(t)$ are hermitian for $t \in I$.

For the non-linear matrix differential system (3.17) one has the following result.

THEOREM 3.3. *Suppose that* $Q(t;\Phi,\Psi)$ *satisfies condition* ($\mathscr{C}$), *and that* $(t_0;\Phi_0,\Psi_0) \in I \times M(n,n) \times M(n,n)$. *Then there exists a unique solution* $\Phi = \Phi(t;t_0,\Phi_0,\Psi_0)$, $\Psi = \Psi(t;t_0,\Phi_0,\Psi_0)$ *of* (3.17) *such that* $\Phi = \Phi_0$, $\Psi = \Psi_0$ *for* $t = t_0$, *and the maximal interval of existence of this solution is the given interval* I; *moreover, the* $n \times n$ *matrix functions* $\Phi^*\Phi + \Psi^*\Psi$ *and* $\Psi^*\Phi - \Phi^*\Psi$ *are constant on* I. *In particular, if* $\Phi^*\Phi + \Psi^*\Psi \equiv E$ *and* $\Psi^*\Phi - \Phi^*\Psi \equiv 0$, *then on* I *the corresponding* $2n \times 2n$ *matrix* (3.12) *is unitary, and also* $\Phi\Phi^* + \Psi\Psi^* \equiv E$, $\Phi\Psi^* - \Psi\Phi^* \equiv 0$. *If the solution* $(\Phi;\Psi)$ *of* (3.17) *is such that* $\Phi(t_0)\Phi^*(t_0) + \Psi(t_0)\Psi^*(t_0) = E$ *and* $\Phi(t_0)\Psi^*(t_0) - \Psi(t_0)\Phi^*(t_0) = 0$, *then* $\Phi\Phi^* + \Psi\Psi^* \equiv E$, $\Phi\Psi^* - \Psi\Phi^* \equiv 0$, *the matrix* (3.12) *is unitary, and also* $\Phi^*\Phi + \Psi^*\Psi \equiv E$, $\Psi^*\Phi - \Phi^*\Psi \equiv 0$ *on* I.

Corresponding to the concept of the "reciprocal equation" associated with a linear homogeneous second order scalar differential equation (II.1.1) as defined in Section II.11, the vector differential equation

$$\mathscr{J}\check{y}'(t) + \check{\mathscr{A}}(t)\check{y}(t) = 0, \tag{3.18}$$

with

$$\check{\mathscr{A}}(t) = \mathscr{J}\mathscr{A}(t)\mathscr{J}^* = \begin{bmatrix} -B(t) & A(t) \\ A^*(t) & C(t) \end{bmatrix} \tag{3.19}$$

is called the "equation reciprocal to (3.1)", (see Ahlbrandt [2, Sec. 2]). Clearly the matrix function $\mathscr{A}(t)$ satisfies hypothesis ($\mathscr{A}$) on an interval I if and only if the matrix $\check{\mathscr{A}}(t)$ satisfies this hypothesis on I. In particular, a vector function $y(t)$ is a solution of (3.1) if and only if $\check{y}(t) = \mathscr{J}y(t)$ is a solution of (3.18). Moreover, if y_1, y_2 are solutions of (3.1), then $\check{y}_1 = \mathscr{J}y_1$, $\check{y}_2 = \mathscr{J}y_2$ are solutions of (3.18) such that $\{\check{y}_1, \check{y}_2\} = \{y_1, y_2\}$, and consequently if y_1, y_2 are conjoined solutions of (3.1) then the corresponding $\check{y}_1, \check{y}_2$ are conjoined solutions of (3.18).

4. Associated Riccati Matrix Differential Equations

The Riccati matrix differential equation

$$\begin{aligned} K[W](t) \equiv W'(t) + A(t)W(t) + W(t)A^*(t) \\ + W(t)B(t)W(t) - C(t) = 0 \end{aligned} \tag{4.1}$$

is associated with the linear differential system (3.1_M) in a manner similar to that of the scalar Riccati differential equation (II.2.4) with the differential equation (II.1.1) and the equivalent first order differential system (II.1.2). In particular, if $Y(t) = (U(t);V(t))$ is an a.c. $2n \times n$ matrix function with $U(t)$ non-singular on a subinterval I_0 of I, then on I_0 we have the identity

$$U^*K[VU^{-1}]U = -U^*\{L_1[U,V] + VU^{-1}L_2[U,V]\}, \tag{4.2}$$

and hence on I_0 the $n \times n$ matrix function $W(t) = V(t)U^{-1}(t)$ is a solution of (4.1), and $U(t)$ is a fundamental matrix solution on I_0 of the corresponding first order homogeneous $n \times n$ matrix differential equation

$$U'(t) = [A(t) + B(t)W(t)]U(t). \tag{4.3}$$

Conversely, if $W(t)$ is a solution of (4.1) and $U(t)$ is defined to be a fundamental matrix solution of (4.3), then $U(t), V(t) = W(t)U(t)$ is a solution of $L_2[U,V] = 0$ on I_0, and from (4.2) it then follows that also $L_1[U,V] = 0$ on I_0. That is, $Y(t) = (U(t);V(t))$ is a solution of (3.1_M) with $U(t)$ non-singular on a subinterval I_0 of I if and only if there is a solution $W(t)$ of (4.1) with $W(t) = V(t)U^{-1}(t)$ on I_0.

In view of the hermitian character of $B(t)$ and $C(t)$ on I, it follows readily that if $W = W_0(t)$ is a solution of (4.1) on a subinterval I_0 of I, then $W = W_0^*(t)$ is also a solution of (4.1) on this subinterval. In particular, $W_0^*(t) \equiv W_0(t)$ on I_0 if there exists a single value $\tau \in I_0$ such that $W_0^*(\tau) = W_0(\tau)$. Now if $W(t)$ is a solution of (4.1) on I_0, and $Y(t) = (U(t);V(t))$ is a solution of (3.1_M) such that $U(t)$ is non-singular and $W(t) = V(t)U^{-1}(t)$, then $W^*(t) - W(t) = U^{*-1}(t)\{Y,Y\}U^{-1}(t)$, and hence $W(t)$ is an hermitian solution of (4.1) if and only if the associated $Y(t) = (U(t);V(t))$ satisfying $W(t) = V(t)U^{-1}(t)$ is a conjoined basis for (3.1).

If $W = W_0(t)$ is a solution of (4.1) on a subinterval I_0 of I, and $c \in I_0$, let $H = H(t,c;W_0)$ and $G = G(t,c,W_0)$ be the solutions of the system

$$\begin{aligned} &(a)\quad H' + H[A(t) + B(t)W_0(t)] = 0, \quad H(c) = E, \\ &(b)\quad G' + [A^*(t) + W_0(t)B(t)]G = 0, \quad G(c) = E, \end{aligned} \tag{4.4}$$

and for s and t on I_0 define $Z(t,s;W_0)$ as

$$Z(t,s;W_o) = \int_s^t H(r,s;W_o)B(s)G(r,s;W_o)ds. \tag{4.5}$$

By direct substitution, it may then be established readily that an $n \times n$ matrix function $W(t)$ is a solution of (4.1) on I_o if and only if the constant matrix $\Gamma = W(s) - W_o(s)$ is such that $E + Z(t,s;W_o)\Gamma$ is non-singular on I_o, and

$$W(t) = W_o(t) + G(t,s;W_o)\Gamma[E+Z(t,s;W_o)\Gamma]^{-1}H(t,s;W_o). \tag{4.6}$$

The solution $W(t)$ is also given by

$$W(t) = W_o(t) + G(t,s;W_o)[E+\Gamma Z(t,s;W_o)]^{-1}\Gamma H(t,s;W_o). \tag{4.6'}$$

In particular, from (4.6) or (4.6') it follows that two solutions $W(t)$ and $W_o(t)$ of (4.1) are such that $W(t) - W_o(t)$ is of constant rank throughout a common interval of existence. Moreover, if Γ is non-singular, (4.6) may be written as

$$W(t) = W_o(t) + G(t,s;W_o)[\Gamma^{-1}+Z(t,s;W_o)]^{-1}H(t,s;W_o), \tag{4.6$_o$}$$

and similarly for (4.6').

If $W = W_o(t)$ is a solution of (4.1) on a subinterval I_o of I, and $Y_o(t) = (U_o(t);V_o(t))$ is a corresponding solution of (3.1_M) such that $W_o(t) = V_o(t)U_o^{-1}(t)$, then it is readily seen that the solution of (4.4a) is given by $H(t,c;W_o) = U_o(c)U_o^{-1}(t)$. One may also establish, (see Reid [35, p. 318]), that the solution of (4.4b) is given by

$$G(t,c;W_o) = H^*(t,c;W_o)U_o^{*-1}(c)T^{*-1}(t,c;U_o)U_o^*(c), \tag{4.7}$$

where $T(t,c;U_o)$ is the solution of the equation (3.9), and consequently

$$Z(t,s;W_o) = U_o(c)S^*(t,c;U_o)U^*(c) \tag{4.8}$$

where $S(t,c;U_0)$ is the matrix function defined on I_0 by (3.10).

In particular, if $W_0(t)$ is an hermitian solution of (4.1) then $G(t,c;W_0) = H^*(t,c;W_0)$, and if $B(t) \geq 0$ a.e. on I the following result is immediate from (4.5) and (4.6_0) whenever $\Gamma > 0$ or $\Gamma < 0$, and the cases of $\Gamma \geq 0$, $\Gamma \leq 0$ follow by a direct limiting argument.

THEOREM 4.1. *Suppose that hypothesis* ($\mathscr{A}$) *is satisfied and* $B(t) \geq 0$ a.e. *on* I, *while* $W = W_0(t)$ *is an hermitian solution of* (4.1) *on a subinterval* (a_0,b_0) *of* I. *If* $s \in (a_0,b_0)$ *and* $W(t)$ *is the solution of* (4.1) *satisfying* $W(s) = W_0(s) + \Gamma$, *then* $W(t)$ *exists and satisfies* $W(t) > W_0(t)$, $\{W(t) \geq W_0(t)\}$ *on* $[s,b_0)$ *if* $\Gamma > 0$, $\{\Gamma \geq 0\}$; *correspondingly*, $W(t)$ *exists and satisfies* $W(t) < W_0(t)$, $\{W(t) \leq W_0(t)\}$, *on* $(a_0,s]$ *if* $\Gamma < 0$, $\{\Gamma \leq 0\}$.

The hermitian Riccati matrix differential equation (4.1) may also be written as

$$W'(t) = (E;W(t))^*\mathscr{A}(t)(E;W(t)) + [W^*(t) - W(t)][A(t) + B(t)W(t)], \tag{4.9}$$

and, in particular, if $W(t)$ is an hermitian matrix solution of (4.1) we have

$$W'(t) = (E;W(t))^*\mathscr{A}(t)(E;W(t)). \tag{4.9_0}$$

Moreover, if $W(t)$ is an hermitian solution of (4.1) and $Y(t) = (U(t);V(t))$ is a conjoined basis for (3.1), then $U^*(t)W'(t)U(t) = Y^*(t)\mathscr{A}(t)Y(t)$. Also, since $\mathscr{J}Y'(t) = -\mathscr{A}(t)Y(t)$ and $\mathscr{A}(t) = \mathscr{A}^*(t)$, we have

$$Y^{*\prime}\mathscr{A}Y + Y^*\mathscr{A}Y' = [-Y^*\mathscr{A}\mathscr{J}]\mathscr{A}Y + Y^*\mathscr{A}[\mathscr{J}\mathscr{A}Y] = 0.$$

Consequently, whenever $\mathscr{A}(t)$ is locally a.c. on I_o we have the equation

$$\{U^*(t)W'(t)U(t)\}' = Y^*(t)\mathscr{A}'(t)Y(t)$$

and hence for $(t,\tau) \in I_o \times I_o$ we have

$$W'(t) = U^{*-1}(t)\left\{U^*(\tau)W'(\tau)U(\tau) + \int_\tau^t Y^*(s)\mathscr{A}'(s)Y(s)ds\right\}U^{-1}(t) \tag{4.10}$$

5. Normality and Abnormality

It is to be emphasized that some of the properties possessed by solutions of a system (2.1) with coefficients satisfying hypothesis ($\mathscr{A}_\Omega$) do not remain valid for solutions of systems (3.1) satisfying hypothesis ($\mathscr{A}$). In particular, for (3.1) there may exist solutions $y(t) = (u(t);v(t))$ with $u(t) \equiv 0$ on a non-degenerate subinterval I_o of I, but $v(t) \not\equiv 0$ on this subinterval. Clearly this phenomenon cannot occur for a system (3.1) in which the matrix $B(t)$ is non-singular a.e. on I; however, this latter condition is not satisfied in some very important cases.

For a non-degenerate subinterval I_o of I, let $\Lambda(I_o)$ denote the vector space of n-dimensional vector functions $v(t)$ which are solutions of $v'(t) + A^*(t)v(t) = 0$ and satisfy $B(t)v(t) = 0$ on I_o; it is to be remarked that in accordance with usage throughout this chapter, this later statement means that $B(t)v(t) = 0$ for t a.e. on I_o. Moreover, clearly $v \in \Lambda(I_o)$ if and only if $y(t) = (u(t) \equiv 0; v(t))$ is a solution of (3.1) on I_o. If $\Lambda(I_o)$ is zero-dimensional the equation (3.1) is said to be *normal* on I_o, or *the order of abnormality on* I_o *is zero*. If $\Lambda(I_o)$ has dimension $d = d(I_o) > 0$, the equation is said to be *abnormal*, with *order of abnormality* d *on* I_o. For $I_o = [a,b]$, the precise symbol $d([a,b])$ is reduced to $d[a,b]$, with similar contractions whenever I_o is of the form (a,b), $(a,b]$, or $[a,b)$. Clearly $0 \le d(I_o) \le n$, if I_o is a non-degenerate subinterval. Moreover, if $I_o \subset I'_o \subset I$, then $d(I_o) \ge d(I'_o)$. Also, if $I_o = [a,b]$ then an elementary continuity argument yields the result that $d[a,b] = d[a,b) = d(a,b] = d(a,b)$. If (3.1) is normal on every non-degenerate subinterval of I, then this equation is said to be *identically normal* on I. It is to be remarked that the system (3.1) specified by (2.16), and which is equivalent to the real self-adjoint differential equation (2.17) of order 2n, is identically normal, although the matrix $B(t)$ has all elements identically equal to zero on I except for the non-zero element $B_{nn}(t)$.

As in the case of the two-dimensional system (II.1.2) equivalent to the second order scalar equation (II.1.1), two distinct points t_1, t_2 of I are said to be *(mutually) conjugate* with respect to (3.1) if there exists a solution $y(t) = (u(t);v(t))$ of this system with $u(t) \not\equiv 0$ on the subinterval with end-points t_1 and t_2, while $u(t_1) = 0 = u(t_2)$. If no two distinct points of a subinterval I_o are conjugate with respect to (3.1), then this equation is said to be *disconjugate* on I_o. It is to be noted that in the discussion of "disconjugacy" some authors automatically restrict attention to identically normal systems. For example, Coppel [2] calls a system (3.1) disconjugate on an interval I_o whenever it admits no non-trivial solution $y(t) = (u(t);v(t))$ with $u(t)$ vanishing at two distinct points of I_o.

If $[a,b]$ is a compact non-degenerate subinterval of I, then the vector space of solutions $y(t) = (u(t);v(t))$ of (3.1) satisfying the end-conditions

$$u(a) = 0 = u(b) \tag{5.1}$$

is denoted by $\Omega_0[a,b]$. If $\kappa[a,b]$ is the dimension of $\Omega_0[a,b]$, then $\kappa[a,b]$ is the index of compatibility of the two-point boundary problem (3.1), (5.1). Also, $\kappa[a,b] \geq d[a,b]$, and $\kappa[a,b] > d[a,b]$ if and only if $t = a$ and $t = b$ are conjugate with respect to (3.1), in which case the positive integer $\kappa[a,b] - d[a,b]$ is the *order of* b *as a conjugate point to* a, and the *order of* a *as a conjugate point to* b.

Now the boundary problem (3.1), (5.1) is self-adjoint, and from the general solvability theorems of differential systems, (see, for example, Reid [35; Theorem III.6.2]), it follows that if $[a,b] \subset I$ and ξ^a, ξ^b are given n-dimensional vectors, then there exists a solution $y_0(t) = (u_0(t);v_0(t))$ of (3.1) satisfying $u_0(a) = \xi^a$, $u_0(b) = \xi^b$ if and only if

$$\begin{gathered} v^*(a)\xi^a - v^*(b)\xi^b = 0, \\ \text{for arbitrary } y(t) = (u(t);v(t)) \in \Omega_0[a,b]. \end{gathered} \tag{5.2}$$

For I_0 a non-degenerate subinterval of I, let $D(I_0)$ denote the set of vector functions $\eta(t)$ which are locally a.c. on I_0, and satisfy with a ζ locally of class $\mathscr{L}^\infty$ the differential equation

$$L_2[\eta;\zeta](t) \equiv \eta'(t) - A(t)\eta(t) - B(t)\zeta(t) = 0. \tag{5.3}$$

The fact ζ is then associated with η is denoted by

$\eta \in D(I_o):\zeta$. Also, if $I_o = [a,b]$ the symbol $D([a,b])$ is abbreviated to $D[a,b]$. Now if $v \in \Lambda(I_o)$, and $\eta \in D(I_o):\zeta$, the identity $0 = v^*L_2[\eta,\zeta] + (v'(t) + A^*(t)v(t))^*\eta = [v^*(t)\eta(t)]'$ implies that the function $v^*(t)\eta(t)$ is constant on I_o. Moreover, if $[a,b] \subset I$ and a and b are not mutually conjugate, then for given n-dimensional vectors ξ^a, ξ^b there exists a solution $y(t) = (u(t);v(t))$ of (3.1) satisfying $u(a) = \xi^a$, $u(b) = 0$, $\{u(a) = 0,\ u(b) = \xi^b\}$, if and only if $v^*(a)\xi^a = 0$, $\{v^*(b)\xi^b = 0\}$, for arbitrary $v \in [a,b]$.

If a point $c \in I$ is such that (3.1) is normal on every subinterval of I for which c is an end-point, and $Y(t;c) = (U(t,c);V(t;c))$ is the solution of (3.1_M) determined by the initial conditions $U(c;c) = 0$, $V(c;c) = E$, then a value $t_o \in I$, $t_o \neq c$, is conjugate to $t = c$, relative to (3.1), if and only if $U(t_o;c)$ is singular, and the order of t_o as a conjugate point to c is equal to $n - r(t_o)$, where $r(t_o)$ is the rank of $U(t_o;c)$. However, if there is a subinterval of I which has c as an end-point, and on which (3.1) is abnormal, then it is no longer true that the points conjugate to c are determined by the condition of singularity of $U(t;c)$.

If I_o is a subinterval of I such that on I_o the equation has a positive index of abnormality $d = d(I_o)$, then for a given $t_o \in I_o$ we shall denote by $\Delta = \Delta(t_o)$ an $n \times d$ matrix $V(t)$ such that the column vectors of $V(t)$ form a basis for $\Lambda(I_o)$. That is, $V'(t) + A^*(t)V(t) = 0$ and $B(t)V(t) = 0$ for $t \in I_o$, while $V(t)$ is such that an arbitrary element $v(t) \in \Lambda(I_o)$ has a unique representation of the form $v(t) = V(t)\xi$, where ξ is a d-dimensional vector. For brevity, this property is indicated by the symbol $\Delta(t_o) \sim \Delta(I_o)$. In particular, the column vectors of $\Delta(t_o)$ may be chosen to be mutually orthogonal, so that $\Delta^*(t_o)\Delta(t_o) = E_d$, and in case $\Delta(t_o)$ has been so selected we write $\Delta(t_o) \approx \Lambda(I_o)$. The following results are basic for the consideration of equations (3.1) which are abnormal, and are easily established from the definition of conjugate point.

LEMMA 5.1. *Suppose that* $[a,b]$ *is a non-degenerate compact subinterval of* I, *and* s *is a point of* $[a,b)$ *such that* $d[a,s] = d[a,b] = d$ *for* $c \in (s,b]$, *and let* $\Delta(a)$ *be an* $n \times d$ *matrix such that* $\Delta(a) \sim \Lambda[a,b]$, *while* N *is an* $n \times (n-d)$ *matrix such that* $[\Delta(a) \quad N]$ *is non-singular. For* $\alpha = 0,1,2,3$ *let* $Y_\alpha(t) = (U_\alpha(t);V_\alpha(t))$ *be solutions of* (3.1_M) *determined by the respective initial conditions*

$$\begin{aligned} Y_0(a) &= (0;\Delta(a)); \; Y_1(a) = (0;N); \\ Y_2(a) &= (\Delta(a);0), \; Y_3(a) = (N;0). \end{aligned} \tag{5.4}$$

Then a value $c \in (s,b]$ *is conjugate to* $t = a$ *relative to* (3.1) *if and only if one of the following conditions is satisfied:*

(i) $U_1(c)$ *has rank less than* $n - d$;

(ii) *the* $n \times n$ *matrix* $[U_2(c) \quad U_1(c)]$ *is singular;*

(iii) *the* $2n \times (2n-d)$ *matrix*

$$\begin{bmatrix} U_1(a) & U_2(a) & U_3(a) \\ U_1(c) & U_2(c) & U_3(c) \end{bmatrix} \tag{5.5}$$

has rank less than $2n - d$.

In particular, if $N^*\Delta(a) = 0$ *then the* $2n \times n$ *matrix* *function*

$$Y(t) = ([U_2(t) \quad U_1(t)]; \quad [V_2(t) \quad V_1(t)]) \tag{5.6}$$

is a conjoined basis for (3.1).

In general, abnormal differential systems (3.1) admit pathologies not present in normal differential systems, and even when the theory is in substantial agreement with that for normal problems the proofs involve details which are complicated in nature. An illustration of this fact is afforded by the above lemma, and more details on abnormal cases are to be found in references [20; 21; 35, Ch. VII, Secs. 3,4,5; 38, Ch. II, Secs. 6,7 and Ch. IV, Sec. 8] of the author. Consequently, throughout the present exposition specific discussion of results will be limited almost entirely to differential systems which possess a property of normality. It is to be noted, however, that in case (3.1) has a fixed order of abnormality $d > 0$ on all non-degenerate subintervals I_0 of I then the determination of solutions $y(t) = (u(t);v(t))$ of (3.1) for which the component vector function $u(t)$ vanishes for some value on I is reducible to the determination of like solutions of a similar system in $(n-d)$-dimensional vector functions.

In general, if $T(t)$ is an $n \times n$ matrix function which is locally a.c. on I, and $y^0(t) = (u^0(t);v^0(t))$ is related to $y(t) = (u(t);v(t))$ by the transformation

$$u^0(t) = T^{-1}(t)u(t), \quad v^0(t) = T^*(t)v(t) \tag{5.7}$$

then the component matrix differential expressions $L_1[u,v]$, $L_2[u,v]$ of (2.5) satisfy the relations

$$\begin{aligned} L_1[u,v](t) &= T^{*-1}(t)L_1^o[u^o,v^o](t), \\ L_2[u,v](t) &= T(t)L_2^o[u^o,v^o](t) \end{aligned} \tag{5.8}$$

where

$$\begin{aligned} L_1^o[u^o,v^o](t) &\equiv -v^{o\prime}(t) + C^o(t)u^o(t) - A^o(t)v^o(t) \\ L_2^o[u^o,v^o](t) &\equiv u^{o\prime}(t) - A^o(t)u^o(t) - B^o(t)v^o(t) \end{aligned} \tag{5.9}$$

and the coefficient matrix functions A^o, B^o, C^o are defined as

$$A^o = T^{-1}(AT - T'), \quad B^o = T^{-1}BT^{*-1}, \quad C^o = T^*CT. \tag{5.10}$$

It is to be noted that if $y_\alpha(t) = (u_\alpha(t);v_\alpha(t))$, $(\alpha = 1,2)$, are solutions of (3.1) and $y_\alpha^o(t) = (u_\alpha^o(t);v_\alpha^o(t))$ are the associated solutions of (5.9) then $\{y_1,y_2\} = \{y_1^o,y_2^o\}$; in particular, y_1 and y_2 are conjoined solutions of (3.1) if and only if y_1^o and y_2^o are conjoined solutions of the system

$$L_1^o[u^o,v^o](t) = 0, \quad L_2^o[u^o,v^o](t) = 0. \tag{5.11}$$

Moreover, if $Y(t) = (U(t):V(t))$ is a solution of (3.1_M) with $U(t)$ non-singular on a subinterval I_o then $Y^o(t) = (U^o(t);V^o(t)) = (T^{-1}(t)U(t);T^*(t)V(t))$ is a solution of the matrix differential system (5.11_M) corresponding to the vector system (5.11) with $U^o(t)$ non-singular on I_o and $V^o(t)U^{o-1} = T^*(t)V(t)U^{-1}(t)T(t)$. Consequently, $W(t) = V(t)U^{-1}(t)$ is a solution of the Riccati matrix differential equation (4.1) if and only if $W^o(t) = T^*(t)W(t)T(t)$ is a solution of the Riccati matrix differential equation

$$\begin{aligned} K^o[W^o](t) \equiv W^{o\prime}(t) &+ W^o(t)A^o(t) + A^{o*}(t)W^o(t) \\ &+ W^o(t)B^o(t)W^o(t) - C^o(t) = 0. \end{aligned} \tag{5.12}$$

In particular, $W(t)$ is an hermitian solution of (4.1) if and only if $W^o(t) = T^*(t)W(t)T(t)$ is an hermitian solution of (5.12).

If $Z(t)$ is a fundamental matrix solution of $Z'(t) + A^*(t)Z(t) = 0$, then $T(t) = Z^{*-1}(t)$ is a fundamental matrix solution of the equation $T'(t) - A(t)T(t) = 0$ and with this choice of $T(t)$ the matrices of (5.10) become

$$A^o(t) \equiv 0, \quad B^o(t) = Z^*(t)B(t)Z(t), \\ C^o(t) = Z^{-1}(t)C(t)Z^{*-1}(t). \tag{5.13}$$

In the terminology used by Reid [29; 38, Sec. III.4] such a $T(t)$ is called a *reducing transformation* of (3.1), and the resulting system (5.11) is termed a *reduced system*. In case (3.1) has order of abnormality equal to d on a subinterval I_o of I, then the fundamental matrix $Z(t)$ of $Z'(t) + A^*(t)Z(t) = 0$ may be so chosen that the last d columns of $Z(t)$ provide a basis for $\Lambda(I_o)$, and with $Z(t)$ thus chosen the transformation $T(t) = Z^{*-1}(t)$ provides for $B^o(t)$ of (5.10) the form $B^o(t) = \operatorname{diag}\{\hat{B}(t);0\}$ on I_o, where $\hat{B}(t)$ is an $(n-d) \times (n-d)$ hermitian matrix function. Again, in the terminology of Reid [29; 38, Sec. III.4], a $T(t)$ of this particular form is called a *preferred reducing transformation* for (3.1). If $C^o(t) = T^*(t)C(t)T(t)$ is written as

$$C^o(t) = \begin{bmatrix} \hat{C}_{11}(t) & \hat{C}_{12}(t) \\ \hat{C}_{21}(t) & \hat{C}_{22}(t) \end{bmatrix}.$$

where $\hat{C}_{11}(t)$ is $(n-d) \times (n-d)$ and hermitian, $\hat{C}_{12}(t) = \hat{C}^*_{21}(t)$ is $(n-d) \times d$, and $\hat{C}_{22}(t)$ is $d \times d$ and hermitian, then in terms of the vector functions $\eta(t) = (u_i(t))$,

$\zeta(t) = (v_i(t))$, $(i = 1,\ldots,n-d)$, and $\rho(t) = (u_{n-d+j}(t))$, $\sigma(t) = (v_{n-d+j}(t))$, $(j = 1,\ldots,d)$, for $t \in I_0$ the vector differential system (3.1) may be written

$$\begin{aligned} -\zeta'(t) + \hat{C}_{11}(t)\eta(t) + \hat{C}_{12}(t)\rho(t) &= 0, \\ -\sigma'(t) + \hat{C}_{21}(t)\eta(t) + \hat{C}_{22}(t)\rho(t) &= 0, \\ \eta(t) - \hat{B}(t)\zeta(t) &= 0, \\ \rho'(t) &= 0. \end{aligned} \tag{5.14}$$

Consequently, if $u^o(t) = (\eta(t);\rho(t))$, $v^o(t) = (\zeta(t);\sigma(t))$ is a solution of (5.14) with $u^o(\tau) = 0$ for some $\tau \in I$ then $\rho(t) \equiv 0$ and $(\eta(t);\zeta(t))$ is a solution of the system

$$\begin{aligned} -\zeta'(t) + C_{11}(t)\eta(t) &= 0, \\ \eta'(t) - B(t)\zeta(t) &= 0, \end{aligned} \tag{5.15}$$

system (5.15), which has been called the truncated preferred reduced system, is normal on I_0. Consequently, in case the original system (3.1) has the same order of abnormality d on all non-degenerate subintervals of I, then (5.15) is identically normal on I.

6. Variational Properties of Solutions of (3.1)

As in the preceding sections, we shall consider a vector differential system whose coefficient functions satisfy hypothesis ($\mathscr{A}$). For a given compact subinterval $[a,b]$ of I, we shall denote by $D[a,b]$ the class of n-dimensional vector functions which are absolutely continuous and for which there exists a $\zeta \in \mathscr{L}^\infty[a,b]$ such that $L_2[\eta,\zeta](t) \equiv \eta'(t) - A(t)\eta(t) - B(t)\zeta(t) = 0$ on $[a,b]$. The subclass of $D[a,b]$ on which $\eta(a) = 0 = \eta(b)$ is denoted by $D_0[a,b]$.

Corresponding to the notation of Section II.8, the fact that $\eta(t)$ is a member of $D[a,b]$ with an associated $\zeta(t)$ is indicated by the symbol $\eta \in D[a,b]:\zeta$ and $\eta \in D_0[a,b]:\zeta$, respectively.

If for $\alpha = 1,2$ the functions $\eta_\alpha(t)$, $\zeta_\alpha(t)$ belong to $\mathscr{L}^\infty[a,b]$, the symbol $J[\eta_1:\zeta_1,\eta_2:\zeta_2;a,b]$ is used to denote the functional

$$J[\eta_1:\zeta_1,\eta_2:\zeta_2;a,b] = \int_a^b \{\zeta_2^*(t)B(t)\zeta_1(t)+\eta_2^*(t)C(t)\eta_1(t)\}dt. \tag{6.1}$$

Since $B(t)$ and $C(t)$ are hermitian matrix functions of class $\mathscr{L}[a,b]$, (6.1) defines an hermitian form on $\mathscr{L}^\infty[a,b] \times \mathscr{L}^\infty[a,b]$.

It is to be noted that if $\eta_\alpha \in D[a,b]:\zeta_\alpha$, $(\alpha = 1,2)$, then in general the vector functions ζ_α are not uniquely determined. The value of the functional is independent of the choice of the ζ_α's, however, and consequently for such η_α's the symbol for this functional is reduced to $J[\eta_1,\eta_2;a,b]$. Also, for $\eta \in D[a,b]:\zeta$ the symbol $J[\eta,\eta;a,b]$ is contracted to $J[\eta;a,b]$. Corresponding to (II.8.4), (II.8.5) and (II.8.6) we now have the following relations.

$$J[\eta_1,\eta_2;a,b] = \eta_2^*\zeta_1\Big|_a^b + \int_a^b \eta_2^*L_1[\eta_1,\zeta_1]dt, \qquad \text{if } \eta_\alpha \in D[a,b]:\zeta_\alpha,\ (\alpha = 1,2); \tag{6.2}$$

$$J[\eta;a,b] = \eta^*\zeta\Big|_a^b + \int_a^b \eta^*L_1[\eta,\zeta]dt, \text{ if } \eta \in D[a,b]:\zeta; \tag{6.3}$$

$$\int_a^b \eta_2^* L_1[\eta_1,\zeta_1]dt - \int_a^b (L_1[\eta_2,\zeta_2])^*\eta_1 dt$$
$$= [\zeta_2^*\eta_1 - \eta_2^*\zeta_1]\Big|_a^b = \{(\eta_1;\zeta_1),(\eta_2;\zeta_2)\}\Big|_a^b, \quad (6.4)$$
$$\textit{if} \quad \eta_\alpha \in D[a,b]:\zeta_\alpha, \qquad (\alpha = 1,2).$$

In particular, if $a \le t_1 \le t_2 \le b$, and the values t_1 and t_2 are conjugate with respect to (3.1), then there exists a solution $y(t) = (u(t);v(t))$ of (3.1) with $u(t) \not\equiv 0$ on $[t_1,t_2]$ and $u(t_1) = 0 = u(t_2)$. Then $(\eta(t);\zeta(t)) = (u(t);v(t))$ for $t \in [t_1,t_2]$, and $(\eta(t);\zeta(t)) = (0;0)$ for $t \in [a,b_1) \cup (t_2,b]$ is such that $\eta \in D_0[a,b]:\zeta$, and with the aid of (6.3) it follows that $J[\eta;a,b] = J[u;t_1,t_2] = u^*v\Big|_{t_1}^{t_2} = 0$. That is, if $[a,b] \subset I$ and there exists a pair of values on $[a,b]$ which are conjugate with respect to (3.1), then there exists an $\eta \in D_0[a,b]$ such that $\eta(t) \not\equiv 0$ on $[a,b]$ and $J[\eta;a,b] = 0$.

Corresponding to Theorem II.8.1, we now have that if $[a,b] \subset I$ then an n-dimensional vector function $u(t)$ has an associated vector function $v(t)$ such that $(u(t);v(t))$ is a solution of (3.1) on $[a,b]$ if and only if $u(t)$ is a.c. and there exists a $v_1 \in \mathscr{L}^\infty[a,b]$ such that $u \in D[a,b]:v_1$ and

$$J[u:v_1,\eta:\zeta;a,b] = 0 \quad \text{for arbitrary } \eta \in D_0[a,b]:\zeta. \quad (6.5)$$

The result of this theorem is somewhat more difficult to establish than the special instance of Theorem II.8.1, and for its proof reference is made to Theorem VII.4.1 of Reid [35]. In particular, it is to be emphasized that in general

(6.5) does not imply that $(u;v_1)$ is a solution of (3.1); however, an associated v such that $(u(t);v(t))$ is a solution of (3.1) satisfies the condition $B[v-v_1] = 0$, where it is to be recalled that this latter equation means that $B(t)[v(t) - v_1(t)] = 0$ for t a.e. on $[a,b]$. Corresponding to Corollary 1 to Theorem II.8.1 we have that if $J[\eta;a,b]$ is non-negative on $D_0[a,b]$, and u is an element of $D_0[a,b]$ satisfying $J[u;a,b] = 0$, then there exists a vector function $v(t)$ such that $(u;v)$ is a solution of (3.1) on $[a,b]$; in particular, if $u(t) \not\equiv 0$ on $[a,b]$ then a and b are conjugate values, relative to (3.1). Also, the analogue of Corollary 2 to Theorem II.8.1 holds, to the effect that if $J[\eta;a,b]$ is non-negative on $D_0[a,b]$, and $(u;v)$ is a solution of (3.1), then for $u_0 \in D[a,b]$ with $u_0(a) = u(a)$, $u_0(b) = u(b)$ we have $J[u_0;a,b] \geq J[u;a,b]$; moreover, if $J[\eta;a,b]$ is positive definite on $D_0[a,b]$ then $J[u_0;a,b] > J[u;a,b]$ unless $u_0(t) \equiv u(t)$ on $[a,b]$.

The results of the following theorem and its corollary are analogues of those presented in Theorem II.8.2 and its corollary, although the context of the identity (6.6) is relatively more general than that of (II.8.11) and (II.8.12). In particular, the Clebsch-von-Escherich transformation of the second variational functional, which was alluded to in Section I.2, is embodied in the given identity.

THEOREM 6.1. *Suppose that* $[a,b] \subset I$ *and* $U(t),V(t)$ *are* $n \times k$ a.c. *matrix functions on* $[a,b]$. *If for* $\alpha = 1,2$ *the vector functions* $\eta_\alpha(t)$ *are* a.c. *on* $[a,b]$, *and the vector functions* $\zeta_\alpha(t)$ *are of class* $\mathcal{L}^\infty[a,b]$, *while there exist* a.c. k-*dimensional vector functions* $h_\alpha(t)$

such that $\eta_\alpha(t) = U(t)h_\alpha(t)$ *on* $[a,b]$, *then on* $[a,b]$ *we have the identity*

$$\begin{aligned}\zeta_2^*B\zeta_1 + \eta_2^*C\eta_1 = &\{\zeta_2 - Vh_2\}^*B\{\zeta_1 - Vh_1\} \\ &- h_2^*V^*L_2[\eta_1,\zeta_1] - (L_2[\eta_2,\zeta_2])^*Vh_1 \\ &+ h_2^*\{V^*L_2[U,V] + U^*L_1[U,V]\}h_1 \\ &- h_2^*\{U^*V - V^*U\}h_1' + \{h_2^*U^*Vh_1\}'.\end{aligned} \tag{6.6}$$

COROLLARY. *If* $[a,b] \subset I$, *and the column vectors of* $Y(t) = (U(t);V(t))$ *form a basis for an* r-*dimensional conjoined family of solutions of* (3.1), *while* $\eta \in D[a,b]:\zeta$ *and there exists an* r-*dimensional* a.c. *vector function* h *such that* $\eta(t) = U(t)h(t)$ *for* $t \in [a,b]$, *then*

$$J[\eta;a,b] = \eta^*Vh\Big|_a^b + \int_a^b [\zeta - Vh]^*B[\zeta - Vh]dt. \tag{6.7}$$

As a ready consequence of (6.7) one has the following basic result.

THEOREM 6.2. *If* $[a,b] \subset I$, $B(t) \geq 0$ *for* t a.e. *on* $[a,b]$, *and there exists a conjoined basis* $Y(t) = (U(t):V(t))$ *for* (3.1) *with* U(t) *non-singular on* $[a,b]$, *then for* $\eta \in D_0[a,b]:\zeta$ *and* $h(t) = U^{-1}\eta(t)$ *we have*

$$J[\eta;a,b] = \int_a^b [\zeta - Vh]^*B[\zeta - Vh]dt \tag{6.8}$$

and $J[\eta;a,b]$ *is positive definite on* $[a,b]$.

If $[a,b] \subset I$, and $J[\eta;a,b]$ is non-negative on $D_0[a,b]$, then $B(t) \geq 0$ for t a.e. on $[a,b]$. This result is essentially the Clebsch condition for variational problems of Lagrange or Bolza type, which is the analogue of the Legendre condition for simpler variational problems of the sort discussed in Section I.2. For a proof of this result under hypotheses of the sort described above, see Reid [12, Theorem 2.1; 35, Theorem VII.4.2].

The following theorem presents a result that corresponds to Theorem II.8.6 for the scalar second order equation.

THEOREM 6.3. *If* $[a,b] \subset I$, *and* $B(t) \geq 0$ *for* t a.e. *on* I, *then the following conditions are equivalent:*

(i) $J[\eta;a,b]$ *is positive definite on* $D_0[a,b]$;

(ii) (3.1) *is disconjugate on* $[a,b]$;

(iii) *there is no point on* $(a,b]$ *conjugate to* $t = a$;

(iv) *there is no point on* $[a,b)$ *conjugate to* $t = b$;

(v) *there exists a conjoined basis* $Y(t) = (U(t);V(t))$ *of* (3.1) *with* $U(t)$ *non-singular on the closed interval* $[a,b]$;

(vi) *there exists an a.c.* $n \times n$ *hermitian matrix function* $W(t)$, $t \in [a,b]$ *which is a solution of the Riccati matrix differential equation* (4.1).

In the general case, the results of the above theorem are a combination of those given in Theorems VII.4.4, VII.4.5 and its Corollary of Reid [35], and presented as Theorem VII.5.1 of that reference. However, it is to be noted that relatively elementary proofs, which are direct analogues of those for the simpler problem of Section II.8, are available under the following normality condition.

($\mathscr{L}_N$) *For* $[a,b]$ *a given subinterval of* I, *system* (3.1) *is normal on each subinterval* $[a,b']$ *with* $b' \in (a,b]$, *and on each subinterval* $[a',b]$ *with* $a' \in [a,b)$.

When ($\mathscr{L}_N$) holds, conditions (iii) and (iv) of Theorem 6.3 are equivalent to the following respective conditions:

(iii') *If* $Y_a(t) = (U_a(t);V_a(t))$ *is a conjoined basis for* (3.1) *determined by the initial conditions* $U_a(a) = 0$, $V_a(a)$ *non-singular, then* $U_a(t)$ *is non-singular for* $t \in (a,b]$.

(iv') *If* $Y_b(b) = (U_b(t);V_b(t))$ *is a conjoined basis for* (3.1) *determined by the initial conditions* $U_b(b) = 0$, $V_b(b)$ *non-singular, then* $U_b(t)$ *is non-singular for* $t \in [a,b)$.

Under the strengthened hypothesis of ($\mathscr{L}_N$), the proof of Theorem 6.3 may be presented as follows: (v) implies (i) by Theorem 6.2; (i) implies (ii) by the last sentence of the paragraph following relation (6.4); (ii) obviously implies (iii'); (iii') implies (iv') with the aid of (v) applied to subintervals $[a',b]$ where $a' \in (a,b)$, and the fact that the non-singularity of $U_a(b)$ implies the non-singularity of $U_b(a)$; (iii') and (iv') imply (v) by noting that a suitable choice of $V_a(a)$ and/or a suitable choice of $V_b(b)$ provides conjoined families $Y_a(t)$ and $Y_b(t)$ satisfying $\{Y_a,Y_b\} \equiv -E$, and with this choice $Y(t) = (U_a(t) + U_b(t); V_a(t) + V_b(t))$ is a conjoined basis for (3.1). Details of this proof are to be found in Reid [12, Theorem 2.1] or Reid [35, Hint to Prob. VII.4.1]. Finally, the equivalence of (v) and (vi) is a direct consequence of equation (4.2).

Corresponding to the result of Theorem II.8.5, one may also show that if $[a,b] \subset I$, $B(t) \geq 0$ for t a.e. on $[a,b]$, and there exists a conjoined basis $Y(t) = (U(t);V(t))$ of (3.1) with $U(t)$ non-singular on the open interval (a,b), then $J[\eta;a,b]$ is non-negative on $D_0[a,b]$, and if $\eta \in D_0[a,b]:\zeta$ is such that $J[\eta;a,b] = 0$ then there exists a constant vector ξ such that $\eta(t) = U(t)\xi$ on $[a,b]$ and

$\zeta(t) - V(t)\xi$ belongs to the abnormality space $\Lambda[a,b]$. In particular, if condition $(\mathscr{C}_N)$ holds then $\zeta(t) = V(t)\xi$ on $[a,b]$. For a proof of this result, see Reid [35, Hint to Prob. VII.4.2], together with Reid [42, 7(a,b) of Sec. 2].

It is to be noted that if condition $(\mathscr{C}_N)$ is not assumed, then whenever $B(t) \geq 0$ a.e. on $[a,b]$ and $J[\eta;a,b]$ is non-negative on $D_0[a,b]$ it is not true in general that there exists a conjoined basis $Y(t) = (U(t);V(t))$ for (3.1) with $U(t)$ non-singular on the open interval (a,b). As shown in Reid [42, Sec. 2], this fact is illustrated by a system (3.1) with $[a,b] = [-\pi,\pi]$, $n = 1$, $A(t) \equiv 0$ and $(B(t),C(t)) = (0,0)$ for $t \in [-\pi, -\frac{1}{2}\pi) \cup (\frac{1}{2}\pi,\pi]$, $(B(t),C(t)) \equiv (1,-1)$ for $t \in [-\frac{1}{2}\pi,\pi]$.

In accord with the terminology of Section II.8 and the first paragraph of the present section, the subclass of $D[a,b]$ on which $\eta(b) = 0$ is denoted by $D_{*0}[a,b]$, and the subclass of $D[a,b]$ on which $\eta(a) = 0$ is designated by $D_{0*}[a,b]$; in particular, $D_0[a,b] = D_{0*}[a,b] \cap D_{*0}[a,b]$. Also, the fact that $\eta(t)$ belongs to $D_{0*}[a,b]$ or $D_{*0}[a,b]$ with an associated $\zeta(t)$ is indicated by the symbol $\eta \in D_{0*}[a,b]:\zeta$ or $\eta \in D_{*0}[a,b]:\zeta$. Corresponding to notations of Section II.8 we set

$$
\begin{aligned}
J_a[\eta_1:\zeta_1,\eta_2:\zeta_2:a,b] &= \eta_2^*(a)\Gamma_a\eta_1(a) \\
&\quad + \int_a^b \{\zeta_2^*(t)B(t)\zeta_1(t) + \eta_2(t)C(t)\eta_1(t)\}dt \qquad (6.8) \\
&= \eta_2^*(a)\Gamma_a\eta_1(a) + J[\eta_1:\zeta_1,\eta_2:\zeta_2:a,b]
\end{aligned}
$$

and

$$
J_b[\eta_1:\zeta_1,\eta_2:\zeta_2:a,b] = \eta_2^*(b)\Gamma_b\eta_1(b)+J[\eta_1:\zeta_1,\eta_2:\zeta_2:a,b], \qquad (6.9)
$$

where the coefficient functions $A(t)$, $B(t)$, $C(t)$ satisfy hypothesis ($\mathscr{A}$) and Γ_a, Γ_b are $n \times n$ hermitian matrices. If $\eta_\alpha \in D[a,b]:\zeta_\alpha$, $(\alpha = 1,2)$, then the values of (6.8) and (6.9) are independent of the particular ζ_α associated with η_α, and the symbols of these functionals are abbreviated to $J_a[\eta_1,\eta_2;a,b]$ and $J_b[\eta_1,\eta_2;a,b]$, and also $J_a[\eta;a,b] = J_a[\eta,\eta;a,b]$, $J_b[\eta;a,b] = J_b[\eta,\eta;a,b]$.

Analogous to Theorem II.8.7, we now have the following result.

THEOREM 6.4. *If* $[a,b] \subset I$, *then an* n*-dimensional* a.c. *vector function* $u(t)$ *has an associated vector function* $v(t)$ *such that* $(u(t);v(t))$ *is a solution of* (3.1) *on* $[a,b]$ *which satisfies the initial condition*

$$\Gamma_a u(a) - v(a) = 0, \tag{6.10a}$$

$$\{\Gamma_b u(b) + v(b) = 0\} \tag{6.10b}$$

if and only if there exists a $v_1(t) \in \mathscr{L}^\infty[a,b]$ *such that* $u \in D[a,b]:v_1$ *and* $J[u:v_1,\eta:\zeta;a,b] = 0$ *for* $\eta \in D_{*0}[a,b]:\zeta$ $\{J[u:v_1,\eta:\zeta;a,b] = 0$ *for* $\eta \in D_{0*}[a,b]:\zeta\}$. *Moreover, if* $J_a[\eta;a,b]$ *is non-negative on* $D_{*0}[a,b]$, $\{J_b[\eta;a,b]$ *is non-negative on* $D_{0*}[a,b]\}$, *and* u *is an element of* $D_{*0}[a,b]$, $\{D_{0*}[a,b]\}$, *such that* $J_a[u;a,b] = 0$, $\{J_b[u;a,b] = 0\}$, *then there exists a vector function* $v(t)$ *such that* $(u;v)$ *is a solution of* (3.1) *satisfying the boundary conditions*

$$\Gamma_a u(a) - v(a) = 0, \quad u(b) = 0, \tag{6.11a}$$

$$\{u(a) = 0, \quad \Gamma_b u(b) + v(b) = 0\}. \tag{6.11b}$$

Relative to the functional (6.9), or relative to the differential system (3.1) with initial condition (6.10a), a value $\tau > a$ on I is a *right-hand focal point* to $t = a$

if there exists a solution $(u(t);v(t))$ of (3.1) satisfying condition (6.10a) and $u(b) = 0$, while $u(t) \not\equiv 0$ on $[a,b]$. Correspondingly, relative to the functional (6.9), or relative to the differential system (3.1) with initial condition (6.10b), a value $\tau < b$ on I is a *left-hand* focal point to $t = b$ in case there exists a solution $(u(t);v(t))$ of (3.1) satisfying condition (6.10b) and $u(a) = 0$, while $u(t) \not\equiv 0$ on $[a,b]$.

Corresponding to Theorem II.8.8, we have the following result.

THEOREM 6.5. *If* $[a,b] \subset I$ *and* $B(t) \geq 0$ *for* t a.e. *on* $[a,b]$, *then* $J_a[\eta;a,b]$ *is non-negative on* $D_{*0}[a,b]$ $\{J_b[\eta;a,b]$ *is non-negative on* $D_{0*}[a,b]\}$ *if and only if the solution* $Y(t) = (U(t);V(t))$ *of* (3.1_M) *determined by the initial conditions*

$$U(a) = E, \quad \Gamma_a - V(a) = 0, \tag{6.12a}$$

$$\{U(b) = E, \quad \Gamma_b + V(b) = 0\} \tag{6.12b}$$

is such that $U(t)$ *is non-singular for* $t \in [a,b)$, $\{t \in (a,b]\}$. *If* $U(b)$ *is non-singular,* $\{U(a)$ *is non-singular*$\}$, *then for* $\eta \in D_{*0}[a,b]:\zeta$, $\{\eta \in D_{0*}[a,b]:\zeta\}$, *and* $h(t) = U^{-1}(t)\eta(t)$ *we have* $J_a[\eta;a,b]$, $\{J_b[\eta;a,b]\}$ *given by the integral* (6.8) *and* $J_a[\eta;a,b]$ *is positive definite on* $D_{*0}[a,b]$, $\{J_b[\eta;a,b]$ *is positive definite on* $D_{0*}[a,b]\}$. *If* $U(b)$ *is singular,* $\{U(a)$ *is singular*$\}$, *however, then* $J_a[\eta;a,b] \geq 0$ *for* $\eta \in D_{*0}[a,b]$, $\{J_b[\eta;a,b] \geq 0$ *for* $\eta \in D_{0*}[a,b]\}$, *and the equality sign holds if and only if there is a constant* n-*dimensional vector* ξ *such that* $\eta(t) = U(t)\xi$ *on* $[a,b]$ *and* $\zeta(t) - V(t)\xi$ *belongs to* $\Lambda[a,b]$.

7. Comparison Theorems

Now consider two systems (3.1^{α}), $(\alpha = 1,2)$, involving $n \times n$ matrix functions A^{α}, B^{α}, C^{α} which satisfy hypothesis $(\mathscr{A})$ on a given interval I on the real line. If $I_o \subset I$ the corresponding classes $D(I_o)$, $D_{*o}(I_o)$, $D_{o*}(I_o)$, $D_o(I_o)$ will be denoted by $D^{\alpha}(I_o)$, $D^{\alpha}_{*o}(I_o)$, $D^{\alpha}_{o*}(I_o)$, $D^{\alpha}_o(I_o)$. Also, for the two problems the classes $\Lambda(I_o)$ will be designated by $\Lambda^{\alpha}(I_o)$, $(\alpha = 1,2)$. In the following theorems we shall be concerned with two systems for which $D^1(I_o) = D^2(I_o)$, and hence also $D^1_{*o}(I_o) = D^2_{*o}(I_o)$, $D^1_{o*}(I_o) = D^2_{o*}(I_o)$ and $D^1_o(I_o) = D^2_o(I_o)$. The class of vector functions $D^1[a,b]$ and $D^2[a,b]$ are clearly equal for arbitrary compact intervals $[a,b] \subset I$ if $A^1(t) \equiv A^2(t)$ and $B^1(t) \equiv B^2(t)$ on I, but are also equal under more general conditions since whenever a vector function η is such that $\eta \in D^1[a,b]:\zeta_1$ and $\eta \in D^2[a,b]:\zeta_2$ it is not required that $\zeta_1 = \zeta_2$. For example, if $B^1(t)$ and $B^2(t)$ are non-singular on $[a,b]$, and each of the matrix functions $(B^2)^{-1}B^1$, $(B^2)^{-1}(A^1 - A^2)$, $(B^1)^{-1}B^2$, and $(B^1)^{-1}(A^2 - A^1)$ is of class $\mathscr{L}^{\infty}[a,b]$, then $\eta \in D^1[a,b]:\zeta_1$ if and only if $\eta \in D^2[a,b]:\zeta_2$, where

$$\zeta_2 = (B^2)^{-1}[B^1\zeta_1 + (A^1 - A^2)\eta],$$
$$\zeta_1 = (B^1)^{-1}[B^2\zeta_2 + (A^2 - A^1)\eta].$$

In particular, for a system (3.1) equivalent under the substitution (2.4) to a second order matrix differential system (2.1) with continuous coefficient matrix functions P, Q, R satisfying $(\mathscr{A}_{\omega})$, the class $D(I)$ consists of vector functions $\eta(t)$ which on arbitrary compact subintervals $[a,b]$ of I_o are Lipschitzian. Similarly, for a system (3.1) equivalent under the substitution (2.12) to a system (2.11) with continuous coefficient matrix functions P, Q, R, ϕ, θ satisfying ($\mathscr{A}_\Omega$), the class $D(I_o)$ consists of vector functions which on arbitrary compact subintervals of I_o are Lipschitzian and satisfy $\Phi(t,\eta(t),\eta'(t)) = 0$.

For $\alpha = 1,2$, we have the corresponding functionals

$$\begin{aligned} &(a) \quad J^\alpha[\eta;a,b] = \int_a^b \{\zeta^*(t)B^\alpha(t)\zeta(t) + \eta^*(t)C(t)\eta(t)\}dt \\ &(b) \quad J_a^\alpha[\eta;a,b] = \eta^*(a)\Gamma_a^\alpha\eta(a) + J^\alpha[\eta;a,b], \\ &(c) \quad J_b^\alpha[\eta;a,b] = \eta^*(b)\Gamma_b^\alpha\eta(b) + J^\alpha[\eta;a,b]. \end{aligned} \tag{7.1}$$

In particular, if $D^1(I) = D^2(I)$ and $[a,b] \subset I$, the difference functionals

$$\begin{aligned} &(a) \quad J^{1,2}[\eta;a,b] = J^1[\eta;a,b] - J^2[\eta;a,b] \\ &(b) \quad J_a^{1,2}[\eta;a,b] = \eta^*(a)[\Gamma_a^1 - \Gamma_a^2]\eta(a) + J^{1,2}[\eta;a,b], \\ &(c) \quad J_b^{1,2}[\eta;a,b] = \eta^*(b)[\Gamma_b^1 - \Gamma_b^2]\eta(b) + J^{1,2}[\eta;a,b], \end{aligned} \tag{7.2}$$

are well-defined for $\eta \in D^1[a,b] = D^2[a,b]$. It is to be emphasized, however, that in the evaluation of any of the functionals of (7.1) or (7.2), for $\alpha = 1,2$ the integrand of the functional $J^\alpha[\eta;a,b]$ involves (η,ζ^α), where ζ^α is a vector function such that $\eta \in D^\alpha[a,b]:\zeta^\alpha$.

Corresponding to Theorem II.1.1 for the case of scalar differential equations, we now have the following basic comparison theorem.

THEOREM 7.1. *Suppose that for* $\alpha = 1,2$ *the matrix functions* $A^\alpha(t)$, $B^\alpha(t)$, $C^\alpha(t)$ *satisfy hypothesis* ($\mathscr{A}$), *while* $B^\alpha(t) \geq 0$ *for* t a.e. *on* I, *and* $D^1(I) = D^2(I)$. *If for a given compact subinterval* $[a,b]$ *of* I *we have that*

$J^{1,2}[\eta;a,b]$ *is non-negative on* $D_0^1[a,b] = D_0^2[a,b]$, *and* $t = b$ *is the first right-hand conjugate point* $t = a$ *for* (3.1^1), *then for a conjoined basis* $Y^2(t) = (U^2(t);V^2(t))$ *of* (3.1^2) *exactly one of the following conditions holds:*

(i) *there exists an* $s \in (a,b)$ *such that* $U^2(s)$ *is singular;*

(ii) $U^2(t)$ *is non-singular for* $t \in (a,b)$ *in which case* $t = b$ *is also the first right-hand conjugate point to* $t = a$ *for* (3.1^2), *and if* $(u;v)$ *is a solution of* (3.1^1) *which determines* $t = b$ *as conjugate to* $t = a$ *with respect to* (3.1^1), *then there is a* v_2 *such that* $(u;v_2)$ *is a solution of* (3.1^2) *which determines* $t = b$ *as conjugate to* $t = a$ *with respect to* (3.1^2); *moreover, there exists a non-zero* n-*dimensional vector* ξ *such that* $u(t) = U^2(t)\xi$ *on* $[a,b]$ *and* $v_2 - V^2\xi \in \Lambda^2[a,b]$.

In particular, if $J^{1,2}[\eta;a,b]$ *is positive definite on* $D_0^1[a,b] = D_0^2[a,b]$, *then conclusion* (i) *holds.*

THEOREM 7.2. *If the coefficient functions of* (3.1) *satisfy hypothesis* $(\mathscr{A})$, *while* $[a,b] \subset I$ *and* $B(t) \geq 0$ *for* t a.e. *on* $[a,b]$, *then this system is disconjugate on* $[a,b]$ *if and only if one of the following conditions holds:*

(1^0) *there exists on* $[a,b]$ *a non-singular* $n \times n$ *matrix function* $U(t)$ *such that* $U \in D[a,b]:V$ *with an* a.c. *matrix function* $V(t)$, *while* $V^*(t)U(t) - U^*(t)V(t) \equiv 0$ *and* $U^*(t)L_1[U,V](t) \geq 0$ *for* t a.e. *on* $[a,b]$;

(2^0) *there exists an* $n \times n$ *hermitian matrix function* $W(t)$ *which is* a.c. *and satisfies* $K[W](t) \leq 0$ *for* t a.e. *on* $[a,b]$.

THEOREM 7.3. *Suppose that for* $\alpha = 1,2$ *the matrix func-*

tions $A^{\alpha}(t)$, $B^{\alpha}(t)$, $C^{\alpha}(t)$ *satisfy hypothesis* ($\mathscr{A}$), $B^{\alpha}(t) \geq 0$ *for* t a.e. *on* I, *and* $D^1(I) = D^2(I)$. *If for a given compact subinterval* $[a,b]$ *of* I, *we have that* $J_a^{1,2}[\eta;a,b]$ *is non-negative on* $D_{*0}^1[a,b] = D_{*0}^2[a,b]$, *and* $t = b$ *is the first right-hand focal point to* $t = a$ *relative to* $J^1[\eta;a,b]$, *then for a conjoined basis* $Y^2(t) = (U^2(t);V^2(t))$ *of* (3.1^2) *satisfying*

$$\Gamma_a^2 U^2(a) - V^2(a) = 0, \quad U^2(a) \text{ non-singular}, \tag{7.3}$$

exactly one of the following conditions holds:

(i) *there exists an* $s \in (a,b)$ *such that* $U^2(s)$ *is singular;*

(ii) $U^2(t)$ *is non-singular for* $t \in [a,b)$, *in which case* $t = b$ *is also the first right-hand focal point to* $t = a$ *relative to* $J^2[\eta;a,b]$, *and if* $(u;v)$ *is a solution of* (3.1^1) *which determines* $t = b$ *as focal to* $t = a$ *relative to* $J_a^1[\eta;a,b]$ *then there is a* v_2 *such that* $(u;v_2)$ *is a solution of* (3.1^2) *which determines* $t = b$ *as focal to* $t = a$ *relative to* $J_a^2[\eta;a,b]$; *moreover, there exists a non-zero* n-*dimensional vector* ξ *such that* $u(t) = U^2(t)\xi$ *on* $[a,b]$, *and* $v_2 - V^2\xi \in \Lambda^2[a,b]$.

In particular, if $J_a^{1,2}[\eta;a,b]$ *is positive definite on* $D_{*0}^1[a,b] = D_{*0}^2[a,b]$, *then conclusion* (i) *holds.*

Results analogous to those of the above theorem hold for functionals $J_b^1[\eta;a,b]$, $J_b^2[\eta;a,b]$ and left-hand focal points to $t = b$, but will not be stated specifically as they should be obvious.

If a system (3.1) is identically normal, and $Y(t) = (U(t);V(t))$ is a conjoined basis for this system, then a value $t = t_0$ is called a *focal point of* $Y(t)$ *of order* k if $U(t_0)$ is of rank $n - k$. The following theorem presents a fundamental property of the set of focal points belonging to a given conjoined basis for such an equation.

THEOREM 7.4. *Suppose that* (3.1) *satisfies* ($\mathscr{A}$) *with* $B(t) \geq 0$ a.e., *is also identically normal on a given interval* I, *and* [a,b] *is a compact subinterval of* I *on which* (3.1) *is disconjugate. If* $Y(t) = (U(t);V(t))$ *is a conjoined basis for* (3.1) *for which* $t = a$, $\{t = b\}$, *is a focal point of order* r, *then on* (a,b], {*on* [a,b)}, *there are at most* $n - r$ *focal points, each focal point being counted a number of times equal to its order. In particular, the focal points of a conjoined basis for* (3.1) *are isolated.*

If (3.1) is identically normal on I, and $Y(t) = (U(t);V(t))$ is a conjoined basis for this equation, then whenever $t = t_0$ is a focal point of $Y(t)$ there exists a non-zero vector ξ such that $U(t_0)\xi = 0$, and $(u(t);v(t)) = (U(t)\xi;V(t)\xi)$ is a solution of (3.1) such that $u(t_0) = 0$ and $u(t)$ is not identically zero throughout any non-degenerate subinterval of I. Moreover, if $t = c$ is a focal point of $Y(t)$ of order $k > 0$, and $\xi = \xi^{(j)}$, $(j = 1,\ldots,k)$, are linearly independent vectors satisfying $U(t_0)\xi = 0$, then $(u^{(j)}(t),v^{(j)}(t)) = (U(t)\xi^{(j)};V(t)\xi^{(j)})$, $(j = 1,\ldots,k)$, are linearly independent solutions of (3.1) such that $u^{(j)}(t),\ldots,u^{(k)}(t)$ are linearly independent vector functions on arbitrary subintervals of I. Now suppose that (3.1) is disconjugate on [a,b], while $Y(t) = (U(t);V(t))$ is a conjoined basis for (3.1) with $U(a)$ of rank $n - r$ and with $n - r + \ell$, $(\ell \geq 1)$, focal points $t_1 \leq t_2 \leq \cdots \leq t_{n-r+\ell}$ on (a,b], where each focal point is counted a number of times equal to its order. Then there exist $n - r + \ell$ solutions $(u^{(j)}(t),v^{(j)}(t)) = (U(t)\xi^{(j)};V(t)\xi^{(j)})$, $(j = 1,\dots,n-r+\ell)$, such that $u^{(j)}(t_j) = 0$, and if $(\eta^{(j)}(t);\zeta^{(j)}(t)) = (u^{(j)}(t);v^{(j)}(t))$ for $t \in [a,t_j]$, $(\eta^{(j)}(t);\zeta^{(j)}(t)) = (0;0)$ for $t \in (t_j,b]$, then $\eta^{(j)} \in D_{*o}[a,b]:\zeta^{(j)}$ and $\eta^{(j)}(t),\dots,\eta^{(n-r+\ell)}(t)$ are linearly independent on $[a,b]$. Also, with the aid of (6.2), one may verify readily that

$$J[\eta^{(j)},\eta^{(j)};a,b] = -\eta^{(j)*}(a)V(a)\xi^{(j)}, \quad (i,j = 1,\dots,n-r+\ell),$$

As $\eta^{(j)}(a) = U(a)\xi^{(j)}$, and $U(a)$ is of rank $n - r$ and $\ell > 0$, there exist constants $d_1,\dots,d_{n-r+\ell}$ not all zero, and such that $\eta(t) = \eta^{(j)}(t)d_1 + \dots + \eta^{(n-r+\ell)}(t)d_{n-r+\ell}$ satisfies the condition $\eta(a) = 0$, and hence

$$J[\eta;a,b] = \sum_{i,j=1}^{n-r+\ell} d_i d_j J[\eta^{(j)},\eta^{(i)};a,b]$$

$$= -\eta^*(a)V(a)[\xi^{(1)}d_1 + \dots + \xi^{(n-r+\ell)}d_{n-r+\ell}] = 0.$$

On the other hand, $\eta(t)$ is a non-identically vanishing element of $D_o[a,b]$, and in view of Theorem 6.3 we have the contradictory result $J[\eta;a,b] > 0$. The proof of the corresponding results for the interval $[a,b)$ are similar, and will be omitted.

Now if $c \in I$ then for the conjoined basis $Y_o(t;c) = (U_o(t;c);V_o(t;c))$ determined by the initial conditions $U_o(c,c) = E$, $V_o(c;c) = 0$ there exists a $\delta > 0$ such that $U_o(t,c)$ is non-singular on $[c-\delta,c+\delta] \cap I$, and in view of the above theorem an arbitrary conjoined basis $Y(t) = (U(t);V(t))$ of (3.1) for which $t = c$ is a focal point has at most $n - 1$ other focal points on each of the subintervals

$(c,c+\delta] \cap I$ and $[c-\delta,c) \cap I$, so that c is an isolated focal point of this conjoined basis.

Results analogous to those of the above theorem hold for functionals $J_b^1[\eta;a,b]$, $J_b^2[\eta;a,b]$ and left-hand focal points to $t = b$, but will not be stated specifically as they should be obvious to the reader.

The results of the preceding sections of this chapter have been presented largely in the terminology of Reid [35], and the above comparison theorems in the context of the author's paper [42]. It is to be emphasized, however, that as basic tools for the study of oscillation phenomena for self-adjoint vector differential equations these results are of long standing, and under varying degrees of generality are to be found in Morse [1, 4, 9], Birkhoff and Hestenes [1], Bliss [7], and Reid [6, 7, 12, 18]. In particular, the systematic use of the family of broken solutions of (3.1) determined by the focal points of a conjoined basis as in the proof of Theorem 7.4, is a central ingredient of the treatments of Morse and Hestenes in the above cited references.

It is to be remarked that in the above theorems there is no generalization of Theorem II.9.4 dealing with the comparative positions of a sequence of right-hand focal points of $t = a$ relative to $J_a^1[\eta;a,b]$ and a sequence of such focal points relative to $J_a^2[\eta;a,b]$. In this connection, it is to be noted that for a differential equation (3.1) we may have s_1 and s_2 right-hand focal points to $t = a$ relative to $J_a[\eta;a,b]$ with $a < s_1 < s_2$, but $J_a[\eta;s_1,s_2]$ be positive definite on $D_o[s_1,s_2]$. For example, if $n = 2$, $\Gamma = 0$, $A(t) \equiv 0$, $B(t) \equiv E_2$, $C(t) = \text{diag}\{-1,-4\}$, then $t = \pi/4$ and $t = \pi/2$ are both focal points to $t = 0$ relative to the functional

$$\int_0^{\pi}\{|\eta_1'|^2 + |\eta_2'|^2 - 4|\eta_1|^2 - |\eta_2|^2\}dt$$

with the conjoined basis (6.12a) given by $U(t) = \operatorname{diag}\{\cos 2t, \cos t\}$, $V(t) = \operatorname{diag}\{-2 \sin 2t, -\sin t\}$, while $U_0(t) = \operatorname{diag}\{\sin[2t - \pi/8], \sin t\}$, $V_0(t) = \operatorname{diag}\{2 \cos[2t - \pi/8], \cos t\}$ is a conjoined basis with $U_0(t)$ non-singular on the closed subinterval $[\pi/4,\pi/2]$.

In this connection, it is to be noted that the paper of Bliss and Schoenberg [1] considers an alternate concept of a "conjugate system of points" associated with a given initial point. The specific system considered in this paper is of the form (2.1) with continuous coefficient matrices and $R(t)$ positive definite, but it may be readily verified that the results remain valid for a more general system (3.1) with $B(t) \geq 0$ for t a.e., and which is identically normal. Described briefly, it is shown that for each initial value s there exists a set of values $t_k(s)$, $k = 0,\pm 1,\pm 2,\ldots$, which may be infinite or finite such that $t_0(s) = s$ and if $t_k(s)$ and $t_{k+1}(s)$ exist then $t_{k+1}(s)$ is the first right-hand conjugate point to $t_k(s)$, and $t_k(s)$ is the first left-hand conjugate point to $t_{k+1}(s)$. For such conjugate systems of points Bliss and Schoenberg obtain a comparison theorem, and also an oscillation theorem for the eigenfunctions of an associated boundary value problem. In particular, two such systems of conjugate points coincide or separate each other in the sense that between two adjacent points of one system there is one and only one point of the second system. In comparing results of this paper with those of others cited above, it is important to realize that in the context of Bliss and Schoenberg [1] each point of a conjugate system of points is counted only once, and not with a multiplicity that may exceed one. Also, it is important to realize that if k is a value distinct from 0 and ± 1 and such that $t_k(s)$ exists, then in general $t = t_k(s)$ is not a point conjugate to $t = t_0(s) = s$. However, with the aid of general oscillation theorems to be discussed later it may be shown that if $k > 1$ and $t_k(s)$ exists, then the k-th right-hand conjugate point $\tau_k^+(s)$ of s does exist and $s < \tau_k^+(s) \leq t_k(s)$; correspondingly, if $k < -1$ and $t_k(s)$ exists, then the k-th left-hand conjugate point $\tau_k^-(s)$ of s does exist and $t_k(s) \leq \tau_k^-(s) < s$.

8. Morse Fundamental Hermitian Forms

Throughout this section we shall suppose that hypothesis ($\mathscr{A}$) holds with $B(t) \geq 0$ a.e., and system (3.1) is identically normal. Corresponding to the terminology of Section II.10, if $[a,b]$ is a compact subinterval of I a partition

$$\Pi: a = t_0 < t_1 < \dots < t_k < t_{k+1} = b \tag{8.1}$$

is called a *fundamental partition of* $[a,b]$, {*relative to* (3.1)}, if this equation is disconjugate on each of the subintervals $[t_{j-1},t_j]$, $(j = 1,\dots,k+1)$. The existence of fundamental partitions is assured by the uniform continuity on $[a,b] \times [a,b]$ of the matrix function $U_0(t;c)$ of the conjoined basis $Y_0(t;c) = (U_0(t;c),V_0(t;c))$ determined by the initial conditions $U_0(c;c) = E$, $V_0(c;c) = 0$.

Analogous to the procedure of Section II.10, for a fundamental partition (8.1) of $[a,b]$ let $X(\Pi)$ denote the totality of sequences of n-dimensional vectors $x = (x^0,x^1,\dots,x^{k+1})$, and let $u = u^{(j)}(t)$ signify the solution of (3.1) determined by the two-point conditions

$$u^{(j)}(t_{j-1}) = x^{j-1}, \quad u^{(j)}(t_j) = x^j, \ (j = 1,\ldots,k+1). \tag{8.2}$$

In view of the disconjugacy of (3.1) on each of the component subintervals $[t_{j-1},t_j]$, the conditions (8.2) determine uniquely a solution $y^{(j)}(t) = (u^{(j)}(t);v^{(j)}(t))$ of (3.1) which is indeed a continuous function of $(t,t_{j-1},x^{j-1},t_j,x^j)$ as long as t_{j-1} and t_j vary in such a manner that (3.1) remains disconjugate on $[t_{j-1},t_j]$. Also, for $x \in X(\Pi)$ let $u_x(t)$ be defined as

$$u_x(t) = u^{(j)}(t) \text{ for } t \in [t_{j-1},t_j], \ (j = 1,\ldots,k+1). \tag{8.3}$$

As in the case of the scalar system considered in Section II.10, if $v_x(t)$ is a piecewise continuous vector function $[a,b]$ satisfying

$$v_x(t) = v^{(j)}(t) \text{ for } t \in (t_{j-1},t_j), \ (j = 1,\ldots,k+1), \tag{8.4}$$

we have that $u_x \in D[a,b]:v_x$. Again, as in the scalar case, the subclass of sequences $x = (x^0,x^1,\ldots,x^k,x^{k+1})$ of $X(\Pi)$ for which $x^{k+1} = 0$ is denoted by $X_{*0}(\Pi)$, the subclass of sequences for which $x^0 = 0$ is denoted by $X_{0*}(\Pi)$, and $X_0(\Pi)$ denotes $X_{*0}(\Pi) \cap X_{0*}(\Pi)$, the subclass of sequences for which $x^0 = 0 = x^{k+1}$. Consequently, the vector function $u_x(t)$ defined by (8.3) belongs to $D_{*0}[a,b]$, $D_{0*}[a,b]$ or $D_0[a,b]$ according as $x = (x^0,\ldots,x^{k+1})$ is a member of the respective classes $X_{*0}(\Pi)$, $X_{0*}(\Pi)$, or $X_0(\Pi)$.

If $x \in X_0(\Pi)$ and $y \in X_0(\Pi)$, from the fact that $u_x \in D_0[a,b]:v_x$ and $u_y \in D_0[a,b]:v_y$ it follows that $J[u_x,u_y;a,b] = \overline{J[u_y,u_x;a,b]}$ and consequently the functional

$$Q^o\{x,y:\Pi\} = J[u_x,u_y;a,b], \text{ for } x \in X_o(\Pi),\ y \in X_o(\Pi) \quad (8.5)$$

is of the form

$$Q^o\{x,y:\Pi\} = \sum_{\alpha,\beta=1}^{k} y^{\alpha *} Q^o_{\alpha\beta}\{\Pi\} x^{\beta},$$

where the $n \times n$ matrices $Q^o_{\alpha\beta}\{\Pi\}$ are such that

$$Q^o_{\alpha\beta}\{\Pi\} = [Q^o_{\beta\alpha}\{\Pi\}]^*, \quad (\alpha,\beta = 1,\ldots,k). \quad (8.7)$$

That is, if $N = kn$, and $\check{x} = (\check{x}_\sigma)$, $\check{y} = (\check{y}_\sigma)$, $(\sigma = 1,\ldots,N)$, where $\check{x}_{(\alpha-1)n+j} = x^\alpha_j$, $\check{y}_{(\alpha-1)n+j} = y^\alpha_j$, $(j = 1,\ldots,n;$ $\alpha = 1,\ldots,k)$, then $Q^o\{x,y:\Pi\}$ is an hermitian form in $\check{x},\check{y}$. Corresponding to equation (II.10.8) we now have

$$Q^o\{x,y:\Pi\} = \sum_{\alpha=1}^{k} y^{\alpha *}[v_x(t^-_\alpha) - v_x(t^+_\alpha)], \quad \text{for } x \in X_o(\Pi),\ y \in X_o(\Pi). \quad (8.8)$$

Moreover, in view of the above stated continuity property of the functions $u^{(j)}(t), v^{(j)}(t)$, if for a fixed positive integer k the symbol T_k denotes the set of values $(t_0,t_1,\ldots,t_{k+1})$ belonging to fundamental partitions (8.1) of compact subintervals $[a,b]$ of I, the coefficient matrices $Q^o_{\alpha\beta}\{\Pi\}$ of the hermitian form (8.6) are continuous functions of $(t_0,t_1,\ldots,t_{k+1})$ on T_k.

Corresponding to Theorem 10.1, we now have the following result which will be stated without proof, as it follows from a direct application of results on general hermitian forms to the particular form $Q^o\{x:\Pi\}$.

THEOREM 8.1 *If* Π *is a fundamental partition* (8.1) *of* $[a,b]$, *then the hermitian form* $Q^o\{x:\Pi\}$ *is singular if and only if* $t = b$ *is conjugate to* $t = a$, *with respect to* (3.1); *moreover, if* $Q^o\{x:\Pi\}$ *is singular then its nullity is equal to the order of* $t = b$ *as a conjugate point to* $t = a$.

Continuing in a manner entirely analogous to that in Section II.10, we have the verbatim analogues of Theorems II.10.2 and II.10.3. Specifically, if Π^1 and Π^2 are two fundamental partitions of the compact subinterval $[a,b]$ of I, then the two associated hermitian forms $Q^o\{x:\Pi^1\}$ and $Q^o\{x:\Pi^2\}$ have the same index and the same nullity. Moreover, if $[a,b_1] \subset [a,b_2] \subset I$, and Π^1,Π^2 are fundamental partitions of $[a,b_1]$ and $[a,b_2]$, respectively, then the indices i_α and nullities n_α of the corresponding hermitian forms $Q^o\{x:\Pi^\alpha\}$, $(\alpha = 1,2)$, are such that $i_1 \leq i_2$ and $i_1 + n_1 \leq i_2 + n_2$.

The basic result on the relationship between the hermitian form $Q^o\{x:\Pi\}$ and the existence of conjugate points is presented in the following theorem, which may be established by the same method of proof indicated for the corresponding Theorem II.10.4 in the scalar case.

THEOREM 8.2. *If* Π *is a fundamental partition of a compact nondegenerate subinterval* $[a,b]$ *of* I, *then the index of* $Q^o\{x:\Pi\}$ *is equal to the number of points on the open interval* (a,b) *which are conjugate to* $t = a$, *where each point conjugate to* $t = a$ *is counted a number of times equal to its index.*

Upon interchanging the roles of $t = a$ and $t = b$ in the arguments leading to Theorem 8.2, one is led to the result that the index of $Q^o\{x:\Pi\}$ is also equal to the number of points on the open interval (a,b) which are conjugate to $t = b$, where each such conjugate point is counted a number of times equal to its index. Hence for a general system

(3.1) that is identically normal one has the following non-trivial result, although its counterpart for the scalar system of Chapter II is a direct consequence of the separation of zeros of two linearly independent real solutions.

THEOREM 8.3. *If* $[a,b]$ *is a compact subinterval of* I, *then the number of points on the interval* (a,b), $\{(a,b]\}$, *conjugate to* $t = a$ *is equal to the number of points on the interval* (a,b), $\{[a,b)\}$, *which are conjugate to* $t = b$, *where in each case a point is counted a number of times equal to its order as a conjugate point.*

If $Q^o\{x:\Pi\}$ is negative definite, {non-positive definite} on a p-dimensional subspace spanned by vectors $x_\rho \equiv (x_{\sigma\rho})$, $(\rho = 1,\dots p)$, where $x_{(\alpha-1)n+j,\rho} = x^\alpha_{j,\rho}$, $(j = 1,\dots,n;$ $\alpha = 1,\dots,k)$, then $J[\eta;a,b]$ is negative definite, {non-positive definite} on a p-dimensional subspace of $D_o[a,b]$ spanned by $u_{x_1}(t),\dots,u_{x_p}(t)$, where $x = (x^o_\rho, x^1_\rho,\dots,x^k_\rho,$ $x^{k+1}_\rho) \in X_o\{\Pi\}$, $\rho = 1,\dots,p$. Conversely, if $J[\eta;a,b]$ is negative definite, {non-positive definite}, on a p-dimensional subspace of $D_o[a,b]$ spanned by $\eta_1(t),\dots,\eta_p(t)$, then for Π a fundamental partition (8.1) of $[a,b]$ and $x^\alpha_\rho = \eta_\rho(t_\alpha)$, $(\rho = 1,\dots,p;\alpha = 0,1,\dots,k+1)$, for arbitrary constants $c_1,\dots,c_p$ we have that

$$\eta(t) = \sum_{\rho=1}^{p} c_\rho \eta_\rho(t), \quad x^\alpha = \sum_{\rho=1}^{p} c_\rho x^\alpha_\rho, \quad (\alpha = 0,1,\dots,k+1)$$

and $x = (x^o,x^1,\dots,x^{k+1})$ are such that $u_x(t) = \sum_{\rho=1}^{p} c_\rho u_{x_\rho}(t)$, and in view of the disconjugacy of (3.1) on each of the subintervals $[t_{j-1},t_j]$, it follows that $J[\eta;t_{j-1},t_j]$ is positive definite on $D_o[t_{j-1},t_j]$ and consequently that $J[\eta;t_{j-1},t_j] \geq J[u_x;t_{j-1},t_j]$, $(j = 1,\dots,k+1)$. Therefore,

$Q^o\{x:\Pi\}$ is correspondingly negative definite, {non-positive definite} on the p-dimensional subspace spanned by the corresponding nk-dimensional vectors x_ρ, $(\rho = 1,\dots,p)$, and we have the following result.

THEOREM 8.4. *If* $[a,b]$ *is a compact subinterval of* I, *and* Π *is a fundamental partition of* $[a,b]$, *then the index, {index plus nullity} of* $Q^o\{x:\Pi\}$ *is equal to the largest non-negative integer* p *such that there is a* p-*dimensional subspace in* $D_o[a,b]$ *on which the functional* $J[\eta;a,b]$ *is negative definite, {non-positive definite}.*

For a given $c \in I$, the set of points on I which are right-hand conjugate points to $t = c$ will be ordered as a sequence $\{t^+_\nu(c)\}$ with $t^+_\nu(c) \le t^+_{\nu+1}(c)$, and each repeated a number of times equal to its order as a conjugate point. If there are no points on I which are right-hand conjugate points to $t = c$ this sequence is vacuous; otherwise, it may be finite or infinite, and in the latter case the final conclusion of Theorem 7.5 implies that only a finite number of points in the sequence lie in a given compact subinterval of I. Correspondingly, the set of points of I which are left-hand conjugate points to $t = c$ are ordered as a sequence $\{t^-_\nu(c)\}$, with $t^-_{\nu+1}(c) \le t^-_\nu(c)$, and similar convention as to repetition according to the order as a conjugate point. A basic result on conjugate points is as follows.

THEOREM 8.5. *If for a given positive integer* ν *the conjugate point* $t^+_\nu(c)$, $\{t^-_\nu(c)\}$, *exists for* $c = c_o$, *then there exists a* $\delta > 0$ *such that* $t^+_\nu(c)$, $\{t^-_\nu(c)\}$, *exists for* $|c - c_o| < \delta$, *and* $t^+_\nu(c)$, $\{t^-_\nu(c)\}$, *is continuous at* $c = c_o$.

For conjugate points one also has the following monotoneity property.

THEOREM 8.6. *If* $c_\alpha \in I$, $(\alpha = 1,2)$, *and* $c_1 < c_2$, *then whenever* $t^+_\nu(c_2)$, $\{t^-_\nu(c_1)\}$, *exists the corresponding* ν-th *conjugate point* $t^+_\nu(c_1)$, $\{t^-_\nu(c_2)\}$, *exists, and* $t^+_\nu(c_1) < t^+_\nu(c_2)$, $\{t^-_\nu(c_2) > t^-_\nu(c_1)\}$.

Also, for comparison theorems of the general type considered in Theorem 7.1, the above Theorems 8.2, 8.3, 8.4, yield the following result.

THEOREM 8.7. *Suppose that for* $\alpha = 1,2$ *the matrix functions* $A^\alpha(t)$, $B^\alpha(t)$, $C^\alpha(t)$ *satisfy hypothesis* $(\mathscr{A})$, $B^\alpha(t) \geq 0$ *for* t a.e. *on* I, *and each of the systems* (3.1^α) *is identically normal; moreover,* $D^1(I) = D^2(I)$, *and for arbitrary compact subintervals* $[a,b]$ *of* I *the functional* $J^{1,2}[\eta;a,b]$ *of* (8.2) *is non-negative on* $D^1_o[a,b] = D^2_o[a,b]$. *If* $t^+_{\nu\alpha}(c)$ *and* $t^-_{\nu\alpha}(c)$, $(\alpha = 1,2,\dots)$, *denote the sequences of right- and left-hand conjugate points to* $t = c$ *relative to the respective system* (3.1^α), *then whenever the conjugate point* $t^+_{\nu1}(c)$, $\{t^-_{\nu1}(c)\}$ *exists the conjugate point* $t^+_{\nu2}(c)$, $\{t^-_{\nu2}(c)\}$, *also exists, and*

$$t^+_{\nu2}(c) \leq t^+_{\nu1}(c), \quad \{t^-_{\nu2}(c) > t^-_{\nu1}(c)\}; \tag{8.9}$$

moreover, if $J^{1,2}[\eta;a,b]$ *is positive definite on* $D^1_o[a,b] = D^2_o[a,b]$ *for arbitrary compact subintervals* $[a,b]$ *of* I, *then strict inequalities hold in* (8.9).

Corresponding to the discussion of focal points in Section II.8, for $[a,b]$ a compact subinterval of I consider the hermitian functional

$$J_a[\eta;a,b] = \eta^*(a)\Gamma_a\eta(a) + J[\eta;a,b], \tag{8.10}$$

where Γ_a is an $n \times n$ hermitian matrix. For Π a fundamental partition (8.1) of $[a,b]$, and x and y elements of $X_{*0}(\Pi)$, the corresponding vector functions $(u_x(t);v_x(t))$ and $(u_y(t);v_y(t))$ defined by (8.3), (8.4) are such that $u_x \in D_{*0}[a,b]:v_x$ and $u_y \in D_{*0}[a,b]:v_y$. Consequently, the functional

$$Q^{*0}_{\alpha\beta}\{x,y:\Pi\} = J[u_x,u_y;a,b] \tag{8.11}$$

is of the form

$$Q^{*0}_{\alpha\beta}\{x,y:\Pi\} = \sum_{\alpha,\beta=0}^{k} y^{\alpha *}Q^{*0}\{\Pi\}x^{\beta}, \tag{8.12}$$

where the $n \times n$ matrices $Q^{*0}_{\alpha\beta}\{\Pi\}$ are such that $Q^{*0}_{\alpha\beta}\{\Pi\} = (Q^{*0}_{\beta\alpha}\{\Pi\})^*$, $(\alpha,\beta = 0,1,\ldots,k)$. Also, corresponding to (8.8) we have for $x \in X_{*0}(\Pi)$ and $y \in X_{*0}(\Pi)$ the relation

$$Q^{*0}\{x,y:\Pi\} = y^{0*}[\Gamma_a x^0 - v_x(a)] + \sum_{\alpha=1}^{k} y^{\alpha *}[v_x(t_\alpha^-) - v_x(t_\alpha^+)].$$

Therefore $Q^{*0}\{x,y:\Pi\}$ is an hermitian form in the $n(k+1)$-dimensional vectors $x = (x_\sigma)$, $y = (y_\sigma)$, where $x_{\beta n+j} = x_j^\beta$, $y_{\beta n+j} = y_j^\beta$, $(j = 1,\ldots,n;\beta = 0,1,\ldots,k)$.

Corresponding to the results of Theorems 8.1, 8.2, and 8.4, we now have the following results.

THEOREM 8.8. *If* Π *is a fundamental partition of* $[a,b]$, *then the hermitian form* $Q^{*0}\{x:\Pi\}$ *is singular if and only if* $t = b$ *is a right-hand focal point to* $t = a$, *relative to the functional* (8.10); *moreover, if* $Q^{*0}\{x:\Pi\}$ *is singular then its nullity is equal to the order of* $t = b$ *as a focal point.*

THEOREM 8.9. *If* Π *is a fundamental partition of a compact subinterval* $[a,b]$ *of* I, *then the index of* $Q^{*0}\{x:\Pi\}$ *is equal to the number of points on the open interval* (a,b) *which are right-hand focal points to* $t = a$ *relative to the* *functional* (8.10), *where each focal point is counted a number of times equal to its order.*

THEOREM 8.10. *If* $[a,b]$ *is a compact subinterval of* I, *and* Π *is a fundamental partition of* $[a,b]$, *then the index, {index plus nullity} of* $Q^{*0}\{x:\Pi\}$ *is equal to the largest non-negative integer* p *such that there exists a* p-*dimensional subspace in* $D_{*0}[a,b]$ *on which the functional* $J[\eta;a,b]$ *is negative definite, {non-positive definite}.*

It is to be commented that for a general conjoined basis $Y_0(t) = (U_0(t);V_0(t))$ of (3.1), the specification of a value c at which $U_0(c)$ is singular as a focal point of this basis is in accord with the above characterization of a focal point with respect to the functional $J_a[\eta;a,b]$ of (8.10). Under the assumption of identical normality of (3.1) the points at which $U_0(t)$ is singular are isolated. If $t = a$ is a value at which $U_0(a)$ is non-singular, then the conjoined character of $Y_0(t)$ implies that $W_0(a) = V_0(a)U_0^{-1}(a)$ is hermitian, and $(U(t);V(t)) = (U_0(t)U_0^{-1}(a); V_0(t)U_0^{-1}(a))$ is the conjoined basis determined by the initial condition $\Gamma_a U(a) - V(a) = 0$ with $\Gamma_a = W_0(a)$. Moreover, a value $t_0 > a$ is a right-hand focal point of order q relative to the functional (8.10) if and only if $U(t_0)$ is singular with rank $n - q$.

For a given $a \in I$, the points on I which are right-hand focal points to $t = a$, relative to the functional $J_a[\eta;a,b]$ of (8.10), will be ordered as a sequence $\tau_\nu^+(\Gamma_a)$, $(\nu = 1,2,\ldots)$, indexed so that $\tau_\nu^+(\Gamma_a) \le \tau_{\nu+1}^+(\Gamma_a)$, and each repeated a number of times equal to its order as a focal point. For focal points we have the following basic separation theorem.

THEOREM 8.11. *Suppose that hypothesis* ($\mathcal{A}$) *is satisfied by the coefficients of* (3.1), *this system is identically normal, and* $B(t) \geq 0$ *for* t a.e. *on* I. *For* $\alpha = 1,2$ *and* $[a,b] \subset I$, *let*

$$J_a^{\alpha}[\eta;a,b] = \eta^*(a)\Gamma_a^{\alpha}\eta(a) + J[\eta;a,b]$$

where Γ_a^1 *and* Γ_a^2 *are* $n \times n$ *hermitian matrices. Also, let* P *and* N *denote the number of positive and negative eigenvalues of the hermitian matrix* $\Gamma_a^1 - \Gamma_a^2$, *where each eigenvalue is repeated a number of times equal to its multiplicity. If for a positive integer* q *the focal point* $\tau_{q+P}^{+}(\Gamma^2)$ *exists, then* $\tau_q^{+}(\Gamma^1)$ *exists and* $\tau_q^{+}(\Gamma^1) \leq \tau_{q+P}^{+}(\Gamma^2)$; *if* $\tau_{q+N}^{+}(\Gamma^1)$ *exists then* $\tau_q^{+}(\Gamma^2)$ *exists and* $\tau_q^{+}(\Gamma^2) \leq \tau_{q+N}^{+}(\Gamma^1)$.

As a consequence of the above theorem, we have the following result.

COROLLARY 1. *For a given subinterval* I_0 *of* I *the number of focal points on* I_0 *of any conjoined basis for* (3.1) *differs from that of any other conjoined basis of this system by at most* n.

Indeed, let $Y_\alpha(t) = (U_\alpha(t);V_\alpha(t))$, $(\alpha = 1,2)$, be two conjoined bases for (3.1), and I_0 a compact subinterval $[a_0,b_0]$ of I. As the focal points of a conjoined basis are isolated, there exists a value $a < a_0$ such that each $U_\alpha(a)$ is non-singular, and the focal points of each $Y_\alpha(t)$ on $[a,b_0]$ all occur on the interval $[a_0,b_0]$. As noted above, $(U_\alpha(t)U_\alpha^{-1}(a);V_\alpha(t)U_\alpha^{-1}(a))$, $(\alpha = 1,2)$, is a conjoined basis for the initial conditions at $t = a$ for the corresponding functional J_a with $\Gamma^\alpha = V_\alpha(a)U_\alpha^{-1}(a)$. As each of the integers P and N or the above theorem for the matrix $\Gamma^1 - \Gamma^2$ does not exceed n, the conclusion of the above theorem implies the result of the Corollary in case I_o is a compact subinterval $[a_o,b_o]$. Moreover, if I_o is an interval of the form $(a,b_o]$, $[a_o,b_o)$ or (a_o,b_o), the result of the Corollary follows from this first result, together with the fact that in each case there is a compact subinterval $[a,b]$ of I_o such that the number of focal points of each family on I_o is equal to the number of focal points of that family on $[a,b]$.

Application of the result of Corollary 1 to a particular conjoined basis which determined points conjugate to $t = a$ yields the following result.

COROLLARY 2. *If* $[a,b] \subset I$, *and relative to* (3.1) *there are* q *conjugate points to* $t = a$ *on the interval* $(a,b]$ *or* (a,b), *then any conjoined basis for* (3.1) *has at most* $q + n$, *and at least* $q - n$ *focal points on this interval.*

In particular, if two conjoined bases $Y_\alpha(t) = (U_\alpha(t);V_\alpha(t))$ of solutions of (3.1) have in common m linearly independent solutions, and $t = a$ is a point of I at which each $U_\alpha(a)$ is non-singular, then for $\Gamma_a^\alpha = V_\alpha(a)U_\alpha^{-1}(a)$ the matrix $\Gamma^1 - \Gamma^2$ has nullity equal to m, and the N and P of Theorem 8.11 satisfy $N + P \leq n - m$. Consequently, the result of the above Corollary may be strengthened as follows.

COROLLARY 3. *If* $Y_\alpha(t)$, $(\alpha = 1,2)$ *are conjoined bases for* (3.1) *which have in common* m *linearly independent solutions, then for a given subinterval* I_o *of* I *the number of focal points of one basis on* I_0 *differs from that of the other basis by at most* $n - m$.

It is to be commented, (see Reid [27]), that subsequent to Morse's initial treatment using "broken extremals" his index theorems have been established by other methods that do not use this device. Certain of these alternative methods are based upon the theory of differential equations and associated boundary problems, as in Hu [1] and Reid [7, 6§4]. Other methods involve the concept of "natural isoperimetric conditions" as introduced by G. D. Birkhoff and M. R. Hestenes [1]. For a brief discussion of this method, with some alteration of details and method of attack, the reader is referred to Hestenes [1]; also, for specific variational problems related detailed treatments appear in various Ph.D. dissertations directed by Hestenes, notably those of Hazard [1], Karush [1], and Ritcey [1]. There are also methods wherein the Morse index theorems appear as a special instance of a more general theory of focal points for an appropriate class of quadratic forms in Hilbert space, as developed by Hestenes [2]; see also Gregory [1,2].

9. Generalized Polar Coordinate Transformations for Matrix Differential Systems

The basic idea that is fundamental for the development of an analogue of the polar coordinate transformation (II.6.1) for self-adjoint matrix differential systems was established by Barrett [3]. Specifically, Barrett considered a real self-adjoint matrix differential equation of the second order

$$[R(t)U'(t)]' - P(t)U(t) = 0, \tag{9.1}$$

where $R(t)$, $P(t)$ are $n \times n$ real, symmetric, continuous matrix functions, and $R(t)$ positive definite. Shortly thereafter, Reid [17] established similar results of somewhat more general character for a differential system of the form (3.1), and still further extensions of both method and range of results were presented in Reid [32]. The general format of the following description of results follows this latter paper.

For $n \times n$ matrix functions $A(t)$, $B(t)$, $C(t)$ satisfying the conditions of $(\mathscr{A})$, let

$$\begin{aligned} \Phi(t;\Phi,\Psi) &= \Psi B(t)\Psi^* + \Psi A(t)\Phi^* + \Phi A^*(t)\Psi^* - \Phi C(t)\Phi^*, \\ M(t;\Phi,\Psi) &= \Phi A(t)\Phi^* + \Psi C(t)\Phi^* + \Phi B(t)\Psi^* - \Psi A^*(t)\Psi^*. \end{aligned} \tag{9.2}$$

With the aid of Theorem 3.3 one may establish the following result on generalized polar coordinate transformations. This theorem contains the result of Reid [17, Theorem 3.1], which extended the result of Barrett [3]. In particular, the methods of both Barrett [3] and Reid [17] were constructive in nature, whereas the presentation of Reid [32] as given below is a direct generalization of the method used in the scalar case.

THEOREM 9.1. *If hypothesis* $(\mathscr{A})$ *is satisfied, and for* $\tau \in I$ *we have a conjoined basis* $Y(t) = (U(t);V(t))$ *of* (3.1) *satisfying* $Y(\tau) = (U_o;V_o)$, *then*

$$U_o^*U_o + V_o^*V_o > 0, \quad V_o^*U_0 - U_0^*V_o = 0. \tag{9.3}$$

Moreover, if Φ_0, Ψ_0, R_0 *are* $n \times n$ *matrices satisfying*

$$R_0^*R_0 = U_0^*U_0 + V_0^*V_0, \quad U_0 = \Phi_0^*R_0, \quad V_0 = \Psi_0^*R_0, \tag{9.4}$$

then

$$\Phi_o\Phi_o^* + \Psi_o\Psi_o^* = E, \quad \Phi_o\Psi_o^* - \Psi_o\Phi_o^* = 0, \tag{9.5}$$

and the solution $(\Phi(t);\Psi(t);R(t))$ *of the differential system*

$$\begin{aligned}
&\text{(a)} \quad L_1^o[\Phi,\Psi](t) \equiv -\Psi'(t) - Q(t;\Phi,\Psi)\Phi(t) = 0, \ \Phi(\tau) = \Phi_o,\\
&\text{(b)} \quad L_2^o[\Phi,\Psi](t) \equiv \Phi'(t) - Q(t;\Phi,\Psi)\Psi(t) = 0, \ \Psi(\tau) = \Psi_o,\\
&\text{(c)} \quad L^o[\Phi,\Psi,R](t) \equiv R'(t)-M(t;\Phi,\Psi)R(t) = 0,\\
&\qquad\qquad R(\tau) = R_o,
\end{aligned} \tag{9.6}$$

where $Q(t;\Phi,\Psi)$ *and* $M(t;\Phi,\Psi)$ *are defined by* (9.2) *is such that*

$$U(t) = \Phi^*(t)R(t), \quad v(t) = \Psi^*(t)R(t) \ \textit{for} \ t \in I. \tag{9.7}$$

Conversely, if $(\Phi;\Psi(t);R(t))$ *is a solution of* (9.6) *where* R_o *is non-singular and* (Φ_o,Ψ_o) *satisfies* (9.5), *then* (9.7) *defines a conjoined basis* $Y(t) = (U(t);V(t))$ *of* (3.1) *with*

$$R^*(t)R(t) = U^*(t)U(t) + V^*(t)V(t), \ \textit{for} \ t \in I. \tag{9.8}$$

It is to be remarked that for a conjoined basis $Y(t) = (U(t);V(t))$ the conditions (9.3), (9.4) imply that R_o is nonsingular and relations (9.5) hold. The result of Theorem 3.3 implies that the solution $(\Phi(t);\Psi(t))$ of the differential system $L_1^o[\Phi,\Psi](t) = 0$, $L_2^o[\Phi,\Psi](t) = 0$ satisfying the initial conditions $\Phi(\tau) = \Phi_o$, $\Psi(\tau) = \Psi_o$ has maximal interval of existence equal to I, and that throughout this interval the identities

$$\begin{aligned}
\Phi\Phi^* + \Psi\Psi^* &\equiv E, \qquad \Phi\Psi^* - \Psi\Phi^* \equiv 0,\\
\Phi^*\Phi + \Psi^*\Psi &\equiv E, \qquad \Psi^*\Phi - \Phi^*\Psi \equiv 0,
\end{aligned} \tag{9.9}$$

hold. If $U(t)$, $V(t)$, $\Phi(t)$, $\Psi(t)$, $R(t)$ are $n \times n$ matrix functions which are a.c. on arbitrary compact subintervals of I, and which are related by equations (9.7), the following identities hold:

$$\begin{aligned} L_1[U,V] &\equiv (L_1^o[\Phi,\Psi])^*R + G_1[\Phi,\Psi]R - \Psi^*L^o[\Phi,\Psi,R], \\ L_2[U,V] &\equiv (L_2^o[\Phi,\Psi])^*R + G_2[\Phi,\Psi]R + \Phi^*L^o[\Phi,\Psi,R], \end{aligned} \tag{9.10}$$

where

$$\begin{aligned} G_1[\Phi,\Psi] &= [E - \Phi^*\Phi - \Psi^*\Psi][C\Phi^* - A^*\Psi^*] \\ &\quad + [\Phi^*\Psi - \Psi^*\Phi][A\Phi^* + B\Psi^*], \\ G_2[\Phi,\Psi] &= [\Phi^*\Psi - \Psi^*\Phi][C\Phi^* - A^*\Psi^*] \\ &\quad - [E - \Phi^*\Phi - \Psi^*\Psi][A\Phi^* + B\Psi^*]. \end{aligned} \tag{9.11}$$

Consequently, if $Y(t) = (U(t);V(t))$ is a conjoined basis for (3.1) with $Y(\tau) = (U_o;V_o)$ and (Φ_o,Ψ_o,R_o) satisfies (9.4), then the solution $(\Phi(t),\Psi(t),R(t))$ of (9.6) is such that $G_1[\Phi,\Psi] \equiv 0$, $G_2[\Phi,\Psi] \equiv 0$ on I, and from (9.10) it follows that the matrix functions $U(t)$, $V(t)$ defined by (9.7) are solutions of (3.1_M) which agree at $t = \tau$ with the given solutions of this system, and therefore are equal to this given solution throughout I. Conversely, if $(\Phi(t), \Psi(t),R(t))$ is a solution of (9.6) with R_o non-singular and the matrices Φ_o,Ψ_o satisfying (9.5), the identities (9.9) are a consequence of Theorem 3.3 for the system $L_1^o[\Phi,\Psi](t) = 0$, $L_2^o[\Phi,\Psi](t) = 0$, so that $G_1[\Phi,\Psi] \equiv 0$, $G_2[\Phi,\Psi] \equiv 0$ on I, and relations (9.10) imply that $Y(t) = (U(t);V(t))$ is a solution of (3.1_M). Moreover, the non-singularity of R_o implies that the column vectors of $Y(t)$ are linearly independent solutions of (3.1), and the fact that $Y(t)$ is a conjoined basis of (3.1) is a direct consequence of the identity $\Phi\Psi^* - \Psi\Phi^* \equiv 0$.

Now if $Y(t) = (U(t);V(t))$ is a conjoined basis for (3.1) and $(\Phi(t),\Psi(t),R(t))$ is the corresponding solution of the differential system (9.6) satisfying relations (9.7), then in view of the identities (9.9) we have that $V(t)U^{-1}(t) = \Psi^*(t)\Phi^{*-1}(t) = \Phi^{-1}(t)\Psi(t)$ if $U(t)$ is non-singular, and $U(t)V^{-1}(t) = \Phi^*(t)\Psi^{*-1}(t) = \Psi^{-1}(t)\Phi(t)$ if $V(t)$ is non-singular. In particular, if $A(t) \equiv 0$, $B(t) \equiv E$, $C(t) \equiv -E$, so that (3.1_M) is equivalent to the second order linear homogeneous differential equation $U'' + U = 0$, and $Y(t,\tau) = (U(t;\tau);V(t;\tau))$ is the solution of (3.1_M) satisfying $U(t;\tau) = 0$, $V(\tau;\tau) = E$ then $(U_0,V_0,R_0) = (0,E,E)$ satisfies (9.4) and $(\Phi(t);\Psi(t);R(t)) = (S(t;\tau);C(t;\tau);E)$ is the solution of the corresponding differential system (9.6). In particular, on a subinterval on which $S(t;\tau)$ is non-singular the matrix function $W(t,\tau) = S^{-1}(t;\tau)C(t;\tau)$ is a solution of the Riccati matrix differential equation

$$W' + W^2 + E = 0 \tag{9.12}$$

Also, on a subinterval on which $C(t;\tau)$ is non-singular the matrix function $W_0(t;\tau) = C^{-1}(t;\tau)S(t;\tau)$ is a solution of the Riccati matrix differential equation

$$W_0' - W_0^2 - E = 0. \tag{9.13}$$

Consequently, in terms of the differential equations satisfied by them individually, the matrix functions $W(t;\tau) = S^{-1}(t;\tau)C(t;\tau)$ and $W_0(t;\tau) = C^{-1}(t;\tau)S(t;\tau)$ are generalizations of the scalar functions $\operatorname{ctn}(t-\tau)$ and $\tan(t-\tau)$, respectively.

In particular, if $\tau \in I$ and $\Phi(t) = S(t;\tau)$, $\Psi(t) = C(t,\tau)$ is the solution of the differential system (9.6a,b) satisfying the initial values $\Phi(\tau) = 0$, $\Psi(\tau) = E$, then whenever (3.1) is normal on arbitrary subintervals of I for which τ is an end-point it follows that relative to (3.1) a value $s \in I$ distinct from τ is a conjugate point to $t = \tau$ of order r if and only if $\Phi(s)$ is of rank $n - r$. Correspondingly, a value $s \in I$ distinct from τ is such that $t = \tau$ is a focal point of $t = s$ relative to (3.1) and the initial condition $v(s) = 0$, (that is, in an obvious extension of the terminology of Picone mentioned in Section I.3, $t = s$ is a pseudo conjugate to $t = \tau$), if and only if $\Psi(s)$ is of rank $n - r$.

For the use of the above described generalized polar coordinate transformation in the study of oscillation and comparison theorems for matrix differential equations of the form (3.1), the reader is referred to Barrett [3; 10, Sec. 5.3], Etgen [1,2,3,4], Kreith [8,10], and Reid [17,32]. In particular, in Reid [32, Secs. 6,7] there is presented a result that is a partial generalization of the results of Section I.4 for a scalar second order equation, and a type of "coupled polar coordinate transformation" for the simultaneous representation of solutions of a given first order linear matrix differential equation and solutions of the corresponding adjoint matrix differential equation.

10. Matrix Oscillation Theory

An alternate generalization of the polar coordinate method for the extension of the Sturmian theory emanates from the work of Lidskiĭ [1]. A special case of the comparison theorem due to Lidskii has been presented by Jakubovič [1], while Atkinson [2, Chapter 10] has placed Lidskiĭ's argument on a rigorous basis, and presented results on separation theorems. Also, Coppel [1] has employed this method to obtain some comparison theorems.

An introduction to this method is afforded by the following theorem.

THEOREM 10.1. *Suppose that the matrix coefficients of* (3.1) *satisfy hypothesis* ($\mathscr{A}$). *If* $Y(t) = (U(t);V(t))$ *is a conjoined basis for this system, then:*

(a) *the matrix functions*

$$\hat{U}(t) = V(t) - iU(t), \quad \hat{V}(t) = V(t) + iU(t) \tag{10.1}$$

are non-singular for $t \in I$;

(b) *the matrix function*

$$\Theta(t|Y) = \hat{V}(t)\hat{U}^{-1}(t) \tag{10.2}$$

is unitary for $t \in I$;

(c) $\Theta(t) = \Theta(t|Y)$ *is a solution of the differential system*

$$\Theta'(t) = i\Theta(t)N(t|Y), \tag{10.3}$$

where $N(t|Y)$ *is the hermitian matrix function*

$$N(t|Y) = -2\hat{U}^{*-1}(t)Y^*(t)\mathscr{A}(t)Y(t)\hat{U}^{-1}(t), \tag{10.4}$$

and $\mathscr{A}(t)$ *is defined by* (3.2).

(d) *if* (3.1) *is identically normal, then* $t = \tau$ *is a focal point of the conjoined basis* $Y(t)$ *if and only if* $\omega = 1$ *is an eigenvalue of* $\Theta(\tau|Y)$.

As $Y(t)$ is a conjoined basis for (3.1), it follows that

$$V^*(t)U(t)-U^*(t)V(t) = 0,\ V^*(t)V(t)+U^*(t)U(t) > 0 \text{ for } t \in I,$$

and conclusion (10.1) follows from the identities

$$\begin{aligned}
\text{(a)}\quad \hat{V}^*(t)\hat{V}(t) &\equiv [V^*(t) - iU^*(t)][V(t) + iU(t)]\\
&\equiv V^*(t)V(t) + U^*(t)U(t);\\
\text{(b)}\quad \hat{U}^*(t)\hat{U}(t) &\equiv [V^*(t) + iU^*(t)][V(t) + iU(t)]\\
&\equiv V^*(t)V(t) + U^*(t)U(t).
\end{aligned} \tag{10.5}$$

Also, these relations imply $\hat{V}^*(t)\hat{V}(t) \equiv \hat{U}^*(t)\hat{U}(t)$, so that $E = \hat{U}^{*-1}(t)\hat{V}^*(t)\hat{V}(t)\hat{U}^{-1}(t) = \Theta^*(t|Y)\Theta(t|Y)$, so that $\Theta(t|Y)$ is unitary for $t \in I$.

By direct computation it follows that $\Theta(t) = \Theta(t|Y)$ satisfies the differential equation (10.3) with

$$\begin{aligned}
iN(t|Y) &= \{[\Theta^*(t|Y) - E]V'(t) + i[\Theta^*(t|Y) + E]U'(t)\}\hat{U}^{-1}(t)\\
&= \hat{U}^{*-1}(t)\{[\hat{V}^*(t)-\hat{U}^*(t)]V'(t) + i[\hat{V}^*(t)+\hat{U}^*(t)]U'(t)\}\hat{U}^{-1}(t),
\end{aligned}$$

from which it follows immediately that $N(t|Y)$ is also given by (10.4).

It is to be remarked that in terms of the matrix functions $Q(t;\Phi,\Psi)$ and $R(t)$ appearing in the trigonometric transform (9.7) of the conjoined basis as defined by (9.2), (9.6), we have the relation

$$N(t|Y) = 2\hat{U}^{*-1}(t)R^*(t)Q(t;\Phi,\Psi)R(t)\hat{U}^{-1}(t). \tag{10.6}$$

Also, since the last members of (10.5a) and (10.5b) are equal to $R^*(t)R(t)$ we have that $E = U^{*-1}(t)R^*(t)R(t)\hat{U}^{-1}(t)$, so that the matrix function $R(t)\hat{U}^{-1}(t)$ is unitary on I and (10.6) yields the result that $N(t|Y)$ and $Q(t;\Phi,\Psi)$ are unitarily similar for each $t \in I$.

Finally, if (3.1) is identically normal then $t = \tau$ is a focal point of $Y(t)$ if and only if $U(\tau)$ is singular and conclusion (d) is a ready consequence of the identity

$$[\Theta(\tau|Y) - \omega E]\theta = [(1-\omega)V(\tau) + i(1+\omega)U(\tau)]\hat{\xi}, \quad \text{for } \hat{\xi} = \hat{U}^{-1}(\tau)\xi. \tag{10.7}$$

From equation (10.7) one also has the following result.

COROLLARY. *If the matrix coefficients of* (3.1) *satisfy hypothesis* ($\mathscr{A}$), *and this system is identically normal, then* $t = \sigma$ *is a value such that* $V(\sigma)$ *is singular if and only if* $\omega = -1$ *is an eigenvalue of* $\Theta(\sigma|Y)$.

As $\Theta(t|Y)$ is unitary, all of its eigenvalues have absolute value equal to 1, and consequently the occurrence of focal points of $Y(t)$ may be determined by the study of the eigenvalues $\omega(t)$ of $\Theta(t|Y)$, and, in particular, by the count of the number of times some eigenvalue $\omega(t)$ of this matrix function has $\arg \omega(t) = 0 \pmod{2\pi}$. Basic results for this problem, and also various other related problems are presented in the following two lemmas, whose results appear in Appendix V of Atkinson [2].

LEMMA 10.1. *If* $\Psi(t)$ *is a continuous* $n \times n$ *matrix function which is unitary for all* $t \in I$, *then there exists an ordering of the eigenvalues* $\omega_1(t),\dots,\omega_n(t)$ *of* $\Psi(t)$ *for* $t \in I$ *such that:*

(a) *the* $\omega_\alpha(t)$, *and their arguments, are continuous functions on* I;

(b) *the* $\omega_\alpha(t)$, $(\alpha = 1,\ldots,n)$, *where each is repeated a number of times equal to its multiplicity, for each* $t \in I$ *appear in a positive order on the unit circle with increasing* α; *that is,*

$$\arg \omega_1(t) \le \arg \omega_2(t) \le \cdots \le \arg \omega_n(t) \le \arg \omega_1(t) + 2\pi. \tag{10.8}$$

For real values θ the Möbius transformation

$$\lambda = \frac{i[e^{i\theta} + \omega]}{e^{i\theta} - \omega} \tag{10.9}$$

maps the positively oriented unit circle in the ω-plane into the positively oriented real axis in the λ-plane. Also, if Ψ is an $n \times n$ unitary matrix and θ is a real value such that $e^{i\theta}$ is not an eigenvalue of Ψ, then a value ω with $|\omega| = 1$ is an eigenvalue of Ψ if and only if the λ given by (10.9) is an eigenvalue of the corresponding $n \times n$ hermitian {Cayley transform} matrix

$$M[\theta] = i\{e^{i\theta}E + \Psi\}\{e^{i\theta}E - \Psi\}^{-1}. \tag{10.10}$$

Indeed, if ζ is an eigenvector of Ψ for the eigenvalue ω then ζ is also an eigenvector of $M[\theta]$ for the corresponding λ, and the index of ω as an eigenvalue of Ψ is equal to the index of the corresponding λ of (10.9) as an eigenvalue of $M[\theta]$.

Now suppose that $\Psi(t)$ is a continuous $n \times n$ unitary matrix function on I, and that for a given $\tau \in I$ the real value θ is such that $e^{i\theta}$ is not an eigenvalue of $\Psi(\tau)$. Then by continuity there exists a subinterval $[a_0,b_0]$ of I

containing τ in its interior and such that $e^{i\theta}$ is not an eigenvalue of $\Psi(t)$ for $t \in [a_0,b_0]$. As the corresponding matrix function

$$M[t;\theta] = i\{e^{i\theta}E + \Psi(t)\}\{e^{i\theta}E - \Psi(t)\}^{-1} \qquad (10.11)$$

is continuous on $[a_0,b_0]$, if $\lambda_1(t) \leq \dots \leq \lambda_n(t)$ denote its eigenvalues arranged in numerical order, and each repeated a number of times equal to its multiplicity, then each $\lambda_\alpha(t)$ is continuous on $[a_0,b_0]$. Consequently on $[a_0,b_0]$ a set of eigenvalues of $\Psi(t)$ is given by

$$\omega_\alpha(t) = e^{i\theta}\frac{[i\lambda_\alpha(t) + 1]}{[i\lambda_\alpha(t) - 1]}, \qquad (\alpha = 1,\dots,n), \qquad (10.12)$$

and each of these functions is continuous on $[a_0,b_0]$; also, $|\omega_\alpha(t)| \equiv 1$ on this interval, and choice of arguments to satisfy (10.8) at a chosen point of this interval insures by continuity that these relations persist throughout this subinterval. As shown in Atkinson [2, Sec. V.5], this process may be continued in a unique fashion to maintain conclusions (a) and (b) throughout I.

Also, by an extension of this continuation process Atkinson established the following result which is of use in the consideration of differential systems involving a parameter.

LEMMA 10.2. *If* $\Psi(t;s)$ *is a continuous* $n \times n$ *matrix function which is unitary for all* $(t,s) \in I \times J$, *where* J *is a subinterval of* $\mathbf{R}$, *and for a given* (t_0,s_0) *the eigenvalues* $\omega_\alpha(t_0;s_0)$ *are ordered to satisfy inequalities of the form* (10.8), *then throughout* $I \times J$ *they are continuable in a unique fashion so that the* $\omega_\alpha(t;s)$ *and their arguments are continuous functions, and the* $\omega_\alpha(t;s)$, *where each is repeated a number of times equal to its multiplicity, satisfy the inequalities*

$$\begin{aligned} \arg \omega_1(t;s) < \arg \omega_2(t;s) \le \dots \le \arg \omega_n(t;s) \\ \le \arg \omega_1(t;s) + 2\pi. \end{aligned} \tag{10.8'}$$

Of particular significance is the behavior of solutions of matrix differential equations of the form

$$\Psi'(t) = i\Psi(t)N(t), \tag{10.13}$$

where $N(t)$ is an hermitian matrix function that is locally Lebesgue integrable on I. If $\Psi(t)$ is an $n \times n$ matrix function which is unitary for $t \in I$, and is locally absolutely continuous, then $N(t) = -i\Psi^*(t)\Psi'(t)$ is an hermitian matrix function which is locally integrable and $\Psi(t)$ satisfies (10.13). Conversely, if $\Psi(t)$ is any matrix function satisfying (10.13) for which there exists a τ such that $\Psi(\tau)$ is unitary, then $W(t) = \Psi^*(t)\Psi(t)$ is the solution of the differential equation $W'(t) = i[W(t)N(t) - N(t)W(t)]$ satisfying $W(\tau) = E$, so that $W(t) \equiv E$ and $\Psi(t)$ is unitary for all $t \in I$.

LEMMA 10.3. *Suppose that* $N(t)$ *is a continuous* $n \times n$ *hermitian matrix function on* I, *and that* $\Psi(t)$ *is a unitary matrix function satisfying the differential equation* (10.13) *with* $\omega_\alpha(t)$, $(\alpha = 1,\dots,n)$ *an ordering of the eigenvalues of* $\Psi(t)$ *satisfying conditions* (a) *and* (b) *of Lemma* 10.1. *If* $\tau \in I$ *and* $e^{i\phi}$ *is an eigenvalue of* $\Psi(\tau)$ *such that* $\zeta^*N(\tau)\zeta > 0$ *for all eigenvectors* ζ *of* $\Psi(\tau)$ *corresponding to the eigenvalue* $e^{i\phi}$, *then for any index value*

$\alpha = \gamma$ *such that* $\arg \omega_\gamma(\tau) = \phi \pmod{2\pi}$ *the function* $\arg \omega_\gamma(t)$ *is a strictly increasing function at* $t = \tau$.

Let α be a real value such that $e^{i\theta}$ is not an eigenvalue of $\Psi(t)$ for t on a subinterval $[a_0,b_0]$ containing $t = \tau$ in its interior. Since the corresponding matrix function $M[t;\theta]$ of (10.11) may be written as $M[t;\theta] = 2ie^{i\theta}\{e^{i\theta}E - \Psi(t)\} - iE$, we have $M'[t;\theta] = 2ie^{i\theta}\{e^{i\theta}E - \Psi(t)\}^{-1}\Psi'(t)\{e^{i\theta}E - \Psi(t)\}^{-1}$. Using (10.13) and the fact that $\Psi^*(t) = \Psi^{-1}(t)$, it then follows that

$$\begin{aligned} M'[t;\theta] &= -2\{E - e^{-i\theta}\Psi(t)\}^{-1}\Psi(t)N(t)\{e^{i\theta}E - \Psi(t)\}^{-1} \\ &= 2\{e^{i\theta}E - \Psi(t)\}^{*-1}N(t)\{e^{i\theta}E - \Psi(t)\}^{-1} \end{aligned} \tag{10.14}$$

Now, as noted above, ζ is an eigenvector of $\Psi(\tau)$ for the eigenvalue $\omega_0 = e^{i\phi}$ if and only if ζ is an eigenvector of $M[\tau;\theta]$ for the corresponding eigenvalue given by (10.9); moreover, $[e^{i\theta}E - \Psi(\tau)]\zeta = [e^{i\theta} - \omega]\zeta$ and hence $\{e^{i\theta}E - \Psi(\tau)\}^{-1}\zeta = [e^{i\theta} - \omega]^{-1}\zeta$. Consequently, from (10.14) it follows that for any eigenvector ζ of $M[\tau;\theta]$ for the value $\lambda_0 = i[e^{i\theta} + \omega_0]/[e^{i\theta} - \omega_0]$ we have that

$$\zeta^*M'[\tau;\theta]\zeta = 2|e^{i\theta} - \omega_0|^2\zeta^*N(\tau)\zeta > 0, \tag{10.15}$$

and, as is well-known, (see, for example, Atkinson [2, Theorem V.3.2]), this implies that the eigenvalues $\lambda(t)$ of $M[t,\theta]$ that are equal to λ_0 for $t = \tau$ are strictly increasing functions at this value, and this latter condition is equivalent to the strictly increasing nature at $t = \tau$ of $\arg \omega_\alpha(t)$.

COROLLARY. *Suppose that* $N(t)$ *is an* $n \times n$ *hermitian matrix function which is locally Lebesgue integrable on* I, *and that* $\Psi(t)$ *is a unitary matrix function satisfying* (10.13) *with* $\omega_\alpha(t)$, $(\alpha = 1,\dots,n)$, *an ordering of the eigenvalues of* $\Psi(t)$ *satisfying conditions* (a) *and* (b) *of Lemma* 10.1.

(a) *If* $N(t)$ *is continuous and* $N(t) > 0$, $\{N(t) \geq 0\}$, *throughout* I, *then for* $\alpha = 1,\dots,n$ *the functions* $\arg \omega_\alpha(t)$ *are strictly increasing,* {*non-decreasing*}, *on* I.

(b) *If* $N(t) \geq 0$ *for* t a.e. *on* I, *then for* $\alpha = 1,\dots,n$ *the functions* $\arg \omega_\alpha(t)$ *are non-decreasing on* I.

The conclusion (a) for the case of $N(t) > 0$ for $t \in I$ is an immediate consequence of Lemma 10.2, since under this hypotheses we have $\zeta^* N(t)\zeta > 0$ for $t \in I$ and ζ an arbitrary non-zero vector. Now if $\Psi(t)$ is a unitary matrix function satisfying (10.13), then for arbitrary $r > 0$ the matrix function $\hat{\Psi}(t) = e^{irt}\Psi(t)$ is a solution of an equation of the form (10.13) with $\hat{N}(t) = N(t) + rE$, and if $\omega_\alpha(t)$ denote the eigenvalues of $\Psi(t)$ ordered to satisfy (10.3), then $\hat{\omega}_\alpha(t) = e^{irt}\omega_\alpha(t)$ are eigenvalues of $\Psi(t)$ similarly ordered and with $\arg \hat{\omega}_\alpha(t) = rt + \arg \omega_\alpha(t)$. If $N(t) \geq 0$ for $t \in I$ then the first part of conclusion (a) implies that the functions $\arg \hat{\omega}_\alpha(t)$ are strictly increasing on I, and the non-decreasing nature of the functions $\arg \omega_\alpha(t)$ follows upon letting r tend to zero.

The proof of conclusion (b) involves a more sophisticated limit process. For $h > 0$ so small that the set $I^h = \{t : [t-h, t+h] \subset I\}$ is non-empty, and $0 \leq s \leq h$, let $N(t;s)$ denote the matrix function

$$N(t;s) = \frac{1}{2s}\int_{t-s}^{t+s} N(\tau)\,d\tau.$$

Under the hypotheses of (b), on $I^h \times (0,h]$ the matrix function $N(t;s)$ is continuous, hermitian and $N(t;s) \geq 0$. Let $\Psi(t;s)$ denote the solution of the matrix differential equation $(\partial/dt)\Psi(t;s) = i\Psi(t;s)N(t;s)$ satisfying the initial condition $\Psi(t_0;s) = \Psi(t_0)$ for some fixed $t_0 \in I^h$. Now $\int_a^b \nu[N(t;s) - N(t)]dt \to 0$ as $s \to 0$ for arbitrary compact subintervals $[a,b]$ of I^h, and in turn this result implies that $\lim_{s\to 0} \Psi(t;s) = \Psi(t)$ uniformly on such subintervals. Therefore, if we set $\Psi(t;0) = \Psi(t)$ then $\Psi(t;s)$ is a continuous unitary matrix function on $I^h \times [0,h]$ with $\Psi(t_0;s) \equiv \Psi(t_0)$. Consequently, by Lemma 10.2 there is an ordering $\omega_\alpha(t;s)$ of the eigenvalues of $\Psi(t;s)$ for which the functions $\omega_\alpha(t;s)$ and $\arg \omega_\alpha(t;s)$ are continuous and inequalities (10.8') hold. Finally, since by conclusion (b) the functions $\arg \omega_\alpha(t;s)$ are non-decreasing functions of t for $s \in (0,h]$, by continuity it follows that the functions $\arg \omega_\alpha(t) = \arg \omega_\alpha(t;0)$ also possess this property. In particular, since the unitary matrix function $\Theta(t|Y)$ of (10.2) satisfies the differential equation (10.3) with $N(t|Y)$ given by (10.4), it follows that the corresponding hermitian matrix function $M(t,\theta|Y) = i\{e^{i\theta}E + \Theta(t|Y)\}\{e^{i\theta}E - \Theta(t|Y)\}^{-1}$ of (10.11) satisfies the differential equation

$$M'(t,\theta|Y) = -2G^*(t,\theta|Y)\mathscr{A}(t)G(t,\theta|Y) \tag{10.16}$$

where

$$G(t,\theta|Y) = Y(t)\hat{U}^{-1}(t)\{e^{i\theta}E - \Theta(t|Y)\}^{-1}. \tag{10.17}$$

Various results on the differential system (3.1) involve the following hypotheses.

($\mathscr{A}_1'$) *The matrix coefficients* $A(t)$, $B(t)$, $C(t)$ *of* (3.1) *are continuous on* I.

($\mathscr{A}_2'$) *The matrix function* $\mathscr{A}(t)$ *of* (3.2) *is negative definite for* $t \in I$.

It is to be noted that condition ($\mathscr{A}_2'$) implies that the $n \times n$ matrix functions $B(t)$ and $-C(t)$ are positive definite for $t \in I$. Moreover, the positive definiteness of $B(t)$ implies the identical normality of the system (3.1), and the positive definiteness of $-C(t)$ implies the identical normality of the associated reciprocal system.

In general, if (3.1) is identically normal on I and for a given $a \in I$ the conjoined basis $Y(t) = (U(t);V(t))$ of (3.1) is specified by the initial conditions $U(a) = 0$, $V(a)$ non-singular, then the values conjugate to $t = a$, which are the values $t \neq a$ at which $U(t)$ is singular, are isolated. Under the assumption of identical normality of (3.1), let τ_j^+, $(j = 1,2,\ldots)$, denote the right-hand conjugate points to $t = a$, ordered in non-decreasing manner and each repeated a number of times equal to its order as a conjugate point; similarly let τ_j^-, $(j = 1,2,\ldots)$, be the left-hand conjugate points to $t = a$, ordered in non-increasing manner, and with the same convention as to repetition. The values t at which $V(t)$ is singular define points which are called pseudoconjugate to $t = a$, in accordance with the terminology introduced by Picone [1], and these pesudoconjugates are isolated whenever the reciprocal equation is identically normal. Under the assumption of identical normality of the reciprocal equation, let σ_j^+, $(j = 1,2,\ldots)$ and σ_j^-, $(j = 1,2,\ldots)$, denote the sequences of right- and left-hand pseudoconjugates to $t = a$, with similar conventions as to order and repetitions. It is to be emphasized that any of the sequences $\{\tau_j^+\}$, $\{\tau_j^-\}$, $\{\sigma_j^+\}$, $\{\sigma_j^-\}$ may be finite or vacuous.

As a value $t = s$ is a point conjugate to $t = a$ if and only if there is an eigenvalue ω of $\Theta(s|Y)$ with $\arg \omega = 0 \pmod{2\pi}$, and $t = s$ is a point pseudoconjugate to $t = a$ if and only if there is an eigenvalue of $\Theta(s|Y)$ with $\arg \omega = \pi \pmod{2\pi}$, the following separation results follow from the above remarks.

THEOREM 10.2. *Suppose that the coefficient matrix functions of* (3.1) *satisfy hypothesis* ($\mathscr{A}_1'$), *and the matrix functions* $B(t)$ *and* $-C(t)$ *are positive definite on* I. *Then*

(i) *if the right-hand conjugate* τ_j^+, {*left-hand conjugate* τ_j^-} *of* $t = a$ *exists, then the right-hand pseudoconjugate* σ_j^+, {*left-hand pseudoconjugate* σ_j^-} *of* $t = a$ *exists and* $a < \sigma_j^+ < \tau_j^+$, $\{\tau_j^- < \sigma_j^- < a\}$.

(ii) *if a closed subinterval* $[a_0,b_0]$ *of* I *contains* $n + 1$ *points which are pseudoconjugate,* {*conjugate*} *to* $t = a$, *then* $[a_0,b_0]$ *contains at least one point which is conjugate,* {*pseudoconjugate*} *to* $t = a$.

In particular, conclusions (i) *and* (ii) *hold whenever hypotheses* ($\mathscr{A}_1'$) *and* ($\mathscr{A}_2'$) *are satisfied.*

For systems (3.1) in the reduced form with $A(t) \equiv 0$, the results of Theorem 10.2 are given in Atkinson [2; Theorem 10.3.1, 10.3.2], and the result of Theorem 10.2 is Atkinson's Theorem 10.3.3.

11.   Principal Solutions

Corresponding to the discussion of Section IV.3, we
shall now consider the concept of principal solutions for a
system (3.1) which satisfies hypothesis ($\mathscr{H}$) on an open interval I = (a_o,b_o), $-\infty \le a_o < b_o \le \infty$, on the real line.   For
systems (3.1) equivalent to the special example (2.1) of Section 2, this concept was initially formulated by Hartman
[10], and shortly thereafter extended by Reid [15].   For the
case of a system (3.1) which is identically normal the
definition and characterization of such solutions is much
simpler than in the more general abnormal situation, and
introduction of this concept will be limited to this case.
That is, we shall assume in our textual discussion that (3.1)
is a system which satisfies on (a_o,b_o) *hypothesis* ($\mathscr{H}$) *and
is identically normal.*

For s $\in$ I, let Y(t;s) = (U(t;s);V(t;s)) be the solution of (3.1_M) satisfying the initial conditions

$$U(s;s) = 0, \quad V(s;s) = E. \qquad (11.1)$$

Then a value t $\in$ I conjugate to s is a value at which
U(t;s) is singular, and the order of t as a conjugate
point to s is equal to r if U(t;s) is of rank n - r.
In particular, the first left-hand conjugate point to s,
t = $\tau_1^-(s)$, is the largest value on (a_o,s) at which U(t;s)
is singular.   Now in view of Theorem 8.6, $\tau_1^-(s)$ is a strictly
monotone increasing function of s, and as s → b_o we have
that $\tau_1^-(s)$ tends to a limit $\tau_1^-(b_o)$, and is called the
first (left-hand) conjugate point to b_o on I.   As noted in
the case of scalar second order equations in Section IV.3,

b_o may be its own first conjugate point on I, in which case the system (3.1) is oscillatory on arbitrary neighborhoods (t_o,b_o) of b_o. In the contrary case, $\tau_1^-(b_o) < b_o$ and (3.1) is disconjugate on the open interval $(\tau_1^-(b_o),b_o)$. We shall assume that such is the case, and proceed to obtain a "principal solution" of (3.1) at b_o which in the scalar case reduces to that already determined in Section IV.3.

An initial result in this direction is that of the following theorem, which appears specifically as Theorem VII.3.1 of Reid [35], and which was essentially proved in Reid [15].

LEMMA 11.1. *Suppose that* (3.1) *is disconjugate on a subinterval* $I_o = (c,b_o)$ *of* $I = (a_o,b_o)$, *and that* $Y_o(t) = (U_o(t);V_o(t))$ *is a solution of* (3.1_M) *with* $U_o(t)$ *non-singular on* I_o. *Then for* $s \in I_o$ *the matrix function* $S(t,s;U_o)$ *defined by* (3.10) *is non-singular for* $t \in I_o$, $t \neq s$; *moreover, if there exists a* $s \in I_o$ *such that* $S^{-1}(t,s;U_o) \to 0$ *as* $t \to b_o$, *then* $S^{-1}(t,b;U_o) \to 0$ *as* $t \to b_o$ *for arbitrary* $b \in I_o$.

In view of the last conclusion of this lemma, if an equation (3.1) is identically normal on an open interval $I = (a_o,b_o)$, and disconjugate on a subinterval $I_o = (c_o,b_o)$, then a solution $Y_o(t) = (U_o(t);V_o(t))$ of (3.1_M) has been called by Reid [15] a *principal solution* at b_o of (3.1_M), or of (3.1), if $U_o(t)$ is non-singular for t on some subinterval $I\{Y_o\} = (d_o,b_o)$ of I, and $S^{-1}(t,s;U_o) \to 0$ as $t \to \infty$ for at least one $s \in I\{Y_o\}$, and consequently for all such values of s. The concept of a *principal solution* at a_o is defined in a similar fashion. In case $Y_o(t)$ is a conjoined basis for (3.1) for which $U_o(t)$ is non-singular on some neighborhood of b_o, this definition of principal solution agrees with that given by Hartman [10] for a system (3.1) specified by an equation (2.1) satisfying hypothesis $(\mathcal{A}_\omega)$. It is to be noted also that for a scalar equation of the sort discussed in Chapter IV this definition of principal solution reduces to that introduced by Hartman and Wintner. Moreover, one has the following result, which shows that if (3.1) is disconjugate on a neighborhood of b_o then a solution $Y_o(t) = (U_o(t);V_o(t))$ which is principal at b_o in the sense defined above also possesses a property which was used as a definitive property by Morse and Leighton [1] for a real scalar second order differential equation.

THEOREM 11.2. *Suppose that* (3.1) *is identically normal on* $I = (a_o,b_o)$, $-\infty \le a_o < b_o \le \infty$. *If this equation is disconjugate on a subinterval* $I_o = (c_o,b_o)$, *then a solution* $Y_o(t) = (U_o(t);V_o(t))$ *of* (3.1_M) *is a principal solution at* b_o *if* $U_o(t)$ *is non-singular on a subinterval* $I\{Y_o\} = (d_o,b_o)$ *of* I, *and there exists a solution* $Y_1(t) = (U_1(t);V_1(t))$ *of* (3.1_M) *with* $U_1(t)$ *non-singular on some interval* $I\{Y_1\} = (d_1,b_o)$, *and such that for some* $s \in (d_o,b_o)$ *we have*

$$U_1^{-1}(t)U_o(t)T(t,s;U_o) \to 0 \quad as \quad t \to b_o; \tag{11.2}$$

moreover, $\{Y_1,Y_o\}$ *is non-singular for any such* $Y_1(t)$. *Conversely, if* (3.1) *is disconjugate on a subinterval* (c_o,b_o), *and* $Y_o(t) = (U_o(t);V_o(t))$ *is a principle solution at* b_o *with* $U_o(t)$ *non-singular on* (d_o,b_o), *then any solution* $Y_1(t) = (U_1(t);U_2(t))$ *of* (3.1_M) *with* $\{Y_2,Y_1\}$ *non-singular is such that* $U_1(t)$ *is non-singular on some subinterval* $I\{Y_1\} = (d_1,b_0)$ *and* (11.2) *holds for arbitrary* $s \in (d_0,b_0)$.

Now in view of the discussion of Section 4, there exists a solution $Y_0(t) = (U_0(t);V_0(t))$ on $I = (a_0,b_0)$ with $U_0(t)$ non-singular on a subinterval $I\{Y_0\} = (d_0,b_0)$ if and only if on this subinterval there is a solution $W = W_0(t)$ of the Riccati matrix differential equation (4.1) with $W_0(t) = V_0(t)U_0^{-1}(t)$; moreover, $U = U_0(t)$ is a fundamental matrix solution of the first order homogeneous matrix differential equation

$$U'(t) = [A(t) + B(t)W_0(t)]U(t). \tag{11.3}$$

Now for a solution $W = W_0(t)$ of (4.1) on (d_0,b_0), let $H = H(t,s;W_0)$, $G = G(t,s;W_0)$ and $Z(t,s;W_0)$ be determined by equations (4.4) and (4.5). As established in equation (4.8) we have that $Z(t,s;W_0) = U_0(s)S^*(t,s;U_0)U_0^*(s)$. Also, as given in No. 1 of the Topics and Exercises at the end of this Chapter, if both $W_0(t)$ and $W_1(t)$ are solutions of (4.1) on a subinterval (d_0,b_0) of I, then for $s \in (d_0,b_0)$ the matrix $\Gamma = W_0(s) - W_1(s)$ is such that $E + \Gamma Z(t,s;W_1)$ is non-singular for $t \in (d_0,b_0)$ and $Z(t,s;W_0) = Z(t,s;W_1)[E + \Gamma Z(t,s;W_1)]^{-1}$, so that if $Z(t,s;W_0)$ is non-singular then $Z(t,s;W_1)$ is also non-singular and

$$Z^{-1}(t,s;W_1) = Z^{-1}(t,s;W_0) + W_0(s) - W_1(s). \tag{11.4}$$

In particular, if $Z^{-1}(t,s;W_0) \to 0$ as $t \to b_0$ then $Z^{-1}(t,s;W_1) \to W_1(s) - W_0(s)$ as $t \to b_0$. Consequently, if also $Z^{-1}(t,s;W_1) \to 0$ it follows that $W_1(s) = W_0(s)$, and therefore $W_1(t) \equiv W_0(t)$ on (d_0,b_0).

Now if $W_0(t)$ is any solution of (4.1) on a subinterval (d_0,b_0) of (a_0,b_0), then for a value $s \in (d_0,b_0)$ the condition that $Z^{-1}(t,s;W_0) \to 0$ as $t \to b_0$ is equivalent to the condition that $S^{-1}(t,s;U_0) \to 0$ as $t \to b_0$, where $Y_0(t) = (U_0(t);V_0(t))$ is a corresponding solution of (3.1_M) with $U_0(t)$ non-singular and $W_0(t) = V_0(t)U_0^{-1}(t)$. Since by Theorem 11.2 this latter limit relation holds for one value $s \in (d_0,b_0)$ only if it holds for all s on this interval, the same is true for the limit relation involving $Z^{-1}(t,s;W_0)$. If $W_0(t)$ is a solution of (4.1) on a subinterval (d_0,b_0) such that $Z^{-1}(t,s;W_0) \to 0$ as $t \to b_0$ for at least one $s \in (d_0,b_0)$, and consequently for all such s, $W_0(t)$ is called a *distinguished solution* of (4.1) at b_0. Indeed, in view of the remark of the preceding paragraph, if there exists such a solution of (4.1) then it is unique, and may properly be called *the distinguished solution* of (4.1) at b_0. That is, if (3.1) is identically normal on $I = (a_0,b_0)$, $-\infty \leq a_0 < b_0 \leq \infty$, and disconjugate on a subinterval (c_0,b_0) of I, then there exists a principal solution $Y_0(t) = (U_0(t);V_0(t))$ of (3.1_M) at b_0 if and only if the Riccati matrix differential equation (4.1) has a distinguished solution $W = W_0(t)$ at b_0 and $W_0(t) = V_0(t)U_0^{-1}(t)$ on a subinterval (d_0,b_0) of I. In case such a distinguished solution of (4.1) at b_0 exists then it is unique, and the most general principal solution of (3.1_M) at b_0 is of the form $Y(t) = Y_0(t)K$, where K is a non-singular $n \times n$ matrix.

Now hypothesis ($\mathscr{U}$) requires the matrix functions $B(t)$ and $C(t)$ to be hermitian on I, and therefore if $W = W_0(t)$ is a solution of the Riccati matrix differential equation

(4.1) on a subinterval I_o of I then $W = W_o^*(t)$ is also a solution of (4.1) on I_o. Moreover, from the definitions of $H(t,s;W_o)$ and $G(t,s;W_o)$ as solutions of the differential systems (4.4) it follows that $\{H(t,s;W_o)\}^* = G(t,s;W_o^*)$, $\{G(t,s;W_o)\}^* = H(t,s;W_o^*)$, and therefore $\{Z(t,s;W_o)\}^* = Z(t,s;W_o^*)$. In particular, if $W = W_o(t)$ is a distinguished solution of (4.1) at b_o or at a_o, then $W = W_o^*(t)$ is also a distinguished solution of (4.1) at the same end-point. As a consequence of the above established results on uniqueness of distinguished solutions when they exist, and the equivalence of the existence of a distinguished solution of (4.1) with the existence of a principal solution of (3.1_M) at the same end-point of I, we have that if $Y_o(t) = (U_o(t);V_o(t))$ is a principal solution of (3.1_M) at b_o, {at a_o}, in the sense of Reid [15] then $Y_o(t)$ is a conjoined basis for (3.1) with $U_o(t)$ non-singular on some subinterval $I\{Y_o\} = (d_o,b_o)$, $\{I\{Y_o\} = (a_o,d_o)\}$, of I.

A still further characterization of a principal solution is given in Theorem VII.3.4 of Reid [35], to the effect that if (3.1) is identically normal on $I = (a_o,b_o)$ and disconjugate on $I_o = (c_o,b_o)$, then for $W(t)$ a solution of (4.1) on I_o the function $Z(t,s;W)$ is non-singular for $(t,s) \in I_o \times I_o$, $t \neq s$ and

$$Z^{-1}(d,s;W) = W(s) - W_d(s) \quad \text{for} \quad d \in I_o, \quad d \neq s,$$

where $W_d(t) = V_d(t)U_d^{-1}(t)$ and $Y_d(t) = (U_d(t);V_d(t))$ is the solution of (3.1_M) determined by the initial conditions

$$U_d(d) = 0, \quad V_d(d) = E; \tag{11.5}$$

in particular, there exists a principal solution $Y_{b_o}(t) = (U_{b_o}(t);V_{b_o}(t))$ of (3.1_M) at b_o if and only if the corresponding solution $W_d(t) = V_d(t)U_d^{-1}(t)$ of (4.1) is such that $W_{b_o}(t) = \lim_{d\to b_o} W_d(t)$ exists for $t \in I_o$ and $W(t) = W_{b_o}(t)$ is a solution of (4.1) on I_o, and in this case $W_{b_o}(t)$ is necessarily the distinguished solution of (4.1) at b_o.

Of particular interest is the case of self-adjoint systems (3.1) which on the open interval $I = (a_o,b_o)$ are identically normal and $B(t) \geq 0$ for t a.e. on this interval. For such systems we have the following existence theorem of construction type, and for which the proof will be given in order to emphasize the pertinent role played by extremizing properties of solutions of this differential system.

THEOREM 11.3. *Suppose that hypothesis* ($\mathscr{A}$) *holds for the system* (3.1) *on* $I = (a_o,b_o)$, *while on this interval the system is identically normal and* $B(t) \geq 0$ *for* t a.e. *on* I, *and that there is a subinterval* $I_o = (c_o,b_o)$ *on which* (3.1) *is disconjugate. For* s *and* b *distinct values on* I_o, *let* $Y_{sb}(t) = (U_{sb}(t);V_{sb}(t))$ *be the solution of* (3.1_M) *satisfying the boundary conditions* $U_{sb}(s) = E$, $U_{sb}(b) = 0$. *Then* $Y_{sb_o}(t) = (U_{sb_o}(t);V_{sb_o}(t)) = \lim_{b\to b_o} Y_{sb}(t)$ *exists and is a principal solution of* (3.1_M) *at* b_o, *with* $U_{sb_o}(t)$ *non-singular on* I_o; *in particular,* $Y_{sb_o}(t)$ *is a conjoined basis for* (3.1) *on* I_o *and* $Y_{cb_o}(t) = Y_{sb_o}(t)U_{cb_o}(s)$ *for* $(s,c,t) \in I_o \times I_o \times I_o$.

As (3.1) is assumed to be identically normal and disconjugate on I_o, if s and b are distinct values on this subinterval then there is a unique solution $Y_{sb}(t) = (U_{sb}(t);V_{sb}(t))$ of (3.1_M) satisfying the boundary conditions $U_{sb}(s) = E$, $U_{sb}(b) = 0$. Clearly $Y_{sb}(t)$ is a conjoined basis for (3.1), and consequently $U^*_{sb}(t)V_{sb}(t)$ is hermitian for $t \in I$; in particular, $V_{sb}(s)$ is hermitian. For a given $s \in I_o$, let a and b be values on I_o satisfying $a < s < b$, and for an arbitrary non-zero n-dimensional vector ξ let $\eta(t)$ be the vector function on $[a,b]$ defined as $\eta(t) = U_{sa}(t)\xi$ for $t \in [a,s]$, $\eta(t) = U_{sb}(t)\xi$ for $t \in [s,b]$. Then $\eta \in D_o[a,b]$, and as (3.1) is disconjugate on $[a,b]$ and $B(t) \geq 0$ for t a.e. on this subinterval, from (6.3) and the remark following Theorem 6.2 it then follows that $0 < J[\eta;a,b] = \xi^*U^*_{sa}(s)V_{sa}(s)\xi - \xi^*U^*_{sb}(s)V_{sb}(s)\xi = \xi^*[V_{sa}(s) - V_{sb}(s)]\xi$. Since this inequality holds for arbitrary non-zero $\xi \in \mathbf{C}_n$, we have that matrix inequality

$$V_{sb}(s) < V_{sa}(s) \text{ for } (a,s,b) \in I_o \times I_o \times I_o, \quad a < s < b. \tag{11.6}$$

For $s < b < d < b_o$, and ξ an arbitrary non-zero vector in $\mathbf{C}_n$, let $(u(t);v(t)) = (U_{sd}(t)\xi;V_{sd}(t)\xi)$ and $\eta_1(t) = U_{sb}(t)\xi$ for $t \in [s,b]$, $\eta_1(t) = 0$ for $t \in [b,d]$. Then $(u(t);v(t))$ is a solution of (3.1) and $\eta_1 \in D[s,d]$ with $\eta_1(s) = u(s)$, $\eta_1(d) = u(s)$; moreover, since $\eta_1(t) \equiv 0$ on $[b,d]$ and (3.1) is identically normal, it then follows that $\eta_1(t) \not\equiv u(t)$ on $[s,d]$. Consequently, in view of Theorem 6.3 and the discussion preceding Theorem 6.1 we have $-\xi^*V_{sd}(s)\xi = J[u;s,d] < J[\eta_1;s,d] = J[\eta_1;s,b] = -\xi^*V_{sb}(s)\xi$; that is,

$$V_{sb}(s) < V_{sd}(s), \text{ for } (s,b,d) \in I_o \times I_o \times I_o, \; s<b<d. \quad (11.7)$$

Using a similar argument, one obtains

$$V_{sr}(s) < V_{sa}(s), \text{ for } (r,a,s) \in I_o \times I_o \times I_o, \; r<a<s. \quad (11.8)$$

Inequalities (11.6), (11.7) imply that for fixed $s \in I_o$ the one parameter family of hermitian matrices $V_{sd}(s)$, $d \in (s,b_o)$, is monotone and bounded, and hence there is an hermitian matrix V_{sb_o} such that $\{V_{sd}(s)\} \to V_{sb_o}$ as $d \to b_o$. Also, in view of (11.6), (11.7), (11.8), it follows that

$$V_{sd}(s) < V_{sb_o} < V_{sr}(s), \text{ for } (r,s,d) \in I_o \times I_o \times I_o, \quad r < s < d. \quad (11.9)$$

If $Y_{sb_o}(t) = (U_{sb_o}(t);V_{sb_o}(t))$ is the solution of (3.1_M) determined by the initial condition $Y_{sb_o}(t) = (E;V_{sb_o})$, then clearly $\{Y_{sd}(t)\} \to Y_{sb_o}(t)$ as $d \to b_o$ for $t \in I$, and indeed the convergence is uniform on arbitrary compact subintervals of I. Also, in view of the result of Theorem 7.3, inequality (11.9) implies that $U_{sb_o}(t)$ is non-singular on each subinterval $[r,d]$ of I_o with $r < s < d$, and consequently $U_{sb_o}(t)$ is non-singular on I_o. Now let $Y_d(t) = (U_d(t);V_d(t))$ be the solution of (3.1_M) determined by the initial conditions (11.5). Then for $(s,d) \in I_o \times I_o$ and $s \neq d$ we have $Y_d(t) = Y_{sd}(t)V_{sd}^{-1}(d)$, and $W_d(t) = V_d(t)U_d^{-1}(t)$ is identical with $W_{sd}(t) = V_{sd}(t)U_{sd}^{-1}(t)$ for $(s,d,t) \in I_o \times I_o \times I_o$, $t \neq d$. As $U_{sb_o}(t)$ is non-singular on I_o, the matrix function $W_{sb_o}(t) = V_{sb_o}(t)U_{sb_o}^{-1}(t)$ is a solution of the Riccati matrix differential equation (4.1), which on arbitrary compact subintervals of I_o is the uniform limit of $W_{sd}(t) = W_d(t)$ as $d \to b_o$. From comments preceding the statement of Theorem 11.3 it then follows that $W_{sb_0}(t)$ is the distinguished solution of (4.1) at b_0, and $Y_{sb_0}(t)$ is a principal solution of (3.1_M) at this end-point; in particular, $Y_{sb_0}(t)$ is a conjoined basis for (3.1). The final conclusion of the theorem follows from the fact that if s,c and d belong to I_0 then $Y_{cd}(t)$ and $Y_{sd}(t)U_{cd}(s)$ are individually solutions of (3.1_M) with the corresponding matrix functions $U_{cd}(t)$ and $U_{sd}(t)U_{cd}(s)$ equal for $t = s$ and $t = d$, so that $Y_{cd}(t) = Y_{sd}(t)U_{cd}(s)$ for arbitrary $t \in I$, and hence $Y_{cb_0}(t) = Y_{sb_0}(t)U_{cb_0}(s)$ for $(s,c,t) \in I_0 \times I_0 \times I_0$.

Other basic properties of principal solutions are presented in the following theorem; for proofs, reference is to Reid [15, Th. 6.1 and Corollary, Th. 6.2; 35, Th. VII.5.5 and Corollary, Prob. VII.5.7].

THEOREM 11.4. *Suppose that the hypotheses of Theorem 11.3 are satisfied.*

(a) *If* $Y(t) = (U(t);V(t))$ *is a solution of* (3.1_M) *with* $U(t)$ *non-singular on a subinterval* $I\{Y\} = (c,b_0)$, *then for* $s \in I_0 \cap I\{Y\}$, *the matrix* $M(s;U) = \lim_{t\to\infty} S^{-1}(t,s,;U)$ *exists and is finite.*

(b) *If* $Y_1(t) = (U_1(t);V_1(t))$ *is a solution of* (3.1_M) *with* $U_1(t)$ *non-singular on a subinterval* $I_1 = (c_1,b_0)$ *of* I, *and* $s \in I_1$, *then it is not true that* $T(t,s;U_1) \to 0$ *as* $t \to b_0$.

(c) *If* $Y_0(t) = (U_0(t);V_0(t))$ *is a principal solution of* (3.1_M) *at* b_0, *then for a* $2n \times n$ *dimensional solution* $Y_2(t) = (U_2(t);V_2(t))$ *of the system the* $n \times n$ *constant matrix* $\{Y_0,Y_2\}$ *is non-singular if and only if* $U_2(t)$ *is non-singular for* t *on some subinterval* $I\{Y_2\} = (c_2,b_0)$ *and* $U_2^{-1}(t)U_0(t) \to 0$ *as* $t \to b_0$*; moreover, if* $\{Y_0,Y_2\}$ *is non-singular then for* s *on some subinterval* (a,b_0) *of* I *the matrix* $\lim_{t\to b_0} S(t,s;U_2)$ *exists and is non-singular.*

(d) *If* $s \in I_0$, *and* $Y_{sb_0}(t) = (U_{sb_0}(t);V_{sb_0}(t))$ *is the solution of* (3.1_M) *determined in Theorem* 11.3*, then for* $Y(t) = (U(t);V(t))$ *a solution of* (3.1_M) *such that* $U(t)$ *is non-singular on* I_0 *and* $S(b_0,c;U) = \lim_{t\to b_0} S(t,c;U)$ *exists and is finite for some value* $c \in I_0$*, then for arbitrary* $s \in I_0$ *we have*

(i) $S(b_0,s;U)$ *exists, and* $S(b_0,s;U) = T(s,c;U)[S(b_0,c;U) - S(s,c;U)]$;

(ii) $\{Y_{sb_0},Y\}$ *is non-singular;*

(iii) $U^{-1}(t)U_{sb_0}(t) \to 0$ *as* $t \to b_0$;

(iv) $\{Y_{sb_0},Y\} - \{Y,Y\}U^{-1}(s)$ *is non-singular, and* $T(b_0,s;U) = \lim_{t\to b_0} T(t,s;U)$ *exists and is equal to the non-singular matrix*

$$\{Y,Y_{sb_0}\}^{-1}[\{Y,Y_{sb_0}\} - U^{*-1}(s)\{Y,Y\}];$$

(v) $U_{sb_0}(t) = U(t)S(b_0,t;U)\{Y_{sb_0},Y\}$.

12. Comments on Systems (3.1) Which are not Identically Normal

For systems (3.1) which are not identically normal, various aspects of the theory treated in the earlier sections of this chapter become decidedly more complicated, and sometimes quite altered. As already noted in Lemma 5.1, if $a \in I$ and the system is not normal on subintervals having $t = a$ as an end-point, then there does not exist a single conjoined basis $Y(t) = (U(t);V(t))$ such that the points conjugate to $t = a$ are the values at which $U(t)$ becomes singular. An even more complicated aspect of oscillation phenomena is that when identical normality is absent then for a given joined basis $Y(t)$ there are in general whole intervals of values, each of which is a focal point of $Y(t)$. As a result, in order to develop an oscillation theory analogous to that explored in the earlier sections, the whole of such an interval of focal values should be considered as a focal element, so that one has "conjugate intervals" and "focal intervals". This altered definition of conjugate point and focal point that preserved the result of the Morse index theorem without the assumption of identical normality was initially considered in the context of certain types of self-adjoint integro-differential equations in Reid [7]. For the accessory differential systems occurring in varied types of variational problems this aspect was discussed in depth in University of Chicago dissertations by Hazard [1], Karush [1] and Ritcey [1], written under the direction of M. R. Hestenes.

For a system of the form (3.1) with continuous real-valued matrix coefficients and $B(t) \geq 0$, and with an altered definition of principal solution to be discussed later in this section, Stokes [1, Ch. 3] has recently defined the first right-hand {left-hand} conjugate point to an initial point $t = a$ as the first value $b > a$, $\{b < a\}$, at which the component matrix $U_a(t)$ belonging to a principal solution $Y_a(t) = (U_a(t);V_a(t))$ at $t = a$ has an expanding null-space. For the purpose of his discussion this definition suffices; however, in the larger context wherein one is dealing with the general problem of oscillation for abnormal systems (3.1), such a point is an end-point of the first right- or left-hand conjugate interval to an initial focal interval containing $t = a$.

A second aspect of the theory that becomes altered when the condition of identical normality is absent is that of the concept of a principal solution at a point b_0 which is the end-point of an interval $(a_0,b_0) \subset I$.

For systems (3.1) that are disconjugate on a subinterval $[c,b_0)$ of the given interval I of definition, and which are not identically normal, Reid [21] has defined a principal solution $Y_{b_0}(t) = (U_{b_0}(t);V_{b_0}(t))$ at b_0 by a method which maintains in large measure the procedure introduced in [15], and with the E. H. Moore generalized inverse of a matrix replacing the ordinary inverse in the definition. Indeed, the presented definition of a principal solution is not restricted to self-adjoint systems of the form (3.1). Moreover, for systems (3.1) satisfying ($\mathscr{U}$), with $B(t) \geq 0$ a.e. on I, and disconjugate on a subinterval $[c,b_0)$ of I, the existence of a principal solution is established by extremizing properties corresponding to those occurring in the proof given in Theorem 11.3 for the identically normal case. In connection with the definition of principal solution used by Reid [21], two properties are to be noted. Firstly, whenever a principal solution $Y(t) = (U(t);V(t))$ exists at a value b_0, then the component matrix $U(t)$ is non-singular in an open interval for which b_0 is an end-point. Secondly, principal solutions when existent are in general not unique, although when principal solutions exist a particular one may be uniquely determined by requiring the corresponding solution $W(t) = V(t)U^{-1}(t)$ of the Riccati differential equation to be orthogonal in a certain sense to the abnormal solutions of (3.1) at an arbitrarily selected value in a suitable neighborhood of b_o. In the case of a system (3.1) which satisfies hypothesis ($\mathscr{A}$), with $B(t) \geq 0$ and disconjugate on a subinterval $[c,b_o)$ of I, Theorem 5.3 of Reid [21] establishes the existence of a principal solution $Y_{b_o}(t) = (U_{b_o}(t);V_{b_o}(t))$, by extremizing properties corresponding to those occurring in the proof of Theorem 11.3; moreover, the principal solution thus determined is normalized in the sense referred to above.

The concept of a principal solution $Y(t) = (U(t);V(t))$ as b_o has been defined by Ahlbrandt [3] as a conjoined basis for (3.1) for which there exists another conjoined basis $Y_o(t) = (U_o(t);V_o(t))$ of (3.1) for which the constant matrix $\{Y_o,Y\}$ is non-singular, while $U_o(t)$ is non-singular on a neighborhood (c,b_o) of b_o and $U_o^{-1}(t)U(t) \to 0$ as $t \to \infty$. Whenever $B(t) \geq 0$ a.e., such a principal solution $Y(t)$ is given by the $Y_1(t) = (U_1(t);V_1(t))$ defined by (3.10) in the statement of Theorem 3.2. It is to be noted, however, that there is no assurance that the matrix function $U(t)$ is non-singular in a neighborhood of b_o.

A somewhat similar definition of principal solution is given by Stokes [1, Ch. III]. Starting with an arbitrary conjoined basis $Y_o(t) = (U_o(t);V_o(t))$ of (3.1) with $U_o(t)$ non-singular in a subinterval (a,b_o) of I, for $c \in (a,b_o)$ let $Y_c(t) = (U_c(t);V_c(t))$ be the solution of (3.1_M)

determined by $U_c(c) = 0$, $V_c(c) = -U_0^{*-1}(c)$. Then $Y_c(t)$ is a conjoined basis for (3.1) with $\{Y_c, Y_0\} = E$. According to Stokes, as $c \to b_0$ the conjoined basis $Y_c(t)$ tends to a limiting $Y(t) = (U(t);V(t))$, which he defines to be the principal solution of (3.1) at b_0. Moreover, if $W_1(t)$ is any hermitian solution of (4.1) which exists on a subinterval (a,b_0) of (a_0,b_0), then $U^*(t)V(t) \leq U^*(t)W_1(t)V(t)$ on (a,b_0); correspondingly, if $Y_2(t) = (U_2(t);V_2(t))$ is a solution of (3.1) for which $U_2(t)$ is non-singular in a neighborhood of b_0 then $\{Y_2,Y\}U_2^{-1}(t)U(t) \to 0$ as $t \to b_0$, so that if $\{Y_2,Y\}$ is non-singular then $U_2^{-1}(t)U(t) \to 0$ as $t \to b_0$. Again, as in the definition of Ahlbrandt, for a principal solution $Y(t) = (U(t);V(t))$ in the sense of Stokes there is no assurance that $U(t)$ is non-singular on a subinterval (c,b_0) of (a_0,b_0). Indeed, in the extreme case wherein $A(t) \equiv B(t) \equiv C(t) \equiv 0$ on $(-\infty,\infty)$, then $Y(t) = (0;K)$, with K non-singular, is a principal solution at ∞ in the sense of either Ahlbrandt or Stokes. In this example, however, the most general principal solution at ∞ in the sense of Reid [21] is $Y(t) = (K_0,KK_0)$, where K is hermitian and K_0 is invertible, and the normalized principal solution as determined in Theorem 5.3 of this paper is $Y(t) = (K_0,0)$, where K_0 is invertible. As is noted in Stokes [1], the definition of principal solution used therein corresponds to the extremal hermitian solutions of the corresponding Riccati differential equation, which in the particular example given above are $W_\infty(t) = (-\infty)E$ and $W_{-\infty}(t) = (+\infty)E$ in the context of infinite valued matrix functions.

13. Comments on the Literature on Oscillation Theory for Hamiltonian Systems (3.1)

Specific comparison and/or oscillation criteria for Hamiltonian systems (3.1) are to be found in the papers of Hartman and Wintner [11], Barrett [3, 10§5], Etgen [1,2,3, 4,6], Atkinson [2-Ch. 10], Kreith [2,4], Coppel [2-Ch. 2], and Reid [12, 31, 34, 36, 37, 42].

For the extension of results on singular quadratic functionals of the type obtained by Morse and Leighton [1], (see Sec. IV. 4), to higher dimensional problems involving n-dimensional vector functions, initial results on points conjugate to the singular end-point were obtained by Chellevold [1]. For this problem a necessary and sufficient condition for the minimum limit among $\mathscr{A}$-admissible arcs in terms of the singularity condition was derived by Tomastik [1]. Also, in his 1971 UCLA dissertation, J. Stein considered the problem of singular quadratic functionals in the general context of the quadratic form theory in Hilbert space of Hestenes [1]. Stein considers two approaches to the problem: the first for unbounded functionals over an enlarged class of functionals, in which there is derived a characterization of a relative singularity condition similar to the singularity condition of Morse and Leighton in terms of an index of the quadratic functional over certain linear manifolds; the second for bounded functionals, wherein there are established results on weakly lower semi-continuous, compact, and positively elliptic quadratic forms. In his recent paper [8], Morse surveyed the interrelations between this type of problem and singular Meyer fields for the accessory problem associated with a simple integral non-parametric problem in

(1+n)-space, and for the singular quadratic functional on (0,b] with single end-point at $t = 0$ derived the necessary and sufficient condition for a minimum limit in the class of $\mathscr{A}$-admissible arcs in terms of the non-existence on [0,b) of a point conjugate to $t = 0$ and the fulfillment of the singularity condition. Morse also discusses this problem briefly in Section 37 of [9], and formulates without proof some theorems on which a general oscillation theory might be built.

In his 1951 dissertation, Sternberg [1] extended to real symmetric systems (3.1) a number of oscillation and/or disconjugacy criteria given by Hille [1] and Wintner [4,5,6] for second order scalar differential equations. In particular, in the case of non-singular $B(t)$ he obtained a necessary and sufficient condition for disconjugacy for large t in terms of the existence of a solution for a Riccati matrix integral equation of the type presented by Hille [1] in the scalar case. Recently, Ahlbrandt [1] has extended significantly the general results of Sternberg. In particular, in Ahlbrandt [1 - Sec. 4, 2] there are established duality relations between disconjugacy criteria for a given equation (3.1), and the corresponding reciprocal equation (3.18). Ahlbrandt [3] has called $Y(t) = (U(t):V(t))$ a coprincipal solution of (3.1) at a value $t = c$ if $\mathscr{J}Y(t) = (-V(t);U(t))$ is a principal solution of (3.18) at this value, and the major results of [3] and subsequent work of Ahlbrandt are concerned with conditions under which a solution is principal if and only if it is coprincipal.

The boundary problem criterion of Nehari, (Theorem IV.5.1), for nonoscillation on a non-compact interval was extended by Reid [19-Ths. 3.1, 3.2] to systems (3.1), and applied to derive a criterion for non-oscillation of a higher order self-adjoint differential equation.

For Hamiltonian systems (3.1) there have not been developed to date a multiplicity of special criteria for oscillation and non-oscillation on a non-compact interval, corresponding to the situation for scalar second order differential equations as discussed in Chapter IV. There have been presented a few such criteria, however, involving comparison of the general system with a scalar equation, (see, for example, Reid [12-Th. 4.1, 17-Th. 5.1]), or the projection of the space of admissible n-dimensional vector functions on a subspace determined by a scalar function, (see, for example, Tomastik [1], Howard [3,4], Kreith [6], Noussair and Swanson [1], Simons [1]). These particular types of criteria are presented in a general form as [25] and [26] of the list of Topics and Exercises of Section 15. In connection with disconjugacy criteria, the dissertation of Stokes [1] on the Riccati matrix differential equation and associated differential inequalities is of particular interest. In addition to showing that these equations are the only type of symmetric matrix equations which preserve ordering of solutions as functions of the independent variable, and a discussion of the relation of such equations to disconjugacy, for these equations there are presented a continued fraction expansion and results on the asymptotic behavior of such solutions of the Riccati equation and the associated linear system. In particular, using the sequence of approximants of the continued fraction expansion there is derived a sequence of increasingly critical necessary conditions for disconjugacy. The sequence of approximants is obtained by the solution of a sequence of Riccati matrix differential equations, and in this regard the details of this procedure are similar to those occurring in the work of Willett [3], although the exact relationship has not been determined by the author.

14. Higher Order Differential Equations

Recently much attention has been devoted to extending elements of the Sturmian theory to higher order self-adjoint differential operators. Considerable impetus for the study of boundary problem involving real even order self-adjoint scalar differential equations was due to the early work of Kamke [6]. In particular, within recent years a great deal of investigation on fourth order equations has emanated from the pioneering study of Leighton and Nehari [1] in 1958. A relationship between the study of real self-adjoint linear differential equations and the functional approach of this chapter was indicated briefly at the end of Section 2 of this Chapter. At this point this discussion will be elaborated in a somewhat more general context. For simplicity, however, attention will still be limited to the case of equations with real coefficients; a corresponding discussion of equations with complex coefficients is to be found in Section 4 of Reid [19].

Suppose that $p_j(t)$, $(j = 0,1,\ldots,n)$ are real-valued functions, with $p_n(t) > 0$ and $p_0(t),\ldots,p_{n-1}(t)$, $1/p_n(t)$ locally integrable on a given non-degenerate interval I. Then the differential system

$$\begin{aligned} u_\alpha'(t) &= u_{\alpha+1}(t), \quad (\alpha = 1,\dots,n-1), \\ u_n'(t) &= [1/p_n(t)]v_n, \\ v_1'(t) &= p_0(t)u_1(t), \\ v_{\alpha+1}'(t) &= p_\alpha(t)u_{\alpha+1}(t) - v_\alpha(t), \quad (\alpha = 1,\dots,n-1) \end{aligned} \tag{14.1}$$

is of the form (3.1) with coefficient matrices given by $A_{\alpha,\alpha+1}(t) \equiv 1$, $(\alpha = 1,\dots,n-1)$, $A_{\alpha\beta}(t) \equiv 0$ otherwise; $B_{\alpha\beta}(t) \equiv 0$ for $(\alpha,\beta) \neq (n,n)$, $B_{nn}(t) = 1/p_n(t)$; $C_{\beta\beta}(t) = p_{\beta-1}(t)$, $(\beta = 1,\dots,n)$, $C_{\alpha\beta}(t) \equiv 0$ otherwise. It follows readily that such a system satisfies hypothesis $(\mathscr{A})$, and $B(t) \geq 0$; also, this system is identically normal on I. Moreover, $(u;v)$ is a solution of (14.1) if and only if there exists a scalar function $z(t)$ of class $\mathscr{C}^{[n-1]}$ with $z^{[n-1]}(t)$ and functions $v_j(t) = v_j(t|z)$, $(j = 1,\dots,n)$, locally a.c. and satisfying

$$\begin{aligned} u_j(t) &= z^{[k-1]}(t), \quad (k = 1,\dots,n) \\ v_n(t) &= p_n(t)z^{[n]}(t), \\ v_1'(t) &= p_0(t)z(t), \\ v_{\alpha+1}'(t) &= p_\alpha(t)z^{[\alpha]}(t) - v_\alpha(t), \quad (\alpha = 1,\dots,n-1). \end{aligned} \tag{14.2}$$

For a subinterval I_0 of I let $D(I_0)$ denote the class of real-valued functions z of class $\mathscr{C}^{[n-1]}$ on I_0, with $z^{[n-1]}$ locally a.c. and $p_n z^{[n]}$ locally of class $\mathscr{L}^\infty$ on I_0. For a compact subinterval $[a,b]$ of I_0, let $D_{0*}[a,b]$ and $D_{*0}[a,b]$ denote the subclasses of functions belonging to $D[a,b]$ such that respectively we have $z^{[k-1]}(a) = 0$, $(k = 1,\dots,n)$ and $z^{[k-1]}(b) = 0$, $(k = 1,\dots,n)$; also, let $D_0[a,b] = D_{00}[a,b] = D_{0*}[a,b] \cap D_{*0}[a,b]$. It follows readily that a vector function

$\eta \doteq (\eta_k(t))$ belongs to the class $D(I_o)$, $D[a,b]$, $D_{o*}[a,b]$, $D_{*o}[a,b]$ relative to the system (14.1), as defined in Section 6, if and only if $\eta_k(t) = \xi^{[k-1]}(t)$, $(k = 1,\ldots,n)$, where $\xi(t)$ is a function belonging to the corresponding class $D(I_o)$, $D[a,b]$, $D_{o*}[a,b]$, $D_{*o}[a,b]$, $D_o[a,b]$. Moreover, if $\eta(t) = (\eta_k(t))$ is a vector function with $\eta_1(t) = \xi(t) \in D[a,b]$ then the value of the functional $J[\eta;a,b]$ of (6.1) is equal to $J[\xi;a,b]$ defined by

$$J[\xi;a,b] = \int_a^b \Big[\sum_{j=0}^{n} p_j(t)(\xi^{[j]}(t))^2\Big]dt. \tag{14.3}$$

In general, for a solution $z(t)$, $v_j(t)$ of a system (14.2) the function $z(t)$ does not possess higher order differentiability beyond that specified above, and this system is not equivalent to an ordinary type linear differential equation in ξ. However, (14.2) is equivalent to a quasi-differential equation, (see Bôcher [5], Achieser-Glasmann [1, Anhang II], Reid [14, p. 451]), Bradley [1] of the form

$$D^{<2n>}z = 0, \tag{14.4}$$

where for $n = 1,\ldots,2n$ and a function z of class $\mathscr{C}^{[n-1]}$ with $z^{[n-1]}(t)$ a.c. we use the following modification of the formal derivative operator $D = d/dt$, as determined by the set of coefficient functions $p_j(t)$: $D^{<k>} = D^k$, $(k = 0,1,\ldots,n)$, $D^{<n>} = p_n D^{[n]}$, $D^{<n+j>} = DD^{<n+j-1>} - (-1)^j p_{n-j} D^{n-j}$, $(j = 1,\ldots,n-1)$, and $D^{<2n>} = DD^{<2n-1>}$. If for $j = 0,1,\ldots,n$ the function $p_j(t)$ is of class $\mathscr{C}^{[j]}$ on I, then (14.2) is satisfied by a set $z, v_1,\ldots,v_n$ if and only if $z \in \mathscr{C}^{[2n]}$ and in this case the equation $D^{<2n>}z = 0$ may be written as

$$\sum_{j=0}^{n} (-1)^{n-j} \{p_j(t) z^{[j]}(t)\}^{[j]} = 0. \qquad (14.5)$$

Moreover, in this case for $j = 1,\ldots,n$ we have

$$v_j(t) = \sum_{\alpha=j}^{n} (-1)^{\alpha-j} \{p_\alpha(t) z^{[\alpha]}(t)\}^{[\alpha-j]}. \qquad (14.6)$$

Employing the terminology of the preceding sections for the first order system (3.1), a pair of distinct points t_1, t_2 will be said to be (mutually) conjugate relative to (14.3) provided there exists a non-identically vanishing solution $z(t)$ of this equation satisfying $z^{[k-1]}(t_1) = z^{[k-1]}(t_2)$, $(k = 1,\ldots,n)$. Correspondingly, a value t_2 will be said to be a focal point to $t = t_1$ relative to the equation (14.4) and the initial condition $v_j(t_1|z) = 0$, $(j = 1,\ldots,k)$, if there exists a non-identically vanishing solution $z(t)$ of this equation satisfying $v_j(t_1|z) = 0$ and $z^{[j-1]}(t_2) = 0$, $(j = 1,\ldots,k)$. Clearly the more general concept of focal points as discussed in Section 6 are applicable to the system (14.1), and thus results for such systems are translatable into corresponding results for the equation (14.4) in the sense that they are results for the system (14.1). In particular, in this terminology, the equation (14.4) is called disconjugate on an interval, or oscillatory for large t, provided the system (14.1) possesses this property.

The literature on higher order linear differential equations may be separated into that dealing with fourth order equations and that concerned with general higher order equations, usually with real coefficients. Employing the terminology introduced above by way of the auxiliary system (14.1), specific consideration of comparison and/or oscillation phenomena for fourth order equations formed the central theme of the papers by Howard [1], Barrett [6,7,8, 9,10, Secs. III, IV], Hinton [1,4], Bradley [2], Kreith [3], Gustafson [2], Kuks [1], Lewis [2] and W. B. Miller [1]. Employing a technique initiated by Whyburn [7] in the case of real self-adjoint equations of the fourth order, the papers of Kreith [11-15], Cheng [1] and Edelson and Kreith [1] consider in the corresponding context oscillation phenomena for a pair of second order linear differential equations to which the general real non-self-adjoint linear differential equation of the fourth order is reducible. Corresponding references for general linear self-adjoint differential equations of higher order are Reid [19, 34, 36], Hunt [1,2], Sherman [1], Glazman [1-Chs. II, III], Hinton [2,4], Leighton [11], Gustafson [3], Kreith [4], Ridenhour [1], and Lewis [1,3]. For fourth order equations and for higher order equations there is much related literature wherein the central role is assumed by boundary problem theory, and which will be referenced in the following chapter.

It is to be emphasized that the above discussed definitions of conjugate point, disconjugacy and oscillation are the more restrictive ones arising from the corresponding concepts for the first order system (14.1) in the context of the earlier discussion of this chapter, and do not have the generality corresponding to the definitions introduced by Leighton and Nehari [1] for fourth order differential equations

$$[r(t)u''(t)]'' + q(t)u(t) = 0, \qquad (14.7)$$

with $r(t)$, $q(t)$ real-valued, continuous functions having

$r(t) > 0$ and $q(t)$ of constant sign on an interval I of the form $[a,\infty)$. Specifically, in the terminology employed by Leighton and Nehari, a value $t_2 > t_1$ is the k-th (right-hand) conjugate point to t_1 if t_2 is the minimum value of a_{k+3} such that there exists a non-identically vanishing solution $u(t)$ of (14.7) with zeros at $t = a_j$, $(j = 1,\ldots, k+3)$, where $t_1 = a_1 \leq a_2 \leq \ldots \leq a_{k+3}$, and zeros are counted according to multiplicities. This concept of conjugate point was extended by Sherman [1] to real n-th order linear differential equations of the form

$$\ell_n[u](t) \equiv u^{[n]}(t) + p_{n-1}(t)u^{[n-1]}(t) + \ldots + p_0(t)u(t) = 0, \tag{14.8}$$

with continuous coefficients on $I = (a,\infty)$, $-\infty \leq a < \infty$. Subsequently an extensive literature in this area has evolved, largely through the works of Sherman [2,3], Hinton [3], Hartman [14,15,16,18], A. Ju. Levin [4], Gustafson [1,2], W. J. Kim [1], Keener [1,2], and Ridenhour [3]. An excellent survey of this area is presented in Chapter 3 of Coppel [2].

An equation (14.8) has been termed *oscillatory* on an interval $[c,\infty)$, $-\infty < c < \infty$, if there exists a non-identically vanishing solution whose set of zeros is unbounded above, and non-oscillatory otherwise. Such an equation has been called *conjugate on a subinterval* I_0 if there exists a non-identically vanishing solution which has on I_0 at least n zeros, where each zero is counted according to its multiplicity, and *disconjugate* on I_0 in the contrary case. Also, (14.8) has been called *eventually disconjugate* on $[c,\infty)$ if there is a value t_0 such that it is disconjugate on $[t_0,\infty)$. A non-identically vanishing solution $u(t)$ is said to have

$(i_1,\ldots,i_k)$-zeros on I_0 if there exist k values $s_1 < s_2 < \ldots < s_k$ on I_0 such that $u(t)$ has at $t = s_j$ a zero of multiplicity at least i_j, $(j = 1,\ldots,k)$. For positive integers $i_1,\ldots,i_k$ with $i_1 + \ldots + i_k \geq n$ the symbol $r_{i_1 i_2 \ldots i_k}(s)$ has been used to denote the infimum of values $\sigma > s$ such that (14.8) has on $[s,\sigma]$ a solution with $(i_1,\ldots,i_k)$-zeros, where by convention the infimum of the empty set is taken to be ∞. Also, the condition $r_{i_1 i_2 \ldots i_k}(s) = \infty$ for all $s \in I$ is expressed by $r_{i_1 i_2 \ldots i_k} = \infty$. If $s \in [a,\infty)$ and k is a positive integer, then the k-th (right-hand) *conjugate point* $\eta_k(s)$ to s is defined to be the infimum of the set of values $\sigma > s$ such that there is a solution of (14.8) with a zero at s and at least $n + k - 1$ zeros on $[s,\sigma]$. When $\eta_k(s) < \infty$ this infimum is a minimum; that is, there exist non-identically vanishing solutions of (14.8) with a zero at s and at least $n + k - 1$ zeros on $[s,\eta_k(s)]$; all such solutions have been termed *extremal solutions* for $\eta_k(s)$.

For an equation (14.7) with $q(t) > 0$ it follows readily that $r_{22} = \infty$, and in this case Leighton and Nehari [1] showed that an extremal solution for the k-th conjugate point $\eta_k(s)$ is unique except for a multiplicative constant and has a triple zero at $t = s$. Subsequently, Peterson [2] proved that for the general self-adjoint fourth order equation with $r_{22} = \infty$ this characterization of extremal solutions remained valid. For the case of an equation (14.7) with $q(t) < 0$ Leighton and Nehari showed that $r_{13} = r_{31} = r_{121} = \infty$, and using only these specific properties they proceeded to show that an extremal solution for the k-th conjugate point $\eta_k(s)$ has a double zero at each of the values $t = s$, $t = \eta_k(s)$, and exactly $k - 1$ simple zeros on the open interval $(s, \eta_k(s))$. For other special fourth order equations this result was generalized by Peterson [1] and Pudei [1,2]. Subsequently Ridenhour and Sherman [1] established for the general fourth order equation $\ell_4[u](t) = 0$ of the form (14.8) that the condition $r_{121} = \infty$ implies also $r_{13} = \infty = r_{31}$, and in turn that for this general equation the Leighton-Nehari characterization of extremal solutions for the k-th conjugate points remains valid.

In the general discussion of oscillation phenomena in the sense of Leighton and Nehari for linear equations of order $2m$, some results have been obtained on the existence and/or non-existence of solutions with two distinct zeros, each of order at least m, (see, for example, Ridenhour [1]). Since for self-adjoint equations of order $2m$ such results are immediately interpretable in terms of conjugate and focal points in the sense emanating from the associated Hamiltonian system (14.1), for such equations there persist relationships between the two concepts of oscillation. This connection is not fully understood at present, however, and in the opinion of the author this area of oscillation theory is far from closed at the present time.

15. Topics and Exercises

1. If $W(t)$ is a solution of the Riccati matrix differential equation (4.1) on a non-degenerate subinterval I_0 of I, and for $s \in I_0$ the matrix functions $H(t) = H(t,s;W)$, $G(t) = G(t,s;W)$ and $Z(t) = Z(t,s;W)$ are defined as in equations (4.4), (4.5) then $W(t)$, $H(t)$, $G(t)$, $Z(t)$ are

characterized by the property that the $2n \times 2n$ matrix function

$$\mathscr{W}(t) = \begin{bmatrix} W(t) & G(t) \\ H(t) & -Z(t) \end{bmatrix}, \quad \text{with } \mathscr{W}(s) = \begin{bmatrix} W(s) & E \\ E & 0 \end{bmatrix}, \qquad (15.1)$$

is the solution of the Riccati matrix differential equation

$$\mathscr{W}'(t) + \mathscr{W}(t)\mathscr{A}(t) + \mathscr{A}^*(t)\mathscr{W}(t) + \mathscr{W}(t)\mathscr{B}(t)\mathscr{W}(t) - \mathscr{C}(t) = 0, \qquad (15.2)$$

where $\mathscr{A}(t) = \operatorname{diag}\{A(t),0\}$, $\mathscr{B}(t) = \operatorname{diag}\{B(t),0\}$, and $\mathscr{C}(t) = \operatorname{diag}\{C(t),0\}$. If $W_0(t)$ is a second solution of (4.1) on I_0, and $\mathscr{W}_0(t)$ the corresponding solution of (15.2) then relations (4.6) and (4.6') for the solutions $\mathscr{W}(t)$ and $\mathscr{W}_0(t)$ yield the following relations, where $\Gamma = W(s) - W(s_0)$:

$$G(t,s;W) = G(t,s;W_0)[E + \Gamma Z(t,s;W_0)]^{-1},$$

$$H(t,s;W) = [E + Z(t,s;W_0)\Gamma]^{-1}H(t,s;W_0),$$

$$Z(t,s;W) = [E + Z(t,s;W_0)\Gamma]^{-1}Z(t,s;W_0),$$

$$= Z(t,s;W_0)[E + \Gamma Z(t,s;W_0)]^{-1}.$$

In particular, if $Z(t,s;W_0)$ is non-singular for a fixed s and t on a subinterval I_1 of I_0, then for the same s the matrix $Z(t,s;W)$ is also nonsingular for $t \in I_1$, and $Z^{-1}(t,s;W) = Z^{-1}(t,s;W_0) + \Gamma$. {Reid [20, 35-Prob. II.11]}.

2. For M_α, $(\alpha = 1,2,3,4)$, $n \times n$ matrices with $M_3 - M_2$ and $M_4 - M_1$ non-singular, define $\{M_1,M_2,M_3\}$ as $(M_3 - M_1)(M_3 - M_2)^{-1}$ and $\{M_1,M_2,M_3,M_4\}$ as $\{M_1,M_2,M_3\}\{M_2,M_1,M_4\}$; thus $\{M_1,M_2,M_3,M_4\}$ is a direct generalization of the scalar cross, or anharmonic, ratio. If $W_0(t)$, $W_\alpha(t)$, $(\alpha = 1,2,3,4)$, are solutions of (4.1) on a non-degenerate subinterval I_0 of I with $W_3 - W_2$ and $W_4 - W_1$

non-singular on this subinterval, then for $(t,s) \in I_0 \times I_0$ we have the identity

$$\{W_1(t),W_2(t),W_3(t),W_4(t)\}$$
$$= \Lambda(t,s;W_0,W_1)\{W_1(s),W_2(s),W_3(s),W_4(s)\}\Lambda^{-1}(t,s;W_0,W_1),$$

where $\Lambda(t,s;W_0,W_1) = G(t,s;W_0)[E + \Gamma_1 Z(t,s;W_0)]^{-1}$ and $\Gamma_1 = W_1(s) - W_0(s)$. That is, if $W_\alpha(t)$, $(\alpha = 1,2,3,4)$, are four solutions of (4.1) on I_0 such that for some value $t = \tau$ on I_0 the two matrix functions $W_3(t) - W_2(t)$ and $W_4(t) - W_1(t)$ are non-singular then these matrix functions are non-singular on I_0, and throughout this subinterval the generalized cross ratio $\{W_1(t),W_2(t),W_3(t),W_4(t)\}$ is similar to a constant matrix. {Sandor [1]; J. J. Levin [1]; Reid [20, 35-Prob. II.10]}.

3. Suppose that $Y_1(t) = (U_1(t);V_1(t))$ and $Y_2(t) = (U_2(t);V_2(t))$ are $2n \times n$ solutions of (3.1_M) such that the $2n \times 2n$ matrix $Y_1^* \mathcal{J} Y_2$ is non-singular, then the $2n \times 2n$ matrix $[Y_1(t) \quad Y_2(t)]$ is a fundamental matrix for (3.1) and

$$\mathcal{J}[Y_1 \quad Y_2]^* \mathcal{J}^* [Y_1 \quad Y_2] = \begin{bmatrix} Y_2^* \mathcal{J} Y_1 & Y_2^* \mathcal{J} Y_2 \\ Y_1^* \mathcal{J} Y_1 & Y_2^* \mathcal{J} Y_2 \end{bmatrix}.$$

If $Y_2(t)$ is a conjoined basis for (3.1) then there exists an associated conjoined basis $Y_1(t)$ such that $Y_1^*(t)\mathcal{J}^* Y_2(t) \equiv E$. If $Y_1(t)$ and $Y_2(t)$ are such conjoined bases, then an arbitrary $2n \times n$ matrix solution $Y(t)$ of (3.1_M) is of the form $Y(t) = Y_1(t)C_1 + Y_2(t)C_2$, where C_1 and C_2 are $n \times n$ matrices uniquely determing as $C_2 = Y_1^* \mathcal{J}^* Y$ and $C_1 = Y_2^* \mathcal{J} Y$; moreover, $Y^* \mathcal{J}^* Y = C_1^* C_2 - C_2^* C_1$. {These results for real Hamiltonian systems arising as accessory systems for variational problems are classic and fundamental although not described in these specific terms, (see Bliss [7, Chs. VIII, IX]; Morse [4-Chs. III, IV; 9-Ch. 5]; Radon [1,2]; Ahlbrandt [3]}.

4. Suppose that hypothesis $(\mathscr{A})$ holds, and $A(t) \equiv 0$, $B(t) \geq 0$ for t a.e. on I. If $[a,b]$ is a compact subinterval of I, then the following conditions are equivalent:

(i) the differential system

$$-v'(t) + C(t)u(t) = 0, \quad u'(t) - B(t)v(t) = 0, \qquad (3.1_0)$$

has order of abnormality d on $[a,b]$;

(ii) the linear manifold of constant n-dimensional vectors π satisfying $B(t)\pi = 0$ a.e. on $[a,b]$ has dimension d;

(iii) the rank of the hermitian matrix $\int_a^b B(t)dt$ is $n - d$.

5. Suppose that for $\alpha = 1,2$ the matrix functions A^α, B^α, C^α satisfy hypothesis $(\mathscr{A})$ on I, and denote by L_1^α and L_2^α the corresponding differential expressions defined by (2.5). Also, suppose that $U_1(t)$, $V_1(t)$ are $n \times r$ locally a.c. matrix functions, while $U_2(t)$, $V_2(t)$ are $n \times n$ locally a.c. matrix functions with $U_2(t)$ non-singular on I. For brevity, let $W_2(t) = V_2(t)U_2^{-1}(t)$. Corresponding to (6.3) we have the differential identity for $Y_1(t) = (U_1(t);V_1(t))$ and $Y_2(t) = (U_2(t);V_2(t))$.

$$V_1^*B_1V_1 + U_1^*C_1U_1 = (U_1^*V)' + U_1^*L_1^1[U_1,V_1] - (L_2^1[U_1,V_1])^*V_1.$$

Combining this relation with one corresponding to (6.6) with $(\eta_\alpha,\zeta_\alpha)$, $(\alpha = 1,2)$, replaced by the matrix functions (U_1,V_1) one obtains the following matrix identity

$$\begin{aligned}\{U_1^*W_2U_1 - U_1^*V_1\}' &+ V_1^*[B_1 - B_2]V_1 + U_1^*[C_1 - C_2]U_1 \\ &+ [V_1 - W_2U_1]^*B_2[V_2 - W_2U_1] \\ &= -U_1^*L_1^2[U_2,V_2]U_2^{-1}U_1 + U_1^*L_1^1[U_1,V_1] \\ &\quad - (L_2^1[U_1,V_1])^*V_1 - U_1^*W_2^*L_2^2[U_1,V_1] \\ &\quad - (L_2^2[U_1,V_1])^*W_2U_1 + U_1^*W_2L_2^2[U_2,V_2]U_2^{-1}U_1 \\ &+ U_1^*U_2^{*-1}[Y_2^*JY_2][U_2^{-1}U_1]'.\end{aligned}$$

This identity is a direct matrix generalization of the Picone identity (3.2). In particular, whenever $L_2^1[U_1,V_1] = 0 = L_2^2[U_1,V_1]$ and $L_2^2[U_2,V_2] = 0$ then $U_2^*V_2 - V_2^*U_2 = 0$ on I and the right hand member of this identity reduces to the sum of its first two terms, and these are zero if also $L_1^1[U_1,V_1] = 0$, $L_1^2[U_2,V_2] = 0$. In particular, these conditions hold whenever the two systems $L^\alpha[y] = 0$, $(\alpha = 1,2)$, have in common the equation $L_2[u,v] = 0$, while $Y_2(t)$ is a conjoined basis for $L^2[y] = 0$ and $L_2[U_1,V_1] = 0$. The above identity includes as special cases the "Picone identities" of Coppel [1] and Kreith [5]. Also, when applied to linear systems (3.1) equivalent to higher order self-adjoint scalar differential equations one obtains a "Picone identity" for such equations, including as special instances those of Cimmino [1,2] and Eastham [1]. It is to be remarked that for differential systems in form more general than (3.1), and in general not self-adjoint, one has an identity similar to (6.6), an integral form of which is to be found in Reid [22-Th. 3.4] for a class of generalized differential equations. With the aid of this relation one may derive a relation corresponding to the above, but more complicated in nature since the formal adjoints of the L^α are also involved, and with the aid of such a relation one may establish "Picone identities" for non-self-adjoint differential systems and scalar equations.

6. For each integer k, $0 \le k \le n$, let I_k denote the subset of $[a,b]$ on which the matrix $B(t)$ is of rank $n-k$. Each set I_k is measurable, since it is the union of subsets on which individual minors of $B(t)$ of order $n-k$ are non-singular, while minors of higher order are singular. Moreover, in view of this characterization of I_k, it is readily seen that on I_k there exists an $n \times k$ measurable matrix $\pi(t;k)$ such that $B(t)\pi(t;k) = 0$, $\pi^*(t;k)\pi(t;k) \equiv E_k$, and the $(n+k) \times (n+k)$ matrix

$$M(t;k) = \begin{bmatrix} B(t) & \pi(t,k) \\ \pi^*(t;k) & 0 \end{bmatrix}, \quad t \in I_k$$

is non-singular. For $t \in I_k$, let $R(t;k)$ denote the $n \times n$ matrix such that

$$M^{-1}(t;k) = \begin{bmatrix} R(t;k) & \pi(t,k) \\ \pi(t;k) & 0 \end{bmatrix}, \quad t \in I_k.$$

The matrix $R(t;k)$ is hermitian, or rank $n - k$, measurable on I_k, and is the E. H. Moore generalized inverse of $B(t)$. In particular, $R(t;k)$ satisfies on I_k the equation $R(t;k)B(t)R(t;k) = R(t;k)$. Now define on I_k matrix functions $\pi(t)$, $M(t)$ and $R(t)$ as equal to $\pi(t;k)$, $M(t;k)$ and $R(t;k)$, respectively. The matrix $R(t)$ is $n \times n$ throughout $[a,b]$, but $M(t)$ is $(n+k) \times (n+k)$ and $\pi(t)$ is $n \times k$ for $t \in I_k$, with the understanding that if $k = 0$ then the matrices $\pi(t;k)$ are deleted and $M(t;0) = B(t)$, $R(t,0) = B^{-1}(t)$. Now if $\eta \in D[a,b]:\zeta$, then $\check{\zeta}(t) = \zeta(t) - \pi(t)\pi^*(t)\zeta(t)$ is such that $\eta \in D[a,b]:\check{\zeta}$; moreover, writing $L[\eta](t)$ for $\eta'(t) - A(t)\eta(t)$, we have

$\check{\zeta}(t) = R(t)L[\eta](t)$ and $\zeta^*(t)B(t)\zeta(t) = \{L[\eta](t)\}^*R(t)L\,\eta\,(t)$ for $t \in [a,b]$. It is to be emphasized that the conditions of hypothesis ($\mathscr{A}$) do not imply that the measurable matrix $R(t)$ is integrable on $[a,b]$. In case $B(t)$ is of class $\mathscr{L}^\infty[a,b]$ then the elements of $M(t;k)$ belong to $\mathscr{L}^\infty(I_k)$, and it follows readily that the matrix function $R(t)$ belongs to $\mathscr{L}^\infty[a,b]$ under the following additional hypothesis.

($\mathscr{A}'[a,b]$) *There exists a positive constant* $\kappa = \kappa[a,b]$ *such that* $|\det M(t)| \geq \kappa$ *for* t a.e. *on* $[a,b]$.

Now $|\det M(t)|$ is the absolute value of the product of the non-zero roots of the characteristic equation $\det[\lambda E - B(t)] = 0$, and for $B(t) \in \mathscr{L}^\infty[a,b]$ the above condition ($\mathscr{A}'[a,b]$) is equivalent to the condition that there exists a positive constant κ_0 such that $B^2(t) \pm \kappa_0 B(t) \geq 0$ for t a.e. on $[a,b]$. If $B(t) \geq 0$ this latter condition reduces to the existence of a positive κ_0 such that $B^2(t) - \kappa_0 B(t) \geq 0$ for t a.e. on $[a,b]$. In particular, if the hermitian matrix function $B(t)$ is continuous and of constant rank on $[a,b]$, then the matrix function $\pi(t)$ may be chosen as continuous on this interval; moreover, in this case the matrix $R(t)$ is continuous and condition ($\mathscr{A}'[a,b]$) holds. {Reid [25-§3]}.

7. If hypothesis ($\mathscr{A}$) is satisfied, while on a compact subinterval $[a,b]$ of I the matrix function $B(t)$ is such that $B(t) \geq 0$, and there exists a positive constant κ_0 satisfying $B^2(t) - \kappa_0 B(t) \geq 0$ for t a.e. on $[a,b]$, then there exists a positive constant κ such that

$$J[\eta;a,b] \leq \kappa\int_a^b \{|\eta'(t)|^2 + |\eta(t)|^2\}dt, \text{ for } \eta \in D[a,b].$$

8. If $R(t)$ is a non-singular hermitian matrix function such that $R(t)$ and $R^{-1}(t)$ belong to $\mathscr{L}^\infty[a,b]$, with $R(t) > 0$ for t a.e. on $[a,b]$, then for η an arbitrary n-dimensional vector function with $\eta' \in \mathscr{L}^2[a,b]$ the inequality

$$\int_a^b \eta^{*\prime}(t)R(t)\eta'(t)dt \geq 4\eta^*(c)\left[\int_a^b R^{-1}(s)ds\right]^{-1}\eta(c)$$

holds for $c \in [a,b]$. Moreover, the strict inequality holds if $\eta(t) \not\equiv 0$ and for each $s \in (a,b)$ with $|\eta(s)| \neq 0$ there is a neighborhood $(s-\delta,s+\delta)$ on which there is defined a continuous function ζ satisfying $\zeta(t) = R(t)\eta'(t)$ for t a.e. on this neighborhood. {Reid [12, Th. 2.3]. It is to be noted that if A, B are hermitian matrices satisfying $A > 0$ and $B - A > 0$, then the hermitian matrix $A^{-1} + (B-A)^{-1} - 4B^{-1}$ is non-negative definite}.

9. Consider two systems (3.1^α), $(\alpha = 1,2)$, involving matrix functions $\mathscr{A}^\alpha(t)$ of the form (3.2), which satisfy hypothesis ($\mathscr{A}$) on I and with $B^\alpha(t) \geq 0$ a.e. on this interval.

(i) If (3.1^1) is disconjugate on a compact subinterval $[a,b]$ of I, then in view of the equivalence of (ii) and (v) of Theorem 6.1, and known continuity properties of solutions of (3.1) as functions of initial data and coefficient functions, (see, for example, Reid [24-Comments in Introduction, and Th. 4.1]) it follows that there exists a $\delta > 0$ such that if $\int_a^b |\mathscr{A}^1(s) - \mathscr{A}^2(s)|ds < \varepsilon$ then (3.1^2) is also disconjugate on $[a,b]$. That is, in the Lebesgue space of matrix functions $\mathscr{A}(t)$ of the form (3.2) satisfying ($\mathscr{A}$) and with norm $\int_a^b |\mathscr{A}(s)|ds$ the set of such matrices with $B(t) \geq 0$ and (3.1) disconjugate is a relatively open set. Indeed, the cited theorem of Reid [24] provides results involving weak limits in place of the strong limits of the above statements.

(ii) If for a given compact subinterval $[a,b]$ of I the classes of functions $D^{\alpha}[a,b]$, $(\alpha = 1,2)$, for these two problems are the same, and each of the equations (3.1^{α}) is disconjugate on $[a,b]$, then in view of the equivalence of (i) and (ii) of Theorem 6.3 it follows that for $\theta \in [0,1]$ the equation (3.1) with $\mathscr{A}(t) = (1-\theta)\mathscr{A}^1(t) + \theta\mathscr{A}^2(t)$ is also disconjugate on $[a,b]$. That is, in the linear space of matrix functions of the form (3.2) satisfying $(\mathscr{A})$, the set of such matrices with $B(t) \geq 0$ and (3.1) disconjugate is a convex set.

10. (i) Suppose that $A^j(t)$, $B^j(t)$, $C^j(t)$, $(j = 1,2,\ldots)$ are continuous $n \times n$ matrix functions on $(-\infty,\infty)$ with the $B^j(t)$, $C^j(t)$ hermitian and $B^j(t) > 0$, and there are matrix functions $A(t)$, $B(t)$, $C(t)$ such that the sequences $\{A^j(t)\}$, $\{B^j(t)\}$, $\{C^j(t)\}$ converge uniformly on arbitrary compact subintervals to $A(t)$, $B(t)$, $C(t)$, respectively. If each system (3.1^j) with coefficient matrix functions A^j, B^j, C^j is disconjugate on $(-\infty,\infty)$, and $B(t) > 0$ for $t \in (-\infty,\infty)$, then the limit system (3.1) with coefficient matrix functions A, B, C is also disconjugate on $(-\infty,\infty)$.

(ii) Suppose that $A(t)$, $B(t)$, $C^{\alpha}(t)$, $(\alpha = 1,\ldots,r)$, are continuous $n \times n$ matrix functions on $(-\infty,\infty)$, with $B(t) > 0$. Consider the system (3.1:x) with coefficient matrix functions $A(t)$, $B(t)$, $C(t) = \sum_{\alpha=1}^{r} x_{\alpha}C^{\alpha}(t)$, where $x_1,\ldots,x_r$ are real parameters. Then the set of values $x = (x_{\alpha})$ in real r-space

R^r for which (3.1:x) is disconjugate on $(-\infty,\infty)$ is a closed convex set. {Special cases of these results for systems equivalent to scalar equations of the second order were used by Markus and Moore [1] in considering problems of oscillation and disconjugacy for such equations with almost periodic coefficients}.

11. Let θ_a and θ_b be $n \times r_a$ and $n \times r_b$ matrices with $0 \leq r_a \leq n$, $0 \leq r_b \leq n$, it being understood that if either $r_a = 0$ or $r_b = 0$ the respective matrix θ_a or θ_b does not occur, and that if $0 < r_a \leq n$ or $0 < r_b \leq n$ then the corresponding θ_a or θ_b has rank r_a or r_b. For θ_a and θ_b given $n \times n$ hermitian matrices, let

$$J^{\theta}[\eta;a,b] = \eta^*(a)\Gamma_a\eta(a) + \eta^*(b)\Gamma_b\eta(b) + J[\eta;a,b],$$

where $A(t)$, $B(t)$, $C(t)$ satisfy $(\mathscr{A})$, and denote by $D^{\theta}[a,b]$ those $\eta \in D[a,b]$ which satisfy $\theta_a^*\eta(a) = 0$, $\theta_b^*\eta(b) = 0$. In particular, $D^{\theta}[a,b] = D_o[a,b]$, $\theta_a = \theta_b = E_n$, $D^{\theta}[a,b] = D[a,b]$ if θ_a and θ_b are non-existent, $D^{\theta} = D_{*o}[a,b]$ if $\theta_b = E_n$ and θ_a non-existent, while $D^{\theta} = D_{o*}[a,b]$ if $\theta_a = E_n$ and θ_b non-existent. Then $J^{\theta}[\eta;a,b]$ is positive definite on $D^{\theta}[a,b]$ if and only if $B(t) \geq 0$ for t a.e. on $[a,b]$ and there exists a conjoined basis $Y(t) = (U(t);V(t))$ of (3.1) such that $U(t)$ is non-singular on $[a,b]$ and there exists an associated constant k such that $W(t) = V(t)U^{-1}(t)$ satisfies the conditions $\Gamma_a + k\theta_a\theta_a^* - W(a) > 0$, $\Gamma_b + k\theta_b\theta_b^* + W(b) > 0$. {Reid [22; Th. 7.2]}.

12. Suppose that for $\alpha = 1,2$ the matrix function $A^{\alpha}(t)$, $B^{\alpha}(t)$, $C^{\alpha}(t)$ satisfy hypotheses $(\mathscr{A})$, $B^{\alpha}(t) \geq 0$ for t a.e. on I, and $D^1(I) = D^2(I)$, while for $[a,b]$ a given compact subinterval of I there exists a solution

$y(t) = (u(t);v(t))$ of (3.1^1) with $u(t) \not\equiv 0$ on $[a,b]$ and

$$\Gamma_a u(a) - v(a) = 0, \quad \Gamma_b u(b) + v(b) = 0,$$

where Γ_a and Γ_b are $n \times n$ hermitian matrices. Moreover, suppose that $J^{1,2}[\eta;a,b]$ is non-negative on $D^1[a,b] = D^2[a,b]$, and there exists a conjoined basis $Y^2(t) = (U^2(t);V^2(t))$ of (3.1^2) with $U^2(a)$ and $U^2(b)$ non-singular, and

$$\Gamma_a - V^2(a)[U^2(a)]^{-1} \geq 0, \quad \Gamma_b + V^2(b)[U^2(b)]^{-1} \geq 0, \tag{15.3}$$

and that either $J^{1,2}[\eta;a,b]$ is positive definite on $D^1[a,b] = D^2[a,b]$ or the equality does not hold in one of the relations (15.3). Then there exists an $s \in [a,b]$ such that $U^2(s)$ is singular. {This result is a direct corollary of the preceding Exercise. For systems (3.1^α), $(\alpha = 1,2)$, of the form (2.1) with $Q(t) \equiv 0$ this result is equivalent to Theorem 1 of Kreith [2]}.

13. Consider the Riccati differential equation (4.1) with coefficients satisfying hypothesis $(\mathscr{A})$ and $B(t) \geq 0$ a.e. on I.

(i) If $W = W_2(t)$ is an hermitian solution of (4.1) on a subinterval $[c,d_0)$ of I, and $W = W_4(t)$ is an hermitian solution of this equation satisfying $W_4(c) > W_2(c)$, $\{W_4(c) \geq W_2(c)\}$, then Theorem 4.1 implies that $W_4(t)$ exists on $[c,d_0)$ and $W_4(t) > W_2(t)$, $\{W_4(t) \geq W_2(t)\}$, on this interval. Moreover, if $W_4(c) > W_2(c)$, and $Y_4(t) = (U_4(t);V_4(t))$ is a conjoined basis for (3.1) such that $U_4(t)$ is non-singular and $W_4(t) = V_4(t)U_4^{-1}(t)$ on $[c,d_0)$, then the hermitian matrix function $S(t,c;U_4) = \int_c^t U_4^{-1}(s)B(s)U_4^{*-1}(s)ds$ is non-decreasing on $[c,b_0)$ and satisfies

$$0 \leq S(t,c;U_4) \leq U_4^{-1}(c)[W_4(c) - W_2(c)]^{-1}U_4^{*-1}(c).$$

In particular, $S(d_o,c;U_4) = \lim_{t\to d_o} S(t,c;U_4)$ is a finite-valued hermitian matrix. This latter result is a consequence of relation (4.6_o) with $W = W_2$ and $W_o = W_4$, together with the remark that since $W_4(t)$ is hermitian the matrix function $T(t,c;U_4)$ defined by (3.7) is identically equal to E, and by (4.8) the matrix function $Z(t,c;W_4)$ is equal to $U_4(c)S(t,c;U_4)U_4^*(c)$.

(ii) Suppose that $Y(t) = (U(t);V(t))$ and $Y_1(t) = (U_1(t);V_1(t))$ are conjoined bases for (3.1) with $U(t)$ non-singular on a subinterval $[c,d_o)$ of I, with $S(d_o,c;U)$ a finite-valued hermitian matrix and equation (3.10) holds; that is,

$$U_1(t) = U(t)S(d_o,t;U), \quad V_1(t) = V(t)S(d_o,t;U)+U^{*-1}(t) \quad (15.4)$$

for $t \in [c,d_o)$. If $W = W_3(t)$ is an arbitrary hermitian matrix function which exists on $[c,d_o)$, and $Y_3(t) = (U_3(t);V_3(t))$ is a conjoined matrix solution of (3.1) with $U_3(t)$ non-singular on $[c,d_o)$ and $W_3(t) = V_3(t)U_3^{-1}(t)$, then

(a) $U_1^*(t)V_1(t) - U_1^*(t)W_3(t)U_1(t) \leq 0$, for $t \in [c,d_o)$;

(b) $\{Y_3,Y_1\}U_3^{-1}(t)U_1(t) \to 0$, as $t \to d_o$. (15.5)

In order to establish these conclusions, for $d \in (c,d_o)$ let

$$\begin{aligned} U_{1d}(t) &= U(t)S(d,t;U), \\ V_{1d}(t) &= V(t)S(d,t;U) + U^{*-1}(t) = W(t)U_{1d}(t) + U^{*-1}(t). \end{aligned} \quad (15.6)$$

Then $Y_{1d} = (U_{1d}(t);V_{1d}(t))$ is a conjoined basis for (3.1) with $\{Y_{1d},Y\} = E$, and $Y_{1d}(t) \to Y_1(t)$ uniformly on arbitrary compact subintervals of $[c,d_o)$. For $Y_3(t) = (U_3(t);V_3(t))$ a conjoined basis with $U_3(t)$ non-singular on $[c,d_o)$, and $W_3 = V_3U_3^{-1}$ we then have

$$\begin{aligned} \text{(a)}\quad U_{1d}(t) &= U_3(t)S(d,t;U_3)\{Y_{1d},Y_3\}, \\ \text{(b)}\quad V_{1d}(t) &= V_3(t)S(d,t;U_3)\{Y_{1d},Y_3\} - U_3^{*-1}(t)\{Y_{1d},Y_3\} \\ &= W_3(t)U_{1d}(t) - U_3^{*-1}(t)\{Y_{1d},Y_3\}, \end{aligned} \tag{15.7}$$

and consequently

$$U_{1d}(t)V_{1d}(t) - U_{1d}^*(t)W_3(t)V_{1d}(t) = -\{Y_{1d},Y_3\}^*S(d,t;U_3)\{Y_{1d},Y_3\}.$$

Since $S(d,t;U_3) \geq 0$ for $d \geq t$, it then follows that $U_{1d}^*(t)V_{1d}(t) - U_{1d}^*(t)W_3(t)V_{1d}(t) \leq 0$ for $d \geq t$, and (15.5a) follows upon letting $d \to d_o^-$.

Now suppose that $Y_4(t) = (U_4(t);V_4(t))$ is a conjoined basis for (3.1) with $U_4(t)$ non-singular on $[c,d_o)$ and $S(d_o,c;U_4)$ finite-valued. Then for $(t,d) \in [c,d_o) \times [c,d_o)$ relations (15.7) hold with $Y_3(t) = (U_3(t);V_3(t))$ replaced by $Y_4(t) = (U_4(t);V_4(t))$ and $W_3(t)$ replaced by $W_4(t) = V_4(t)U_4^{-1}(t)$, and upon letting $d \to d_o^-$ we have that

$$\begin{aligned} \text{(a)}\quad U_1(t) &= U_4(t)S(d_o,t;U_4)\{Y_1,Y_4\} \\ \text{(b)}\quad V_1(t) &= V_4(t)S(d_o,t;U_4)\{Y_1,Y_4\} - U_4^{*-1}\{Y_1,Y_4\}. \end{aligned} \tag{15.8}$$

In particular, it then follows that $U_4^{-1}(t)U_1(t) \to 0$ as $t \to d_o$. Moreover, equations (15.8) imply that for $t \in [c,d_o)$ we have

$$E = \{Y_1,Y\} = [\{Y_4,Y\}S(d_o,t;U_4)+U^*(t)U_4^{*-1}(t)]\{Y_1,Y_4\}$$

so that $\{Y_1,Y_4\}$ is non-singular, and $U_4^{-1}(t)U(t) \to [\{Y_1,Y_4\}^*]^{-1} = -\{Y_4,Y_1\}^{-1}$ as $t \to d_o^-$.

Finally, for a given hermitian solution $W_3(t)$ of (4.1) on $[c,d_o)$, let $W_4(t)$ be any hermitian solution of this equation with $W_4(c) > W_3(c)$. By conclusion (i) we then have $W_4(t) > W_3(t)$ on $[c,d_o)$, and by a result just established $U_4^{-1}(t)U_1(t) \to 0$. Then

$$\{Y_3,Y_1\}U_3^{-1}(t)U_1(t) = V_1^*(t)U_1(t) - U_1^*(t)W_3(t)U_1(t) \leq 0 \quad \text{on } [c,d_o).$$

Moreover, since $U_1^*(t)V_1(t) - U_1^*(t)W_4(t)U_1(t) = \{Y_4,Y_1\}U_4^{-1}(t)U_1(t)$ we have

$$0 \geq U_1^*(t)V_1(t) - U_1^*(t)W_3(t)U_1(t)$$
$$= \{Y_4,Y_1\}U_4^{-1}(t)U_1(t) + U_1^*(t)[W_4(t) - W_3(t)]U_1(t),$$

and since $U_4^{-1}(t)U_1(t) \to 0$ and $U_1^*(t)[W_4(t) - W_3(t)]U_1(t) \geq 0$ it follows that $U_1^*(t)[W_4(t) - W_3(t)]U_1(t) \to 0$ as $t \to d_o^-$, and hence also $\{Y_3,Y_1\}U_3^{-1}(t)U_1(t) \to 0$ as $t \to d_o^-$. {The principal results of the above are those of Stokes [1-Th. 1 of Ch. 3], with details of proof organized in the spirit of Reid [15, 20]}.

14. Suppose that the coefficient matrix functions in (3.1) are constant $n \times n$ matrices A, B, C with B and C hermitian. Then

(i) (3.1) is identically normal on $(-\infty,\infty)$ if and only if the $n \times n^2$ matrix $[B \;\; AB \; \ldots \; A^{n-1}B]$ has rank n;

(ii) if (3.1) is identically normal on $(-\infty,\infty)$ and $B \geq 0$, then this system is disconjugate on $(-\infty,\infty)$ if and only if there exists an hermitian constant matrix W satisfying the algebraic matrix equation

$$WA + A^*W + WBW - C = 0. \tag{15.9}$$

If such a system is disconjugate on $(-\infty,\infty)$, then the distinguished solutions $W_{\infty}(t)$, $W_{-\infty}(t)$ of (4.1) at ∞ and $-\infty$, respectively, are constant matrices W_{∞}, $W_{-\infty}$. Moreover, if $(U_0(t);V_0(t))$ is the solution of the corresponding system (3.1_M) satisfying the initial conditions $U_0(0) = 0$, $V_0(0) = E$, then $W_0(t) = V_0(t)U_0^{-1}(t)$ converges to W_{∞} and $W_{-\infty}$ as $t \to -\infty$ and $t \to \infty$, respectively. Furthermore, if W is any hermitian solution of the equation (15.9) then $W_{\infty} \leq W \leq W_{-\infty}$. {References: (i) Reid [35-Prob. VII.5.12]; (ii) Reid [20-Theorem 7.2; 34-Prob. VII.5.12]. For the case of $\mathscr{A}$ real, Coppel [3] presents an excellent discussion of the theory of solutions of (15.9), and various applications. In particular, he presents a greatly simplified proof of the following result due to Molinari [1,2]: If $\mathscr{A}$ is real, and $B \leq 0$, then the following statements are equivalent: (i) (A,B) is controllable, (i.e., the $n \times n^2$ matrix $[B \quad AB \ \ldots \ A^{n-1}B]$ is of rank n), and $\det[\lambda E_{2n} - \mathscr{J}\mathscr{A}]$ has no pure imaginary zeros; (ii) (15.9) has a solution W_1 such that $A + BW_1$ is stable, (i.e., all zeros of $\det[\lambda E_n - A - BW_1]$ have negative real parts), and a solution W_2 such that $-(A + BW_2)$ is stable}.

15. Suppose that on (a_0,b_0) the system (3.1) satisfies hypothesis ($\mathscr{A}$), if identically normal, and also $C(t) \geq 0$ for t a.e. If $Y_{b_0}(t) = (U_{b_0}(t);V_{b_0}(t))$ is a principal solution of (3.1_M) at b_0, then there exists a $c \in I$ such that the non-negative matrix function $V_{b_0}^* BV_{b_0} + U_{b_0}^* CU_{b_0}$ is integrable on $[s,b_0)$ for $s \in [c,b_0)$, and for such values of s,

$$-U^*_{b_o}(s)W_{b_o}(s)U_{b_o}(s)$$
$$= \int_s^{b_o} \{V^*_{b_o}(t)B(t)V_{b_o}(t) + U^*_{b_o}(t)C(t)U_{b_o}(t)\}dt;$$

also, $U^*_{b_o}(s)V_{b_o}(s) \leq 0$ for $s \in [c,b_o]$ and $U^*_{b_o}(s)V_{b_o}(s) \to 0$ as $s \to b_o$. Furthermore, if $C(t)$ is such that for arbitrary $c \in I$ there exists a value $\tau(c) > c$ such that $\int_c^t C(r)dr > 0$ for $t \geq \tau(c)$, then $W_{b_o}(s) < 0$ for $s \in [c,b_o)$. {This result may be proved by the argument used to establish Th. 8.1 of Reid [15]}.

16. Suppose that hypothesis ($\mathscr{A}$) holds, with $A(t) \equiv 0$, and $B(t) \geq 0$ for t a.e. on $I = (a_o,b_o)$.

(a) The differential system (3.1_o) is disconjugate on I whenever there exists hermitian $n \times n$ matrix functions $M(t)$ and $C_1(t)$ which are respectively Lipschitzian and of class $\mathscr{L}^\infty$ on arbitrary compact subintervals of I and satisfy the conditions

$$M'(t) + C(t) = C_1(t), \quad C_1(t) \geq M(t)B(t)M(t), \quad t \text{ a.e. on } I.$$

In particular, this condition holds in each of the following cases:

1°. $I = [0,1]$, $B(t) \equiv E$, $C(t) \leq -4\left[\int_0^t C(s)ds\right]^2$ for t a.e. on I;

2°. $I = (0,\infty)$, $B(t) \equiv E$, $\int_1^\infty C(s)ds = \lim_{t\to\infty} \int_1^t C(s)ds$

exists and is finite, and the matrix function $M_1(t) = -\int_t^\infty C(s)ds$ satisfies either (α): $C(t) \leq -4M_1^2(t)$ for t a.e. on I, or (β): $-3E \leq 4tM_1(t) \leq E$ for t a.e. on I.

(b) If, in addition, on a compact subinterval $[a,b]$ of I the matrix function $B(t)$ is non-singular, $B^{-1}(t) \in \mathscr{L}^\infty[a,b]$, and $B(t) > 0$ for t a.e. on $[a,b]$, then (3.1_0) is disconjugate on $[a,b]$ if there exists a non-negative real-valued function $\theta(t)$ of class $\mathscr{L}^\infty$ such that $\theta(t)E + C(t) \geq 0$ for t a.e. on $[a,b]$, and the constant hermitian matrix $4\left[\int_a^b B(t)dt\right]^{-1} - \left(\int_a^b \theta(t)dt\right)E$ is non-negative definite. {References: Reid [12, Sec. 4]: These results extend results of Wintner [7] for scalar equations}.

(c) If $B(t) > 0$ and (3.1_0) is non-oscillatory on a subinterval $[c,b_0)$ of I, then there exists a conjoined basis $Y_1(t) = (U_1(t);V_1(t))$ of (3.1_0) such that $U_1(t)$ is non-singular on a subinterval $[a,b_0)$ of I_0, and the hermitian matrix integral $\int_a^{b_0} U_1^{-1}(t)B(t)U_1^{*-1}(t)dt$ is convergent; also, for any such conjoined basis the scalar integral $\int_a^{b_0}\{|B(t)|/|U_1(t)|^2\}dt$ is convergent. In particular, in case solutions of (3.1_0) remain bounded as $t \to b_0$ we have $\int_c^{b_0}|B(t)|dt < \infty$ for $c \in I$. Moreover, in view of the non-negative hermitian character of $B(t)$, this latter condition is equivalent to each of the following: (i) $\int_c^{b_0} \operatorname{Tr} B(t)dt < \infty$ for $c \in I$; (ii) the matrix integral $\int_c^{b_0} B(t)dt$ is convergent, for $c \in I$. {Reid [17-Th. 5.2]}.

(d) Suppose that on $I = (a_0,b_0)$, $-\infty \leq a_0 < b_0 \leq \infty$ the $n \times n$ matrix function $Q(t)$ is hermitian, locally of class $\mathscr{L}$, and $Q(t) > 0$ a.e. on I. Then there exists a subinterval $I = [a,b_0)$ of I such that the system

$$-v'(t) - Q(t)u(t) = 0, \quad u'(t) - Q(t)v(t) = 0$$

is disconjugate on I_0 if and only if $\int_s^{b_0}|Q(t)|dt < \infty$ for

$s \in I$. Moreover, this latter condition is equivalent to each of the following: (i) $\int_c^{b_0} \operatorname{Tr} Q(t)dt < \infty$ for $c \in I$; (ii) the matrix integral $\int_c^{b_0} Q(t)dt$ is convergent for $c \in I$. {Reid [17-Th. 5.3]}.

17. Consider the vector differential equation $L[u](t) \equiv [R(t)u'(t)]' + P(t)u(t) = 0$, where $R(t)$ and $P(t)$ are $n \times n$ real symmetric matrix functions on $[0,\infty)$ with $P(t)$ continuous, $R(t)$ continuously differentiable and positive definite. If $\int_0^\infty [1/|R(s)|]ds = \infty$, $P(t) > 0$ on $[0,\infty)$, and $L[u] = 0$ is disconjugate on $[a,\infty)$, then

$$(t-a)\left|\int_t^\infty P(s)ds\right| < M\{|R|;a,t\},\ (t-a)\int_t^\infty |P(s)ds| < nM\{|R|;a,t\};$$

$$\frac{1}{t-a}\left|\int_a^t (s-a)^2P(s)ds\right| < M\{|R|;a,t\},\ \frac{1}{t-a}\int_a^t (s-a)^2|P(s)|ds < nM\{|R|;a,t\},$$

where $M\{|R|;a,t\}$ denotes the maximum of $|R(s)|$ for $a \leq s \leq t$. Furthermore, if $\int_a^\infty s^\nu|R(s)|ds < \infty$ then $\left|\int_a^\infty s^{\nu+2}P(s)ds\right| < \infty$, $\int_a^\infty s^{\nu+2}|P(s)|ds < \infty$. Also, if $\int_a^\infty |R(s)|^2 ds < \infty$ then $(t-a)\left|\int_t^\infty (s-a)P(s)ds\right|^2 < (25/16)\int_a^\infty |R(s)|^2 ds$ and $(t-a)\left[\int_t^\infty (s-a)|P(s)|ds\right]^2 < (25/16)n^2\int_a^\infty |R(s)|^2 ds$. {Simons [1]}.

18. Suppose that hypothesis ($\mathscr{A}$) holds for a system (3.1) on an interval I, and let $\Phi(t)$ be a fundamental matrix solution of $\Phi'(t) = A(t)\Phi(t)$. Then under the substitution

$$y(t) = \begin{bmatrix} \Phi(t) & 0 \\ 0 & \Phi^{*-1}(t) \end{bmatrix} \text{y}(t)$$

the vector differential equation (3.1) is transformed into

$$\mathcal{J}y'(t) - A(t)y(t) = 0 \tag{15.10}$$

where

$$\mathcal{J} = \begin{bmatrix} 0 & -E \\ E & 0 \end{bmatrix}, \quad A(t) = \begin{bmatrix} F(t) & 0 \\ 0 & G(t) \end{bmatrix}, \tag{15.11}$$

with $F(t) = -\Phi^*(t)C(t)\Phi(t)$ and $G(t) = \Phi^{-1}(t)B(t)\Phi^{*-1}(t)$. In particular, $F(t)$ and $G(t)$ are hermitian on I; moreover, $G(t) \geq 0$ if and only if $B(t) \geq 0$ and $F(t) \geq 0$ if and only if $C(t) \leq 0$.

If $y(t)$ is a solution of (15.10), then $y_0(t) = \mathcal{J}y(t)$ is a solution of the differential system

$$\mathcal{J}y_0'(t) - A_0 y_0(t) = 0, \tag{15.12}$$

where

$$A_0(t) = \begin{bmatrix} G(t) & 0 \\ 0 & F(t) \end{bmatrix}.$$

In accord with the terminology introduced at the end of Section 3, equation (15.12) is called the equation *reciprocal to* (15.10).

For a non-negative hermitian matrix function $K(t)$ which is locally integrable on I, the symbol $N^+(K)$ is used to denote the condition that for arbitrary $b \in I$ there exists a $d \in I^+\{b\} = \{t : t \in I,\ t > b\}$ such that $\int_b^d K(s)ds > 0$. If for a given $a \in I$ the hermitian matrix function $\int_a^t K(s)ds$ is positive definite for each $t \in I^+\{a\}$, then $K(t)$ is said to satisfy condition $N^{++}(K|a)$. For an $n \times n$ hermitian matrix H the eigenvalues of H are denoted by $\lambda_j(H)$, $j = 1,\dots,n$, with $\lambda_1(H) \leq \dots \leq \lambda_n(H)$; also, we write $\lambda_{Min}(H)$ for $\lambda_1(H)$ and $\lambda_{Max}(H)$ for $\lambda_n(H)$.

The following results hold for systems (15.10) and (15.12) when on $I = [a,\infty)$ the *hermitian matrix functions*

F *and* G *are locally integrable, and both are non-negative definite.*

1. If also the conditions $N^+(F)$ and $N^+(G)$ hold on I, then the following ten conditions are equivalent:

(i) (15.10) is disconjugate for large t;

(ii) (15.12) is disconjugate for large t;

(iii) {(iii)'} for large t there exists an $n \times n$ hermitian, {non-singular hermitian}, solution of the Riccati differential equation

$$W'(t) - F(t) - W(t)G(t)W(t) = 0; \qquad (15.13)$$

(iv) {(iv)'} for large t there exists an $n \times n$ hermitian, {non-singular hermitian}, solution of the Riccati differential equation

$$W'(t) - G(t) - W(t)F(t)W(t) = 0; \qquad (15.14)$$

(v) {(v)'} there exists a $2n \times n$ matrix of conjoined solutions $(U(t);V(t))$ of (15.10) such that $U(t)$, $\{V(t)\}$, is non-singular for large t.

(vi) {(vi)'} there exists a $2n \times n$ matrix of conjoined solutions $(U_0(t);V_0(t))$ of (15.12) such that $U_0(t)$, $\{V_0(t)\}$, is non-singular for large t. {Ahlbrandt [2, Th. 3.1]}.

2. If condition $N^+(F)$ holds, and there is a value $c \in I$ such that $\lambda_{Min}\left[\int_c^t G(s)ds\right] \to \infty$ as $t \to \infty$, then (15.10) is disconjugate for large t if and only if the matrix integral $\int_a^\infty F(s)ds$ is finite, and for large t there exists a continuous $n \times n$ hermitian matrix function $W(t)$ such that the integral $\int_t^\infty W(s)G(s)W(s)ds$ exists and

$$W(t) = \int_t^\infty F(s)ds + \int_t^\infty W(s)G(s)W(s)ds. \qquad (15.15)$$

In particular, if $N^+(F)$ holds, and there is a value $a \in I$ such that (15.10) is disconjugate on $[a,\infty)$, while condition $N^{++}(G|a)$ also holds, and $\lambda_{Min}\left[\int_a^t G(s)ds\right] \to \infty$ as $t \to \infty$, then for $Y(t;a) = (U(t;a),V(t;a))$ the solution of (15.10) determined by $U(a;a) = 0$, $V(a;a) = E$, and $W(t;a) = V(t;a)U^{-1}(t;a)$, we have

$$0 < \int_t^\infty F(s)ds \le W(t;a) \le \left[\int_a^t G(s)ds\right]^{-1}, \text{ for } t \in (a,\infty). \qquad (15.15)$$

In turn, (15.16) implies that

$$\left[\lambda_i\left(\int_t^\infty F(s)ds\right)\right]\lambda_{n-i+1}\left(\int_a^t G(s)ds\right)\right] \le 1, \quad (i = 1,\dots,n). \qquad (15.17)$$

{Ahlbrandt [1, Th. 4.1; 2, Th. 3.2]. In particular, this result is an extension to equations (15.10) of results of Hille [1; p. 243] for second order scalar equations, and earlier results of Sternberg [1, p. 316] for self-adjoint systems}.

3. Suppose that conditions $N^+(F)$ and $N^+(G)$ hold on $I = [a,\infty)$, and there exists a value $c \in [a,\infty)$ and a real-valued function $\alpha(t)$ of class $\mathscr{L}'$ on $[c,\infty)$ with $\alpha(t)$ non-zero, $\alpha'(t)$ positive on $[c,\infty)$, and $\int_c^\infty \alpha(t)G(t)dt$ convergent, while $[1/\alpha'(t)]F(t)$ is essentially bounded on $[c,\infty)$. Then for each real number λ the system

$$u'(t) = G(t)v(t), \quad v'(t) = -\lambda F(t)u(t) \qquad (15.18)$$

is disconjugate for large t. {Ahlbrandt [2, Th. 3.3]. The conditions of this result form a partial dual to the sufficient conditions for disconjugacy for large t given in Corollary 1 to Theorem 4.1 of Ahlbrandt [1]}.

4. Suppose that hypothesis $N^+(G)$ holds on $[a,\infty)$, and there exists a value $c \in [a,\infty)$ such that $\int_c^t F(s)ds$ is a strictly increasing matrix function on $[c,\infty)$ and $\lambda_{Min}\left[\int_c^t F(s)ds\right] \to \infty$ as $t \to \infty$. Then (15.10) is disconjugate for large t if and only if there exists a value $a_0 \in [a,\infty)$ such that for each $b_0 \in (a_0,\infty)$ the functional

$$J_0[\eta;a_0,b_0] = \int_{a_0}^{b_0}\{\zeta^*(t)F(t)\zeta(t) - \eta^*(t)G(t)\eta(t)\}dt$$

is positive definite on $D^0_{0*}[a_0,b_0]$, the class of a.c. vector functions η on $[a_0,b_0]$ satisfying $\eta(a_0) = 0$, and for which there exists a $\zeta \in \mathscr{L}^\infty[a_0,b_0]$ such that $\eta'(t) = F(t)\zeta(t)$. Applied to the transform of the general system (3.1), this result yields the following criterion:

A. Suppose that on $I = [a,\infty)$ hypothesis ($\mathscr{A}$) is satisfied, while $B(t) \geq 0$ and $C(t) \leq 0$ a.e. on I, and (3.18), the reciprocal system to (3.1), is identically normal on I. Moreover, if $\Phi(t)$ is a fundamental matrix solution of $\Phi'(t) = A(t)\Phi(t)$ then

(i) for each $b \in [a,\infty)$ there exists a $c \in (b,\infty)$ such that $\int_b^c \Phi^{-1}(t)B(t)\Phi^{*-1}(t)dt > 0$;

(ii) if $c \in [a,\infty)$, and π is an arbitrary non-zero n-dimensional vector, then $\pi^*\left[\int_c^t \Phi^*(s)C(s)\Phi(s)ds\right]\pi \to -\infty$ as $t \to \infty$. Then (3.1) is disconjugate for large t if and only if there exists a value $a_0 \in [a,\infty)$ such that for each $b_0 \in (a_0,\infty)$ the functional

$$J_0[\eta;a_0,b_0] = \int_{a_0}^{b_0}\{\zeta^*(t)C(t)\zeta(t) + \eta^*(t)B(t)\eta(t)\}dt$$

is negative definite on the class $D^1_{0*}[a_0,b_0]$ of a.c. vector functions η on $[a,b]$ satisfying $\eta(a_0) = 0$ and for which there exists a corresponding $\zeta \in \mathscr{L}^\infty[a_0,b_0]$ such that $\eta' + A^*(t)\eta + C(t)\eta(t) = 0$. {Ahlbrandt [2; Th. 3.4]. The result of A is a partial dual of the criterion of Reid [19; Corollary 1 to Th. 3.2], which provides necessary and sufficient conditions for disconjugacy on $[a,\infty)$}.

5. For $c \in [a,\infty)$, let $t_1^+(c)$ be the first right-hand conjugate point to $t = c$, relative to system (15.10), and $\tau_1^+(c)$ the first right-hand focal point to $t = c$, relative to this system; that is, if $(U(t;c);V(t;c))$ is the matrix solution satisfying $U(c,c) = 0$, $V(c,c) = E$, then $t_1^+(c)$ and $\tau_1^+(c)$ are the smallest values greater than c at which, respectively, $U(t)$ and $V(t)$ are singular. Similarly, let $t_{1*}^+(c)$ and $\tau_{1*}^+(c)$ denote the first right-hand conjugate and focal points to $t = c$, relative to the system (15.12). Moreover, suppose that there exists a $c \in [a,\infty)$ such that condition $N^{++}(G|c)$ holds. If $t_1^+(c)$ exists, then $\tau_1^+(c)$ and $\tau_{1*}^+(c)$ exist and lie in the interval $(c,t_1^+(c)]$. If, in addition, condition $N^{++}(F|c)$ holds, then the existence of either $t_1^+(c)$ or $t_{1*}^+(c)$ implies the existence of both $\tau_1^+(c)$ and $\tau_{1*}^+(c)$. {Ahlbrandt [2, Theorem 4.1]. This result provides an extension of Ths. 1.1, 2.1 of Barrett [5] and Th. 4.1 of Hunt [1]}.

6. Suppose that condition $N^+(F)$ holds on $[a,\infty)$ and there exists a $c \in [a,\infty)$ such that $N^{++}(G|c)$ holds. If $\lambda_{Min}\left[\int_a^t G(s)ds\right] \to \infty$ as $t \to \infty$, and $\tau_1^+(c)$ exists, then $t_1^+(c)$ and $\tau_{1*}^+(c)$ exist. {Ahlbrandt [2, Th. 4.2]. This result is a consequence of 5 above, and Th. 3.3 of Ahlbrandt [1]}.

7. Suppose that system (15.10) and (15.12) are identically normal on $[a,\infty)$, and $c \in [a,\infty)$ is such that

$\lambda_{Min}\left[\int_c^t F(s)ds\right] \to \infty$ as $t \to \infty$. If (15.10) is disconjugate on $[c,\infty)$, and $(U(t;c);V(t;c))$ is the solution of the corresponding matrix system satisfying $U(c;c) = 0$, $V(c;c) = E$, then $V(\ ;c)$ has exactly n points of singularity on (c,∞), where singularities of order k are counted k times. {Ahlbrandt [2; Th. 4.3]. This result is an extension of an observation of Barrett [4, Lemma 3.1] for the case $n = 1$}.

8. Suppose that there exists a $c \in [a,\infty)$ such that condition $N^{++}(G|c)$ holds. If $\tau_1^+(c)$ does not exist, then $\int_c^\infty F(t)dt < \infty$, and there exists an $n \times n$ hermitian positive definite solution $W(t)$ of the Riccati matrix differential equation $W'(t) + F(t) + W(t)G(t)W(t) = 0$ on (c,∞), such that $\int_t^\infty F(s)ds \le W(t) \le \left[\int_a^t G(s)ds\right]^{-1}$ for $t \in (c,\infty)$. {Ahlbrandt [2, Th. 4.4]}.

19. Suppose that: (i) $F(t)$ is an $n \times n$ hermitian matrix function continuous on $[a,\infty)$; (ii) $E(t,M)$ is for arbitrary $n \times n$ matrices M an $n \times n$ matrix function on $[a,\infty)$ which is such that $E(t,M(t))$ is a continuous hermitian matrix function on $[a,\infty)$ whenever $M(t)$ is of class $\mathscr{C}'$ and hermitian on this interval; (iii) $W(t)$ is an $n \times n$ hermitian matrix function on $[a,\infty)$ such that $E(t,W(t)) \ge 0$ on this interval, and there exists a constant $c > 0$ and a positive continuous function ϕ satisfying $\int_a^\infty \phi(s)ds = +\infty$, and for arbitrary unit vectors π we have

$$\pi^*[E(t,W(t)) + F(t)]\pi \ge \phi(t)|\pi^*W(t)\pi|^{1+c}, \text{ for } t \in [a,\infty).$$

Then $W(t)$ is a solution of the matrix differential equation

$$W'(t) + E(t,W(t)) + F(t) = 0 \tag{15.19}$$

on $[a,\infty)$ if and only if $W(t) \geq 0$ for $t \in [a,\infty)$, $W(t) \to 0$ as $t \to \infty$, and the improper matrix integrals $\int_a^\infty E(t,W(t))dt$, $\int_a^\infty F(t)dt$ exist and are finite, so that $W(t)$ satisfies the integral equation

$$W(t) = \int_t^\infty E(s,W(s))ds + \int_t^\infty F(s)ds, \quad t \in [a,\infty).$$

Moreover, if $W(t)$ is such a solution of (15.19), and $\psi(t) = \int_a^t \phi(s)ds$, then for all unit vectors π we have

$$0 \leq \limsup_{t\to\infty} [\psi(t)]^{1/c}\pi^*\left(\int_t F(s)ds\right)\pi$$

$$\leq \limsup_{t\to\infty} [\psi(t)]^{1/c}\pi^*W(t)\pi \leq (1/c)^{1/c},$$

and for each r satisfying $0 \leq r < 1/c$ the improper integral $\int_a^\infty [\psi(t)]^r F(t)dt$ exists and is finite. {Sternberg [2]}.

20. If $Q(t)$ is an $n \times n$ continuous real-valued symmetric matrix function on $[0,\infty)$, then the equation

$$u''(t) + Q(t)u(t) = 0 \tag{15.20}$$

is oscillatory for large t if there exists a positive function $g:[0,\infty) \to R$ of class $\mathscr{C}'$ and such that:

(a) $\int_a^\infty [1/g(s)]ds = +\infty$; (b) the matrix

$$K(t) = \int_a^t \left\{g(s)Q(s) - \frac{1}{4}\frac{[g'(s)]^2}{g(s)} E\right\}ds + \frac{1}{2} g'(t)E$$

is such that $\pi^*K(t)\pi \to \infty$ as $t \to \infty$ for arbitrary non-zero vectors π. {Howard [3, Th. 2]}.

21. Suppose that $Q(t)$, $g(t)$ and $K(t)$ are as in Exercise 20, and $h_0 \geq 0$ is such that $K(t) \geq -h_0E$ on a compact interval $[a,b]$. Moreover, suppose there exist on $[a,b]$ continuous scalar functions ψ and ℓ, and a continuously

differentiable function ϕ satisfying on this interval the conditions $\psi^2(t) \geq [\phi(t) + \ell(t)]^2$, $\phi'(t) \geq \psi^2(t)/g(t)$, $\ell(t)E \geq H(t)$, and $\phi(a) = t_o + 2h_o$, with $t_o \geq 0$. If $U(t)$ is an $n \times n$ matrix function with column vectors solutions of (15.19) which is such that $(U(t);U'(t))$ is a conjoined solution of the corresponding matrix system (3.1_M), while $U(a)$ is non-singular and satisfies $(t_o + 2h_o)E \geq -g(a)U'(a)U^{-1}(a) \geq -t_oE$, then $U(t)$ is non-singular throughout the interval $[a,b]$. {Howard [3; modified Th. 6]}.

22. Suppose that $r(t)$, $p(t)$, $q(t)$ are real-valued functions which on an interval $[c,\infty)$ are such that $r(t) > 0$ and $r(t)$, $1/r(t)$, $p(t)$ and $q(t)$ are of class $\mathscr{L}^\infty[a,b]$ on arbitrary compact subintervals $[a,b]$ of $[c,\infty)$, and the scalar differential equation

$$\ell[u] \equiv [r(t)u'(t)+q(t)u(t)]' - [q(t)u'(t)+p(t)u(t)] = 0$$

is oscillatory on arbitrary intervals $[a,\infty) \subset [c,\infty)$. If $R(t)$, $P(t)$, $Q(t)$ are $n \times n$ matrix functions satisfying hypothesis $\mathscr{A}_\omega$ on $[c,\infty)$, while the $2n \times 2n$ matrix function

$$\begin{bmatrix} r(t)E_n - R(t) & q(t)E_n - Q(t) \\ q(t)E_n - Q^*(t) & p(t)E_n - P(t) \end{bmatrix}$$

is non-negative definite, then for any conjoined basis $Y(t) = (U(t);V(t))$ of (2.5) the matrix function $U(t)$ is singular at some point on arbitrary $[a,\infty) \subset [c,\infty)$.

23. Consider a system (3.1) with continuous matrix coefficients and $B(t) > 0$ for $t \in I$, and let $X(t)$ be an $n \times r$, $(0 < r \leq n)$, matrix function of class $\mathscr{C}'$ and of rank r on I. Then the $r \times r$ matrix functions

$R(\ |X) = X^*B^{-1}X$, $Q(\ |X) = X^*B^{-1}(X' - AX)$, $P(\ |X) = X^*CX + (X' - AX)^*B^{-1}(X' - AX)$ are continuous, with $R(\ |X) > 0$ on I. Moreover, if $\xi(t)$ is an r-dimensional vector function which is continuous and has a piecewise continuous derivative on a compact subinterval $[a,b]$ of I, then relative to the functional $J[\eta;a,b]$ the vector function $\eta(t) = X(t)\xi(t)$ is of class $D[a,b]$ with $\zeta = B^{-1}[X\xi' + (X' - AX)\xi]$. Also, the value of $J[\eta;a,b]$ is equal to

$$\check{J}[\xi;a,b] = \int_a^b \{\xi'[R(\ |X)\xi' + Q(\ |X)\xi] + \xi^*[Q^*(\ |X)\xi' + P(\ |X)\xi]\}dt, \tag{15.21}$$

for which the corresponding system (3.1) may be written as the linear second order vector differential equation

$$[R(t\ |X)\xi'(t) + Q(t\ |X)\xi(t)]' - [Q^*(t\ |X)\xi'(t) + P(t|X)\xi(t)] = 0. \tag{15.22}$$

In particular, for $t_0 \in I$ let $t_j^+(t_0)$, $t_j^-(t_0)$, $(j = 1,2,\ldots)$, be the set of right- and left-hand conjugate points to $t = t_0$ relative to (3.1), and $t_j^+(t_0)$, $t_j^-(t_0)$, $(j = 1,2,\ldots)$, the set of right- and left-hand conjugate points to $t = t_0$ relative to the equation (15.22), where in each case these points are indexed according to order. In view of the characterization of conjugate points in Theorems 8.3 and 8.4, we have that if for a given positive integer k the conjugate point $t_k^+(t_0)$, $\{t_k^-(t_0)\}$ exists then the k-th conjugate point $t_k^+(t_0)$, $\{t_k^-(t_0)\}$, exists and $t_0 < t_k^+(t_0) \leq t_k^+(t_0)$, $\{t_k^-(t_0) \leq t_k^-(t_0) < t_0\}$. Corresponding results hold relative to focal points, where if the initial condition at $t = a$

relative to (3.1) is $\Gamma_a u(a) - v(a) = 0$ then the corresponding initial condition at $t = a$ relative to (15.22) is $X^*(a)\Gamma_a X(a)\xi(a) - [R(a|X)\xi'(a) + Q(a|X)\xi(a)] = 0$. In particular, if I is of the form (a,∞) and (15.22) is oscillatory for large t, then (3.1) is also oscillatory for large t.

Of special interest is the case of $A(t) \equiv 0$, which may always be obtained by a subsidiary transformation as in Section 5, and $r = 1$ with $X(t)$ a constant non-zero n-dimensional vector x, since then the coefficients of the associated system are real scalars $R(\ |x) = x^*B^{-1}x$, $Q(\ |x) \equiv 0$, $p(\ |x) = x^*Cx$, and (15.22) is the real scalar equation

$$[r(t|x)\xi'(t)]' - p(t|x)\xi(t) = 0. \qquad (15.23)$$

In particular, whenever (3.1) is non-oscillatory for large t the scalar equation (15.23) is also non-oscillatory for large t, and hence any sufficient condition for (15.23) to be oscillatory immediately yields a criterion for (3.1) to be oscillatory. For example, by the Wintner-Leighton criterion, if $I = [a,\infty)$ and there exists a non-zero vector x such that $\int_a^\infty [1/r(s|x)]ds = +\infty$ and $-\int_a^\infty p(s|x)ds = -\lim_{t\to\infty}\int_a^t p(s|x)ds = +\infty$, then (15.23) is oscillatory for large t, and consequently (3.1) is oscillatory for large t. Now if $B(t)$ is such that its smallest eigenvalue $\lambda_{Min}[B(t)]$ satisfies $\int_a^\infty \lambda_{Min}[B(s)]ds = +\infty$, then for any non-zero vector x the coefficient function $r(t|x) = x^*B^{-1}(t)x$ does not exceed $1/\lambda_{Min}[B(s)]$, and therefore $\int_a^\infty [1/r(s|x)]ds = +\infty$. Consequently, in this case if (3.1) is non-oscillatory for large t it follows that for each non-zero vector x we cannot have $-\int_a^\infty p(s|x)ds = +\infty$. In particular, if $C(t) \le 0$ for large t, then the integral $\int_a^\infty C(s)ds = \lim_{t\to\infty} \int_a^t C(s)ds$ must exist and be finite. Special cases of this simple criterion have appeared in Reid [12, 17], Tomastik [1], Howard [3] and Kreith [6]. Further comments on its use will appear in the Topics and Exercises at the end of the next chapter on boundary problems.

24. Suppose that $p_j(t)$, $(j = 0,1,\dots,n)$ are real-valued continuous functions on $[a,\infty)$ with $p_n(t) > 0$ on this interval, and let $F(t|\pi)$ denote the quadratic form $F(t|\pi) = \sum_{\beta=1}^{n} p_{\beta-1}(t)\pi_\beta^2$. Moreover, suppose that $\int_a^\infty [1/p_n(s)]ds = +\infty$.

(i) If there exists a non-negative continuous function $q(t)$ which is not identically zero for large t, and such that on some subinterval $[c,\infty)$ we have $F(t|\pi) + q(t)\pi_1^2 \le 0$ for arbitrary real $\pi = (\pi_\beta)$, $(\beta = 1,\dots,n-1)$, then the equation (13.4) is disconjugate on a subinterval $[a_0,\infty)$ if and only if for arbitrary $b_0 > a_0$ the functional

$$J[\xi;a_0,b_0] = \int_{a_0}^{b_0} \left\{\sum_{j=0}^{n} p_j(t)(\xi^{[j]}(t))^2\right\}dt$$

is positive definite on the class of functions $D_{0*}[a_0,b_0]$.

(ii) If $F(t|\pi)$ is non-positive for $t \in [a,\infty)$, and (13.4) is non-oscillatory for large t, then each of the integrals $\int^\infty p_\beta(t)t^{2n-2\beta-2}dt$, $(\beta = 0,1,\dots,n-1)$, is convergent.

{Reid [19, Ths. 4.1, 4.2]. For $n = 1$, the result of (i) was established by Leighton [4], while for $n = 2$ and $p_2(t)$ bounded on $[a,\infty)$ the result is a special case of a criterion for analogous fourth order matrix differential equations by Sternberg and Sternberg [1]. The above results are established in Reid [19] as consequences of a more general theorem extending to self-adjoint differential systems the boundary value problem criterion for oscillation and non-oscillation derived initially by Nehari [2] for second order scalar equations}.

25. Consider the self-adjoint differential equation

$$(r(t)u^{[n]})^{[n]} + (-1)^{n+1}p(t)u(t) = 0, \qquad (15.24)$$

where $r(t)$ and $p(t)$ are positive functions which are continuous on $[0,\infty)$. Let $\eta_1(0)$ denote the smallest right-hand conjugate point to $t = 0$ with respect to (15.24); that is, $t_2 = \eta_1(0)$ is the smallest positive value such that there is a non-identically vanishing solution of this equation satisfying $u^{[\alpha-1]}(0) = u^{[\alpha-1]}(t_2) = 0$, $(\alpha = 1,\ldots,n)$. Also, let $\mu_1(0)$ be the smallest positive value such that there is a non-identically vanishing solution of (15.24) satisfying $u^{[\alpha-1]}(0) = v_n^{[\alpha-1]}(t_2) = 0$, $(\alpha = 1,\ldots,n)$, where $v_n(t) = v_n(t\ u) = r(t)u^{[n]}(t)$. Corresponding to the Picone terminology, $\mu_1(0)$ may be called the first right-hand pseudo-conjugate to $t = 0$.

(i) If $\eta_1(0)$ exists, then $\mu_1(0)$ exists and $0 < \mu_1(0) \leq \eta_1(0)$.

(ii) If $\int^{\infty} t^{2n-2}p(t)dt = \infty$, then $\mu_1(0)$ exists.

(iii) If $I_n(t|p)$ denotes the n-th iterated integral of $p(t)$ on the interval $[0,t]$, and $\int^{\infty} p(t)[I_n(t|p)]^2 dt = +\infty$, then $\mu_1(0)$ exists.

(iv) If $\mu_1(0)$ exists and $\int^{\infty}[1/r(t)]dt = \infty$, then $\eta_1(0)$ exists. {Hunt [1; Ths. 4.1, 4.2, 4.3, 5.2]}.

26. Consider the self-adjoint differential equation (15.24) where $r(t)$ and $q(t)$ are real-valued continuous functions with $r(t) > 0$ on $[a,\infty)$, but with no restriction as to the algebraic sign of $q(t)$.

(i) If there exists a positive continuous function $h(t)$ on $[a,\infty)$ such that as $t \to \infty$ we have (i): $\int_a^t s^{n-1}h(s)ds \to \infty$, and (ii): $\liminf J(t) = -\infty$, where

$$J(t) = \left[\int_a^t \left\{r(s)[(n-1)!h(s)]^2 - p(s)\left[\int_s^t (\xi-s)^{n-1}h(\xi)d\xi\right]^2\right\}ds\right] \times \left[\int_a^t s^{n-1}h(s)ds\right]^{-2},$$

then there exists a $b > a$ such that (15.24) is oscillatory on $[a,b]$. This conclusion still holds if condition (ii) is replaced by the condition that

$\limsup_{t\to\infty} \left\{\int_a^t r(s)h^2(s)ds\right\}\left\{\int_a^t s^{n-1}h(s)ds\right\}^{-2} < \infty$ and $t^{1-n}\int_a^t p(s)[t-s]^{n-1}ds \to \infty$ as $t \to \infty$; in particular, the choice of $h(t) = 1/r(t)$ yields the result that there exists a $b > a$ such that (15.24) is oscillatory on $[a,b]$ whenever $\int_a^\infty [t^{n-1}/r(t)]dt = \infty$ and $t^{1-n}\int_a^t p(s)(t-s)^{n-1}ds \to \infty$ as $t \to \infty$. {Hinton [2]}.

27. Consider the differential equation

$$L_{2n}[u](t) \equiv \sum_{j=0}^{n} (-1)^{n-j}\{p_j(t)u^{[n-j]}(t)\}^{[n-j]} \tag{15.25}$$

where $p_j(t)$, $(j = 0,1,\ldots,n)$, is a real-valued function of class $\mathscr{C}^{[n-j]}$ on $(0,\infty)$ and $p_0(t) > 0$ on this interval.

(i) Suppose $p_0(t) \equiv 1$, and for $P_j^0(t) = p_j(t)$ set $P_j^m(t) = \int_t^\infty P_j^{m-1}(s)ds$, $m = 1,2,\ldots$ whenever $\int_t^\infty P_j^{m-1}(s)ds$ exists and is finite for $t \in (0,\infty)$. If for $k = 1,\ldots,n$

and $t \geq a$ we have that $\int_a^\infty P_k^m(t)dt$ exists and is finite for $m = 0,1,\ldots,k-1$, and $t^k|P_k^k(t)| \leq a_k$ for $t \in [a,\infty)$, while $\sum_{j=1}^n a_k M_k \leq 1$, where $M_k = k!2^{4k-1}/(2k)!$, then $L_{2n}[u] = 0$ is non-oscillatory on $[a,b]$ for all $b > a$. {Lewis [1, Th. 1.1]}.

28. Consider the differential equation

$$L_4[u](t) \equiv [r(t)u''(t)]'' - [q(t)u'(t)]'+p(t)u(t) = 0 \quad (15.26)$$

where p, q, r are real-valued functions on $(0,\infty)$ that are respectively continuous, continuously differentiable, and of class $\mathscr{C}''$, and with $r(t) > 0$. Then $L_4[u] = 0$ is oscillatory on $(0,\infty)$ in each of the following cases:

(i) there exist constants N, M such that $0 < r(t) \leq N$, $q(t) \leq M$ for $t \in (0,\infty)$, and $\int_1^\infty p(s)ds = \lim_{t\to\infty} \int_1^t p(s)ds = -\infty$;

(ii) there exists a constant M such that $0 < r(t) \leq M$ for $t \in (0,\infty)$, while one of the following conditions holds:

(a) $\int^\infty q(s)ds = -\infty$, and $\int^\infty s^2|p(s)|ds < \infty$,

(b) $\int^\infty q(s)ds = -\infty$, and $\int^\infty p(s)ds = -\infty$.

(c) $-\infty < \int^\infty p(s)ds < \infty$, $\int^\infty P_1(s)ds = -\infty$, $\int^\infty |q(s)|s^{-1}ds < \infty$ and $q(t) \to 0$ as $t \to \infty$.

{Lewis [1, Ths. 2.1, 2.2, 2.3, 2.4]}.

29. Consider the differential equation

$$L_{2n}[u](t) \equiv (-1)^n\{r(t)u^{[n]}(t)\}^{[n]} + p(t)u(t) = 0 \quad (15.27)$$

where $p(t)$ and $r(t)$ are real-valued functions on $(0,\infty)$ which are respectively continuous and of class $\mathscr{C}^{[n]}$, with $r(t) > 0$ on this interval. Then $L_{2n}[u](t)$ is non-oscillatory on $(0,\infty)$ in each of the following cases:

(i) $p(t) \leq 0$, there exist numbers M and α such that $0 < r(t) \leq Mt^{\alpha}$ on $(0,\infty)$ with $\alpha < 2n - 1$, and

$$\limsup_{t \to \infty} t^{2n-1-\alpha}\int_t^{\infty} |p(s)|ds > MA_n^2,$$

where $A_n^{-1} = \frac{\sqrt{2n - 1}}{(n-1)!} \sum_{j=1}^{n} (-1)^{j-1}\binom{n-1}{j-1}(2n - j)^{-1}$;

(ii) There exist numbers M and α such that $0 < r(t) \leq Mt^{\alpha}$ on $(0,\infty)$, and for some $\nu > 1$ and A_n as in (i) we have

$$\lim_{t\to\infty} \left\{Kt^{\alpha-2n+1} + \int_a^t p(s)ds\right\} = -\infty,$$

where $K = MA_n^2\nu^{\alpha}/(\nu-1)^{2n-1}$. {Lewis [1; Ths. 3.1, 3.2]}.

30. The differential equation (15.25) is oscillatory for large t provided there exists a number α such that $\int^{\infty} t^{\alpha}p_n(t)dt = -\infty$, and $\int^{\infty} t^{\alpha-2n+2j}|p_j(t)|dt < \infty$ for $0 \leq j \leq n - 1$. In particular, if there exists a value $\alpha < 2n - 1$ such that $\int^{\infty} t^{\alpha}p(t)dt = -\infty$, then the differential equation $(-1)^n u^{[2n]}(t) + p(t)u(t) = 0$ is oscillatory for large t, with no sign restrictions on $p(t)$. Moreover, for this latter result the bound on α is sharp, since the Euler differential equation $(-1)^n u^{[2n]}(t) + ct^{-2n}u(t) = 0$ is not oscillatory for large t whenever $c \geq -\sigma_n^2$, where $\sigma_n = [1\cdot3\cdot5...(2n-1)]/2^n$, whereas for $c < -\sigma_n^2$ the Euler equation is oscillatory for large t. {Lewis [3; Th. 2]}.

31. Consider the self-adjoint fourth order linear differential equation (15.26) where p, q, r are real-valued functions on $(0,\infty)$ which are respectively continuous, continuously differentiable, and of class $\mathscr{C}''$. This equation is oscillatory for large t in each of the following cases:

(i) there exists a constant M such that $0 < r(t) \leq M$ for $t \in (0,\infty)$, while $\int^{\infty} q^{+}(s)ds < \infty$ and $\int^{\infty} s^2 p(s)ds = -\infty$;

(ii) there exists a constant M such that $0 < r(t) \leq M$ on $(0,\infty)$, while $q(t) \leq 0$ for $t \in (0,\infty)$, $-\infty < \int^{\infty} p(s)ds < \infty$, and there exists a constant $\alpha \in [0,1]$ such that $\int^{\infty} t^{\alpha} P_1(t)dt = -\infty$.

(iii) $r(t)$ is bounded, $q(t)$ is bounded above, $p(t) \geq 0$ on $(0,\infty)$ and $\int^{\infty} [s^2 p(s) + q(s)]ds = -\infty$;

(iv) there exist constants M and α with $\alpha < 1$ such that $0 < r(t) \leq Mt^{\alpha}$, $p(t) \leq 0$, $q(t) \leq 0$ on $(0,\infty)$, and $\liminf_{t \to \infty} t^{1-\alpha} \int_t^{\infty} [s^2 p(s) + q(s)]ds < -4M$. {Lewis [2; Ths. 2.3,4,5]}.

32. Consider the n-th order linear differential equation $\ell_n[u](t) \equiv u^{[n]}(t) + p_{n-1}(t)u^{[n-1]}(t) + \dots + p_0(t)u(t) = 0$, where the $p_j(t)$, $(j = 0,1,\dots,n-1)$, are real-valued continuous functions on a given interval I. If $p_i(t) \leq 0$, $(i = 0,1,\dots,n-2)$, then every extremal solution for $[s,\eta_1(s)]$ has a zero of order 2 at $\eta_1(s)$ and not more than $n - 2$ zeros on $[s,\eta_1(s)]$. Similarly, if $(-1)^{n-i} p_i(t) \leq 0$, $(i = 0,1,\dots,n-2)$, then every extremal solution for $[s,\eta_1(s)]$ has a zero of order 2 at s and not more than $n - 2$ zeros on $[s,\eta_1(s)]$. {W. J. Kim [1-Th. 3.1]}.

33. Consider the n-th order linear differential equation $Lu \equiv u^{[n]} + \sum_{k=0}^{n-1} p_k(t)u^{[k]} = 0$, where $p_k(t)$, $(k = 0, 1,\dots,n-1)$ are real-valued functions of class $\mathscr{C}^k$ on $[a,b]$, so that the corresponding adjoint equation is $L^*v = \sum_{k=0}^{n-1} (-1)^{n-k}[p_k(t)v]^{(k)} = 0$.

(i) If $u(t)$ is a non-identically vanishing solution of $Lu = 0$ on $[a,b]$ with $u^{[n-1]}(a)u^{[n-1]}(b) < 0$, and zeros of order $n-1$ at $t = a$ and $t = b$, then any real solution $v(t)$ of $L^*v = 0$ vanishes at least once in $[a,b]$.

(ii) An equation $Lu = 0$, (or the operator Lu), is called a *separator* on a ray $[c,\infty)$ if for each $[c_0,\infty) \subset [c,\infty)$ there exists a corresponding solution $u = u(t;c_0)$ and values a, b on $[c_0,\infty)$ with $u^{[j]}(a) = 0 = u^{[j]}(b)$, $(j = 0,1,\dots, n-2)$, and $u^{[n-1]}(a)u^{[n-1]}(b) < 0$. Moreover, an equation, (or operator), is called *strongly oscillatory* on $[c,\infty)$ if all solutions have infinitely many zeros on this ray. Then: (a) the adjoint of a separator is strongly oscillatory; (b) self-adjoint even order equations are strongly oscillatory; (c) suppose that $Lu = 0$ is a self-adjoint even order differential equation such that for $c < \infty$ there exists values a, b on $[c,\infty)$ such that the boundary problem $Lu = 0$, $u^{[j]}(a) = u^{[j]}(b)$, $(j = 0,1,\dots,n-2)$ has a non-trivial non-negative solution. Then this equation is a separator, and all solutions of $Lu = 0$ are oscillatory. {Gustafson [4]}.

34. If the functions $p_j(t)$, $(j = 0,1,\dots,n)$ of (14.1) are real-valued and continuous on $[a,b]$, with $p_n(t) > 0$ on this interval, then (14.4) is disconjugate on $[a,b]$ in the sense of the Hamiltonian system (14.1) if and only if there exist real-valued continuous functions $r_0(t),\dots,r_n(t)$ on $[a,b]$ with $r_n(t) > 0$, and such that if L_0 is the n-th order differential operator defined by $(L_0z)(t) = \sum_{j=0}^{n} r_j(t)z^{[j]}(t)$, then the quasi-differential operator $D^{<2n>}$ of (14.4) has the factorization $D^{<2n>} = (-1)^n L_0^* L_0$, where the adjoint operator L_0^* is interpreted in an analogous system sense. Moreover, L_0z has the determinantal representation

$$L_o z = \frac{[p_n(t)]^{1/2}}{W[t:z_1,\dots,z_n]} \begin{vmatrix} z_1(t) & \dots & z_n(t) & z \\ \cdot & \dots & \cdot & \cdot \\ \cdot & \dots & \cdot & \cdot \\ z_1^{[n]}(t) & \dots & z_n^{[n]}(t) & z^{[n]} \end{vmatrix}$$

where $z_1,\dots,z_n$ are solutions of (14.4) such that the $(z_j^{[\alpha-1]}(t);v_\alpha(t|z_j))$, $(j = 1,\dots,n)$ form a conjoined basis for (14.1) with Wronskian determinant $W[t:z_1,\dots,z_n]$ non-zero on $[a,b]$. {Heinz [1], Kegley [1-§3]; for other work on factorization of linear differential operators, see Polya [1], Barrett [10], Zettl [1,2,3], Coppel [2]}.

CHAPTER VI.
SELF-ADJOINT BOUNDARY PROBLEMS

1. Introduction

Corresponding to the two-point boundary problems considered in Chapter III, we now consider vector boundary problems associated with a system of the form (V.3.1). *It will be assumed that the coefficient matrix functions* $A(t)$, $B(t)$, $C(t)$ *satisfy hypothesis* ($\mathcal{U}$) *of Section* V.3. The boundary problem to be considered is then of the form

$$(\mathcal{B})\quad \begin{array}{ll} \text{(i)} & \begin{array}{l} L_1[u,v](t) \equiv -v'(t) + C(t)u(t) - A^*(t)v(t) = 0, \\ L_2[u,v](t) \equiv u'(t) - A(t)u(t) - B(t)v(t) = 0, \end{array} \\ & \qquad\qquad t \in [a,b] \\ \text{(ii)} & M_1u(a) + M_2v(a) + M_3u(b) + M_4v(b) = 0, \end{array}$$

where the coefficient matrices M_1, M_2, M_3, M_4 are of dimension $2n \times n$, and such that the $2n \times 4n$ matrix

$$\mathcal{M} = [M_1 \quad M_2 \quad M_3 \quad M_4] \tag{1.1}$$

is of rank $2n$. In terms of the 2n-dimensional vector function $y(t) = (u(t);v(t))$, for which $y_\alpha(t) = u_\alpha(t)$, $y_{n+\alpha}(t) = v_\alpha(t)$, $(\alpha = 1,\ldots,n)$, the system ($\mathcal{B}$) may be written as

$$(\mathscr{B}'') \quad \begin{aligned} \mathscr{L}[y](t) &\equiv \mathscr{J}y'(t) + \mathscr{A}(t)y(t) = 0, \quad t \in [a,b], \\ s[y] &\equiv \mathscr{M}\hat{y} = 0, \end{aligned}$$

where the hermitian matrix function $\mathscr{A}(t)$ and the constant skew matrix $\mathscr{J}$ are as in (V.3.2), and $\hat{y}$ denotes the 4n-dimensional vector with $y_j = y_j(a)$ and $y_{2n+j} = y_j(b)$, $(j = 1,\ldots,2n)$. One may verify readily that for $y(t)$ and $z(t)$ arbitrary 2n-dimensional vector functions which are a.c. on $[a,b]$ we have the identity

$$\begin{aligned} \int_a^b z^*(t)\mathscr{L}[y](t)dt - \int_a^b \{\mathscr{L}[z](t)\}^* y(t)dt &= z^*(t)\mathscr{J}y(t)\Big|_{t=a}^{t=b} \\ &= \hat{z}^*[\operatorname{diag}\{-\mathscr{J},\mathscr{J}\}]\hat{y}. \end{aligned} \tag{1.2}$$

Consequently, the above defined boundary problem $(\mathscr{B})$ is formally self-adjoint if and only if

$$\hat{z}^*[\operatorname{diag}\{-\mathscr{J},\mathscr{J}\}]\hat{y} = 0, \text{ whenever } \mathscr{M}\hat{y} = 0 \text{ and } \mathscr{M}\hat{z} = 0. \tag{1.3}$$

In turn, it follows readily that condition (1.3) holds if and only if $\mathscr{M}$ satisfies the matrix equation

$$\mathscr{M}[\operatorname{diag}\{-\mathscr{J},\mathscr{J}\}]\mathscr{M}^* = 0; \tag{1.4}$$

a matrix equation equivalent to (1.4) is

$$\mathscr{P}^*[\operatorname{diag}\{-\mathscr{J},\mathscr{J}\}]\mathscr{P} = 0, \tag{1.4'}$$

where $\mathscr{P}$ is a $4n \times 2n$ matrix such that $\mathscr{M}\mathscr{P} = 0$.

For brevity, let $\Delta_{\sigma\tau}$ denote the $2n \times 2n$ matrix

$$\Delta_{\sigma\tau} = [M_\sigma \quad M_\tau], \quad (\sigma,\tau = 1,2,3,4) \tag{1.5}$$

and set

$$D = \operatorname{diag}\{-E_n, E_n\}, \quad N = \Delta_{24}D. \tag{1.6}$$

By direct verification it may be established that the matrix equation (1.4) may be written as

$$\Delta_{13}N^* - N\Delta_{13}^* = 0, \tag{1.4''}$$

which states that the $2n \times 2n$ matrix $\Delta_{13}N^* = \Delta_{13}D\Delta_{24}^*$ is hermitian. Moreover, since $\Delta_{24} = ND$, the boundary conditions (ii) of ($\mathscr{B}$) may be written vectorially as

$$\Delta_{13}\hat{u} + ND\hat{v} = 0. \tag{1.7}$$

THEOREM 1.1. *A differential system* ($\mathscr{B}$) *is self-adjoint if and only if there exists a* $2n \times 2n$ *hermitian matrix* $Q = Q[\mathscr{B}]$, *and a linear subspace* $S = S[\mathscr{B}]$ *of* $\mathbf{C}_{2n}$ *such that* u,v *satisfies the boundary conditions* ($\mathscr{B}$ii) *if and only if*

$$\hat{u} \in S, \quad T[u,v] \equiv Q\hat{u} + D\hat{v} \in S^{\perp}, \tag{1.8}$$

where $S^{\perp} = S^{\perp}[B]$ *denotes the orthogonal complement of* $S = S[\mathscr{B}]$ *in* $\mathbf{C}_{2n}$.

The proof indicated for this theorem is essentially the same as that of Theorem III.1.1, made more concise through the use of matrix algebra. If the matrix N is of rank zero, then the boundary conditions reduce to $\Delta_{13}\hat{u} = 0$ with Δ_{13} non-singular, so that (1.7) is of the form (1.8) with S the zero-dimensional subspace of $\mathbf{C}_{2n}$. If N is of rank $2n$, then (1.7) is equivalent to $N^{-1}\Delta_{13}\hat{u} + D\hat{v} = 0$, and since $N^{-1}\Delta_{13}$ is hermitian by (1.4') it follows that (1.7) is of the form (1.8) with $S = \mathbf{C}_{2n}$, $Q = N^{-1}\Delta_{13}$.

Now suppose that N has rank r, $1 \leq r < 2n - 1$, and let ψ be a $2n \times (2n - r)$ matrix of rank $2n - r$ such that $\psi^*N = 0$, and let χ be a $2n \times r$ matrix such that the $2n \times 2n$ matrix $[\psi \quad \chi]$ is non-singular. The boundary conditions (1.7) are then equivalent to the conditions

$$\begin{aligned} \chi^*\Delta_{13}\hat{u} &= 0, \\ \chi^*\Delta_{13}\hat{u} + \chi^*ND\hat{v} &= 0 \end{aligned} \tag{1.7'}$$

and as the matrix $\mathscr{M}$ is of rank $2n$ it follows that the $(2n - r) \times 2n$ matrix $\psi^*\Delta_{13}$ is of rank $2n - r$, and the $r \times 2n$ matrix χ^*N is of rank r. Moreover, by (1.4") we have that

$$(\psi^*\Delta_{13})(\chi^*N)^* = \psi^*\Delta_{13}N^*\chi = (\psi^*N)\Delta_{13}^*\psi = 0,$$

and as the matrices χ^*N and N are each of rank r we have that $\psi^*\Delta_{13}N^* = 0$, and consequently any $q \times 2n$ matrix K satisfying $KN^* = 0$ is of the form $K = \mu\psi^*\Delta_{13}$ for some $q \times (2n - r)$ matrix μ.

Now let H be the E. H. Moore generalized inverse of the hermitian matrix $\Delta_{13}N^*$. The matrix H is an hermitian $2n \times 2n$ matrix satisfying the equation

$$0 = \Delta_{13}N^* - (\Delta_{13}N^*)H(\Delta_{13}N^*) = (\Delta_{13} - N\Delta_{13}^*H\Delta_{13})N^*,$$

and in view of the above comment there exists a $2n \times (2n - r)$ matrix Ω such that

$$\Delta_{13} - N\Delta_{13}^*H\Delta_{13} = \Omega\psi^*\Delta_{13}. \tag{1.9}$$

The boundary conditions (1.7') are equivalent to the conditions

$$\begin{aligned} \psi^*\Delta_{13}\hat{u} &= 0, \\ \chi^*N[\Delta_{13}^*H\Delta_{13}\hat{u} + D\hat{v}] &= 0, \end{aligned}$$

which is a set of conditions of the form (1.8) with S the r-dimensional subspace of C_{2n} defined as $\{\hat{u}:\psi^*\Delta_{13}\hat{u} = 0\}$, and Q the hermitian matrix $\Delta_{13}^*H\Delta_{13}$.

If the $2n \times 2n$ matrix $Q = Q[\mathscr{B}]$ has been determined as in the above theorem, we write $Q[\eta_1,\eta_2] = Q[\eta_1,\eta_2:\mathscr{B}]$ for the hermitian form $\eta_2^*Q\eta_1$, and for $\eta_\alpha \in D[a,b]:\zeta_\alpha$, $(\alpha = 1,2)$, the corresponding functional $J[\eta_1,\eta_2;\mathscr{B}]$ is defined as

$$J[\eta_1,\eta_2:\mathscr{B}] = Q[\eta_1,\eta_2:\mathscr{B}] + \int_a^b \{\zeta_2^*(t)B(t)\zeta_1(t) + \eta_2^*(t)C(t)\eta_1(t)\}dt. \tag{1.10}$$

Corresponding to earlier abbreviated notations, whenever $\eta \in D[a,b]:\zeta$ the symbol $J[\eta,\eta:\mathscr{B}]$ is contracted to $J[\eta:\mathscr{B}]$. Analogous to the formula II.3.5, the results presented in Section III.1, we have that if $\eta_\alpha \in D[a,b]:\zeta_\alpha$, $(\alpha = 1,2)$, then

$$\text{(i)}\quad J[\eta_1,\eta_2:\mathscr{B}] = \hat{\eta}_2^*T[\eta_1,\zeta_1] + \int_a^b \eta_2^*L_1[\eta_1,\zeta_1]dt,$$

$$\text{(ii)}\quad J[\eta_1:B] = \hat{\eta}_1^*T[\eta_1,\zeta_1] + \int_a^b \eta_1^*L_1[\eta_1,\zeta_1]dt. \tag{1.11}$$

The symbol $D_e[\mathscr{B}]$ will be used to denote the set $\{\eta:\eta \in D[a,b],\ \hat{\eta} \in S[\mathscr{B}]\}$. Also, if $\eta \in D_e[\mathscr{B}]$, and $L[\eta] = B\zeta$, we write $\eta \in D_e[\mathscr{B}]:\zeta$. Finally, for n-dimensional vector functions $\eta_\alpha(t)$, $(\alpha = 1,2)$ which are continuous on the compact interval $[a,b]$, we introduce the notations

$$\text{(i)}\quad N[\eta_1,\eta_2] = \hat{\eta}_2^*\hat{\eta}_1 + \int_a^b \eta_2^*(t)\eta_1(t)dt,$$

$$\text{(ii)}\quad N[\eta_1] = N[\eta_1,\eta_1] = |\hat{\eta}_1|^2 + \int_a^b |\eta_1(t)|^2dt. \tag{1.12}$$

As noted above, if the dimension of the end-space S is equal to zero then the boundary conditions of (1.8) reduce to $\hat{u} = 0$; that is, the second condition $T[u,v] \in S^{\perp}$ imposes no additional restriction upon the end vectors $\hat{u}$ and $\hat{v}$. On the other hand, if S is of dimension r, with $1 \leq r \leq 2n$, then there exists a $2n \times r$ matrix Φ whose column vectors form a basis for S, and in terms of an r-dimensional vector ξ the condition that $\eta \in S$ is expressible parametrically as $\eta = \Phi\xi$. Also, the condition $T[u,v] \in S^{\perp}$ is expressible as $\Phi^*T[u,v] = 0$, so that an alternate form of the end-conditions (1.8) is

$$\hat{u} = \Phi\xi, \quad Q\xi + \Phi^*D\hat{v} = 0, \tag{1.8'}$$

where $Q = Q[\mathscr{B}]$ is given by

$$Q = \Phi^*Q\Phi.$$

Also, for $\eta_\alpha \in D[a,b]:\zeta_\alpha$, $(\alpha = 1,2)$, there exists parametric vectors ξ_1 and ξ_2 such that $\hat{\eta}_\alpha = \Phi\xi_\alpha$ and an alternate form of (1.10) is

$$J[\eta_1,\eta_2:\mathscr{B}] = \xi_2^*Q\xi_1 + \int_a^b \{\zeta_2^*(t)B(t)\zeta_1(t) + \eta_2^*(t)C(t)\eta_1(t)\}dt. \tag{1.10'}$$

In particular, if $\eta \in D[a,b]:\zeta$ and $\hat{\eta} = \Phi\xi$ then we have

$$J[\eta:\mathscr{B}] = \xi^*Q\xi + \int_a^b \{\zeta^*(t)B(t)\zeta(t) + \eta^*(t)C(t)\eta(t)\}dt. \tag{1.10''}$$

Theoretically, the forms (1.8) and (1.8') are entirely equivalent and it becomes largely a matter of personal preference as to which is used for general considerations, although in a particular problem there may be a definite advantage of one form over the other. In particular, the parametric form has been consistently preferred by Morse.

A basic result on the solvability of certain non-homogenous differential systems associated with $(\mathscr{B})$ is presented in the following theorem.

THEOREM 1.2. *If* $(\mathscr{B})$ *is self-adjoint and has only the identically vanishing solution* $(u(t);v(t)) \equiv (0;0)$, *then there exist* $n \times n$ *matrix functions* $G(t,s)$, $G_0(t,s)$ *for* $(t,s) \in \square = [a,b] \times [a,b]$ *possessing the following properties:*

(a) $G(t,s)$ *is continuous in* (t,s) *on* $\square$, *is* a.c. *in each argument on* $[a,b]$ *for fixed values of the other argument, and* $G(t,s) \equiv [G(s,t)]^*$ *on* $\square$.

(b) $G_0(t,s)$ *is continuous in* (t,s) *on each of the triangular domains* $\Delta_1 = \{(t,s)\colon (t,s) \in \square,\ s < t\}$ *and* $\Delta_2 = \{(t,s)\colon (t,s) \in \square,\ t < s\}$, *is bounded on* $\square$, *and the restriction of* G_0 *to* Δ_α, $(\alpha = 1,2)$, *has a finite limit at each* (t,t) *with* $t \in [a,b]$.

(c) *if* $s \in (a,b)$, *and* ξ *is an arbitrary* n-*dimensional vector, then* $(u(t);v(t)) = (G(t,s)\xi;G_0(t,s)\xi)$ *is a solution of the differential system* $(\mathscr{B}$-i) *on each of the subintervals* $[a,s)$ *and* $(s,b]$, *and also satisfies the boundary conditions* $(\mathscr{B}$-ii); *in particular,* $u \in D[\mathscr{B}]:v$.

(d) *if* $f \in \mathscr{L}_n[a,b]$, *then the unique solution of the differential system*

$$\begin{aligned} &\text{(i)} \quad L_1[u,v](t) = f(t), \quad L_2[u,v](t) = 0, \quad t \in [a,b], \\ &\text{(ii)} \quad u \in S[\mathscr{B}], \quad T[u,v] \in S^{\perp}[\mathscr{B}], \end{aligned} \tag{1.13}$$

is given by

$$u(t) = \int_a^b G(t,s)f(s)ds, \quad v(t) = \int_a^b G_0(t,s)f(s)ds. \tag{1.14}$$

(e) *if* $\eta \in D[\mathscr{B}]:\zeta$, *and* $(u(t);v(t))$ *is the unique solution of the differential system*

$$\text{(i)}\quad L_1[u,v](t) = \eta(t),\quad L_2[u,v](t) = 0,\quad t \in [a,b],$$
$$\text{(ii)}\quad u \in S[\mathscr{B}],\quad T[u,v] - \eta \in S^{\perp}[\mathscr{B}], \tag{1.15}$$

then there exists a $k > 0$ *such that* $N[u] \le k^2 N[\eta]$; *also*,

$$J[u,\eta:\mathscr{B}] = N[\eta],\quad J[u:\mathscr{B}] = N[\eta,u]. \tag{1.16}$$

Conclusions (a)-(d) of Theorem 1.2 are immediate consequences of properties of the Green's matrix for an incompatible boundary problem involving two-point boundary conditions, (see, for example, Reid [35, Section III.7]). If $Y(t) = (U(t);V(t))$ is a fundamental matrix of the differential equation of ($\mathscr{B}$-i), then there exists a $2n \times 2n$ matrix R such that the solution of (1.15) is given by the integral transforms

$$\begin{aligned} u(t) &= U(t)R\eta + \int_a^b G(t,s)\eta(s)ds, \\ v(t) &= V(t)R\eta + \int_a^b G_o(t,s)\eta(s)ds. \end{aligned} \tag{1.17}$$

With the aid of elementary inequalities it then follows that there exists a constant c_1 such that

$$|u(t)|^2 \le c_1 N[\eta] \quad \text{for} \quad t \in [a,b], \tag{1.18}$$

and in turn it follows that $N[u] \le c_1[(b-a) + 2]N[\eta]$. Also, if $(u(t);v(t))$ is the solution of the differential system (1.15), with the aid of relation (1.11-i) it follows that

$$J[u,\eta_1:\mathscr{B}] = \hat{\eta}_1^*\hat{\eta} + \int_a^b \eta_1^*(t)\eta(t)dt = N[\eta,\eta_1],$$

from which the relations of (1.16) follow upon setting $\eta_1 = \eta$ and $\eta_1 = u$, respectively.

2. Normality and abnormality of boundary problems

As in Section V.5, let $\Lambda[a,b]$ denote the vector space of n-dimensional vector functions which are solutions of the differential equation $v'(t) + A^*(t)v(t) = 0$ and satisfy $B(t)v(t) = 0$ on $[a,b]$. That is, v belongs to $\Lambda[a,b]$ if and only if $u(t) \equiv 0$, $v(t)$ is a solution of the vector differential system (Bi). Following the notation of Reid [35, Sec. VII.9], for a problem $(\mathscr{B})$ involving a subspace of S of $\mathbf{C}_{2n}$ let $\Lambda\{S\}$ denote the subspace of $\Lambda[a,b]$ on which the 2n-dimensional vector $D\hat{v}$ belongs to $S^{\perp}$. Obviously, $u(t) \equiv 0$, $v(t)$ is a solution of the boundary problem $(\mathscr{B})$ if and only if $v \in \Lambda\{S\}$. The boundary problem $(\mathscr{B})$ is said to be *normal* if $\dim \Lambda\{S\} = 0$, and to be *abnormal* with order of abnormality equal to k if $\dim \Lambda\{S\} = k$. If $\dim S = r$, then in terms of the matrices introduced in Section 1 the column vectors of the $2n \times (2n - r)$ matrix $P = \Delta^*_{13}\psi$ form a basis for $S^{\perp}$. Also, if $V(t)$ is a $n \times d$ matrix whose column vectors form a basis for $\Lambda[a,b]$, then a boundary problem $(\mathscr{B})$ is normal if and only if the $2n \times (2n - r + d)$ matrix

$$[P \qquad D\hat{V}] \tag{2.1}$$

is of rank $2n - r + d$. If $(\mathscr{B})$ has order of abnormality equal to $k > 0$, then the matrix (2.1) has rank $2n - r + d - k$, and there is a $(2n - r) \times (2n - r - k)$ matrix R such that the $2n \times (2n - r - k + d)$ matrix

$$[PR \qquad D\hat{V}] \tag{2.1'}$$

is of rank $2n - r - k + d$, and consequently if the linear subspace S_ν of $\mathbf{C}_{2n}$ is defined as $\{w : w \in \mathbf{C}_{2n}, R^*P^*w = 0\}$

then S_ν is of dimension $r + k$ and the boundary problem

$$(\mathscr{B}_\nu) \quad \begin{array}{ll} L_1[u,v](t) = 0, \quad L_2[u,v](t) = 0, & t \in [a,b] \\ \hat{u} \in S_\nu, & T[u,v] \in S_\nu^\perp, \end{array}$$

is normal. Moreover, since the column vectors of $V(t)$ form a basis for $\Lambda[a,b]$ we have that $V^*(t)\eta(t)$ is constant for each $\eta \in D[a,b]$, it follows that an n-dimensional vector function $\eta(t)$ belongs to $D[\mathscr{B}]$ if and only if $\eta(t)$ belongs to $D[\mathscr{B}_\nu]$. Also, $(\mathscr{B}_\nu)$ is a normal problem equivalent to $(\mathscr{B})$ in the following sense. If $(u(t);v(t))$ is a non-identically vanishing solution of $(\mathscr{B}_\nu)$ then $u(t) \not\equiv 0$ on $[a,b]$, and $(u(t);v(t))$ is a solution of $(\mathscr{B})$. Moreover, if $(u(t);v(t))$ is a solution of $(\mathscr{B})$ then there exists a unique d-dimensional vector ρ such that $(u(t);v(t) + V(t)\rho)$ is a solution of $(\mathscr{B}_\nu)$.

For brevity, let $\mathscr{S}_a$ denote the $(2n - d)$-dimensional subspace of $\mathbf{C}_{2n}$ defined as

$$\mathscr{S}_a = \{\hat{\eta}: V^*D\hat{\eta} = 0\}. \tag{2.2}$$

For the particular boundary problem $(\mathscr{B}^0)$ involving the differential equations $(\mathscr{B}\text{-i})$, and with S the zero-dimensional subspace of $\mathbf{C}_{2n}$, then a corresponding normal problem $(\mathscr{B}_\nu^0)$ determined by the above process is $(\mathscr{B})$ with $S = \mathscr{S}_a^\perp$; that is, the system involving the differential equations $(\mathscr{B}\text{-i})$ and the boundary conditions

$$(\mathscr{B}_\nu^0\text{-ii}) \qquad u \in \mathscr{S}_a^\perp, \qquad T[u,v] \in \mathscr{S}_a.$$

For a normal problem $(\mathscr{B})$ the condition that the matrix (2.1) has rank $2n - r + d$ is equivalent to the condition that

$$\dim \{S[\mathscr{B}] \cap \mathscr{S}_a^{\perp}\} = \dim S^{\perp}[\mathscr{B}] + \dim \mathscr{S}_a^{\perp} = (2n-r) + d,$$

so that $\dim \{S[\mathscr{B}] \cap \mathscr{S}_a\} = r - d$. Now consider with $(\mathscr{B})$ a second boundary problem $(\mathscr{B}_0)$ involving the same differential equation $(\mathscr{B}_0 - i)$ as $(\mathscr{B}\text{-i})$, and boundary conditions

$(\mathscr{B}_0\text{-ii})$ $\qquad u \in \mathscr{S}_0, \quad T[u,v] \in \mathscr{S}_0^{\perp},$

where $\mathscr{S}_0$ is a second subspace of $\mathbf{C}_{2n}$. If $\dim \mathscr{S} = r$, $\dim \mathscr{S}_0 = r_0$, and each of the systems $(\mathscr{B})$, $(\mathscr{B}_0)$ is normal, then $\dim [S \cap \mathscr{S}_a] = r - d$ and $\dim [S_0 \cap \mathscr{S}_a] = r_0 - d$. If $S_0 \cap \mathscr{S}_a \subset S \cap \mathscr{S}_a$ then $r \geq r_0$ and $(\mathscr{B}_0)$ is called a *subproblem of* $(\mathscr{B})$ *of dimension* $r - r_0$. If $r > r_0$ then there exists a $2n \times (r - r_0)$ matrix θ such that

$$S_0 \cap \mathscr{S}_a = \{\hat{\eta}: \hat{\eta} \in S \cap \mathscr{S}_a, \ \Theta^*\hat{\eta} = 0\}. \tag{2.3}$$

In particular, if $(\mathscr{B})$ is normal then the problem $(\mathscr{B}_\nu^o)$ involving the differential equations $(\mathscr{B}\text{-i})$ and the above boundary conditions $(\mathscr{B}_0\text{-ii})$ is a subproblem of $(\mathscr{B})$ of dimension $r - r_0$.

3. Self-adjoint boundary problems associated with $(\mathscr{B})$

Attention will now be devoted to differential systems of the form

$$\begin{aligned} &\text{(i)} \quad L_1[u,v](t) = \lambda K(t)u(t), \quad L_2[u,v](t) = 0, \\ &\qquad\qquad t \in [a,b], \\ &\text{(ii)} \quad \hat{u} \in S, \quad T[u,v] \equiv Q\hat{u} + D\hat{v} \in S^{\perp}, \end{aligned} \tag{3.1}$$

involving the parameter λ, and under the following conditions.

$(\mathscr{A}_{\mathscr{B}})$ *On a given nondegenerate compact interval* $[a,b]$ *on the real line the* $n \times n$ *matrix functions* $A(t)$, $B(t)$, $C(t)$ *satisfy hypothesis* $(\mathscr{A})$, S *is a subspace of* $\mathbf{C}_{2n}$ *such that the system* (3.1) *is normal,* Q *is any hermitian* $2n \times 2n$ *matrix,* $D = \text{diag}\{-E_n, E_n\}$, *and* $K(t)$ *is an hermitian matrix function of class* $\mathscr{L}_{nn}[a,b]$ *such that* $K(t) \not\equiv 0$ *for* t *on a subset of* $[a,b]$ *of positive measure.*

For brevity, the notations

$$K[\eta_1,\eta_2] = \int_a^b \eta_2^*(t)K(t)\eta_1(t)dt, \quad K[\eta] = K[\eta,\eta] \tag{3.2}$$

are introduced for η_1,η_2 and η functions of class $\mathscr{L}^2[a,b]$; clearly $K[\eta_1,\eta_2]$ is an hermitian form on $\mathscr{L}^2[a,b] \times \mathscr{L}^2[a,b]$.

A general system (3.1) has in common with $(\mathscr{B})$ of Section 1 the equation $L_2[u,v](t) = 0$ and the subspace S of $\mathbf{C}_{2n}$, so that for (3.1) the classes of vector functions $D[a,b]$, $D_e[\mathscr{B}]$ and $D_o[\mathscr{B}] = D_o[a,b]$ have the same meaning as for the system $(\mathscr{B})$. Moreover, the condition that (3.1) is normal is the same as the condition that $(\mathscr{B})$ is normal; that is, if $u(t) \equiv 0$, $v(t)$ is a solution on $[a,b]$ of (3.1) for any one value of λ, and consequently for all values of λ, then $v(t) \equiv 0$ on $[a,b]$.

Let $(\mathscr{A}_K)$ denote the following condition.

$(\mathscr{A}_K)$ *Condition* $(\mathscr{A}_B)$ *holds, and if* $(u(t);v(t))$ *is an eigenvector of* (3.1) *corresponding to an eigenvalue* λ *then* $K[u] > 0$.

Whenever condition $(\mathscr{A}_K)$ holds, all eigenvalues of this system are real, and solutions $y(t) = (u(t);v(t))$, $y_o(t) = (u_o(t);v_o(t))$ corresponding to distinct eigenvalues

λ and λ_o are K-orthogonal in the sense that $K[u_o,u] = 0$; moreover, if λ is an eigenvalue of index m then the linear vector space of solutions of (3.1) for this value λ has a basis $y_\alpha(t) = (u_\alpha(t);v_\alpha(t))$, $(\alpha = 1,\ldots,m)$, which is K-orthogonal in the sense that $K[u_\alpha,u_\beta] = \delta_{\alpha\beta}$, $(\alpha,\beta = 1,\ldots,m)$.

If the subspace $\mathscr{S}_a$ of $\mathbf{C}_{2n}$ is defined by (2.2) and the $n \times n$ matrix functions $A(t)$, $B(t)$, $C(t)$ satisfy hypothesis $(\mathscr{A})$, then for an arbitrary $2n \times 2n$ hermitian matrix Q the boundary problem

$$\begin{gathered} L_1[u,v](t) = u(t), \quad L_2[u,v](t) = 0, \quad t \in [a,b] \\ \hat{u} \in \mathscr{S}_a^{\perp}, \qquad T[u,v] \equiv Q\hat{u} + D\hat{v} \in \mathscr{S}_a^{\perp}, \end{gathered} \tag{3.3}$$

is a special case of (3.1) with $K(t) \equiv E_n$. Moreover, this problem is normal, so that this system satisfies the above hypothesis $(\mathscr{A}_{\mathscr{B}})$. In particular, (3.3) has only real eigenvalues, and consequently the set of all eigenvalues of this system is at most denumerably infinite. Let $\lambda = \lambda_o$ be a real value which is not an eigenvalue of (3.3), so that for $\lambda = \lambda_o$ the system (3.1) has only the identically vanishing solution, and consequently for arbitrary $w \in \mathbf{C}_{2n}$ the non-homogeneous system

$$\begin{gathered} L_1[u,v](t) = \lambda_o u(t), \quad L_2[u,v](t) = 0 \\ \hat{u} - w \in \mathscr{S}_a^{\perp}, \qquad T[u,v] \equiv Q\hat{u} + D\hat{v} \in \mathscr{S}_a^{\perp}, \end{gathered} \tag{3.4}$$

has a unique solution $(u(t);v(t))$. Since $u \in D[a,b]:v$, it follows that $\hat{u} \in \mathscr{S}_a$. Consequently, if $w \in \mathscr{S}_a$ then $\hat{u} - w \in S_a \cap \mathscr{S}_a^{\perp}$, and hence $\hat{u} = w$. In view of this result, and the fact that $v^*(t)\eta(t)$ is constant on $[a,b]$ for arbitrary $v \in \Lambda[a,b]$ and $\eta \in D[a,b]:\zeta$, it follows that a vector $w \in C_{2n}$ is such that there exists a vector function $\eta \in D[a,b]$ with $\hat{\eta} = w$ if and only if $w \in \mathscr{S}_a$.

For $N[\eta]$ defined by (1.12), let $D_N[\mathscr{B}]$ denote the class of vector functions $\{\eta : \eta \in D_e[\mathscr{B}], N[\eta] = 1\}$. Corresponding to the results of Lemmas III.2.1 and III.2.2 for the scalar boundary problem, we have the following auxiliary results.

LEMMA 3.1. *If* ($\mathscr{B}$) *is normal, and the hermitian functional* $J[\eta:\mathscr{B}]$ *is non-negative on* $D_e[\mathscr{B}]$, *then the infimum of* $J[\eta:\mathscr{B}]$ *on* $D_N[\mathscr{B}]$ *is zero if and only if there exists a non-identically vanishing solution* $(u(t);v(t))$ *of* ($\mathscr{B}$); *moreover, if* $u \in D_e[\mathscr{B}]$ *and* $J[u:\mathscr{B}] = 0$, *then there exists a vector function* $v(t)$ *such that* $(u(t);v(t))$ *is a solution of* ($\mathscr{B}$).

Now let $\mathscr{F}$ denote a finite set $\mathscr{F} = [f_1,\dots,f_r]$ of n-dimensional vector functions which are of class $\mathscr{L}_n[a,b]$, and for which there exists an associated set of vector functions $w_i \in D_e[\mathscr{B}]$, $(i = 1,\dots,r)$, such that the $r \times r$ matrix

$$\left[\int_a^b w_i^*(t) f_j(t)\,dt\right], \quad (i,j = 1,\dots,r), \qquad (3.5)$$

is non-singular; in particular, this assumption requires that the set of vector functions $\mathscr{F}$ is linearly independent on $[a,b]$. Also, corresponding to the notation of Section III.2, the class of functions η in $D_e[\mathscr{B}]$ and $D_N[\mathscr{B}]$ which satisfy the integral conditions

$$\int_a^b f_i^*(t)\eta(t)\,dt = 0, \quad (i = 1,\dots,r), \qquad (3.6)$$

will be denoted by $D_e[\mathscr{B}|\mathscr{F}]$ and $D_N[\mathscr{B}|\mathscr{F}]$, respectively.

LEMMA 3.2. *Suppose that* $(\mathscr{B})$ *is normal, and that* $J[\eta:\mathscr{B}]$ *is non-negative on* $D_e[\mathscr{B}|\mathscr{F}]$, *where* $\mathscr{F}$ *is a set of vector functions possessing the above-described properties. Then the infimum of* $J[\eta:\mathscr{B}]$ *on* $D_N[\mathscr{B}|\mathscr{F}]$ *is zero if and only if there exist constants* $k_1,\ldots,k_r$ *such that the system*

$$\begin{aligned}
&\text{(a)}\quad L_1[u,v](t) + \sum_{j=1}^{r} k_j f_j(t) = 0,\\
&\qquad L_2[u,v](t) = 0, \quad t \in [a,b],\\
&\text{(b)}\quad \hat{u} \in S, \quad T[u,v] \in S^{\perp},\\
&\text{(c)}\quad \int_a^b f_i^*(t)u(t)dt = 0, \quad (i = 1,\ldots,r)
\end{aligned} \tag{3.7}$$

has a non-identically vanishing solution $(u(t);v(t))$. *Moreover, if* $u_o \in D_e[\mathscr{B}|\mathscr{F}]$ *and* $J[u_o:\mathscr{B}] = 0$, *then* u_o *is a solution of the system* (3.7) *for suitable constants* $k_1,\ldots,k_r$. *In particular, if the vector functions* f_i *are such that there exist constants* γ_i *and vector functions* $(u_i(t);v_i(t))$ *satisfying for* $i = 1,\ldots,r$ *the differential system*

$$\begin{aligned}
&L_1[u_i,v_i](t) + \gamma_i f_i(t) = 0, \quad L_2[u_i,v_i](t) = 0,\\
&\qquad\qquad t \in [a,b],\\
&\hat{u}_i \in S, \qquad T[u_i,v_i] \in S^{\perp},
\end{aligned} \tag{3.8}$$

and the $r \times r$ *matrix* $\left[\int_a^b u_i^*(t)f_j(t)dt\right]$ *is non-singular, then whenever* $(u(t);v(t))$ *is a solution of* (3.7) *with constants* $k_1,\ldots,k_r$ *then necessarily each* k_i, $(k = 1,\ldots,r)$, *is equal to zero.*

As indicated in Section III.2 for the corresponding scalar problems, the results of the above two lemmas have been established by various methods, indirect and direct, in general centering around the fact that the determined solutions are minimizing elements for an hermitian functional in a certain class of vector functions. One proof is to be found in Reid [35, Sec. VII.10]. Allied results are presented in the set of Exercises at the end of this chapter.

For the treatment of boundary problems of the form (3.1), in addition to $(\mathscr{A}_{\mathscr{B}})$ the following hypothesis is assumed.

$(\mathscr{A}_K')$ (i) *the hermitian matrix function* $K(t)$ *is non-negative definite for* $t \in [a,b]$;

(ii) *there exists a real number* λ_0 *such that* $J[\eta;\lambda_0:\mathscr{B}] = J[\eta:\mathscr{B}] - \lambda_0 K[\eta]$ *is positive definite on* $D_e[\mathscr{B}]$;

(iii) *there exist subspaces of* $D_e[\mathscr{B}]$ *of arbitrarily high dimension on which* $K[\eta]$ *is positive definite.*

It follows readily that hypotheses $(\mathscr{A})$ and $(\mathscr{A}_K'\text{ii})$ imply the following conditions:

(i) $B(t) \geq 0$ *for* t a.e. *on* $[a,b]$;

(ii) *if* $\eta \in D_e[\mathscr{B}]$, $\eta(t) \not\equiv 0$ *on* $[a,b]$, *and* $K[\eta] = 0$, *then* $J[\eta:\mathscr{B}] > 0$;

(iii) $J[\eta:\mathscr{B}]$ *is bounded below on the set* $D_N[\mathscr{B}:K]$, *defined as*

$$D_K[\mathscr{B}:K] = \{\eta:\eta \in D_e[\mathscr{B}],\ K[\eta] = 1\}. \quad (3.9)$$

Let $(\mathscr{A}_{\mathscr{B}}')$ denote the hypothesis $(\mathscr{A}_{\mathscr{B}})$ with the added conditions that $A(t)$ and $B(t)$ are of class $\mathscr{L}_{nn}^{\infty}[a,b]$. By the method of proof suggested for Problem VII.11.1 of Reid [35] one may then show that condition $(\mathscr{A}_2\text{iii})$ is satisfied whenever $(\mathscr{A}_{\mathscr{B}}')$, $(\mathscr{A}_K'\text{i})$, $(\mathscr{A}_2\text{i,ii})$ hold, and also

there exists a constant $\kappa_0 > 0$ *such that*
$$B^2(t) - \kappa_0 B(t) \geq 0 \text{ for } t \text{ a.e. on } [a,b]. \tag{3.10}$$

In particular, condition (3.10) holds for a problem (3.1) specified by (V.2.16), whenever $0 < r_n(t) \leq 1/\kappa_0$ on $[a,b]$.

Corresponding to Theorem III.2.1 for scalar equations, a basic existence theorem for boundary problems (3.1) is as follows.

THEOREM 3.1. *Whenever hypotheses* $(\mathscr{A}_{\mathscr{B}})$ *and* $(\mathscr{A}_K'\text{i,ii,iii})$ *are satisfied there exists for the boundary problem* (3.1) *an infinite sequence of real eigenvalues* $\lambda_1 \leq \lambda_2 \leq \ldots$, *with corresponding eigenvectors* $(u(t);v(t)) = (u_j(t);v_j(t))$ *for* $\lambda = \lambda_j$ *such that:*

(a) $K[u_i,u_j] = \delta_{ij}, \quad (i,j = 1,2,\ldots);$

(b) $\lambda_1 = J[u_1:\mathscr{B}]$ *is the minimum of* $J[\eta:\mathscr{B}]$ *on the class* $D_N[\mathscr{B}:K]$; (3.11)

(c) *for* $j = 2,3,\ldots$ *the class*

$$D_{N_j}[\mathscr{B}:K] = \{\eta : \eta \in D_N[\mathscr{B}:K],\ K[\eta,u_i] = 0,\ i = 1,\ldots,j-1\} \tag{3.12}$$

is non-empty, and $\lambda_j = J[u_j:\mathscr{B}]$ *is the minimum of* $J[\eta:\mathscr{B}]$ *on* $D_{N_j}[\mathscr{B}:K]$;

(d) $\{\lambda_j\} \to +\infty$ *as* $j \to \infty$.

Also, corresponding to the Corollary to Theorem II.2.1, we have the following results.

COROLLARY. *Suppose that hypotheses* $(\mathscr{A}_{\mathscr{B}})$ *and* $(\mathscr{A}_K'\text{-i,ii,iii})$ *are satisfied, and* $\{\lambda_j, y_j(t) = (u_j(t);v_j(t))\}$ *is a sequence of eigenvalues and corresponding eigenvectors as specified in Theorem* 3.1. *Then:*

(a) *if* k *is a positive integer, and* $c_1,\ldots,c_k$ *are constants such that* $|c_1|^2 + \ldots + |c_k|^2 = 1$, *then* $\eta(t) = c_1u_1(t) + \ldots + c_ku_k(t)$ *belongs to* $D_N[\mathscr{B}]$ *and* $J[\eta:\mathscr{B}] \leq \lambda_k$.

(b) MAXIMUM-MINIMUM PROPERTY. *If* $F = \{f_1,\ldots,f_r\}$ *is a set of* n-*dimensional vector functions of class* $\mathscr{L}_n[a,b]$, *and* $\lambda\{F\}$ *denotes the minimum of* $J[\eta:\mathscr{B}]$ *on the set*

$$D_N[\mathscr{B}:K|F] = \left\{\eta:\eta \in D_N[\mathscr{B}:K],\ \int_a^b f_i^*(t)\eta(t)dt = 0,\ (i = 1,\ldots,r)\right\}$$

then λ_{r+1} *is the maximum of* $\lambda\{F\}$, *and this maximum is attained for* $f_i(t) = K(t)u_i(t)$, $(i = 1,\ldots,r)$.

If $\{\lambda_j;y_j(t) = (u_j(t);v_j(t))\}$, $(j = 1,2,\ldots)$, is a sequence of eigenvalues and eigenvectors determined as in Theorem 3.1, the "Fourier coefficients"

$$c_j[\eta] = \int_a^b u_j^*(t)K(t)\eta(t)dt, \quad (j = 1,2,\ldots) \tag{3.13}$$

are well defined for arbitrary vector functions of class $\mathscr{L}_n^2[a,b]$. With the aid of certain basic inequalities resulting from the extremizing properties of the eigenvalues of (3.1), and which are presented as Nos. 8, 9 and 10 of the list of Topics and Exercises at the end of this Chapter, one may establish the following result.

THEOREM 3.2. *If hypotheses* $(\mathscr{A}_{\mathscr{B}}')$ *and* $(\mathscr{A}_K'$-i,ii,iii) *are satisfied and* $\lambda_0 < \lambda_1$, *then for arbitrary* $\eta \in D_e[\mathscr{B}]$ *the infinite series* $\sum_{j=1}^{\infty}|c_j[\eta]|^2$, $\sum_{j=1}^{\infty}(\lambda_j - \lambda_0)|c_j[\eta]|^2$ *converge, and*

$$\sum_{j=1}^{\infty}|c_j[\eta]|^2 = K[\eta], \tag{3.14}$$

$$\sum_{j=1}^{\infty}(\lambda_j - \lambda_0)|c_j[\eta]|^2 \leq J[\eta;\lambda_0:\mathscr{B}]. \tag{3.15}$$

In view of (3.14), inequality (3.15) is equivalent to

$$\sum_{j=1}^{\infty} \lambda_j |c_j[\eta]|^2 \leq J[\eta:\mathscr{B}]. \tag{3.15'}$$

Also, (3.14) clearly implies the relation

$$\sum_{j=1}^{\infty} c_j[\eta_1]c_j[\eta_2] = K[\eta_1,\eta_2], \text{ for } \eta_\alpha \in D_e[\mathscr{B}], \quad \alpha = 1,2. \tag{3.14'}$$

Corresponding to the result (d) of Theorem 1.2, if λ is not an eigenvalue of (3.1) then there exist $n \times n$ matrix functions $G(t,s;\lambda)$, $G_0(t,s;\lambda)$ which satisfy conditions analogous to (a), (b), (c) of Theorem 1.2, and if $f \in \mathscr{L}_n[a,b]$ then the unique solution of the differential system

$$L_1[u,v](t) - \lambda K(t)u(t) = f(t), \quad L_2[u,v](t) = 0, \quad t \in [a,b],$$
$$\hat{u} \in S, \qquad T[u,v] \in S^{\perp}, \tag{3.16}$$

is given by

$$\text{(i)} \quad u(t) = \int_a^b G(t,s;\lambda)f(s)ds,$$
$$\text{(ii)} \quad v(t) = \int_a^b G_0(t,s;\lambda)f(s)ds, \quad t \in [a,b]. \tag{3.17}$$

Also, in view of the self-adjoint nature of (3.1) it follows that $G(t,s;\lambda) = [G(s,t;\lambda)]^*$. In particular, if $\lambda = \lambda_0$ is not an eigenvalue of (3.1) then $(u(t);v(t)) = (u_j(t);v_j(t))$ is the solution of (3.17) for $\lambda = \lambda_0$ and $f(t) = (\lambda_j - \lambda_0)K(t)u_j(t)$, and therefore the corresponding equation (3.17i) gives

$$(\lambda_j - \overline{\lambda}_0)^{-1}u_j^*(t) = \int_a^b u_j^*(s)K(s)G(s,t;\lambda_0)ds, \quad t \in [a,b]. \tag{3.18}$$

As $G(\ ,t;\lambda_0)\xi \in D_e[\mathscr{B}]:G_0(\ ,t;\lambda_0)\xi$ for arbitrary n-dimensional vectors ξ, equation (3.18) provides the values of the j-th Fourier coefficients of the column vectors of $G(\ ,t;\lambda_0)$, so that the following result is a consequence of the relation (3.14').

THEOREM 3.3. *If the hypotheses of Theorem* 3.1 *are satisfied and* λ *is not an eigenvalue of* (3.1), *then for* $(t,\tau) \in [a,b] \times [a,b]$,

$$\sum_{j=1}^{\infty} |\lambda_j - \lambda|^{-2} u_j(\tau)u_j^*(t) = \int_a^b [G(s,\tau;\lambda)]^* K(s) G(s,t;\lambda)ds; \tag{3.19}$$

in particular,

$$\sum_{j=1}^{\infty} |\lambda_j - \lambda|^{-2} |u_j(t)|^2 = \mathrm{Tr}\left\{\int_a^b [G(s,t;\lambda)]^* K(s) G(s,t;\lambda)ds\right\}. \tag{3.20}$$

A stronger result holds whenever condition ($\mathscr{A}_K'$i) is replaced by the following condition.

($\mathscr{A}_K'$i') *The hermitian matrix function* $K(t)$ *is such that* $K[\eta]$ *is a positive definite functional on* $\mathscr{L}_n[a,b]$.

THEOREM 3.4. *Suppose that the boundary problem* (3.1) *satisfies hypotheses* ($\mathscr{A}_{\mathscr{B}}'$), ($\mathscr{A}_K'$-i,ii,iii), *Condition* (3.10), *and* $\{\lambda_j, y_j(t) = (u_j(t);v_j(t))\}$, $(j = 1,2,\ldots)$ *is a sequence of eigenvalues and corresponding eigenvectors as determined in Theorem* 3.1.

(i) *If* $\eta \in D_e[\mathscr{B}]$, *then*

(a) $\sum_{j=1}^{\infty} c_j[\eta]u_j(t)$ *converges to* $\eta(t)$, *uniformly on* $[a,b]$;

(b) $\left\{\int_a^b |\eta'(t) - \sum_{j=1}^{m} c_j[\eta]u_j(t)|^2 dt\right\} \to 0$ *as* $m \to \infty$;

(c) $J[\eta:\mathscr{B}] = \sum_{j=1}^{\infty} \lambda_j |c_j[\eta]|^2.$

(ii) *If* λ *is not an eigenvalue of* (3.1) *then for* $(t,s) \in [a,b] \times [a,b]$ *the corresponding Green's matrix* $G(t,s;\lambda)$ *has the expansion*

$$G(t,s;\lambda) = \sum_{j=1}^{\infty} (\lambda_j - \lambda)^{-1} u_j(t) u_j^*(s), \tag{3.21}$$

and the series in (3.21) *converges uniformly on* $[a,b] \times [a,b]$.

Combining the results of Theorems V.8.4 and the Corollary to Theorem 3.1, then one has the following oscillation theorem.

THEOREM 3.5. *Suppose that the* $n \times n$ *matrix functions* $A(t)$, $B(t)$, $C(t)$ *satisfy hypothesis* ($\mathscr{A}$) *on* $[a,b]$, $B(t) \geq 0$ *for* t a.e. *on this interval, while* $K(t)$ *is an hermitian matrix function of class* $\mathscr{L}^{\infty}_{nn}[a,b]$ *satisfying condition* ($\mathscr{A}_K'$i') *and the differential system* (3.1-i) *is identically normal on* $[a,b]$. *Then the number of conjugate points to* $t = a$ *on* (a,b), {*on* $(a,b]$}, *relative to* (3.1-i) *for a given value* $\lambda = \ell$ *is equal to the number of eigenvalues of the boundary problem*

$$L_1[u,v](t) = \lambda K(t)u(t), \quad L_2[u,v](t) = 0, \quad t \in [a,b],$$
$$u(a) = 0, \qquad u(b) = 0, \tag{3.22}$$

which are less than, {*not greater than*}, ℓ, *where each eigenvalue is counted a number of times equal to its index.*

4. Comparison theorems

A boundary problem (3.1) depends upon the vector differential operator $L[u:\mathscr{B}](t) = u'(t) - A(t:\mathscr{B})u(t)$, $L_2[u,v:\mathscr{B}](t) = L[u:\mathscr{B}](t) - B(t;\mathscr{B})v(t)$, the vector function space $D[a,b]$ consisting of those n-dimensional a.c. vector functions for which there is a corresponding $\zeta \in \mathscr{L}_n[a,b]$ such that $L_2[\eta,\zeta:\mathscr{B}](t) = 0$, the end-manifold $S[\mathscr{B}]$ in $\mathbf{C}_{2n}$, the domain $D_e[\mathscr{B}] = \{\eta:\eta \in D[a,b],\ \eta \in S[\mathscr{B}]\}$, the end-form $Q[\eta:\mathscr{B}] = \eta^*Q[\mathscr{B}]\eta$, and the hermitian functionals on $D[a,b] \times D[a,b]$ defined as

$$\text{(i)}\quad J[\eta:\mathscr{B}] = Q[\eta:\mathscr{B}] + \int_a^b \{\zeta^*(t)B(t;\mathscr{B})\zeta(t) + \eta^*(t)C(t:\mathscr{B})\eta(t)\}dt$$

$$\text{(ii)}\quad K[\eta:\mathscr{B}] = \int_a^b \eta^*(t)K(t:\mathscr{B})\eta(t)dt. \tag{4.1}$$

Throughout this section it will be assumed that the considered problems satisfy hypotheses $(\mathscr{A}_{\mathscr{B}})$ and $(\mathscr{A}'_k\text{i,ii,iii})$ appearing as the hypotheses of Theorem 3.1. In particular, these hypotheses require each considered problem to be normal.

As an immediate consequence of the minimizing properties of the eigenvalues of two problems $\mathscr{B}$ and $\tilde{\mathscr{B}}$ as given specifically in Theorem 3.1, one has the following comparison theorem.

THEOREM 4.1. *Suppose that* $\mathscr{B}$ *and* $\tilde{\mathscr{B}}$ *have in common* $S[\mathscr{B}] = S[\tilde{\mathscr{B}}]$, $D[\mathscr{B}] = D[\tilde{\mathscr{B}}]$, *and* $K(t:\mathscr{B}) \equiv K(t:\tilde{\mathscr{B}}) \equiv K(t)$. *If* $\Delta J[\eta:\mathscr{B},\tilde{\mathscr{B}}] = J[\eta:\tilde{\mathscr{B}}] - J[\eta:\mathscr{B}]$ *is non-negative on* $D[\mathscr{B}] = D[\tilde{\mathscr{B}}]$, *then* $\tilde{\lambda}_j \geq \lambda_j$, $(j = 1,2,\ldots)$; *moreover, if* $\Delta J[\eta:\mathscr{B},\tilde{\mathscr{B}}]$ *is positive defininte on* $D[\mathscr{B}] = D[\tilde{\mathscr{B}}]$, *then* $\tilde{\lambda}_j > \lambda_j$, $(j = 1,2,\ldots)$.

Obviously, $D[\mathscr{B}] = D[\tilde{\mathscr{B}}]$ if $S[\mathscr{B}] = S[\tilde{\mathscr{B}}]$, and the matrix functions $A(t)$, $B(t)$ for the two problems are identical. This latter condition is not necessary, however, in view of the comments in the first paragraph of Section V.7. Indeed, if the matrices $\pi(t) = \pi(t:\mathscr{B})$ and $\pi(t) = \pi(t:\tilde{\mathscr{B}})$ are defined for the respective problems as in No. 6 of the Topics and Exercises at the end of Chapter V, then $D[\mathscr{B}] = D[\tilde{\mathscr{B}}]$ if and only if $S[\mathscr{B}] = S[\tilde{\mathscr{B}}]$ and for arbitrary n-dimensional a.c. vector functions $\eta(t)$ the condition $\pi^*(t:\mathscr{B})L[\eta:\mathscr{B}](t) = 0$ for $t \in [a,b]$ holds if and only if $\pi^*(t:\tilde{\mathscr{B}})L[\eta:\tilde{\mathscr{B}}](t) = 0$ for $t \in [a,b]$.

Now suppose that problems $(\mathscr{B})$ and $(\tilde{\mathscr{B}})$ differ only in the end-forms $Q[\eta:\mathscr{B}]$ and $Q[\eta:\tilde{\mathscr{B}}]$; that is, $A(t:\mathscr{B}) \equiv A(t:\tilde{\mathscr{B}})$, $B(t:\mathscr{B}) \equiv B(t:\tilde{\mathscr{B}})$, $C(t:\mathscr{B}) \equiv C(t:\tilde{\mathscr{B}})$, $K(t:\mathscr{B}) \equiv K(t:\tilde{\mathscr{B}})$ and $S[\mathscr{B}] = S[\tilde{\mathscr{B}}]$; in particular, $D[\mathscr{B}] = D[\tilde{\mathscr{B}}]$. As for the corresponding scalar problem in Section III.3, let $Q^{\Delta}[\eta] = Q[\eta:\mathscr{B}] - Q[\eta:\tilde{\mathscr{B}}] = \eta^* Q^{\Delta}\eta$. Also, let M be a $2n \times (2n - r)$ matrix whose column vectors form a basis for $S^{\perp}[\mathscr{B}] = S^{\perp}[\tilde{\mathscr{B}}]$, and $V_0(t)$ a $n \times d$ matrix function whose column vectors form a basis for $\Lambda[a,b]$. Since each of the problems $\mathscr{B}$ and $\tilde{\mathscr{B}}$ is assumed to be normal, the $2n \times (2n - r + d)$ matrix $[M \quad D\hat{V}_0]$, where $D = \operatorname{diag}\{-E_n, E_n\}$, has rank $2n - r + d$.

Now consider the polynomial $P(\mu:\mathscr{B},\tilde{\mathscr{B}})$ of degree $r - d$ defined as follows: if $r = d$, then $P(\mu:\mathscr{B},\tilde{\mathscr{B}}) \equiv 1$; if $1 \le r - d \le 2n$, then $P(\mu:\mathscr{B},\tilde{\mathscr{B}})$ is the determinant of the matrix $M(\mu:\mathscr{B},\tilde{\mathscr{B}})$ defined as:

$$M(\mu:\mathscr{B},\tilde{\mathscr{B}}) = \begin{bmatrix} E_{2n} - Q^{\Delta} & M & D\hat{V}_o \\ M^* & 0 & 0 \\ \hat{V}_o^* D & 0 & 0 \end{bmatrix}, \text{ if } 1 \le r-d \le 2n-1;$$

$$= \mu E_{2n} - Q^{\Delta}, \quad \text{if} \quad r - d = 2n.$$

If I is a finite interval on the real line, the symbol $\sigma(I:\mathscr{B})$ will denote the number of eigenvalues of $\mathscr{B}$ on I, each counted a number of times equal to its multiplicity. Also, corresponding to a real number x, the symbol $V_x(\mathscr{B})$, $\{W_x(\mathscr{B})\}$ denotes the number of eigenvalues of $\mathscr{B}$ which are less, {not greater}, than x; that is, $V_x(\mathscr{B}) = \sigma((-\infty,x):\mathscr{B})$ and $W_x(\mathscr{B}) = \sigma((-\infty,x]:\mathscr{B}]$.

Corresponding to results of Theorem III.3.3 and its Corollary, we now have the following result.

THEOREM 4.2. *If problems* $\mathscr{B}$ *and* $\tilde{\mathscr{B}}$ *differ only in the end-forms* $Q[\eta:\mathscr{B}]$ *and* $Q[\eta:\tilde{\mathscr{B}}]$, *while* n_Q *and* p_Q *denote, respectively, the number of negative and positive zeros of* $P(\mu:B,\tilde{\mathscr{B}})$, *each zero being counted a number of times equal to its multiplicity, then*

$$\tilde{\lambda}_{j+n_Q} \ge \lambda_j, \quad \text{and} \quad \lambda_{j+p_Q} \ge \tilde{\lambda}_j. \tag{4.2}$$

COROLLARY. *Under the hypotheses of Theorem* 4.2, *for each real number* x *we have*

$$\begin{aligned} V_x(\mathscr{B}) - p_Q \le V_x(\tilde{\mathscr{B}}) \le V_x(\mathscr{B}) + n_Q; \\ W_x(\mathscr{B}) - p_Q \le W_x(\tilde{\mathscr{B}}) \le W_x(\mathscr{B}) + n_Q; \end{aligned} \tag{4.3}$$

moreover, $|\sigma(I:\mathscr{B}) - \sigma(I:\tilde{\mathscr{B}})| \le n_Q + p_Q$ *for every bounded interval* I *on the real line.*

Now consider two problems $(\mathscr{B})$ and $(\tilde{\mathscr{B}})$ that differ only in the end-manifolds $S[\mathscr{B}]$ and $S[\tilde{\mathscr{B}}]$, which are of respective dimensions r and $\tilde{r}$. As each of the problems is assumed to be normal, from the discussion of Section 2 it follows that $\dim\{S[\mathscr{B}] \cap \mathscr{S}_a\} = r - d$, and $\dim(S[\tilde{\mathscr{B}}] \cap \mathscr{S}_a\} = \tilde{r} - d$. Also, $(\tilde{\mathscr{B}})$ is a subproblem of $(\mathscr{B})$ whenever $S[\tilde{\mathscr{B}}] \cap \mathscr{S}_a \subset S[\mathscr{B}] \cap \mathscr{S}_a$, in which case $r \geq \tilde{r}$, and $(\tilde{\mathscr{B}})$ is said to be a subproblem of $(\mathscr{B})$ of dimension $r - \tilde{r}$. Also, if $r > \tilde{r}$ there exists a $2n \times (r - \tilde{r})$ matrix Θ such that $S[\tilde{\mathscr{B}}] \cap \mathscr{S}_a = \{\hat{\eta}:\hat{\eta} \in S[\mathscr{B}] \cap \mathscr{S}_a,\ \Theta\hat{\eta} = 0\}$. As a consequence of these remarks and the extremizing properties of the eigenvalues of $(\mathscr{B})$ and $(\tilde{\mathscr{B}})$, we have the following result.

THEOREM 4.3. *If the problems* $(\mathscr{B})$ *and* $(\tilde{\mathscr{B}})$ *differ only in the end-manifolds* $S[\mathscr{B}]$, $S[\tilde{\mathscr{B}}]$, *and* $(\tilde{\mathscr{B}})$ *is a subproblem of* $(\mathscr{B})$ *of dimension* $\rho = r - \tilde{r}$, *then* $\lambda_{j+\rho} \geq \tilde{\lambda}_j \geq \lambda_j$, $(j = 1,2,\ldots)$. *Also, for each real number* x *we have*

$$V_x(\mathscr{B}) - \rho \leq V_x(\tilde{\mathscr{B}}) \leq V_x(\mathscr{B}),\quad W_x(\mathscr{B}) - \rho \leq W_x(\tilde{\mathscr{B}}) \leq W_x(\mathscr{B});$$

and $|\sigma(I:\mathscr{B}) - \sigma(I:\tilde{\mathscr{B}})| \leq \rho$ *for every bounded subinterval* I *of the real line.*

5. Treatment of self-adjoint boundary problems by matrix oscillation theory

In notation closely related to that of the preceding chapters, we shall consider a vector differential system of the form

$$Sy'(t) - H(t)y(t) = 0 \tag{5.1}$$

where S is a non-singular $m \times m$ matrix which is skew-hermitian, (i.e., $S^* = -S$), and $H(t)$ is an $m \times m$ hermitian matrix function of class $\mathscr{L}[a,b]$ on arbitrary compact subintervals $[a,b]$ of the interval I. In particular, equation (V.3.1) is of this form with $m = 2n$, S the real skew matrix $\mathscr{J}$ of (V.3.2), and $-H(t)$ the matrix $\mathscr{A}(t)$ as defined by (V.3.2).

With (5.1) we consider two-point boundary conditions

$$My(a) + Ny(\tau) = 0, \quad \tau > a, \quad [a,\tau] \in I, \tag{5.2_τ}$$

where M and N are $m \times m$ matrices such that the $m \times 2m$ matrix $[M \quad N]$ is of rank m, and $\tau \in I_a^+ = \{t : t \in I, t > a\}$. If P and Q are $m \times m$ matrices such that the $m \times 2m$ matrix $[P^* \quad Q^*]$ is of rank m and

$$MP + NQ = 0, \tag{5.3}$$

then in parametric form the boundary conditions (5.2_τ) may be written as

$$y(a) = P\xi, \qquad y(\tau) = Q\xi, \tag{5.2'}$$

where ξ is an n-dimensional parameter vector. Moreover, (see, for example, Reid [35, Ch. III]), the boundary problem (5.1), (5.2'), is self-adjoint if and only if the matrices S, P, Q satisfy the algebraic condition

$$P^*SP - Q^*SQ = 0, \tag{5.4}$$

and *throughout the following discussion it will be assumed that this condition is satisfied.*

If $Y(t)$ is a solution of the corresponding matrix differential equation

$$SY'(t) - H(t)Y(t) = 0, \tag{5.1_M}$$

then it follows readily that the matrix function $Y^*(t)SY(t)$

is constant on I. Consequently, if $Y = F(t)$ is the solution of (5.1_M) determined by the initial condition $F(a) = E$, then $F^*(t)\mathcal{S}F(t) \equiv \mathcal{S}$ on I.

In terms of these matrix functions, now define

$$U(t) = \mathcal{S}[F(t)\mathcal{P} - \mathcal{Q}], \quad V(t) = F(t)\mathcal{P} + \mathcal{Q}. \tag{5.5}$$

Since $\mathcal{S}^* = -\mathcal{S}$, it follows that

$$\begin{aligned} U^*V - V^*U &= -[\mathcal{P}^*F^* + \mathcal{Q}^*]\mathcal{S}[F\mathcal{P} - \mathcal{Q}] - [\mathcal{P}^*F^* + \mathcal{Q}^*]\mathcal{S}[F\mathcal{P} - \mathcal{Q}] \\ &= 2\{\mathcal{Q}^*\mathcal{S}\mathcal{Q} - \mathcal{P}^*F^*\mathcal{S}F\mathcal{P}\} \\ &= 2\{\mathcal{Q}^*\mathcal{S}\mathcal{Q} - \mathcal{P}^*\mathcal{S}\mathcal{P}\} = 0. \end{aligned} \tag{5.6}$$

Also, it may be verified readily that $Y(t) = (U(t);V(t))$ is a solution of the matrix differential system $(V.3.1_M)$ with

$$A = \frac{1}{2}\,\mathcal{H}\mathcal{S}^{-1}, \quad B = \frac{1}{2}\,\mathcal{H}, \quad C = -\frac{1}{2}\,\mathcal{S}^{*-1}\mathcal{H}\mathcal{S}^{-1}. \tag{5.7}$$

Consequently, on any subinterval on which $U(t)$ is nonsingular, the matrix function $W(t) = V(t)U^{-1}(t)$ is an hermitian solution of the Riccati matrix differential equation

$$W'(t) + \frac{1}{2}[W(t) + \mathcal{S}^{*-1}]\mathcal{H}(t)[W(t) + \mathcal{S}^{-1}] = 0. \tag{5.8$_1$}$$

Correspondingly, on any subinterval on which $V(t)$ is nonsingular, the matrix function $\hat{W}(t) = U(t)V^{-1}(t)$ is an hermitian solution of the Riccati matrix differential equation

$$\hat{W}'(t) - \frac{1}{2}[\mathcal{S}^* + \hat{W}(t)]\mathcal{S}^{*-1}\mathcal{H}(t)\mathcal{S}^{-1}[\mathcal{S} + \hat{W}(t)] = 0. \tag{5.8$_2$}$$

If as in Section V.10 we set

$$\hat{U}(t) = V(t) - iU(t), \quad \hat{V}(t) = V(t) + iU(t), \tag{5.9}$$

then in view of (5.6) it follows that

$$\hat{U}^*(t)\hat{U}(t) = \hat{V}^*(t)\hat{V}(t) = \hat{U}^*(t)\hat{U}(t) + \hat{V}^*(t)\hat{V}(t). \tag{5.10}$$

The right-hand member of (5.10) is a non-negative hermitian matrix, and if for some $s \in I$ this matrix were singular then there would exist a non-zero m-dimensional vector ζ such that $U(s)\zeta = 0$, $V(s)\zeta = 0$, and consequently $F(s)P\zeta = 0$ and $Q\zeta = 0$; as $F(s)$ is non-singular, and the $m \times 2m$ matrix $[P^* \quad Q^*]$ is of rank m, this would imply the contradictory result $\zeta = 0$. Hence $\hat{U}(t)$ and $\hat{V}(t)$ are non-singular for all $t \in I$, and the matrix functions

$$\Phi_1(t|Y) = \hat{V}(t)\hat{U}^{-1}(t), \quad \Phi_2(t|Y) = \hat{U}(t)\hat{V}^{-1}(t) \tag{5.11}$$

are well-defined throughout the interval I. Moreover, as for the analogous matrix function in Section V.10, the relation $\hat{U}^*(t)\hat{U}(t) = \hat{V}^*(t)\hat{V}(t)$ of (5.10) implies that $\Phi_1(t|Y)$ and $\Phi_2(t|Y)$ are unitary for $t \in I$.

Corresponding to (V.10.3), we now have that $\Phi_1(t) = \Phi_1(t|Y)$ is a solution of the differential equation

$$\Phi_1'(t) = i\Phi_1(t)N_1(t|Y), \tag{5.12}$$

where

$$N_1(t|Y) = 4\hat{U}^{*-1}(t)P^*F^*(t)H(t)F(t)P\hat{U}^{-1}(t), \tag{5.13}$$

and $\Phi_2(t) = \Phi_2(t|Y)$ is a solution of the differential equation

$$\Phi_2'(t) = i\Phi_2(t)N_2(t|Y) \tag{5.14}$$

where

$$N_2(t|Y) = -4\hat{V}^{*-1}(t)P^*F^*(t)H(t)F(t)P\hat{V}^{-1}(t). \tag{5.15}$$

For $\tau \in I_a^+$, it follows that there is a non-identically vanishing solution $y(t)$ of (5.1), (5.2_τ) if and only if the matrix $U(\tau)$ is singular. Indeed, the most general solution of (5.1) which satisfies the boundary condition $y(a) = P\xi$ is for the form $y(t) = F(t)P\xi$, and the second boundary condition $y(\tau) = \mathcal{Q}\xi$ is satisfied if and only if $[F(\tau)P - \mathcal{Q}]\xi = 0$, which is equivalent to $U(\tau)\xi = 0$. Also, this solution $y(t)$ is identically zero if and only if $P\xi = 0 = \mathcal{Q}\xi$, which holds if and only if $\xi = 0$. Consequently, the boundary problem (5.1), (5.2_τ) has index of compatibility equal to k if and only if the rank of $U(\tau)$ is $m - k$. Also, it follows readily that $U(\tau)\xi = 0$ if and only if $\zeta = V(\tau)\xi = \hat{V}(\tau)\xi = \hat{U}(\tau)\xi$ satisfies $\Phi_1(\tau)\zeta = \hat{V}(\tau)\xi = \zeta$, and consequently the boundary problem (5.1), (5.2_τ) has index of compatibility equal to k if and only if $\mu = 1$ is an eigenvalue of the unitary matrix $\Phi_1(\tau)$ of index k.

If τ is a value on I greater than $t = a$ which is such that the boundary problem (5.1), (5.2_τ) has a non-identically vanishing solution, then τ is said to be a *companion point* of $t = a$, relative to this boundary problem. In the terminology of Atkinson [2, Chapter 10] such points are said to be "conjugate to $t = a$, relative to the boundary problem". To the present author it seems more desirable to restrict "conjugate" to its more classical usage, however, and hence the introduction of the term "companion point".

Corresponding to the procedure of Section V.10, let the eigenvalues of the unitary matrix function $\Phi_1(t|Y)$ be arranged as continuous functions $\omega_\alpha(t)$, $(\alpha = 1,\ldots,m)$, with arguments continuous and satisfying

$$\arg \omega_1(t) \leq \ldots \leq \arg \omega_m(t) \leq \arg \omega_1(t) + 2\pi. \quad (5.16)$$

In particular, the eigenvalues are uniquely specified by this procedure whenever the initial values at $t = a$ are prescribed according to (5.16). As

$$\Phi_1(a|Y) = [(P + Q)i + iS(P - Q)][(P + Q) - iS(P - Q)]^{-1},$$

the initial values of the $\arg \omega_\alpha(t)$ depend upon the values of P and Q; however, for simplicity of notation this dependency is not shown. Now if $\omega = e^{i\beta}$ is an eigenvalue of $\Phi_1(t|Y)$ and ζ is a corresponding eigenvector, then $F(t)P\hat{U}^{-1}(t)\zeta = S^{-1}[(e^{i\beta} + 1)S - i(e^{i\beta} - 1)E]$. Moreover, since S is a non-singular skew matrix and $i(e^{i\beta} - 1)/(e^{i\beta} + 1)$ is real, for arbitrary real β the matrix

$$(e^{i\beta} + 1)S - i(e^{i\beta} - 1)E = 2e^{i\beta/2}[\cos \tfrac{\beta}{2}S + \sin \tfrac{\beta}{2}E] \quad (5.17)$$

is non-singular and $F(t)P\hat{U}^{-1}(t)\zeta$ is a non-zero vector. Consequently, if the hermitian matrix function $H(t)$ is continuous and positive definite on I, for such an eigenvector ζ we have $\zeta^* N_1(t|Y)\zeta > 0$, and it follows from Lemma V.10.3 that the $\arg \omega_\alpha(t)$, $(\alpha = 1,\ldots,m)$, are strictly increasing functions of t on I.

Moreover, if ζ is an eigenvector of $\Phi_1(t|Y)$ for an eigenvalue $\omega = e^{i\beta}$, and $\xi = \hat{U}^{-1}(t)\zeta$, we have $[V(t) + iU(t)]\xi = e^{i\beta}[V(t) - iU(t)]\xi$, which is equivalent to $\sin \frac{\beta}{2} V(t)z = \cos \frac{\beta}{2} U(t)z$, or

$$[\cos \tfrac{\beta}{2}S - \sin \tfrac{\beta}{2}E]F(t)P\xi = [\cos \tfrac{\beta}{2}S + \sin \tfrac{\beta}{2}E]Q\xi. \quad (5.18)$$

From (5.17), with β replaced by $-\beta$, it follows that the matrix $\cos \frac{\beta}{2}S - \sin \frac{\beta}{2}E$ is non-singular, and hence (5.18) is equivalent to $F(t)P\xi = Q(\beta)\xi$, with

$$Q(\beta) = [\cos \tfrac{\beta}{2}S - \sin \tfrac{\beta}{2}E]^{-1}[\cos \tfrac{\beta}{2}S + \sin \tfrac{\beta}{2}E]Q. \quad (5.19)$$

The fact that the $m \times 2m$ matrix $[P^* \quad \mathfrak{Q}^*]$ is of rank m clearly implies that for arbitrary real β the $m \times 2m$ matrix $[P^* \quad \mathfrak{Q}(\beta)]$ is of rank m. Also, from the above relations it follows that if $\tau \neq a$ then $e^{i\beta}$ is an eigenvalue of $\Phi_1(\tau|Y)$ if and only if there exists a non-identically vanishing solution of the boundary problem

$$Sy'(t) + H(t)y(t) = 0, \quad y(a) = P\xi, \quad y(\tau) = \mathfrak{Q}(\beta)\xi, \qquad (5.20^{\beta}_{\tau})$$

where $\mathfrak{Q}(\beta)$ is defined by (5.19). Indeed, the mulitplicity of $e^{i\beta}$ as an eigenvalue of $\Phi_1(\tau|Y)$ is equal to the index of compatibility of the boundary problem (5.20^{β}_{τ}). If an eigenvalue $\omega_\alpha(t)$ of $\Phi_1(t|Y)$ makes a complete circuit of the unit circle as t increases from a value t_1 to a value t_2 then the argument of this eigenvalue passes through a range of 2π, and hence assumes each intermediate value of β at least once. Consequently, we have the following result, (Atkinson [2, Theorem 10.8.2]).

THEOREM 5.1. *If the continuous hermitian matrix function* $H(t)$ *is positive definite for* $t \in I$, *then whenever a closed subinterval* $[a_0,b_0]$ *of* I_a^+ *contains* $n > m$ *companion points of* $t = a$ *relative to the boundary problem* (5.1), (5.2), *then for a given real value* β *this interval contains at least* $n - m$ *compansion points to* $t = a$, *relative to the boundary problems* (5.20^{β}_{τ}).

For the boundary problem (5.1), (5.2_τ), let $(\mathscr{A}_{\mathfrak{Q}}(\tau))$ denote the following condition:

$(\mathscr{A}_{\mathfrak{Q}}(\tau))$ *If* ξ *is an* m*-dimensional vector such that* $\mathfrak{Q}\xi \neq 0$, *then*

$$\xi^*\mathfrak{Q}^*H(\tau)\mathfrak{Q}\xi > 0.$$

The following result has been established by Coppel [1].

THEOREM 5.2. *If* $\tau \in I_a^+$ *is such that the continuous hermitian matrix function* $H(t)$ *satisfies* $(\mathscr{A}_{\mathfrak{a}}(\tau))$, *and* $\omega = 1$ *is an eigenvalue of* $\Phi_1(\tau|Y)$, *then for any value* γ *such that* $\omega_\gamma(t)$ *of the set of eigenvalues chosen above to satisfy* (5.16) *has* $\omega_\gamma(\tau) = 1$ *the function* $\arg \omega_\gamma(t)$ *is a strictly increasing function of* t *at* $t = \tau$. *In particular, if* $(\mathscr{A}_{\mathfrak{a}}(\tau))$ *holds for* $\tau \in I_a^+$, $b \in I_a^+$, *and* $c(b)$ *denotes the number of values* $\tau \in (a,b]$ *which are companion points to* $t = a$ *relative to the boundary problem* (5.1), (5.2_τ) *with each companion point counted a number of times equal to its index as an eigenvalue, then* $c(b)$ *is finite.*

Indeed, if $\omega_\gamma(\tau) = 1$, and ζ is an eigenvector of $\Phi_1(\tau|Y)$ corresponding to this eigenvalue, then for $\xi = \hat{U}^{-1}(\tau)\zeta$ we have $\xi \neq 0$ and $U(\tau)\xi = 0$, so that $F(\tau)P\xi = \mathfrak{a}\xi$. Consequently, $\mathfrak{a}\xi \neq 0$ and

$$\zeta^*N_1(\tau|Y)\zeta = 4\xi^*P^*F^*(\tau)H(\tau)F(\tau)P\xi = \xi^*\mathfrak{a}^*H(\tau)\mathfrak{a}\xi,$$

so that $\zeta^*N_1(\tau|Y)\zeta > 0$ in view of hypothesis $(\mathscr{A}_{\mathfrak{a}}(\tau))$. From the basic result of Lemma V.10.2 it then follows that $\arg \omega_\gamma(t)$ is a strictly increasing function at $t = \tau$ for any index γ such that $\omega_\gamma(\tau) = 1$. In particular, this result implies that the points on a given compact subinterval $[a,b]$ of I_a^+ which are companions to $t = a$ relative to (5.1), (5.2_τ) are individually isolated points, and consequently the number of such points is finite.

The same general procedure may be used to study the dependence of solutions upon a parameter. Consider the vector differential equation

$$Sy'(t) - [H(t) + \lambda K(t)]y(t) = 0 \tag{5.21}$$

where S and H satisfy the conditions given above, $K(t)$ is an $m \times m$ hermitian matrix function of class $\mathscr{L}^\infty[a,b]$ on arbitrary compact subintervals of I, and λ is a real parameter. The fundamental matrix $Y = F(t;\lambda)$ determined by the initial condition $F(a;\lambda) = E$ is then analytic as a function of the real variable λ, and the corresponding functions $U(t;\lambda)$, $V(t;\lambda)$ of (5.5) possess similar properties and satisfy (5.6). Similarly, the functions $U(t;\lambda)$, $V(t;\lambda)$ defined by (5.9) are non-singular for all real λ, and the matrix functions $\Phi_1(t;\lambda|Y) = \hat{V}(t;\lambda)\hat{U}^{-1}(t;\lambda)$ and $\Phi_2(t;\lambda|Y) = \hat{U}(t;\lambda)\hat{V}^{-1}(t;\lambda)$ defined by (5.11) are analytic functions of the real variable λ. Now by direct computation one may show that, (see, for example, Atkinson [2, p. 331]), $\Phi_1 = \Phi_1(t;\lambda|Y)$ satisfies the differential equation

$$\frac{\partial}{\partial\lambda} \Phi_1(t;\lambda|Y) = i\Phi_1(t,\lambda|Y)N_1^+(t,\lambda|Y), \tag{5.22}$$

where

$$N_1^+ = 2(V^* + iU^*)^{-1}(V^*U_\lambda - U^*V_\lambda)(V - iU)^{-1}.$$

Moreover, $V^*U_\lambda - U^*V_\lambda = 2P^*F^*SF_\lambda P$, and

$$F^*(t;\lambda)SF_\lambda(t;\lambda) = \int_a^t F^*(s;\lambda)K(s)F(s;\lambda)ds,$$

so that

$$N_1^+(t,\lambda|Y) = 4U^{*-1}(t,\lambda)P^*\left\{\int_a^t F^*(s;\lambda)K(s)F(s;\lambda)ds\right\}U^{-1}(t;\lambda). \tag{5.23}$$

In particular, N_1^+ is an hermitian matrix function on I and $N_1^+(t,\lambda|Y) \geq 0$ for $t \in I_a^+$ if $K(t) \geq 0$ for $t \in I$.

Correspondingly, $\Phi_2(t,\lambda|Y)$ is a solution of the matrix differential equation

$$\frac{\partial}{\partial\lambda}\,\Phi_2(t;\lambda|Y) = i\Phi_2(t;\lambda|Y)N_2^+(t;\lambda|Y), \tag{5.24}$$

where $N_2^+(t;\lambda|Y)$ is the hermitian matrix function

$$N_2^+(t;\lambda|Y) = -N_1(t;\lambda|Y). \tag{5.25}$$

Now let the eigenvalues $\omega_\alpha(t;\lambda)$ of the unitary matrix function $\Phi_1(t;\lambda|Y)$ be fixed at $t = a$ subject to (5.16), and continued throughout the (t,λ)-plane in the unique fashion to obtain values which are continuous and satisfy inequalities corresponding to (5.16) for all (t,λ), (see Atkinson [2, Appendix V]). Considering the $\omega_\alpha(t;\lambda)$ as functions of λ, it follows that $N_1^+(\tau;\lambda|Y) > 0$ for $\tau \in I_a^+$ and the $\omega_\alpha(\tau;\lambda)$ move in a strictly increasing positive fashion on the unit circle as λ increases, whenever the following condition is satisfied.

($\mathscr{A}^\lambda$) *On* I *the matrix function* $\mathsf{K}(t)$ *is continuous, satisfies* $\mathsf{K}(t) \geq 0$ *and if* $\tau \in I_a^+$ *and* $y(t)$ *is a non-identically vanishing solution of* (5.21) *on* $[a,\tau]$ *for any value* $\lambda \in \mathsf{R}$, *then*

$$\int_a^\tau y^*(s)\mathsf{K}(s)y(s)ds > 0. \tag{5.26}$$

In particular, hypothesis ($\mathscr{A}^\lambda$) implies that the matrix function $N_1^+(t,\lambda|Y)$ of (5.23) is positive definite for $t \in I_a^+$, and one may establish the following result, (see Atkinson [2, Sec. 10.9]).

Theorem 5.3. *If* $\tau \in I_a^+$ *and hypotheses* ($\mathscr{A}$), ($\mathscr{A}^\lambda$) *are satisfied, whenever a closed interval* J *of the real* λ*-line contains* $p > m$ *eigenvalues of* (5.21), (5.2_τ), *then for a given real value* β *this interval contains at least* $p - m$ *eigenvalues of* (5.21), (5.20_τ).

Coppel [2] has employed a related method of proof to establish the following comparison theorem for two systems involving the respective differential equations

$$Sy'(t) - H_\rho(t)y(t) = 0, \quad (\rho = 1,2) \tag{5.1$_\rho$}$$

and the same set of boundary conditions (5.2_τ), and where each of the systems (5.1_ρ), (5.2_τ) satisfies condition $(\mathscr{A}_{\mathfrak{a}}(\tau))$ for $\tau \in I_a^+$. If $\tau \in I_a^+$ then the number of companion points to $t = a$ on $(a,\tau]$, relative to the system (5.1_ρ), (5.2_τ), is denoted by $c_\rho(\tau)$. The result of Theorem 4 of Coppel [2] is as follows.

THEOREM 5.4. *Suppose that* $H_1(t)$ *and* $H_2(t)$ *are two continuous hermitian matrix functions satisfying* $(\mathscr{A}_{\mathfrak{a}}(\tau))$ *and* $H_2(t) \geq H_1(t)$ *for* $t \in I_a^+$. *Then for* $\tau \in I_a^+$ *we have* $c_2(\tau) \geq c_1(\tau)$. *Moreover,* $c_2(\tau^-) \geq c_1(\tau)$ *if* $y(t) \equiv 0$ *is the only solution of* (5.1_1), (5.2_τ) *for which* $H_1(t)y(t) \equiv H_2(t)y(t)$ *on* $[a,\tau]$.

For the proof of Theorem 5.4, Coppel's method involves the character of the auxiliary system

$$Sy'(t) - [H_1(t) + \lambda K(t)]y(t) = 0,$$
$$\text{with } K(t) = H_2(t) - H_1(t), \tag{5.27}$$

subject to the boundary conditions (5.2_τ), where λ is a real parameter with values on the unit interval $[0,1]$. In view of the assumption that $H_2(t) \geq H_1(t)$, for this auxiliary system the matrix function (5.23) satisfies $N_1^+(t;\lambda|Y) \geq 0$ for $(t,\lambda) \in I \times [0,1]$. Consequently, if $\Phi_1(t;\lambda|Y)$ is the solution of the corresponding equation (5.22), with eigenvalues $\omega_\alpha(t;\lambda)$ ordered in the usual fashion as used above, it follows that $\arg \omega_\alpha(t;1) \geq \arg \omega_\alpha(t;0)$ for $t \in I_a^+$ and $\alpha = 1,\ldots,m$, and these relations imply $c_2(\tau) \geq c_1(\tau)$ for arbitrary $\tau \in I_a^+$. If for $\tau \in I_a^+$ the only solution of (5.1_1), (5.2_τ) satisfying $H_1(t)y(t) \equiv H_2(t)y(t)$ on $[a,\tau]$ is $y(t) \equiv 0$, one may show that $\zeta^* N_1^+(\tau;0|Y) > 0$ for all eigenvectors ζ of $\Phi_1(\tau,0|Y)$ corresponding to the eigenvalue 1. From this it follows that $\arg \omega_\alpha(\tau;1) >$ $\arg \omega_\alpha(\tau;0)$ for every α for which $\arg \omega_\alpha(\tau,0)$ is a multiple of 2π, and in turn this implies that $c_2(\tau^-) \geq c_1(\tau)$.

With the aid of an auxiliary system involving linearly a parameter in the boundary condition, Coppel also established the following comparison theorem for two systems involving the same differential equation and different boundary conditions.

THEOREM 5.5. *Let* P_ρ, Q_ρ, $(\rho = 1,2)$ *be* $m \times m$ *constant matrices such that each of the* $m \times 2m$ *matrices* $[P_\rho^* \quad Q_\rho^*]$ *is of rank* m, *the matrix* $D = P_1^* S P_2 - Q_1^* S Q_2$ *is non-netative hermitian, and for all* $\mu \geq 0$ *the matrix* $P_2 - Q_2 + \mu[P_1 - Q_1]$ *is non-singular. Also, for* $\rho = 1,2$ *let* (5.2_τ^ρ) *denote the boundary conditions* (5.2_τ) *with* $P = P_\rho$, $Q = Q_\rho$ *and for* $\tau \in I_a^+$ *let* $c_\rho(\tau)$ *denote the number of companion points to* $t = a$ *on* $(a,\tau]$, *relative to the system* (5.1), (5.2_τ^ρ). *If* $H(t)$ *satisfies* $(\mathscr{H}_{Q_\rho}(\tau))$ *for* $\alpha = 1,2$ *and all* $\tau \in I_a^+$, *then for such* τ *we have* $c_2(\tau) \geq c_1(\tau)$. *Moreover,* $c_2(\tau^-) \geq c_1(\tau)$ *if* $y(t) \equiv 0$ *is the only solution of* (5.1) *satisfying the boundary conditions* $y(a) = P_1\xi$, $y(\tau) = Q_1\xi$ *with* $D\xi = 0$.

6. Notes and comments on the literature

For a simple integral variational problem of Lagrange or Bolza type the so-called accessory boundary problem is a real system of the form (3.1) with $K(t)$ the identity matrix, (in passing, it is to be commented that equally well one might consider the case of $K(t)$ a positive definite matrix function). For a discussion of such problems and related matters in the strictly variational context the following references of the relatively early literature on the subject are pertinent: Bolza [2-Ch. 12], Radon [1,2], Bliss [5,7-Chs. VIII and IX], Morse [2,3,4-Chs. 1 and 2], Boerner [2,3], Hölder [1,2], Reid [2,3,5,6]. The presentation of the Jacobi condition for a variety of variational problems of this general category is to be found in the Chicago dissertations of Cope [1], Bamforth [1], Hickson [1], Jackson [1] and Wiggin [1]. For a variational problem wherein the supposedly minimizing arcs possesses corners, and hence the boundary conditions of the accessory boundary problem apply at more than two points, a presentation of the Jacobi condition in terms of a boundary problem is found in Reid [4; see also the much later paper 33]. Other early papers dealing with related boundary problems, and also directly concerned with index theorems and oscillation and comparison results, are Morse [1, 4-Ch. 4], Carathéodory [2,3], Hu [1], Birkhoff and Hestenes [1], Reid [7], Hazard [1], Karush [1] and Ritcey [1].

It is to be commented that the general "accessory systems" considered by Reid [2,7] and Birkhoff and Hestenes [1] involve boundary conditions which contain linearly the characteristic parameter. There are also additional papers in this and associated fields, but intimately related to the area of "definitely self-adjoint boundary problems" as formulated by Bliss [2], and these are referenced in the following chapter.

At this stage it seems desirable to elaborate on the procedure of Birkhoff and Hestenes [1], both to indicate its relation to other methods and to serve as an introduction to the quadratic form theory in Hilbert space as developed later by Hestenes [2], and which will be discussed further in Chapter VIII. The presentation follows closely that of Hestenes [1-Sec. 10], and is a combination of results obtained by Birkhoff and Hestenes [1] and Hazard [1]. The specific system there considered may be described as that associated with a real symmetric functional of the form (1.10) with continuous matrix coefficients and $B(t) \geq 0$ on $[a,b]$, although the statements are equally applicable with complex-valued matrix coefficient functions and weaker conditions than continuity.

For $D = D[a,b]$ and $J[\eta_1,\eta_2] = J[\eta_1,\eta_2:\mathscr{B}]$ as defined by (1.10), two arcs $\eta_\alpha \in D$, $(\alpha = 1,2)$, are called J-*orthogonal* if $J[\eta_1,\eta_2] = 0$. If $\hat{D} \subset D$ and $\eta_1 \in D$ is such that η_1 is J-orthogonal to all $\eta \in \hat{D}$, the arc η_1 is said to be J-orthogonal to $\hat{D}$. Also, if $\hat{D}_1 \subset \hat{D}_2 \subset D$ the class of $\eta \in \hat{D}_2$ and J-orthogonal to $\hat{D}_1$ is termed the J-*complement* of $\hat{D}_1$ in $\hat{D}_2$. For example, the statement in Section V.6 leading to the condition (V.6.5) may be phrased as follows: an arc $u \in D$ belongs to a solution $(u;v)$ of the differential system ($\mathscr{B}$i) if and only if u is J-orthogonal to D_0, where $D_0 = D_0[a,b]$ is the class of $\eta \in D$ with zero boundary vector $\hat{\eta}$. By the J-index of a subspace $\hat{D}$ of D is meant the largest integer k such that there is a k-dimensional subspace of $\hat{D}$ on which J is negative definite.

For $\hat{D}$ a subspace of D, let $\hat{D}_{\perp}$ denote the set of $\eta \in \hat{D}$ which are also J-orthogonal to $\hat{D}$. In the terminology of Hestenes [1], a subspace $\hat{E}$ of D is called a *special subclass* of D if every $\eta \in D$ that is orthogonal to $\hat{E}$ may be written as $\eta = \eta_1 + \eta_2$, where $\eta_1 \in \hat{E}$ and η_2 is J-orthogonal to $\hat{E}$. For example, D_o is a special subclass of D. This fact is equivalent to the condition that if $\eta \in D$ and η is J-orthogonal to all u belonging to solutions $(u;v)$ of ($\mathscr{B}$i) with $u(a) = 0 = u(b)$, then condition (V.5.2) holds and there is a solution $(u_o;v_o)$ of ($\mathscr{B}$i) satisfying $u_o(a) = \eta(a)$, $u_o(b) = \eta(b)$. Indeed, any subspace of D containing D_o is a special subclass in this sense; in particular, if S is a subspace of $\mathbf{C}_{2n}$ as in Theorem 1.1, then $D_e = \{\eta : \eta \in D, \hat{\eta} \in S\}$ is a special subclass.

Let $\hat{E}$ be a special subclass of D_e, and denote by m the dimension of a maximal subspace of $\hat{E}_{\perp}$ whose non-null arcs are not J-orthogonal to D_e. If the J-indices of $\hat{E}$ and its J-complement in D_e are denoted by k and k', respectively, then the J-index of D_e is equal to $\hat{k} + k' + m$. Following the discussion of Hestenes [2-Sec. 10], for simplicity assume that ($\mathscr{B}$i) is identically normal on $[a,b]$ and consider the problem of determining the J-index of the subspace D_o of D. Let m denote the number of points on (a,b) conjugate to $t = a$, each conjugate point counted a number of times equal to its multiplicity. Then there are m linearly independent arcs $\eta_1,\dots,\eta_m$ of D_o such that for $j = 1,\dots,m$ we have $\eta_j \equiv 0$ on $(t_j,b]$, where $t_j \in (a,b)$ is a value conjugate to $t = a$ and there exists a solution

$y_j = (u_j;v_j)$ of ($\mathscr{B}$i) determining t_m as conjugate to $t = a$ with $u_j(t) = \eta_j(t)$ on $[a,t_j]$. (In a suggestive terminology recently used by Morse [9-Ch. 6], the arc η_j is called an *axial extension* of $u_j(t)$, $t \in [a,t_j]$, and the linear subspace $\hat{E}$ of D_o generated by $\eta_1,\ldots,\eta_m$ is designated a *nuclear subspace*). The functional $J[\eta]$ is zero for each $\eta \in \hat{E}$, so that its J-index on this class is zero. Now no arc $\eta \not\equiv 0$ of $\hat{E}$ is J-orthogonal to D_o; moreover, the index of the J-complement of $\hat{E}$ in D_o is also zero. Consequently, the J-index of D_o is equal to the dimension m of $\hat{E}$, which is the sum of the orders of the conjugate points to $t = a$ between a and b.

As far as the class D_e is concerned, its J-index is the sum of three quantities, the first of which is the J-index of D_o just described. The second quantity is the J-index of the class of u belonging to solutions $(u;v)$ of ($\mathscr{B}$i) with $u \in D_e$; that is, the J-complement of D_o in D_e. The third quantity is the dimension of a maximal subspace of solutions $(u;v)$ of ($\mathscr{B}$i) with $u \in D_o$ and which contains no solution $(u;v)$ that satisfies the transversality, or natural boundary, condition $T[u,v] \equiv Q\hat{u} + D\hat{v} \in S^{\perp}$.

As a further example, consider the class $\hat{E}$ of all $\eta \in D$ forming the J-complement of the class of all arcs vanishing at a finite set of values $t_o = a_1 < t_1 < \ldots < t_{r-1} < t_r = b$. In particular, there are such sets of values with the maximum of the values $t_j - t_{j-1}$, $(j = 1,\ldots,r)$, arbitrarily small; consequently, the index of the J-complement of $\hat{E}$ is zero, and every arc in $\hat{E}$ that is J-orthogonal to $\hat{E}$ is also orthogonal to D_e. The J-index of D_e is therefore identical with that of $\hat{E}$, so that $\hat{E}$ is composed of the totality of "broken solutions" $(u;v)$ of $(\mathscr{B}i)$ with corners at the points $t_1,\ldots,t_{r-1}$ and the J-index in this sense is equivalent to that introduced by Morse.

Associated boundary problems (3.1) appear through the consideration of two functionals, the $J[\eta]$ of (1.10), and a second functional $J_0[\eta]$, which is usually taken to be of the form

$$J_0[\eta] = \hat{\eta}^* Q_0 \hat{\eta} + \int_a^b \eta^*(t) K(t) \eta(t)\,dt. \tag{6.1}$$

For the simplification of further discussion it will be assumed that $J_0[\eta]$ is positive definite on D, although this assumption can be avoided. In the terminology of Hestenes [1], a non-identically vanishing $u \in D_e$ is called a characteristic arc if its J_0-complement in D_e is identical with its J-complement in D_e. If u is a characteristic arc then there is a characteristic value (eigenvalue) λ such that $J[u,\eta] = \lambda J_0[u,\eta]$ for all $\eta \in D_e$. Indeed, $\lambda = J[u]/J_0[u]$, and u has an associated function v such that $(u;v)$ is a solution of the differential equations of (3.1) for this value λ, while

$$\hat{u} \in S, \quad T[u,v|\lambda] \equiv [Q - \lambda Q_0]\hat{u} + D\hat{v} \in S^{\perp}. \tag{6.2}$$

The number of characteristic arcs in a maximal linearly independent set determining the same eigenvalue λ is called the order of λ, and the basic index theorem is that the J-index of D_e is equal to the sum of the orders of the negative eigenvalues.

7. Topics and Exercises

Throughout this section it will be supposed that the coefficient matrix functions $A(t)$, $B(t)$, $C(t)$ of the cited differential systems ($\mathscr{B}$i), (1.13i), or (1.3i) satisfy hypothesis ($\mathscr{A}$) of Section V.3 unless stated otherwise.

1. Let ($\mathscr{B}$) be a given self-adjoint problem (3.1) with boundary conditions in the form (1.8) involving the linear subspace $S = S[\mathscr{B}]$ of $\mathbf{C}_{2n}$. Then there exists an orthonormal system of vectors ϕ_j, $(j = 1,\ldots,2n)$, such that these boundary conditions are given in the non-parametric form

$$\begin{aligned} &\phi_\sigma^* u = 0, \quad (\sigma = 1,\ldots,2n-r), \\ &\phi_\tau^* T[u,v] \equiv \phi_\tau^*[Q\hat{u} + D\hat{v}] = 0, \quad (\tau = 2n-r+1,\ldots,2n) \end{aligned} \tag{7.1|B}$$

where $r = \dim S[\mathscr{B}]$, with the usual convention that the symbols $\sigma = 1,\ldots,0$ and $\tau = 2n+1,\ldots,2n$ denote empty sets.

Also, if as in Theorem 4.3 two self-adjoint problems ($\mathscr{B}$) and ($\tilde{\mathscr{B}}$) differ only in the end-manifolds $S[\mathscr{B}]$, $S[\tilde{\mathscr{B}}]$, and ($\tilde{\mathscr{B}}$) is a subproblem of ($\mathscr{B}$) of dimension $\rho = r - \tilde{r}$, then there is an orthonormal set of vectors ϕ_j, $(j = 1,\ldots,2n)$, such that the boundary conditions of ($\mathscr{B}$) are given by (7.1|B), and the boundary conditions of ($\tilde{\mathscr{B}}$) are given by similar equations

$$\begin{aligned} &\phi_{\sigma'}^* u = 0, \quad (\sigma' = 1,\ldots,2n-\tilde{r}), \\ &\phi_{\tau'}^* T[u,v] \equiv \phi_{\tau'}^*[Q\hat{u} + D\hat{v}] = 0, \quad (\tau' = 2n-\tilde{r}+1,\ldots,2n). \end{aligned} \tag{7.1|$\tilde{B}$}$$

Consequently, if for $j = 0,1,\ldots,$ we denote by ($\mathscr{B}^j$) the self-adjoint boundary problems differing only in the end-manifolds $S[\mathscr{B}^j]$ which are specified by the equations

$$\phi_s^*\hat{u} = 0, \qquad (s = 1,\dots,2n-r+j), \tag{7.1|B^j}$$
$$\phi_t^* T[u,v] \equiv \phi_t^*[Q\hat{u} + D\hat{v}] = 0, \quad (t = 2n-r+j+1,\dots,2n)$$

we have that $(\mathscr{B}^0) = (\mathscr{B})$, $(\mathscr{B}^\rho) = (\tilde{\mathscr{B}})$ and for $i = 0,1,\dots,\rho-1$ the problem $(\mathscr{B}^{i+1})$ is a subproblem of $(\mathscr{B}^i)$ of dimension 1.

2. Suppose that $Y_1(t) = (U_1(t);V_1(t))$ and $Y_2(t) = (U_2(t);V_2(t))$ are two conjoined bases for $(\mathscr{B}i)$ with $\{Y_2,Y_1\} \equiv V_1^*U_2 - U_1^*V_2 \equiv E$. Then also $U_1U_2^* - U_2U_1^* \equiv 0$, $V_1U_2^* - U_1V_2^* \equiv E$, and on $I \times I$ the matrix functions Λ, Λ_o defined as

$$\Lambda(t,s) = U_1(t)U_2^*(s) \text{ for } t \le s, \ \Lambda(t,s) = U_2(t)U_1^*(s) \text{ for } t \ge s;$$
$$\Lambda_o(t,s) = V_1(t)U_2^*(s) \text{ for } t < s, \ \Lambda_o(t,s) = V_2(t)U_1^*(s) \text{ for } t > s \tag{7.2}$$

possess continuity and differentiability properties corresponding to those of G, G_o in conclusions (a), (b), (c) of Theorem 1.2. Moreover, if f is locally of class $\mathscr{L}$ on I, and $[a,b]$ is a compact subinterval of I, then $y(t) = (u(t);v(t))$ defined as

$$u(t) = \int_a^b \Lambda(t,s)f(s)ds, \quad v(t) = \int_a^b \Lambda_o(t,s)f(s)ds$$

is a solution of the differential system (1.13i).

3. For $s \in I$ let Q_s be an $n \times n$ hermitian matrix, S_s a linear subspace of $\mathbf{C}_n$, and denote by $S_s^\perp$ the orthogonal complement of S_s in $\mathbf{C}_n$. Then the one-point conditions

$$u(s) \in S_s, \quad Q_s u(s) + v(s) \in S_s^\perp \tag{7.3}$$

determine a conjoined basis $Y_s(t) = (U_s(t);V_s(t))$ of the differential system ($\mathscr{B}$i) such that $y(t) = (u(t);v(t))$ is a solution of this system satisfying (7.3) if and only if there is an n-dimensional vector ξ such that $y(t) = Y_2(t)\xi$. In particular, if d_s denotes the dimension of S_s then (7.3) becomes: $u(s) = 0$ if $d_s = 0$; $Q_s u(s) + v(s) = 0$ if $d_s = n$; $\theta_s^* u(a) = 0$, $\psi_s^*[Q_s u(s) + v(s)] = 0$ if $1 \le d_s \le n - 1$ and θ_s is an $n \times (n - d_s)$ matrix of rank $n - d_s$, ψ_s is an $n \times d_s$ matrix of rank d_s with $\theta_s^*\psi_s = 0$ and such that the column vectors of ψ_s form a basis for S_s.

4. Suppose that $[a,b]$ is a compact subinterval of I and S is a linear subspace of $\mathbf{C}_{2n}$ such that there exist linear subspaces S_a and S_b of $\mathbf{C}_n$ with the property that $\eta = (\eta(a);\eta(b)) \in S$ if and only if $\eta(a) \in S_a$ and $\eta(b) \in S_b$. Moreover, suppose that Q is a $2n \times 2n$ hermitian matrix of the form $Q = \mathrm{diag}\{-Q_a, Q_b\}$, where Q_a and Q_b are $n \times n$ hermitian matrices. If $Y_a(t) = (U_a(t);V_a(t))$ and $Y_b(t) = (U_b(t);V_b(t))$ are conjoined bases determined as in the above Exercise 3 for $s = a$ and $s = b$, then the boundary problem ($\mathscr{B}$) with boundary conditions in the form (1.8) has only the identically vanishing solution $(u(t);v(t)) \equiv (0;0)$ if and only if the constant matrix function $\{Y_b, Y_a\}$ is non-singular, in which case these conjoined bases may be so chosen that $\{Y_b, Y_a\} \equiv E_n$, and with such a choice the results of the above Exercises 2, 3 imply that the partial Green's matrix functions $G(t,s)$, $G_o(t,s)$ of Theorem 1.2 are given by $G(t,s) = U_a(t)U_b^*(s)$ for $t \in [a,s]$, $G(t,s) = U_b(t)U_a^*(s)$ for $t \in [s,b]$ with corresponding $G_o(t,s) = V_a(t)U_b^*(s)$ for $t \in [a,s)$, $G_o(t,s) = V_b(t)U_a^*(s)$ for $t \in (s,b]$.

5. Suppose that the boundary problem $(\mathscr{B})$ is normal, and the vector space of solutions of this problem is of dimension d. If $Y_0(t) = (U_0(t);V_0(t))$ is a $2n \times d$ matrix function whose column vector functions form a basis for the space of solutions of $(\mathscr{B})$, while $\Theta(t)$ and $\Psi(t)$ are $n \times d$ matrix functions of class $\mathscr{L}$ on $[a,b]$ and such that the $d \times d$ matrices $\int_a^b \Theta^*(s)U_0(s)ds$, $\int_a^b U_0^*(s)\Psi(s)ds$ are non-singular, then for $(\mathscr{B})$ there exist unique generalized partial Green's matrix functions $G\#(t,s|\Theta,\Psi)$, $G_0\#(t,s|\Theta,\Psi)$ such that the conditions (a), (b), (c) of Theorem 1.2 are satisfied when G, G_0 are replaced by $G\#$, $G_0\#$, respectively, while conclusion (d) of that theorem is replaced by:

(d') If $f \in \mathscr{L}_n[a,b]$, then the unique solution of the differential system

$$\text{(i)}\quad \begin{aligned} L_1[u,v](t) &= f(t) - \Psi(t)\left[\int_a^b U_0^*(s)\Psi(s)ds\right]\int_a^b U_0^*(t)f(t)dt, \\ L_2[u,v](t) &= 0, \end{aligned} \tag{7.4}$$

$$\text{(ii)}\quad \hat{u} \in S, \quad T[u,v] \in S^{\perp}, \quad \int_a^b \Theta^*(s)u(s)ds = 0,$$

is given by

$$\begin{aligned} &\text{(i)} \quad u(t) = \int_a^b G\#(t,s|\Theta,\Psi)f(s)ds, \\ &\text{(ii)} \quad v(t) = \int_a^b G_0\#(t,s|\Theta,\Psi)f(s)ds. \end{aligned} \tag{7.5}$$

In case f is such that the system (1.13) has a solution $(u;v)$, then (7.5) provides the solution of this system which satisfies the integral condition of (7.4ii).

6. Suppose that the boundary problem (3.1) satisfies hypotheses $(\mathscr{A}_{\mathscr{B}})$, $(\mathscr{A}_K' i,ii,iii)$, Condition (3.10), and $(\lambda_j, y_j(t)) = (u_j(t); v_j(t))$, $(j = 1,2,\ldots)$, is a seuqnece of eigenvalues and associated eigenvectors as determined in Theorem 3.1. For $\lambda = \mu$ an eigenvalue of this problem of multiplicity k, let $Y(t|\mu) = (U(t|\mu); V(t|\mu))$ be a $2n \times k$ matrix function whose column matrix functions form a basis for the linear space of solutions of (3.1) for $\lambda = \mu$ chosen as in Theorem 3.1, so that $\int_a^b U^*(s|\mu)B(s)U(s|\mu)ds = E_k$. Then from Theorem 3.4 and the above Exercise 5 it follows that for $\Theta(t) = \Psi(t) = K(t)U(t|\mu)$ the corresponding partial generalized Green's matrix functions $G\#(t,s;\mu|\Theta,\Psi)$, $G_0\#(t,s;\mu|\Theta,\Psi)$ are such that

$$G\#(t,s;\mu|\Theta,\Psi) = \sum_{\lambda_j \neq \mu} (\lambda_j - \mu)^{-1} u_j(t) u_j^*(s). \quad (7.6)$$

7. Suppose that the hypotheses $(\mathscr{A}_{\mathscr{B}})$ and $(\mathscr{A}_K' i,ii,iii)$ are satisfied and $\mathscr{F} = \{f_1,\ldots,f_r\}$ is a finite set of n-dimensional vector functions of class $\mathscr{L}_n[a,b]$, and for which there exists an associated set of vector functions $w_\alpha \in D_e[\mathscr{B}]$, $(\alpha = 1,\ldots,r)$, such that the $r \times r$ matrix $\left[\int_a^b w_\alpha^*(t) f_\beta(t)dt\right]$ is non-singular. As in Section 3, let $D_e[\mathscr{B}|\mathscr{F}]$ denote the class of functions η in $D_e[\mathscr{B}]$ satisfying (3.6). Then for the boundary problem consisting of the differential system

$$\text{(a)} \quad L_1[u,v](t) + \sum_{\alpha=1}^{r} k_\alpha f_\alpha(t) = \lambda K(t)u(t), \; L_2[u,v](t) = 0.$$

$$\text{(b)} \quad \hat{u} \in S, \qquad T[u,v] \in S^{\perp}, \qquad (7.7)$$

$$\text{(c)} \quad \int_a^b f_\alpha^*(t)u(t)dt = 0, \quad (\alpha = 1,\ldots,r),$$

there exists an infinite sequence of real eigenvalues $\tilde{\lambda}_1 \leq \tilde{\lambda}_2 \leq \cdots$ with corresponding eigensolutions $(\tilde{u}(t);\tilde{v}(t);\tilde{k}_\alpha) = (\tilde{u}_j(t);\tilde{v}_j(t);\tilde{k}_{\alpha j})$ for $\lambda = \tilde{\lambda}_j$ such that

(a) $K[\tilde{u}_i,\tilde{u}_j] = \delta_{ij}$, $(i,j = 1,2,\ldots)$;

(b) $\tilde{\lambda}_1 = J[\tilde{u}_1:\mathscr{B}]$ is the minimum of $J[\eta:\mathscr{B}]$ on the class $D_N[\mathscr{B}|\mathscr{F}:K] = \{\eta:\eta \in D_e[\mathscr{B}|\mathscr{F}],\ K[\eta] = 1\}$;

(c) for $j = 2,3,\ldots$ the class

$D_{N_j}[\mathscr{B}|\mathscr{F}:K] = \{\eta:\eta \in D_N[\mathscr{B}|\mathscr{F}:K],\ K[\eta,u_i] = 0,$ $(i = 1,\ldots,j-1)$

is non-empty, and $\tilde{\lambda}_j = J[\tilde{u}_j:\mathscr{B}]$ is the minimum of $J[\eta:\mathscr{B}]$ on $D_{N_j}[\mathscr{B}|\mathscr{F}:K]$. Moreover, $\lambda_j \leq \tilde{\lambda}_j \leq \lambda_{j+r}$, $(j = 1,2,\ldots)$, where $\{\lambda_j\}$ are the eigenvalues of (3.1) as in Theorem 3.1.

8. If the hypotheses of Theorem 3.2 are satisfied, and $\lambda_0 < \lambda_1$, where λ_1 is the smallest eigenvalue of (3.1), then there exists a value $\kappa > 0$ such that

$$J[\eta;\lambda_0:\mathscr{B}] \geq \kappa\Big[|\eta(a)|^2 + |\eta(b)|^2 + |\eta(t)|^2 + \int_a^b \{|\eta'(t)|^2 + |\eta(t)|^2\}dt\Big],$$

for $t \in [a,b]$, and $\eta \in D_e[\mathscr{B}]$.

Hints. Note that since $\lambda_0 < \lambda_1$ the functional $J[\eta;\lambda_0:\mathscr{B}]$ is non-negative on $D_e[\mathscr{B}]$, and from the result of Lemma 3.1 applied to $J[\eta;\lambda_0:\mathscr{B}]$ conclude that there exists a $\kappa_1 > 0$ such that

$$J[\eta;\lambda_0:\mathscr{B}] \geq \kappa_1\Big[|\eta(a)|^2 + |\eta(b)|^2 + \int_a^b |\eta(t)|^2 dt\Big]$$

for $\eta \in D_e[\mathscr{B}]$.

Also, by a matrix argument similar to that used in the scalar case to establish inequality (III.2.14), show that there exist constants $\kappa_0 > 0$, $\kappa_2 \geq 0$ such that

$$J[\eta;\lambda_0:\mathscr{B}] \geq \kappa_0 \int_a^b |\eta'(t)|^2 dt - \kappa_2 \int_a^b |\eta(t)|^2 dt, \text{ for } \eta \in D_e[\mathscr{B}],$$

and combine these inequalities to conclude that for $\kappa' = \kappa_0\kappa_1/(\kappa_0 + \kappa_1 + \kappa_2)$ we have for $\eta \in D_e[\mathscr{B}]$ the inequality

$$J[\eta;\lambda_0:\mathscr{B}] \geq \kappa' \; |\eta(a)|^2 + |\eta(b)|^2 + \int_a^b \{|\eta'(t)|^2 + |\eta(t)|^2\} dt \; .$$

Proceed to establish the stated inequality, noting that if $\eta \in D_e[\mathscr{B}]$, then $\eta(t) = \eta(a) + \int_a^t \eta'(s)ds = \eta(b) - \int_t^b \eta'(s)ds$, and elementary inequalities yield the result

$$\tfrac{1}{2}|\eta(t)|^2 \leq |\eta(a)|^2 + |\eta(b)|^2 + (b-a)\int_a^b |\eta'(s)|^2 ds$$

for $t \in [a,b]$.

9. If $z_i(t)$, $(i = 1,2,\ldots)$, are n-dimensional a.c. vector functions on $[a,b]$ with $z_i' \in \mathscr{L}_n^2[a,b]$, while $\{z_i(t)\} \to 0$ for $t \in [a,b]$, and $\left\{\int_a^b |z_j'(t) - z_i'(t)|^2 dt\right\} \to 0$ as $i,j \to \infty$, then $\{z_i(t)\} \to 0$ uniformly on $[a,b]$, and $\left\{\int_a^b |z_i'(t)|^2 dt\right\} \to 0$ as $i \to \infty$.

10. If the hypotheses of Theorem 3.2 are satisfied, and $\eta \in D_e[\mathscr{B}]$, then the sequence $\eta_p(t) = \eta(t) - \sum_{j=1}^{p} c_j[\eta]u_j(t)$, $p = 1,2,\ldots$, converges uniformly on $[a,b]$ to a continuous vector function $\eta_0(t)$ which is such that $K[\eta_0] = 0$; also,

$$\left\{\int_a^b |\eta_p'(t) - \eta_q'(t)|^2 dt\right\} \to 0 \quad \text{as} \quad p,q \to \infty.$$

11. If the hypotheses of Theorem 3.2 are satisfied, and $\eta_p(t) \in D_e[\mathscr{B}]$, $(p = 1,2,\ldots)$, with $K[\eta_p,\eta_q] = \delta_{pq}$, $(p,q = 1,2,\ldots)$, then the sequence $\{J[\eta_p:\mathscr{B}]\}$ is unbounded.

Hint. With the aid of the above Problem 8, show that the assumption that $\{J[\eta_p:\mathscr{B}]\}$ is bounded implies that the vector functions $\{\eta_p(t)\}$ are uniformly bounded and equicontinuous, and hence there is a subsequence $\{\eta_{p_j}(t)\}$ which converges uniformly on $[a,b]$ to a continuous vector function $\eta_0(t)$, which yields the contradictory results $K[\eta_0] = \lim_{j\to\infty} K[\eta_{p_j}] = 1$, $K[\eta_0] = \lim_{p\to\infty} K[\eta_{p_j},\eta_{p_{j+1}}] = 0$.

12. Suppose that hypothesis $(\mathscr{A})$ of Section V.3 holds, with $B(t) \geq 0$ for t a.e., and that the system $(\mathscr{B}i)$ is normal on every subinterval $[a,s]$ and $[s,b]$ of $[a,b]$.

(i) If $Z(t) = Z(t;s)$ is the fundamental matrix for $z'(t) + A^*(t)z(t) = 0$ satisfying $Z(s) = E$, and $N(t;s) = \int_a^t Z^*(r;s)B(r)Z(r;s)dr$, $t \in [a,b]$, then for $\eta \in D_0[a,b]:\zeta$ and $s \in (a,b)$, we have

$$\int_a^b \zeta^*(t)B(t)\zeta(t)dt \geq 4\eta^*(s)N^{-1}(b,s)\eta(s).$$

(ii) If the system $(\mathscr{B}i)$ is normal on arbitrary subintervals on $[a,b]$, and there exists a pair of distinct values on $[a,b]$ which are mutually conjugate with respect to $(\mathscr{B}i)$, then

$$\int_a^b \lambda_{Max}[C^+(t)]dt > 4/[Max\{\lambda_{Max}[N(b;s)]:a \leq s \leq b\}],$$

where $C^+(t) = \frac{1}{2}[|C(t)| + C(t)]$ with $|C(t)|$ the non-negative definite hermitian square root matrix of $C^2(t)$, and $\lambda_{Max}[N(b;s)]$ denotes the largest eigenvalue of the hermitian matrix $N(b;s)$. {Reid [30-Ths. 3.1, 3.2]}.

13. Suppose that hypotheses $(\mathscr{A}_{\mathscr{B}})$, $(\mathscr{A}_K')$ are satisfied, and the differential system (3.1i) is identically normal on [a,b]. Then condition $(\mathscr{A}_K'\text{iii})$ is satisfied; indeed, there exist subspaces of $D_0[a,b]$ of arbitrarily high dimension on which $K[\eta]$ is positive definite. {A result equivalent to this is established in Reid [47]}.

14. If the boundary problem (3.1) satisfies hypotheses $(\mathscr{A}_{\mathscr{B}}')$, $(\mathscr{A}_K'\text{i,ii,iii})$, Condition (3.10), and $\{\lambda_j, y_j(t) = (u_j(t);v_j(t))\}$, is a system of eigenvalues and eigenvectors satisfying the conclusions of Theorem 3.1, then for λ_0 a real value less than λ_1 we have the trace formula

$$\int_a^b \operatorname{Tr} G(s,s;\lambda_0)K(s)ds = \sum_{j=1}^{\infty} (\lambda_j - \lambda_0)^{-1}. \tag{7.8}$$

If condition $(\mathscr{A}_K'\text{i})$ is satisfied then (7.8) is a ready consequence of conclusion (ii) of Theorem 3.4. In general, consider the auxiliary problem involving the functional $J[\eta;\lambda_0:\mathscr{B}]$ and the modified function $K_\varepsilon(t) = \varepsilon E_n + K(t)$ for $\varepsilon > 0$, for which the cited results yield the existence of eigenvalues and eigenvectors $\{\lambda_{j\varepsilon}, y_{j\varepsilon}(t) = (u_{j\varepsilon}(t);v_{j\varepsilon}(t))\}$ satisfying the conclusions of Theorem 3.1 and formula corresponding to (7.8). As $\varepsilon \to 0^+$ an elementary comparison theorem implies that the individual $\lambda_{j\varepsilon}$ increases monotonically to $\lambda_j - \lambda_0$, and hence $1/\lambda_{j\varepsilon}$ decreases monotonically to $1/(\lambda_j - \lambda_0)$, so that (7.8) remains valid by a continuity argument. {Reid [47]}.

15. Suppose that in addition to the conditions of hypothesis $(\mathscr{A}_{\mathscr{B}}')$ the boundary problem $(\mathscr{B})$ satisfies the following conditions: (i) the differential system (3.1i) is identically normal on [a,b]; (ii) $J[\eta:\mathscr{B}]$ is positive definite on $D_e[\mathscr{B}]$. If $H(t)$ is an hermitian matrix

function of class $\mathscr{L}$ on $[a,b]$ such that the system

$$\begin{aligned} &L_1[u,v](t) = H(t)u(t), \quad L_2[u,v](t) = 0 \\ &\hat{u} \in S, \quad T[u,v] \equiv Q\hat{u} + [\operatorname{diag}\{-E_n, E_n\}]\hat{v} \in S^{\perp} \end{aligned} \tag{7.9}$$

has a non-identically vanishing solution, then

$$\int_a^b \operatorname{Tr} G(s,s)H^{+}(s)ds > 1 \tag{7.10}$$

where $G(t,s)$ belongs to the partial Green's matrix functions $G(t,s)$, $G_0(t,s)$ of Theorem 1.2, and $H^{+}(t)$ denote the hermitian matrix function $\frac{1}{2}[H(t) + |H(t)|]$, where $|H(t)|$ is the unique non-negative definite hermitian matrix function satisfying $|H(t)|^2 = H(t)H(t)$.

{Reid [47]. Under the stated conditions, and with the aid of results of the above Exercises 13, 14 the boundary problem

$$\begin{aligned} L_1[u,v](t) = \lambda H^{+}(t)u(t), \quad L_2[u,v](t) = 0, \\ \hat{u} \in S, \qquad T[u,v] \in S^{\perp} \end{aligned}$$

is shown to have infinitely many eigenvalues, all positive and with the smallest one less than 1. This generalized Liapunov inequality extends an inequality for second order linear homogeneous scalar equations due to Hartman and Wintner [5] and Nehari [1]; see also Hartman [13-Ch. XI, Th. 5.1]}.

16. Consider a self-adjoint boundary problem defined by (3.1), where in addition to the stated conditions of $(\mathscr{A}_{\mathscr{B}})$ we require that the $n \times n$ matrix functions $A(t)$, $B(t)$, $C(t)$, $K(t)$ are continuous on $[a,b]$; moreover, $B(t) > 0$ and $K(t) > 0$ for $t \in [a,b]$. In particular, conditions $(\mathscr{A}_K'\text{i,ii,iii})$ of Section 3 are satisfied by such a system, and

the system of differential equations (3.1i) is identically normal on $[a,b]$. Moreover, the hypotheses of Theorem 3.1 are satisfied, so for this system there exists a sequence of eigenvalues and eigenfunctions satisfying the conclusions of that theorem. Let $X(t)$ be an $n \times r$, $(0 < r \leq n)$, matrix of class $\mathscr{C}'$ and rank r on $[a,b]$, as in Exercise V.15.23. Then under the substitution $\eta(t) = X(t)\xi(t)$ the functional $J[\eta:\mathscr{B}]$ associated with (3.1) becomes

$$\hat{\xi}^*\check{Q}\hat{\xi} + \check{J}[\xi;a,b], \tag{7.11}$$

where $\check{J}[\xi;a,b]$ is given by (V.15.21), and $\check{Q}$ is the $2r \times 2r$ hermitian matrix corresponding to the $2n \times 2n$ matrix Q of (3.1ii) by

$$\check{Q} = \begin{bmatrix} X^*(a)Q_{11}X(a) & X^*(a)Q_{12}X(b) \\ X^*(b)Q_{21}X(a) & X^*(b)Q_{22}X(b) \end{bmatrix}, \text{ where } Q = \begin{bmatrix} Q_{11} & Q_{12} \\ Q_{21} & Q_{22} \end{bmatrix},$$

and $Q_{\alpha\beta}$, $(\alpha,\beta = 1,2)$, are $n \times n$ matrices. Let $\check{S}$ denote the subspace of $\mathbf{C}_{2r}$ consisting of values $\xi = (\xi(a);\xi(b))$ such that $(X(a)\xi(a);X(b)\xi(b)) \in S$, where S is the subspace of $\mathbf{C}_{2n}$ of (3.1ii). Finally, let $K(t|X) = X^*(t)K(t)X(t)$. Associated with (7.11) and the essential boundary condition $\hat{\xi} \in \check{S}$ one then has the boundary problem

$$\begin{aligned} &\text{(i)} \quad L[\xi](t) = \lambda K(t|X)\xi(t) \\ &\text{(ii)} \quad \hat{\xi} \in \check{S}, \ (\check{Q}\hat{\xi} + [\operatorname{diag}\{-E_r,E_r\}]\hat{\zeta}) \in \check{S}^{\perp}, \end{aligned} \tag{7.12}$$

where $\zeta = R(\ |X)\xi' + Q(\ |X)\xi$, and

$$L[\xi] = -[R(\ |X)\xi' + Q(\ |X)\xi]' + [Q^*(\ |X)\xi' + P(\ |X)\xi]$$

in the notations of Exercise V.15.23. Then the system (7.12) has a sequence of eigenvalues $\check{\lambda}_1 \leq \check{\lambda}_2 \leq \ldots$ with corresponding eigenvectors $\xi_j(t)$ for $\check{\lambda}_j$ satisfying the conclusions of Theorem 3.1, and in view of the extremum properties of the eigenvalues we have $\check{\lambda}_j \geq \lambda_j$, $(j = 1,2,\ldots)$.

17. Consider a system (3.1) satisfying $(\mathscr{A}'_{\mathscr{B}})$, $(\mathscr{A}'_K\text{-i,ii,iii})$, and $\{\lambda_j, y_j = (u_j(t); v_j(t))\}$ as in Theorem 3.1. For $i = 1,2,\ldots$ and $r = 0,1,2,\ldots$ let D_r denote any r-dimensional subspace of $D_e[\mathscr{B}]$, and for $D_r^{\perp}[\mathscr{B}] = \{\eta : \eta \in D_e[\mathscr{B}],\ K[\eta,\hat{\eta}] = 0 \text{ for } \hat{\eta} \in D_r\}$, let $D_i^{\#}$ denote any i-dimensional subspace of $D_r^{\perp}[\mathscr{B}]$. Then

$$\lambda_{i+r} = \operatorname*{Max}_{D_r} \operatorname*{Min}_{D_i^{\#}} \operatorname*{Max}_{\substack{\eta \in D_i^{\#} \\ \eta \neq 0}} (J[\eta]/K[\eta]). \tag{7.13}$$

{For $i = 1$, (7.13) reduces to the maximum-minimum property of the Corollary to Theorem 3.1, and the result for $r = 0$ is established readily with the aid of this property. For $r \geq 1$, let $\mathscr{F} = \{f_1,\ldots,f_r\}$ be a basis for D_r, and if $\tilde{\lambda}_1 = \lambda_{1,1}$, $\tilde{u}_1 = u_{1,1}$ is the smallest eigenvalue and associated eigensolution of (7.7), then $\lambda_1 \leq \lambda_{1,1} \leq \lambda_{r+1}$. If $\mathscr{F}_1 = \{f_1,\ldots,f_r,u_{1,1}\}$, and $\tilde{\lambda}_1 = \lambda_{1,2}$, $\tilde{u}_1 = u_{1,2}$ is the smallest eigenvalue and associated eigensolution of (7.7), then $\lambda_{1,1} \leq \lambda_{1,2} \leq \lambda_{2,1} \leq \lambda_{r+2}$. Proceed by induction to obtain a class $\mathscr{F}_{i-1} = \{f_1,\ldots,f_r,u_{1,1},u_{1,2},\ldots,u_{1,i-1}\}$ such that the smallest eigenvalue $\tilde{\lambda}_1 = \lambda_{1,i}$ of the corresponding problem (7.7) satisfies $\lambda_{1,i} \leq \lambda_{r+i}$, and show that this result implies (7.13). The Max-min-max principle (7.13) is due to Stenger [2], (see also Weinstein and Stenger [1, Sec. 3.6]), who presented it in the context of self-adjoint linear operators in Hilbert space}.

CHAPTER VII.
A CLASS OF DEFINITE BOUNDARY PROBLEMS

1. Introduction

Antedating the work of Morse on the extension of the Sturmian theory to self-adjoint differential systems, Bliss [2] considered a real two-point boundary problem which in terms of an n-dimensional vector function $y(t) = (y_\alpha(t))$, $(\alpha = 1,\dots,n)$, may be written as

$$\begin{aligned} &(a) \quad y'(t) = A(t)y(t) + \lambda B(t)y(t), \quad t \in [a,b]. \\ &(b) \quad s[y] \equiv My(a) + Ny(b) = 0. \end{aligned} \tag{1.1}$$

For such a system he introduced the concept of "self-adjointness under a real non-singular transformation $z = T(t)y$", and considered in detail a special class of such problems which he called "definitely self-adjoint". This class of problems included the so-called accessory boundary problem for a non-singular simple integral variational problem involving no auxiliary differential equations as restraints, but included the accessory problem for a variational problem of Lagrange or Bolza type only in case the condition of identical normality held. In 1938 Bliss [6] gave a new definition of definite self-adjointness with the involved normality assumption considerably weaker than in the original definition, so that the class of systems (1.1) which are definitely self-adjoint according to this modified definition does include the accessory boundary problem for a Lagrange or Bolza type problem that is normal, but wherein the condition of identical normality is not necessarily satisfied. It is to be commented that the proof presented by Bliss of the existence of eigenvalues for the considered system (1.1) followed the general pattern used by E. Schmidt [1] in the treatment of linear integral equations of the second kind with real symmetric kernel, and thus in basic idea rested on the method introduced by Schwarz [1].

In both the original and modified definitions of Bliss the definiteness property of the real system is possessed by the matrix $S(t) = T^*(t)B(t)$. Subsequently, Reid [9] considered a differential system (1.1) satisfying the conditions of Bliss [6] aside from the definiteness condition, and with this condition replaced by a suitable condition of definiteness on the functional

$$\int_a^b y^*(t)T^*(t)[y'(t) - A(t)y(t)]dt. \tag{1.2}$$

In particular, the class of problems considered in this paper includes as special instances systems which are equivalent to two-point boundary problems involving scalar differential equations of higher order and which are definite in the sense considered by Kamke [6]. Also, extension of the results to systems with complex coefficients was discussed in Reid [9; Sec. 12]. At about the same time, E. Hölder [3] studied a real system (1.1) with $B(t)$ of constant rank on $[a,b]$ and satisfying the hypotheses of Bliss [6], with the exception of the normality condition. By the consideration of a related "canonical system" of twice the dimension of (1.1), Hölder reduced the determination of "normal solutions" of this problem to a pair of adjoint vector integral equations, and thus obtained results on the existence of normal eigenvalues and associated expansions in terms of the normal eigenfunctions of (1.1).

In Reid [16] the results of the earlier paper [9] were extended in various manners. Firstly, the system considered was of the more general form

$$L[y](t) \equiv A_1(t)y'(t) + A_0(t)y(t) = \lambda B(t)y, \ t \in [a,b].$$
$$s[y] \equiv My(a) + Ny(b) = 0, \tag{1.3}$$

with the $n \times n$ coefficient matrix $A_1(t)$ non-singular on $[a,b]$, as the results for such systems possess certain niceties of character not present for the restricted case (1.1). Secondly, the matrix coefficient functions $A_0(t)$, $A_1(t)$, $B(t)$, and the matrices M, N were allowed to be complex valued.

In terms of a non-singular transformation

$$z(t) = T(t)y(t), \quad t \in [a,b], \tag{1.4}$$

the conditions for (1.3) corresponding to those of Bliss [6] are as follows:

(a) (1.3) *is equivalent to its adjoint*

$$[-A_1^*(t)z(t)]' + A_0^*(t)z(t) = \lambda B^*(t)z(t), \quad t \in [a,b]$$
$$P^*z(a) + Q^*z(t) = 0, \tag{1.5}$$

under the transformation (1.4);

(b) $S(t) \equiv T^*(t)B(t)$ *is hermitian on* $[a,b]$;

(c) $S(t)$ *is non-negative definite on* $[a,b]$;

(d) *if* $L[y](t) = 0$, $s[y] = 0$, *and* $B(t)y(t) \equiv 0$, *then* $y(t) \equiv 0$ *on* $[a,b]$.

For (1.3) the conditions corresponding to those of Reid [9] are the above conditions (a), (b), (d), while (c) is replaced by the following condition:

(c') $\int_a^b y^*(t)T^*(t)L[y](t)dt > 0$ *for arbitrary* $y(t)$ *satisfying with an associated vector function* $g(t)$ *the conditions*

$$L[y](t) = B(t)g(t),\ s[y] = 0,\ B(t)y(t) \not\equiv 0 \text{ on } [a,b]. \tag{1.6}$$

In presenting a detailed analysis of the equivalence of a system (1.3) to its adjoint under a transformation (1.4), it is shown in Reid [16; Sec. 5] that whenever (1.3) satisfies condition (a) with a transformation (1.3), then there exists a second transformation $z = T_1(t)y$ with which (1.3) satisfies both conditions (a) and (b), the matrix $A_1^*(t)T_1(t)$ is skew-hermitian on $[a,b]$, and the corresponding equivalent system

$$\begin{aligned} T_1^*(t)L[y](t) &= T_1^*(t)B(t)y(t), \\ s[y] &\equiv My(a) + Ny(b) = 0, \end{aligned} \tag{1.7}$$

is self-adjoint in the classical Lagrange sense; moreover, if (1.3) satisfies with (1.4) conditions (a) and (b), then there is a real constant k_1 such that $T_1^*(t)B(t) \equiv k_1T^*(t)B(t)$ on $[a,b]$. In connection with this result it is to be remarked that if the coefficient matrices of (1.3) are all real-valued, and (1.3) is equivalent to its adjoint (1.5) under a transformation (1.4) with $T(t)$ real-valued on $[a,b]$, then in general the associated matrix $T_1(t)$ may not be chosen real-valued. Consequently, this result gives added incentive to the consideration of systems (1.3) wherein the coefficient matrix functions are complex-valued. Section 6 of Reid [16] is devoted to an analysis of the normality condition (d) for problems (1.3). In particular, it is shown that if (1.3) satisfies condition (a) with a transformation (1.4), but the "normality condition" (d) does not hold, then for any associated transformation $z = T_1(t)y$ satisfying the above stated properties there is a second system (1.3) which is "equivalent" to the original system, and such that the second system with the transformation $z = T_1(t)y$ satisfies conditions (a), (b) and (d). Finally, for systems (1.3) satisfying conditions (a), (b) with a transformation (1.4) there is considered a condition of definiteness that includes as special instances the above condition (c) and the earlier modification of this condition considered in Reid [9]. In particular, a system that is definite in the considered sense, and which satisfies the normality condition (d), is equivalent to a vector integral equation of the types treated by Wilkins [1] and Zimmerberg [2]. Also, the equivalent integral equation for such a boundary problem is a special case of symmetrizable transformations considered by Zaanen [1,2,3] and Reid [11].

2. Definite self-adjoint boundary problems

The study of the class of problems described in the preceding section has been presented in a still more general context in Sections 5, 6 of Chapter IV of Reid [35]. The system considered is of the form

$$\begin{aligned} &(a) \quad L[y](t) \equiv A_1(t)[A_2(t)y(t)]' + A_o(t)y(t) = \lambda B(t)y(t), \\ &(b) \quad \mu[y] \equiv \mathscr{M}\hat{u}_y = 0, \end{aligned} \tag{2.1}$$

where the $n \times n$ matrix coefficient functions $B(t)$, $A_o(t)$, $A_1(t)$, $A_2(t)$ are continuous on $[a,b]$, with $B(t) \not\equiv 0$ and $A_1(t)$, $A_2(t)$ non-singular on this interval, while $\mathscr{M}$ is an $n \times 2n$ matrix of rank n, $u_y(t) = A_2(t)y(t)$, and $\hat{u}_y$ denotes the 2n-dimensional boundary vector $w = (w_\sigma)$, $(\sigma = 1,\ldots,2n)$, with the component n-dimensional vectors (w_α) and $(w_{n+\alpha})$ given by $u_y(a)$ and $u_y(b)$, respectively. The symbol $\mathscr{D}(L)$ will denote the linear vector space of $y(t) \in \mathscr{L}_n[a,b]$ such that $y = A_2^{-1}u_y$ with $u_y \in \mathscr{L}_n'[a,b]$, $\mathscr{D}_o(L) = \{y{:}y \in \mathscr{D}(L), \hat{u}_y = 0\}$, while $D(L)$ is a linear subspace of $\mathscr{L}_n[a,b]$ satisfying $\mathscr{D}_o(L) \subset D(L) \subset \mathscr{D}(L)$, and determined by the end-conditions $D(L) = \{y{:}y \in \mathscr{D}(L), \mathscr{M}\hat{u}_y = 0\}$.

The (formal) adjoint of the boundary problem (2.1) is

$$\begin{aligned} &(a) \quad L^*[z](t) \equiv -A_2^*(t)[A_1^*(t)z(t)]' + A_o^*(t)z(t) = \lambda B^*(t)z(t), \\ &(b) \quad \nu[z] \equiv \mathscr{P}^*\mathscr{D}\hat{v}_z = 0, \end{aligned} \tag{2.2}$$

where $\mathscr{P}$ is an $n \times 2n$ matrix of rank n satisfying $\mathscr{M}\mathscr{P} = 0$, $\mathscr{D}$ is the $2n \times 2n$ constant matrix $\operatorname{diag}\{-E_n,E_n\}$ and v_z is the 2n-dimensional boundary vector of the n-dimensional vector function $v_z(t) = A_1^*(t)z(t)$. Also, corresponding to the above definition associated with (2.1), the symbol

$\mathscr{D}(L^*)$ denotes the linear vector space of vector functions $z(t) \in \mathscr{C}_n[a,b]$ of the form $z(t) = A_1^{*-1}(t)v_z(t)$, where $v_z \in \mathscr{C}_n^1[a,b]$.

The general theory of differential systems (2.1) and their adjoints (2.2) is discussed in Chapter III, Sections 2-7 and Chapter IV, Sections 1-3 of Reid [34]. Indeed, as partially noted in Problem III.2.8 of this reference, the basic results for such systems are readily extensible to systems of the same form, wherein the matrix functions $A_1(t)$, $A_2(t)$, $A_1^{-1}(t)$, $A_2^{-1}(t)$ all belong to $L_{nn}^\infty[a,b]$, and $A_0(t) \in L_{nn}[a,b]$. For simplicity in describing the major results, however, attention will be limited to the case wherein the matrix coefficient functions are continuous.

For typographical simplicity, the symbol $((x,y))$ is used for the integral

$$((x,y)) = \int_a^b y^*(t)x(t)dt, \tag{2.3}$$

where $x(t)$, $y(t)$ are n-dimensional vector functions on $[a,b]$ of class $\mathscr{L}_n^2[a,b]$.

A boundary problem (2.1) is called (formally) self-adjoint if the following conditions hold:

$$\begin{aligned} &\text{(i)} \quad B(t) \textit{ is hermitian for } t \in [a,b]; \\ &\text{(ii)} \quad ((L[y],z)) = ((y,L[z])), \textit{ for } y \in D(L),\ z \in D(L). \end{aligned} \tag{2.4}$$

In particular, condition (2.4ii) holds if and only if $D(L) = D(L^*)$, and $L[y] = L^*[y](t)$ for $y \in D(L) = D(L^*)$, while $u_y = A_2 y$ and $v_z = A_1^* z$ satisfy the boundary condition

$$\hat{v}_z^* \mathscr{D} \hat{u}_y = 0 \text{ if } \mathscr{M}\hat{u}_y = 0,\ \mathscr{P}^*\mathscr{D}\hat{v}_z = 0. \tag{2.4}$$

For a self-adjoint boundary problem (2.1) one has the basic property that if $\lambda = \lambda_0$ is an eigenvalue of (2.1) of index k then $\lambda = \overline{\lambda}_0$ is an eigenvalue of the adjoint problem (2.2) of index k. Also, if λ_0 is not an eigenvalue of (2.1), and $G(t,s;\lambda_0)$ is the Green's matrix of this system for $\lambda = \lambda_0$ then $\lambda = \overline{\lambda}_0$ is not an eigenvalue of (2.1) and $G(t,s;\overline{\lambda}_0) = [G(s,t;\lambda_0)]^*$; in particular, if λ_0 is real then $G(t,s;\lambda_0) = [G(s,t;\lambda_0)]^*$. Moreover, if $y = y_1(t)$ and $y = y_2(t)$ are solutions of (2.1) for respective values λ_1 and λ_2 with $\lambda_1 \neq \lambda_2$ then

$$\text{(i)} \quad ((By_1,y_2)) = 0, \qquad \text{(ii)} \quad ((Ly_1,y_2)) = 0. \tag{2.5}$$

As noted by Reid [35, Prob. IV.5.1], essentially the only restriction placed upon the eigenvalues of a system (2.1) by the condition of self-adjointness is the conclusion expressed by (2.5).

In the terminology of Reid [11; also, 35, Sec. IV.5], a boundary problem (2.1) is called *fully self-adjoint* if it is self-adjoint and $((By,y)) \neq 0$ for each eigenfunction $y(t)$ of this problem. It is to be remarked that this condition may hold vacuously, as there exist self-adjoint boundary problems (2.1) which possess no eigenvalues. If (2.1) is fully self-adjoint then all eigenvalues are real, and the set of all eigenvalues is at most denumerably infinite with no finite limit point. Also, the index of each eigenvalue is equal to its multiplicity, and if λ is an eigenvalue of index k there exist corresponding eigenfunctions $y^{(1)}(t),\dots,y^{(k)}(t)$ such that

$$((By^{(\alpha)},y^{(\beta)})) = \varepsilon(\lambda)\delta_{\alpha\beta}, \quad (\alpha,\beta = 1,\dots,k), \tag{2.6}$$

where either $\varepsilon(\lambda) = +1$ or $\varepsilon(\lambda) = -1$. Moreover, each eigenvalue is a simple pole of the Green's matrix $G(t,s;\lambda)$ of (2.1), and at an eigenvalue λ the residue of the Green's matrix is

$$-\varepsilon(\lambda) \sum_{\alpha=1}^{k} y^{(\alpha)}(t)[y^{(\alpha)}(s)]^*. \tag{2.7}$$

To illustrate the relation of the above boundary problems to those considered in Chapters V and VI, the following example is noted here.

EXAMPLE 2.1. *Suppose that* $A(t)$, $B(t)$, $C(t)$ *and* $K(t)$ *are* $m \times m$ *matrix functions continuous on* $[a,b]$, *while* $B(t)$, $C(t)$, $K(t)$ *are hermitian on this interval. If* κ^1, μ^1, κ^2, μ^2 *are* $2m \times m$ *constant matrices such that the* $2m \times 4m$ *matrix* $[\kappa^1 \quad -\mu^1 \quad \kappa^2 \quad \mu^2]$ *is of rank* $2m$, *and the* $2m \times 2m$ *matrix* $\kappa^1\mu^{1*} + \kappa^2\mu^{2*}$ *is hermitian, then the* $2m$-*dimensional differential system in* $y(t) = (y_\sigma(t))$, *with* $y_\alpha(t) = u_\alpha(t)$, $y_{n+\alpha}(t) = v_\alpha(t)$, $(\alpha = 1,\ldots,m)$, *defined by*

$$\begin{aligned} L_1[u,v](t) &\equiv -v'(t) + C(t)u(t) - A^*(t)v(t) = \lambda K(t)u(t) \\ L_2[u,v](t) &\equiv u'(t) - A(t)u(t) - B(t)v(t) = 0, \\ &\kappa^1 u(a) - \mu^1 v(a) + \kappa^2 u(b) + \mu^2 v(b) = 0 \end{aligned} \tag{2.8}$$

is self-adjoint.

For boundary problems (2.1) there are certain linear vector spaces that are of basic importance. For such a problem, the set of all n-dimensional vector functions $y(t) \in D(L)$ which satisfy the boundary conditions $\mu[y] = 0$ of (2.1), will be denoted by F_o. For $j = 1,2,\ldots$, the symbol F_j will denote the set of all vector functions $\eta(t) \in D(L)$ for which there is an associated $\zeta(t) \in F_{j-1}$ such that

$$L[\eta](t) = B(t)\zeta(t), \quad t \in [a,b], \quad \mu[\eta] = 0. \tag{2.9}$$

Also, F_∞ will designate the set of all vector functions common to F_j, $j = 0,1,\ldots$. Clearly $F_{j+1} \subset F_j$, $(j = 0,1,\ldots)$, and all eigenfunctions of (2.1) belong to F_∞. Moreover, if $\zeta \in F_\infty$ and there is a $\eta(t) \in D(L)$ satisfying (2.9), then also $\eta \subset F_\infty$.

In the terminology of Reid [35, Sec. V.6], a boundary problem (2.1) is termed F_j-definite, $(j = 0,1,\ldots)$, when the following conditions are satisfied.

(a) *the problem is self-adjoint, as defined above;*

(b) *the only vector function* $y(t) \in F_0$ *which satisfies*

$$L[y](t) = 0, \quad B(t)y(t) \equiv 0, \quad t \in [a,b] \tag{2.10}$$

is $y(t) \equiv 0$ *on* $[a,b]$.

(c) *there exist real constants* k_1, k_2 *not both zero and such that*

$$((k_1 L[y] + k_2 By, y)) \geq 0 \tag{2.11}$$

for arbitrary $y \in F_j$, *and if the equality sign holds for a* $y \in F_j$ *then* $B(t)y(t) \equiv 0$ *on* $[a,b]$.

This class of problems includes those of the type discussed by Bliss [2,6] for the case in which the constants in (2.10c) are $k_1 = 0$, $k_2 = 1$. For the case in which $k_1 = 1$, $k_2 = 0$ the class of problems includes those treated by Reid [9], while the general case contains as special instances problems of the sort considered in Reid [16].

In general, if a problem (2.1) is F_j-definite, then this problem is F_k-definite for $k > j$. An elementary, but basic property of such problems is that if there exists a non-negative integer j such that (2.1) if F_j-definite, then this system if fully self-adjoint, so, in particular, the eigenvalues are all real and corresponding eigenfunctions may be chosen to satisfy (2.6).

Now if $L^{(1)}[y](t) = c_1 L[y](t) + c_2 B(t)y(t)$, $B^{(1)}(t) = c_3 B(t)$, where c_1, c_2, c_3 are constants with $c_1 \neq 0$, $c_3 \neq 0$, then $y(t)$ is a solution of the boundary problem

$$L^{(1)}[y](t) = \lambda B^{(1)}(t)y(t), \quad \mu[y] = 0, \tag{2.12}$$

for $\lambda = \lambda_0$ if and only if $y(t)$ is solution of (2.1) for $\lambda = (\lambda_0 c_3 - c_2/c_1)$. Also, for (2.12) the above defined classes F_j are identical with the corresponding classes for (2.1), as F_0 is clearly identical for the two problems, and $L^{(1)}[y](t) = B^{(1)}(t)z^{(1)}(t)$ if and only if $L[y](t) = Bz(t)$, with $z(t) = [\{c_3 z^{(1)}(t) - c_2\}/c_1]$. Moreover, if c_1, c_2, c_3 are real constants then (2.1) is F_j-definite if and only if (2.12) is F_j-definite. In view of these remarks it follows that if (2.1) is F_j-definite then by a suitable choice of real constants $c_1 \neq 0$, c_2, $c_3 \neq 0$ the boundary problem (2.12) is F_j-definite; moreover, for this problem condition (c) above holds with either $k_1 = 0$, $k_2 = 1$ or $k_1 = 1$, $k_2 = 0$. That is, for the consideration of an F_j-definite system (2.1) it may be assumed without loss of generality that the above condition (c) holds either with $k_1 = 0$, $k_2 = 1$ or $k_1 = 1$, $k_2 = 0$; in the respective cases the system will be called BF_j-definite and LF_j-definite. It is to be noted that for BF_0-definite problems the last statement of (c) is extraneous, since the condition that $((By,y))$ is non-negative for all $y \in F_0$ implies that the continuous hermitian matrix function $B(t)$ is non-negative for

$t \in [a,b]$, and consequently, if $y \in F_0$ and $((By,y)) = 0$ then $B(t)y(t) \equiv 0$ on $[a,b]$.

One may verify readily that if (2.1) is LF_j-definite then $\lambda = 0$ is not an eigenvalue of this problem. Moreover, in view of the above comments on the equivalence of systems (2.1) and (2.12), for a BF_j-definite system there is no loss of generality in assuming that $\lambda = 0$ is not an eigenvalue, as this condition holds for a system (2.12) with $c_1 = c_3 = 1$, and c_2 a real constant such that $\lambda = -c_2$ is not an eigenvalue of the given system.

Now if (2.1) is F_j-definite, and $y = y(t)$ is an eigenfunction of this system for an eigenvalue λ, then for $\eta(t)$ a vector function of F_1 satisfying (2.9) with a $\zeta \in F_0$ we have $((By,\zeta)) = ((y,B\zeta)) = ((y,L[\eta])) = ((L[y],\eta)) = \lambda((By,\eta))$, and consequently,

$$((By,\zeta)) = \lambda((By,\eta)); \tag{2.13}$$

in particular, if $\lambda \neq 0$ then $((By,\zeta)) = 0$ if and only if $((By,\eta)) = 0$. Now for a BF_j-definite system the condition that $((B\eta,\eta)) \geq 0$ for all $\eta \in F_j$ implies the corresponding Cauchy-Bunyakovsky-Schwarz inequality

$$|((B\eta_1,\eta_2))|^2 \leq ((B\eta_1,\eta_1))((B\eta_2,\eta_2)), \text{ for } \eta_\alpha \in F_j, \quad (\alpha = 1,2).$$

Another elementary result which is of frequent use in the consideration of such systems is the fact that if (2.1) is BF_j-definite and $\eta \in F_{j+1}$ then whenever $((B\eta,\eta)) = 0$ we also have $((L[\eta],\eta)) = 0$; moreover, if $((B\eta,\eta)) > 0$ and η_1 is a vector function of F_{j+2} such that

$$L[\eta_1](t) = B(t)\eta(t), \quad \mu[\eta_1] = 0, \tag{2.14}$$

then for y_o equal to at least one of the vector functions $\eta, \eta_1, \eta + \eta_1$ of F_{j+1} we have $((L[y_o], y_o)) \neq 0$ and $((By_o, y_o)) > 0$.

In the treatment of definite problems, the method employed by Reid [35; Ch. IV, Sec. 6] is based upon the procedure introduced by Schwarz [1], and used subsequently in varied situations by Schmidt, Picone and Bliss. As earlier references to this procedure gave no real indication of its details, it is considered worthwhile to describe briefly its application to the present problem. As noted above, for a BF_j-definite system there is no loss of generality in assuming that $\lambda = 0$ is not an eigenvalue, and this we shall do in the following discussion.

LEMMA 2.1. *Suppose that* (2.1) *is* BF_j*-definite, and* $\lambda = 0$ *is not an eigenvalue of this system. If* $y_o(t) \in F_j$, *and* $y_\alpha(t)$, $(\alpha = 1,2,\ldots)$, *are vector functions in* $F_{j+\alpha}$ *defined recursively by*

$$L[y_\alpha](t) = B(t)y_{\alpha-1}(t), \quad t \in [a,b], \; \mu[y_\alpha] = 0, \tag{2.15}$$

then the (Schwarz) constants

$$W_\alpha = ((By_o, y_\alpha)), \quad (\alpha = 0,1,\ldots), \tag{2.16}$$

are real and possess the following properties:

$$W_{\alpha+\beta} = ((By_\alpha, y_\beta)), \quad (\alpha, \beta = 0,1,\ldots); \tag{2.17}$$

$$W_{2\alpha-1}^2 \leq W_{2\alpha}W_{2\alpha-2}, \; W_{2\alpha}^2 \leq W_{2\alpha-2}W_{2\alpha+2}, \; (\alpha = 1,2,\ldots); \tag{2.18}$$

$$\text{if } W_2 \neq 0, \text{ then } W_{2\alpha} > 0 \text{ and } W_{2\alpha} \geq W_o(W_2/W_o)^\alpha, \quad (\alpha = 0,1,\ldots); \tag{2.19}$$

if $W_1 \neq 0$, *then* $W_{2\alpha} > 0$, $(\alpha = 0,1,\ldots,)$,
and $W_o/W_2 \leq W_o^2/W_1^2$. (2.20)

The fundamental existence theorem for boundary problems (2.1) is as follows.

THEOREM 2.1. *Suppose that* (2.1) *if* BF_j*-definite and* $\lambda = 0$ *is not an eigenvalue of this problem. Then either* $((B\eta,\eta)) = 0$ *for all* $\eta \in F_{j+1}$ *and* (2.1) *has no eigenvalues, or there is a proper value* λ_1 *of this problem such that for arbitrary* $\eta \in F_{j+1}$ *and associated* $\zeta \in F_j$ *satisfying* (2.9) *with* η, *we have*

$$|((L[\eta],\eta))| \leq |\lambda_1|^{-1}((B\zeta,\zeta)). \qquad (2.21)$$

If (2.1) *has eigenvalues* $\lambda_1,\ldots,\lambda_k$ *and* $\Delta\{\lambda_1,\ldots,\lambda_k\}$ *denotes the set of* $\eta \in F_{j+1}$ *satisfying* $((B\eta,y)) = 0$ *for all eigenvectors* $y(t)$ *of* (2.1) *corresponding to an eigenvalue of the set* $\lambda_1,\ldots,\lambda_k$, *then either* $((B\eta,\eta)) = 0$ *for all* $\eta \in \Delta\{\lambda_1,\ldots,\lambda_k\}$ *and* (2.1) *has no other eigenvalues, or there exists an eigenvalue* λ_{k+1} *distinct from* $\lambda_1,\ldots,\lambda_k$, *and such that for arbitrary* $\eta \in \Delta\{\lambda_1 \ldots,\lambda_k\}$ *and associated* $\zeta \in F_j$ *satisfying* (2.9) *with* η *we have*

$$|((L[\eta],\eta))| \leq |\lambda_{k+1}|((B\zeta,\zeta)). \qquad (2.22)$$

In view of the fact that a BF_j-definite problem is fully self-adjoint, it follows that if $((B\eta,\eta)) = 0$ for all $\eta \in F_{j+1}$ then the boundary problem (2.1) has no eigenvalues. On the other hand, if $((B\eta,\eta))$ is not zero for all $\eta \in F_{j+1}$ then, as noted above, there exists a $y \in F_{j+1}$ for which $((L[y],y)) \neq 0$ and $((By,y)) > 0$. For $\eta = y$ such a vector function, and $\zeta = z$ an associated vector function of F_j

satisfying (2.9) with y, let $y_o(t) = z(t)$, and define the $y_\alpha(t)$, $(\alpha = 1,2,\ldots)$, recursively by (2.15). From a basic solvability theorem for the solution by successive approximations of a non-homogeneous differential system involving linearly a parameter, (see, for example, Reid [35, Theorem IV.2.2 and its Corollary]), the infinite series

$$y_1(t) + \lambda y_2(t) + \ldots + \lambda^{\alpha-1} y_\alpha(t) + \ldots \tag{2.23}$$

is the Maclaurin expansion for the solution $y(t,\lambda)$ of the homogeneous differential system

$$L[y](t) = \lambda B(t)y(t) + B(t)y_o(t), \quad t \in [a,b], \qquad \mu[y] = 0. \tag{2.24}$$

Moreover, if ρ is a positive constant such that (2.1) has no eigenvalue satisfying $|\lambda| < \rho$, then the series (2.23) converges uniformly in (t,λ) on each set of the form $t \in [a,b]$, $|\lambda| \leq \rho_1 < \rho$. In particular, if $w = w(t;\lambda) = y_o(t) + \lambda y(t,\lambda)$, then for $|\lambda| < \rho$ the value of $((Bw,y_o))$ may be obtained by termwise integration of the series $y_o^*(t)B(t)w(t;\lambda)$, which yields the result that the infinite power series in λ,

$$W_o + \lambda W_1 + \ldots + \lambda^\beta W_\beta + \ldots, \tag{2.25}$$

converges to the value $((Bw,y_o))$. In turn, the convergence of (2.25) implies that the power series

$$W_o + \lambda^2 W_2 + \ldots + \lambda^{2\alpha} W_{2\alpha} + \ldots \tag{2.25'}$$

converges for all λ satisfying $|\lambda| < \rho$, where ρ is a positive constant such that (2.1) has no eigenvalue on the circular disk $|\lambda| < \rho$ in the complex plane.

Now $y(t) = y_1(t)$ is such that $0 \neq ((L[y_1],y_1)) =$

$((By_0,y_1)) = W_1$, and in view of the conclusion (2.20) of Lemma 2.1 it follows that $W_{2\alpha} > 0$, $(\alpha = 0,1,\ldots)$. Consequently, the infinite series (2.25') does not converge for $\lambda = W_0/|W_1|$, and therefore, $\rho \leq W_0/|W_1|$. That is, whenever $((B\eta,\eta))$ does not vanish for all $\eta \in F_{j+1}$, then there exists an eigenvalue λ_1 of (2.1) such that the inequality

$$|\lambda_1| \leq \frac{((B\zeta,\zeta))}{|((L[\eta],\eta))|}, \tag{2.21'}$$

holds for all $\eta \in F_{j+1}$ with $((L[\eta],\eta)) \neq 0$, and where ζ is an associated vector function of F_j with which η satisfies (2.9). Clearly (2.21') for this class of vector functions is equivalent of the inequality (2.21) for arbitrary $\eta \in F_{j+1}$ and associated $\zeta \in F_j$ with which η satisfies (2.9).

The proof of the second portion of the theorem is quite analogous to that of the first part. If $((B\eta,\eta)) = 0$ for all $\eta \in \Delta\{\lambda_1,\ldots,\lambda_k\}$ then with the aid of properties of eigenvalues and eigenfunctions stated above it follows that (2.1) has no eigenvalue distinct from $\lambda_1,\ldots,\lambda_k$. Also, if $\eta \in \Delta\{\lambda_1,\ldots,\lambda_k\}$ and η_1 satisfies with η the system (2.14), then relation (2.13) for $y = \eta_1$ and $\zeta = \eta$ implies that $\eta_1 \in \Delta\{\lambda_1,\ldots,\lambda_k\}$. consequently, it follows that if $((B\eta,\eta))$ does not vanish for all $\eta \in \Delta\{\lambda_1,\ldots,\lambda_k\}$ then there is a vector function y of this set such that $((L[y],y)) \neq 0$ and $((By,y)) > 0$. For $y(t)$ such a vector function, and $z(t)$ an associated vector function of F_j satisfying with $y(t)$ the system (2.9), let $y_0(t) = z(t)$ and define the $y_\alpha(t)$, $(\alpha = 1,2,\ldots)$ recursively by (2.15). In particular, $y_1(t) = y(t)$ and from a relation corresponding to (2.13) it follows that $((By_0,y)) = 0$ for all solution $y(t)$ of (2.1) for the eigenvalues $\lambda = \lambda_j$, $(j = 1,\ldots,k)$.

It then follows that the solution $y(t,\lambda)$ of (2.24) given by (2.23) defines a regular function of λ at the eigenvalues $\lambda_1,\ldots,\lambda_k$, and hence for ρ a positive constant such that the disk $|\lambda| < \rho$ in the complex plane contains no eigenvalue of (2.1) distinct from $\lambda_1,\ldots,\lambda_k$ the series (2.13) converges uniformly in (t,λ) on each set of the form $t \in [a,b]$, $|\lambda| \leq \rho_1 < \rho$. By an argument identical with that employed in the proof of the first portion of the theorem, it then follows that if $((B\eta,\eta))$ does not vanish for all $\eta \in \Delta\{\lambda_1,\ldots,\lambda_k\}$ then there exists an eigenvalue λ_{k+1} distinct from $\lambda_1,\ldots,\lambda_k$ and such that if $\eta \in \Delta\{\lambda_1,\ldots,\lambda_k\}$ with $((L[\eta],\eta)) \neq 0$, and ζ is an associated vector function of F_j which satisfies with η the system (2.9), then we have the inequality

$$|\lambda_{k+1}| \leq \frac{((B\zeta,\zeta))}{|((L[\eta],\eta))|}, \tag{2.22'}$$

which is equivalent to (2.22).

It is to be noted that if $z \in F_j$ and $((Bz,y)) = 0$ for all eigenfunctions y of an F_j-definite system (2.1), then for $y_0(t) = z(t)$ and the $y_\alpha(t)$, $(\alpha = 1,2,\ldots)$, defined recursively by (2.15) we have that the non-homogeneous system (2.24) has a solution $y(t,\lambda)$ for all finite values of λ which is a regular function of λ at each eigenvalue of (2.1). It then follows that the power series (2.25) and (2.25') converge for all values of λ. Since relations (2.19), (2.20) imply that if $W_1 \neq 0$ then $W_0 \neq 0$ and the series does not converge for $\lambda = W_0/|W_1|$, we must have $W_1 = 0$. Also, W_2 must be equal to zero since otherwise (2.19) implies that $W_0 \neq 0$ and the series does not converge for $\lambda = (W_0/W_2)^{1/2}$. Therefore $0 = W_2 = ((By_1,y_1))$, and as $y_1 \in F_{j+1} \subset F_j$ and

$((B\eta,\eta)) \geq 0$ on F_j it follows that $((By_1,u)) = 0$ for arbitrary $u \in F_j$. That is, we have the following result.

COROLLARY. *If* (2.1) *is* BF_j*-definite, and* z *is an element of* F_j *satisfying* $((Bz,y)) = 0$ *for all eigenfunctions* y *of* (2.1), *then the vector function* $\eta = y_1(t)$ *determined as the solution of the system* (2.9) *with* $\zeta = z(t)$ *is such that* $((By_1,u)) = 0$ *for arbitrary* $u \in F_j$; *in particular,* $((By_1,y_1)) = W_2 = 0$, $((L[y_1],y_1)) = ((Bz,y_1)) = 0$, *and* $((By_1,y)) = 0$ *for all eigenfunctions of* (2.1).

If a BF_j-definite system possesses eigenvalues, we shall consider the eigenvalues and corresponding eigenfunctions labelled as a simple sequence

$$\lambda_\alpha, y^{(\alpha)}(t), \quad (\alpha = 1,2,\ldots), \tag{2.26}$$

where it is understood that each eigenvalue is repeated a number of times equal to its index, and the corresponding eigenfunctions are chosen to be B*-orthonormal in the sense that*

$$((By^{(\alpha)},y^{(\beta)})) = \delta_{\alpha\beta}, \quad (\alpha,\beta = 1,2,\ldots). \tag{2.27}$$

Also, the corresponding Fourier coefficients of an n-dimensional vector function η relative to the B-orthogonal set of eigenvectors $y^{(\alpha)}$ of a BF_j-definite system are denote denoted by

$$c_\alpha[\eta] = ((B\eta,y^{(\alpha)})), \quad (\alpha = 1,2,\ldots), \tag{2.28}$$

where it is to be emphasized that the sequence (2.26) may be vacuous, finite, or denumerably infinite. The basic expansion results for a BF_j-definite system are presented in the following theorem.

THEOREM 2.2. *If* (2.1) *is* BF_j*-definite, then:*

(i) *if* $\eta \in F_{j+1}$ *then*

$$\begin{aligned}&\text{(a)}\quad ((L[\eta],\eta)) = \textstyle\sum_\alpha \lambda_\alpha |c_\alpha[\eta]|^2,\\ &\text{(b)}\quad ((B\eta,\eta)) = \textstyle\sum_\alpha |c_\alpha[\eta]|^2;\end{aligned} \tag{2.29}$$

(ii) *if* $\eta \in F_{j+1}$, *and* $\zeta \in F_j$, *then*

$$((B\eta,\zeta)) = \textstyle\sum_\alpha c_\alpha[\eta]\overline{c_\alpha[\zeta]}; \tag{2.30}$$

(iii) *if* $\eta \in F_{j+1}$, *then* $B(t)\eta(t) \equiv 0$ *on* $[a,b]$ *if and only if* $((B\eta,y)) = 0$ *for all eigenfunctions* $y(t)$ *of* (2.1).

Boundary problems (2.1) which are LF_j-definite possess properties similar to those established above for BF_j-definite problems. They may be proved by methods which are quite analogous to those presented above, although the specific results for the two types of problems are appreciably different.

In the first place, as noted above, for any LF_j-definite problem $\lambda = 0$ is not an eigenvalue. Also, for such a problem the non-negativeness of $((L[\eta],\eta))$ of F_j implies the Cauchy-Bunyakovsky-Schwarz inequality

$$|((L[\eta_1],\eta_2))|^2 \leq ((L[\eta_1],\eta_1))((L[\eta_2],\eta_2)), \text{ for } \eta_\beta \in F_j,\ \beta = 1,2.$$

Corresponding to results noted above for BF_j-definite problems, if (2.1) is LF_j-definite and $\eta \in F_j$ with $((L[\eta],\eta)) = 0$, then $((B\eta,\eta)) = 0$; also, if $\eta \in F_{j+1}$ with $((L[\eta],\eta)) > 0$, and ζ is a vector function of F_j satisfying (2.9) with η, then $((By_0,y_0)) \neq 0$ and $((L[y_0],y_0)) > 0$ for y_0 equal to at least one of the vector functions η, ζ, and $\eta + \zeta$ of F_j.

Corresponding to the results of Lemma 2.1 and Theorem 2.1

for BF_j-definite systems, for LF_j-definite boundary problems we have the following results.

LEMMA 2.2. *Suppose that* (2.1) *is* LF_j*-definite, and* $y_o \in F_j$. *For the vector functions* $y_\alpha(t)$ *in* $F_{j+\alpha}$ *defined by* (2.15), *then the (Schwarz) constants*

$$V_\alpha = ((L[y_o],y_\alpha)), \quad (\alpha = 0,1,\ldots), \tag{2.31}$$

are real, and possess the following properties:

$$V_{\alpha+\beta} = ((L[y_\alpha],y_\beta)), \quad (\alpha,\beta = 0,1,\ldots); \tag{2.32}$$

$$V^2_{2\alpha-1} \le V_{2\alpha}V_{2\alpha-2}, \quad V^2_{2\alpha} \le V_{2\alpha-2}V_{2\alpha+2}, \quad (\alpha = 1,2,\ldots); \tag{2.33}$$

if $V_2 \neq 0$, *then* $V_{2\alpha} > 0$ *and* $V_{2\alpha} \ge V_o(V_2/V_o)^\alpha$,

$$(\alpha = 0,1,\ldots); \tag{2.34}$$

if $V_1 \neq 0$, *then* $V_{2\alpha} > 0$, $(\alpha = 0,1,\ldots)$,

$$\textit{and } V_o/V_2 \le V_o^2/V_1^2. \tag{2.35}$$

THEOREM 2.3. *Suppose that the boundary problem* (2.1) *is* LF_j*-definite. Then either* $((B\eta,\eta)) = 0$ *for all* $\eta \in F_j$ *and* (2.1) *has no eigenvalues, or there is an eigenvalue* λ_1 *of this problem such that we have*

$$|((B\eta,\eta))| \le |\lambda_1|^{-1}((L[\eta],\eta)), \text{ for } \eta \in F_j. \tag{2.38}$$

If (2.1) *has eigenvalues* $\lambda_1,\ldots,\lambda_k$ *and* $\Delta_o\{\lambda_1,\ldots,\lambda_k\}$ *denotes the set of all* $\eta \in F_j$ *satisfying* $((B\eta,y)) = 0$ *for all eigenvectors* y *corresponding to an eigenvalue of the set* $\lambda_1,\ldots,\lambda_k$, *then either* $((B\eta,\eta)) = 0$ *for all* $\eta \in \Delta_o\{\lambda_1,\ldots,\lambda_k\}$, *and* (2.1) *has no other eigenvalues, or there exists an eigenvalue* λ_{k+1} *distinct from* $\lambda_1,\ldots,\lambda_k$, *and such that for arbitrary* $\eta \in \Delta_o\{\lambda_1,\ldots,\lambda_k\}$ *we have*

$$|((B\eta,\eta))| \leq |\lambda_{k+1}|^{-1}((L[\eta],\eta)). \tag{2.37}$$

If an LF_j-definite problem (2.1) possesses eigenvalues, again the eigenvalues and corresponding eigenfunctions are labelled as a simple sequence (2.26), with each eigenvalue repeated a number of times equal to its index of compatibility, and the corresponding eigenvectors are chosen as linearly independent solutions for the eigenvalue. Since $0 < ((L[y],y)) = \lambda((By,y))$ for an eigenfunction y corresponding to the eigenvalue λ of a LF_j-definite problem, the eigenfunctions of the sequence (2.26) may now be selected to satisfy the condition

$$((By^{(\alpha)},y^{(\beta)})) = \frac{|\lambda_\alpha|}{\lambda_\alpha}\,\delta_{\alpha\beta}, \quad (\alpha,\beta = 1,2,\ldots), \tag{2.38}$$

and such a choice will be assumed for the further discussion. With this choice, the corresponding Fourier coefficients of an n-dimensional vector function η are given by

$$d_\alpha[\eta] = \frac{|\lambda_\alpha|}{\lambda_\alpha}\,((B\eta,y^{(\alpha)})), \quad (\alpha = 1,2,\ldots), \tag{2.39}$$

where again it is to be emphasized that the sequence of eigenvalues and corresponding eigenvectors may be vacuous, finite, or denumerably infinite.

Corresponding to Theorem 2.2, for LF_j-definite systems we have the following results.

THEOREM 2.4. *If* (2.1) *is* LF_j*-definite, then:*

(i) *if* $\eta \in F_j$, *then*

$$\begin{aligned} &(a) \quad ((L[\eta],\eta)) \geq \textstyle\sum_\alpha |\lambda_\alpha|\,|d_\alpha[\eta]|^2, \\ &(b) \quad ((B\eta,\eta)) = \textstyle\sum_\alpha (|\lambda_\alpha|/\lambda_\alpha)\,|d_\alpha[\eta]|^2; \end{aligned} \tag{2.40}$$

(ii) *if* $\eta \in F_{j+1}$ *and* $\zeta \in F_j$, *then*

$$((L[\eta],\zeta)) = \sum_\alpha |\lambda_\alpha| d_\alpha[\eta]\overline{d_\alpha[\zeta]}; \qquad (2.41)$$

in particular, if $\eta \in F_{j+1}$ *then the equality sign holds in* (2.40a).

(iii) *if* $\eta \in F_{j+1}$, *then* $B(t)\eta(t) \equiv 0$ *on* $[a,b]$ *if and only if* $((B\eta,y)) = 0$ *for all eigenfunctions* y *of* (2.1).

3. Comments on related literature

As noted in the Introduction of this chapter, Bliss initiated in 1926 the study of boundary problems of the form (1.1) which are "definitely self-adjoint under a real non-singular transformation". This investigation was motivated by the desire to develop a theory of two-point boundary problems that would encompass the accessory boundary problem associated with the Jacobi condition for variational problems of Lagrange or Bolza type. In turn, the study of such differential systems engendered the consideration of corresponding types of matrix integral equations in Reid [1], Wilkins [1], Zimmerberg [3,6], and later the work of Zaanen [1,2,3] and Reid [11] on symmetrizable compact linear transformations in Hilbert space.

Also, as noted in Section 6 of Chapter VI, for variational problems of Lagrange or Bolza type with variable end-conditions the accessory boundary problem may be phrased quite naturally in a form wherein the characteristic parameter enters linearly into the boundary conditions. Corresponding generalizations of Bliss' definitely self-adjoint problems involving the parameter linearly in the boundary conditions were explored in the Chicago dissertations of Bobonis [1] and Zimmerberg [1]; this study was later continued by Zimmerberg [4,5]. As mentioned in the Introduction of this chapter, the works of Hölder [3,4] and Reid [9] include as special instances many results on self-adjoint scalar differential equations, such as Kamke [6]; in this regard see also Collatz [1-Ch.]; related considerations of boundary problems associated with higher order scalar equations are to be found in Zimmerberg [2], Schubert [1], and Sloss [1]. This area of study of boundary problems has been greatly generalized by Schäfke and A. Schneider [1], and A. Schneider [1], whose results will now be briefly surveyed.

The general eigenvalue problem considered in paper I of Schäfke and Schneider is written as

$$Fy = \lambda Gy, \quad y \in \mathscr{F}_0, \quad y \notin \mathscr{R}_0, \tag{3.1}$$

where: (i) $\mathscr{G}_0$, $\mathscr{F}_0$, $\mathscr{A}_0$ are complex linear spaces having $\mathscr{G}_0 \supset \mathscr{F}_0$, with $[f,g]$ a complex inner product for f, $g \in \mathscr{A}_0$, while G and S are linear mappings of $\mathscr{G}_0$ into $\mathscr{A}_0$ and F is a linear mapping of $\mathscr{F}_0$ into $\mathscr{A}_0$; (ii) $\mathscr{R}_1 = \{y : y \in \mathscr{F}_0, Fy \in G\,\mathscr{G}_0\}$, $\mathscr{R}_0 = \{y : y \in \mathscr{F}_0, Fy = Gy = 0\}$; (iii) (3.1) is S-hermitian in the sense that $F(u,v) = [Fu,Sv]$ and $G(u,v) = [Gu,Sv]$ are hermitian forms on $\mathscr{R}_1$. Such a problem is called "normal" whenever $G(y,y) \neq 0$ for all eigensolutions of (3.1) not belonging to $\mathscr{R}_0$. The problem (3.1) is called "reducible" whenever $\lambda = 0$ is not an eigenvalue, $F\,\mathscr{F}_0 \supset G\,\mathscr{G}_0$, and there is a subspace $\mathscr{R}_{10}$ such that $\mathscr{R}_1$ is the direct sum of $\mathscr{R}_0$ and $\mathscr{R}_{10}$. Also, (3.1) is of "finite defect" if for each λ its defect indices $\delta_1(\lambda)$, the dimension of the null space of $F - \lambda G$, and $\delta_2(\lambda)$, the dimension of the factor space $\mathscr{A}_0/(G-\lambda G)\,\mathscr{F}_0$, are finite, and

$\delta_1(\lambda) - \delta_2(\lambda)$ is a constant independent of λ. Finally, an S-hermitian problem is called "left-definite" when there exist sub-spaces $\mathscr{R}$, $\mathscr{F}$, $\mathscr{G}$ satisfying $\mathscr{R}_1 \subset \mathscr{R} \subset \mathscr{F} \subset \mathscr{G} \subset \mathscr{G}_0$ such that $F(u,v)$, $G(u,v)$ are hermitian on $\mathscr{F}$ and $\mathscr{G}$, respectively, while $F(u,u) \geq 0$ for $u \in \mathscr{R}$; correspondingly, such a problem is called right-definite when there is a subspace $\mathscr{R}$ satisfying $\mathscr{R}_1 \subset \mathscr{R} \subset \mathscr{G}$ and $G(u,u) \geq 0$ for $u \in \mathscr{R}$. If (3.1) is S-hermitian, normal, reducible, of finite defect, and definite, then Schäfke and Schneider showed that the problem is equivalent to an eigenvalue problem of the form $y = \lambda Ay$ to which the central result of Wielandt [1] on problems with real discrete eigenvalues is applicable to provide a spectral representation theorem from which one may deduce readily results on the existence and extremizing properties of eigenvalues and associated iteration procedures. The general results were applied to a class of differential boundary problems

$$\begin{aligned} C_1(t)y'(t) + D_1(t)y(t) &= \lambda[C_2(t)y'(t) + D_2(t)y(t)], \\ A_1y(a) + B_1y(b) &= \lambda[A_2y(a) + B_2y(b)], \end{aligned} \tag{3.2}$$

in the n-dimensional vector function $y(t)$, where for $t \in [a,b]$ the continuous matrix functions $C_\alpha(t)$, $D_\alpha(t)$, $(\alpha = 1,2)$, are $n \times n$ and $C_1(t) - \lambda C_2(t)$ is non-singular for arbitrary complex λ, while A_α, B_α are $m \times n$ constant matrices. This problem may be placed in the above described algebraic setting, with $\mathscr{F}_0 = \mathop{\times}\limits_{\nu=1}^{n} \mathscr{C}'[a,b] \subset \mathscr{G}_0 \subset \mathop{\times}\limits_{\nu=1}^{n} \mathscr{C}[a,b]$; $\mathscr{H} = \mathop{\times}\limits_{\nu=1}^{n} \mathscr{C}[a,b] \times R^m$; $f = (f_1(t);f_2)$, $g = (g_1(t);g_2)$; $[f,g] = \int_a^b g_1^*(t)f_1(t)dt + g_2^*f_2$; $Fy = (C_1(t)y'(t) + D_1(t)y(t); A_1y(a) + B_1y(b))$; $Gy = (C_2(t)y'(t) + D_2(t)y(t); A_2y(a) + B_2y(b))$, and $Sy = (C_3(t)y'(t) + D_3(t)y(t); A_3y(a) + B_3y(b))$. Included as special instances of such boundary problems are the definitely self-adjoint systems considered by Bliss [2,6], Reid [8,9,16], Zimmerberg [1,4,5], the canonical systems of Hölder [2,3], and consequently the boundary problems for higher order self-adjoint scalar differential equations cited above.

Paper II of Schäfke and Schneider continues the study of boundary problems of the form (3.2), with detailed attention given to the structure of such systems that are S-hermitian in the normal case, in the sense that there exist $n \times n$ matrix functions $C_3(t)$, $D_3(t)$ and $n \times n$ constant matrices A_3, B_3 such that there is a non-singular, continuously differentiable skew-hermitian matrix function $H(t)$ satisfying for all real λ the conditions:

$$\begin{aligned}&[C_3v' + D_3v]^*[(C_1 - \lambda C_2)u' + (D_1 - \lambda D_2)u]\\ &\qquad - [(C_1 - \lambda C_2)v' + (D_1 - \lambda D_2v)]^*[C_3u' + D_3u] \equiv [v^*Hu]',\\ &[A_3v(a) + B_3v(b)]^*[(A_1 - \lambda A_2)u(a) + (B_1 - \lambda B_2)u(b)]\\ &\qquad - [(A_1 - \lambda A_2)v(a) + (B_1 - \lambda B_2)v(b)]^*[A_3u(a) + B_3u(b)]\\ &\qquad\qquad = v^*(a)H(a)u(a) - v^*(b)H(b)u(b).\end{aligned}$$

In terms of such problems there is presented a classification and ordering of various types of boundary problems which have been studied in the past. Finally, the third paper of Schäfke and Schneider, and the paper of A. Schneider [1], are concerned with matters related to the character of the resolvent function for boundary problems of the sort considered in the first two papers of Schäfke and Schneider.

For brevity, no detailed discussion will be included of the relations between the various hypotheses of the papers of Schäfke and Schneider and those mentioned in the preceding sections. It is clear, however, that the "normal" condition of Schäfke and Schneider corresponds to the concept of "full self-adjointness" mentioned in Section 2. Also, the replacement of a given problem by a corresponding non-degenerate one in paper III is an abstraction of the concept of replacing a given boundary problem (V.3.1) by a corresponding normal boundary problem as discussed in Section V.2.

4. Topics and Exercises

1. The boundary problem of Example 2.1 is BF_o-definite whenever the matrix function $K(t)$ is non-negative definite for $t \in [a,b]$ and $K(t) \not\equiv 0$ on arbitrary non-degenerate subintervals of $[a,b]$, while the only solution of $v'(t) + A^*(t)v(t) = 0$ satisfying $B(t)v(t) \equiv 0$ on $[a,b]$ is $v(t) \equiv 0$. In particular, such a system is BF_o-definite if $K(t)$ is positive definite and $B(t)$ is non-singular for $t \in [a,b]$. {Reid [35-Prob. IV.6.3]}.

2. If the boundary problem (2.1) is BF_o-definite, and the matrix function $A_2(t)$ is of class $\mathscr{C}^1[a,b]$, then:

(i) $B(t) \geq 0$ for $t \in [a,b]$;

(ii) if $F^o_o = C_n[a,b]$, and for $j = 1,2,\ldots$ the class F^o_j is defined as the set of n-dimensional vector functions η for which there exists a corresponding $\zeta \in F^o_{j-1}$ satisfying with η the differential system (2.9), then the results of Theorems 2.1, 2.2 hold for $j = 0$ and F_j replaced by F^o_j;

(iii) the expansion (2.30) holds for any $\eta \in \mathscr{L}_n^2[a,b]$ which is such that for an arbitrary $\varepsilon > 0$ there exists an $\eta_\varepsilon \in F_1^0$ satisfying $((B[\eta-\eta_\varepsilon],\eta-\eta_\varepsilon)) < \varepsilon$;

(iv) if $B(t)$ is non-singular for $t \in [a,b]$, then the expansion (2.30) holds for arbitrary $\eta \in \mathscr{L}_n^2[a,b]$; in particular, $c_\alpha[\eta] = 0$, $(\alpha = 1,2,\ldots)$, for such an η if and only if $\eta(t) = 0$ for t a.e. on $[a,b]$. Moreover,

$$((L[\eta],\eta_1)) = \textstyle\sum_\alpha \lambda_\alpha c_\alpha[\eta]\overline{c_\alpha[\eta_1]} \tag{4.1}$$

for $\eta_1 \in \mathscr{L}_n^2[a,b]$, and η a solution of a system (2.9) with $\eta_1 \in \mathscr{L}_n^2[a,b]$.

(v) if λ is not an eigenvalue of (2.1), then the infinite series

$$\textstyle\sum_\alpha |\lambda_\alpha - \lambda|^{-2}|y^{(\alpha)}(t)|^2 \tag{4.2}$$

converges for $t \in [a,b]$, and does not exceed $\int_a^b \mathrm{Tr}\{G(t,s;\lambda)B(s)G(s,t;\lambda)\}ds$, so that for each such λ there is a value $c(\lambda)$ such that the sum of this series does not exceed $c(\lambda)$.

(vi) if η is a solution of a system (2.9) with $\zeta \in \mathscr{L}_n^2[a,b]$, then the infinite series

$$\textstyle\sum_\alpha c_\alpha[\eta]y^{(\alpha)}(t) \tag{4.3}$$

converges uniformly on $[a,b]$; moreover, if $B(t)$ is non-singular for $t \in [a,b]$, then the sum of the series (4.3) is $\eta(t)$. {Reid [35-Prob. IV.6.8]}.

3. With the understanding that each eigenvalue is counted a number of times equal to its index of compatibility, a BF_j-definite problem has at least ρ eigenvalues if and only if $((B\eta,\eta))$ is positive definite on a linear manifold in F_{j+1}

of dimension ρ. {Reid [35-Prob. IV.6.5]}.

4. If (2.1) is BF_j-definite then $((B\zeta,\zeta)) \geq \sum_\alpha |c_\alpha[\eta]|^2$ for arbitrary $\zeta \in F_j$; also, if $\eta \in F_{j+1}$ then $\sum_\alpha \lambda_\alpha^2 |c_\alpha[\eta]|^2$ converges and $((B\zeta,\zeta)) \geq \sum_\alpha \lambda_\alpha^2 |c_\alpha[\eta]|^2$, where ζ is a vector function of F_j which satisfies with η system (2.9). {Reid [35-Prob. IV.6.6]}.

5. For (2.1) BF_j-definite with $\lambda = 0$ not an eigenvalue, consider the sequence (2.26) labelled as

$$\lambda_\beta,\ y^{(\beta)}(t), \quad (\beta = \pm 1, \pm 2, \dots),$$

with

$$\leq \lambda_{-2} \leq \lambda_{-1} < 0 < \lambda_1 \leq \lambda_2 \leq \dots$$

it being understood that the individual sequences $\lambda_1, \lambda_2, \dots$ and $\lambda_{-1}, \lambda_{-2}, \dots$ may be either vacuous or finite. If $\eta \in F_{j+1}$, and ζ is a vector function of F_j satisfying $L[\eta](t) = B(t)\zeta(t)$, $t \in [a,b]$, then whenever $c_\beta[\eta] \equiv ((B\eta, y^{(\beta)})) = 0$ for $-p \leq \beta \leq q$, then

$$[1/\lambda_{-p-1}]((B\zeta,\zeta)) \leq ((L[\eta],\eta)) \leq [1/\lambda_{q+1}]((B\zeta,\zeta)),$$

with suitable interpretation of the inequality whenever there are only a finite number of positive eigenvalues, or only a finite number of negative eigenvalues. {Reid [35-Prob. IV.6.7]}.

6. Suppose that a BF_j-definite problem (2.1) is such that there is a real constant λ_0 such that this system has no eigenvalue less than λ_0, and let the set (2.26) be ordered so that $\lambda_1 \leq \lambda_2 \leq \dots$ If $\Delta_1 = \{\eta : \eta \in F_{j+1}, ((B\eta,\eta)) = 1\}$, and Δ_1 is non-vacuous, then λ_1 exists and is the minimum of $((L[\eta],\eta))$ on Δ_1; moreover, this minimum is attained by an $\eta \in \Delta_1$ if and only if $\eta(t) = y(t) + \eta_0(t)$ where y is

an eigenvector of (2.1) for $\lambda = \lambda_1$ and $\eta_0 \in F_{j+1}$ with $B(t)\eta_0(t) \equiv 0$ on $[a,b]$. In general, if eigenvalues $\lambda_1,\dots,\lambda_k$ exist, and $\Delta_{k+1} = \{\eta : \eta \in F_{j+1}, ((B\eta,\eta)) = 1, c_\alpha[\eta] = ((B\eta,y^{(\alpha)})) = 0, \alpha = 1,\dots,k\}$, then whenever Δ_{k+1} is nonvacuous the eigenvalue λ_{k+1} exists and is the minimum of $((L[\eta],\eta))$ on Δ_{k+1}; moreover, this minimum is attained by an $\eta \in \Delta_{k+1}$ if and only if $\eta(t) = y(t) + \eta_0(t)$, where y is an eigenvector of (2.1) for $\lambda = k+1$, and $\eta_0 \in F_{j+1}$ with $B(t)\eta_0(t) \equiv 0$ on $[a,b]$.

7. Consider a system (2.1) wherein the $n \times n$ matrix coefficient functions $B(t)$, $A_0(t)$, $A_1(t)$, $A_2(t)$ and the $n \times 2n$ matrix $\mathscr{M}$ satisfy the conditions listed in the opening paragraph of Section 2, and also the further conditions.

(i) (2.1) is self-adjoint, so that $\mathscr{D}(L) = \mathscr{D}(L^*)$ and $L[y](t) = L^*[y](t)$ for $y \in \mathscr{D}(L) = \mathscr{D}(L^*)$; moreover, $B(t) \equiv B^*(t)$.

(ii) $B(t) \geq 0$ for $t \in [a,b]$.

(iii) If $[a_0,b_0]$ is a non-degenerate subinterval of $[a,b]$, and $L[y](t) = 0$, $B(t)y(t) = 0$ for $t \in [a_0,b_0]$, then $y(t) \equiv 0$ on $[a,b]$.

(iv) $\lambda = 0$ is not an eigenvalue of (2.1).

Then $(y(t);z(t))$ is a solution of the differential system

$$\begin{aligned} &\text{(a)} \quad L[z](t) = B(t)y(t), \quad L[y](t) = B(t)z(t), \\ &\text{(b)} \quad \mathscr{M}\hat{u}_y = 0, \quad \mathscr{M}\hat{u}_z = 0, \end{aligned} \tag{4.4}$$

if and only if $u = u_y = A_2 y$, $v = v_z = A_1^* z$ is a solution of the system

$$\begin{aligned} &\text{(a)} \quad -v'(t) - A^*(t)v(t) = \lambda K(t)u(t), \\ &\qquad u'(t) - A(t)u(t) - B(t)v(t) = 0, \\ &\text{(b)} \quad \mathscr{M}\hat{u} = 0, \quad P^*[\operatorname{diag}\{-E_n, E_n\}]\hat{v} = 0, \end{aligned} \tag{4.5}$$

where $A = -A_1^{-1}A_0A_2^{-1}$, $B = A_1^{-1}BA_1^{*-1}$, $K = A_2^{*-1}BA_2^{-1}$, and $\mathscr{P}$ is a $2n \times n$ matrix of rank n which satisfies $\mathscr{M}\mathscr{P} = 0$. Consequently, (4.5) is a Hamiltonian system $\mathscr{B}$ of the form (VI.3.1) in $(u;v)$, with S the linear subspace of C_{2n} specified by $\mathscr{M}\hat{u} = 0$, and the $2n \times 2n$ hermitian matrix Q of the boundary conditions (VI.3.1ii) the zero matrix. Condition (iii) above implies that the system of differential equations in (4.5) is identically normal on $[a,b]$, and in view of (ii) and (iv) the corresponding functional $J[\eta:\mathscr{B}] = \int_a^b \zeta^*B\zeta dt$ is positive definite on the class $D_e[\mathscr{B}]$. With the aid of the result of Exercise VI.7.13, it then follows that the boundary problem (4.5) satisfies hypotheses $(\mathscr{A}_{\mathscr{B}})$ and $(\mathscr{A}_K'i,ii,iii)$ of Section VI.3, and consequently there exists an infinite sequence of eigenvalues and eigenvectors $\{\lambda_j, u_j, v_j\}$ satisfying the conclusions of Theorem VI.3.1. In particular, $\lambda_j > 0$ for $j = 1,2,\ldots$ Also, if $(y_1(t);z_1(t))$ is a solution of (4.4) for a value $\lambda > 0$ then $y(t) = y_1(t) + \lambda^{-1/2}z_1(t)$ is a solution of (2.1) for $\lambda = \lambda^{1/2}$, and $y(t) = y_1(t) - \lambda^{-1/2}z_1(t)$ is a solution of (2.1) for $\lambda = -\lambda^{1/2}$. Moreover, λ is an eigenvalue of (4.4) iff either $\lambda = \lambda^{1/2}$ or $\lambda = -\lambda^{1/2}$ is an eigenvalue of (2.1), and the index of λ as an eigenvalue of (4.4) is equal to the sum of the indices of $-\lambda^{1/2}$ and $\lambda^{1/2}$ as eigenvalues of (2.1).

Finally, for $\lambda \neq 0$ let $Y(t,\lambda)$ denote the solution of the matrix differential system $L[Y](t) = \lambda B(t)Y(t)$,

$Y(a) = E$, and set $Y_0(t,\lambda) = (2\lambda)^{-1}[Y(t,\lambda) - Y(t,-\lambda)]$, $Z_0(t,\lambda) = 2^{-1}[Y(t,\lambda) + Y(t,-\lambda)]$, $U_0(t,\lambda) = A_2(t)Y_0(t,\lambda)$, $V_0(t,\lambda) = A_1^*(t)Z_0(t,\lambda)$. Then $U_0(t) = U_0(t,\lambda)$, $V_0(t) = V_0(t,\lambda)$ is a solution of the matrix differential system

$$-V_0'(t) - A^*(t)V_0(t) = \lambda^2 K(t)U_0(t),$$

$$U_0'(t) - A(t)U_0(t) - B(t)V_0(t) = 0$$

$$U_0(a) = 0, \quad V_0(a) = A_1^*(a)$$

and consequently a value $\tau \in (a,b]$ is conjugate to $t = a$ relative to (4.5a) for $\lambda = \lambda^2$ iff $Y_0(\tau,\lambda)$ is singular, and the order of τ as a conjugate point to $t = a$ is equal to k if $Y_0(\tau,\lambda)$ has rank $n - k$. {Reid [35-Prob. VII.12.8]. A corresponding result in the setting of a generalized differential system of the type considered in Section VIII.3, is presented in Th. 7.3 of Reid [25]}.

CHAPTER VIII. GENERALIZATIONS OF STURMIAN THEORY

1. Introduction

This chapter is devoted to a brief survey of certain areas to which the Sturmian theory has been extended, and which are intimately related to the subject matter of the preceding chapters. Briefly stated, we shall mention related problems for integro-differential systems, a type of generalized differential equation that in major instances is equivalent to a Riemann-Stieltjes integral equation, quadratic functional theory in Hilbert space due to Hestenes, interrelations with the Weinstein theory of intermediate problems, oscillation theory in the context of a B*-algebra and the topological interpretation of the Sturm theorems in terms of intersection numbers and deformation cycles that has emanated from the 1956 paper of R. Bott [1].

2. Integro-differential boundary problems

Only recently did the author become aware of a paper published by Liouville [2] in 1837 on the solution of the partial integro-differential equation representing a problem in thermomechanics, and which upon using the usual separation of variables technique led to the auxiliary one-dimensional integro-differential boundary problem which we may write as

$$\begin{aligned} &u''(x) + \rho^2\left[u(x) + \kappa x \int_0^1 su(s)ds\right] = 0, \\ &u(0) = 0, \quad u'(1) + hu(1) = 0. \end{aligned} \tag{2.1}$$

Applying the methods used by himself and Sturm for boundary problems involving second order ordinary differential equations, Liouville determined for (2.1) the asymptotic nature of eigenvalues, eigenfunctions, and the associated expansion in eigenfunctions needed for the solution of the thermo-mechanical problem.

Relatively early in the twentieth century, (see Fubini [1]; Courant [2]), it was realized that for certain variational problems the "accessory problem" was a boundary problem involving a self-adjoint integro-differential equation. In particular, Lichtenstein [2] treated by means of the Hilbert theory of quadratic forms in infinitely many variables a boundary problem involving a single self-adjoint linear integro-differential equation of the second order and a special set of two-point boundary conditions. Under certain conditions he established the existence of infinitely many eigenvalues, together with an expansion theorem for functions in terms of the corresponding eigenfunctions. Subsequently, Lichtenstein [4] used the results of his earlier paper to establish by expansion methods sufficient conditions for a weak relative minimum for a simple integral isoperimetric problem of the calculus of variations. Courant [2-Secs. 5,13] treated by means of difference equations an integro-differential boundary problem similar to that considered by

Lichtenstein [1]. A few years after Lichtenstein's paper [4] the author [7] considered a self-adjoint boundary problem involving a system of integro-differential equations and two-point boundary conditions, and in addition to the proof of existence of infinitely many eigenvalues established comparison and oscillation theorems which are generalizations of such theorems of the classical Sturmian theory, and contain as special instances the comparison and oscillation theorems of the Morse generalizations of such theorems to self-adjoint differential systems. Indeed, as noted in Section 13 of Chapter V, the paper [7] of the author presented for the first time such theorems not involving any assumption of normality on subintervals. Recently the author [46] reconsidered such integro-differential boundary problems in the general context of a Hamiltonian system with two-point boundary conditions, and which in terminology analogous to that of Section 6 of Chapter V may be described as the "Euler-Lagrange" system for the hermitian functional

$$\int_a^b \{\zeta^*B(t)\zeta + \eta^*C(t)\eta\}dt + \int_a^b\int_a^b \eta^*(t)N(t,s)\eta(s)ds \qquad (2.2)$$

subject to the constraints $L_2[\eta,\zeta](t) \equiv \eta'(t) - A(t)\eta(t) - B(t)\zeta(t) = 0$, $\eta \in S$, where the matrix functions A, B, C and subspace S of $\mathbf{C}_{2n}$ are as in the earlier discussion of ordinary differential equations and $N(t,s)$ is an $n \times n$ matrix function of class $\mathscr{L}^\infty$ on $[a,b] \times [a,b]$ satisfying the hermitian condition $N(t,s) \equiv [N(s,t)]^*$. It is to be remarked that for such integro-differential boundary problems there do not exist results on the existence of solutions satisfying given initial data, as hold in the ordinary differential equation case. Moreover, one of the greatest differences between the theory of such self-adjoint integro-differential systems and the corresponding ordinary differential boundary problems occurring when $N(t,s) \equiv 0$ is that for the latter we have the extremely useful concept of conjugate or conjoined solutions, whereas for the general integro-differential system described above there remains only a meager remnant of this concept. Consequently, for boundary problems involving integro-differential equations the methods of Morse using "broken extremals" are no longer available for the derivation of oscillation and comparison theorems. In particular, for such integro-differential systems involving matrix kernel functions of degenerate form the results involving a given problem and its subproblems are equivalent to corresponding problems and subproblems for an ordinary differential equation problem in a higher number of dimensions, and thus in such cases the comparison theorems for the integro-differential systems are deducible from corresponding comparison theorems for the associated enlarged differential system. As may be illustrated by simple examples, however, for such integro-differential systems the problem of focal points is equivalent to a corresponding focal point problem for the associated differential system, although the specific conjugate point problem for the integro-differential system is not the same as the conjugate point problem for the related differential system. In regard to comparison theorems, it appears that a modified Weinstein method in the general character of that used by Weinberger [1] for ordinary differential boundary problems may be of value.

3. A class of generalized differential equations

Various generalizations of the classical differential equation have been studied. The one to be considered specifically here is intimately related to the ordinary differential systems and boundary problems considered in Chapters V, VI, and is equivalent to a type of linear vector Riemann-Stieltjes integral equation. The real scalar generalized second order differential equations occurring in the works of Sz.-Nagy [1], Feller [1], Kac and Krein [1], and Guggenheimer [1] are particular instances of the general system considered here. Various aspects of the theory of generalized differential systems appear in the author's papers [18, 22, 24, 25, 28, 41, 43, 44, 45].

In the following the $n \times n$ matrix functions A_0, A_1, B, C, M are supposed to be measurable on an interval I on the real line, with B, C, M hermitian, A_1 non-singular while $A_1^{-1}A_0$, $A_1^{-1}B$, C are locally of class L and M is locally of bounded variation on I. The symbol $L[y]$ is used to denote the vector differential expression $L[y](t) = A_1(t)y'(t) + A_0(t)y(t)$, and for I_0 a subinterval of I the symbol $D(I_0)$ signifies the class of n-dimensional vector functions y which are locally absolutely continuous, and for which there is a z locally of class $\mathscr{L}^\infty$ and such that $L[y] - Bz = 0$ on I_0; corresponding to the notation for differential systems in Chapters V, VI, this association of z with y is denoted by $y \in D(I_0):z$. Continuing with notation corresponding to that for ordinary differential systems, if $[a,b]$ is a compact subinterval of I then the subclass of $D[a,b]$ on which $y(a) = 0$ is denoted by D_{0*}, the subclass of $D[a,b]$ on which $y(b) = 0$ is denoted by D_{*0}, and $D_0 = D_{0*} \cap D_{*0}$, with corresponding meanings of the symbols $y \in D_{0*}[a,b]:z$, $y \in D_{*0}[a,b]:z$, and $y \in D_0[a,b]:z$. Attention is restricted to operators with domain $\mathcal{D}$ on linear manifolds satisfying $D_0 \subset \mathcal{D} \subset D$. In particular, $D_0[a,b]$ is the subspace of D on which the 2n-dimensional boundary vector $\hat{y} = (y(a);y(b))$ is zero. In general, if S denotes the set of 2n-dimensional vectors ξ for which there exists an $y \in \mathcal{D}$ satisfying $\hat{y} = \xi$, then S is a subspace of C_{2n} and $\mathcal{D} = \{y:y \in D, \hat{y} \in S\}$. Finally, the symbol D^* is used to denote the class of n-dimensional vector functions z that are locally of class $\mathscr{L}^\infty$ and for which there exists a v_z locally of bounded variation and satisfying $z = (A_1^*)^{-1}v_z$.

The generalized differential system to be considered may then be written as

$$\Delta[y,z](t) \equiv -dv_z(t) + [C(t)y(t)+A_0^*(t)z(t)] + [dM(t)]y(t) = 0,$$

$$L[y,z](t) \equiv A_1(t)y'(t) + A_0(t)y(t) - B(t)z(t) = 0. \qquad (3.1)$$

By a solution $(y;z)$ of (3.1) is meant a pair of n-dimensional vector functions y, z with y locally absolutely continuous, $z \in D^*$ and satisfying with y the ordinary differential equation $L[y,z](t) = 0$ and the Riemann-Stieltjes integral equation

$$v_z(t) = v_z(\tau) + \int_\tau^t \{C(s)y(s) + A_0^*(s)z(s)\}ds + \int_\tau^t [dM(s)]y(s) \qquad (3.2)$$

for $(t,\tau) \in I \times I$. In general, $n \times r$ matrix functions Y, Z are a solution of the corresponding matrix generalized differential system

$$\Delta[Y,Z](t) = 0, \quad L[Y,Z](t) = 0 \tag{3.1$_M$}$$

if each column vector of the $2n \times r$ matrix function $(Y(t);Z(t))$ is a solution of (3.1).

In case $M(t)$ is locally absolutely continuous and $P(t)$ is matrix function locally integrable and such that $M(t) = M(\tau) + \int_\tau^t P(s)ds$, then $(y;z)$ is a solution of (3.1) if and only if it is a solution of the ordinary differential equation system

$$\begin{aligned} L^*[z](t) + [C(t) + P(t)]y(t) &= 0, \\ L[y](t) - B(t)z(t) &= 0, \end{aligned} \tag{3.2}$$

where $L^*[Z]$ is the adjoint operator $-[A^*(t)z(t)]' + A_o^*(t)z(t)$. Another frequently occurring case is $M(t) = G(t) + \int_\tau^t P(s)ds$, where $P(t)$ is locally integrable and $G(t)$ is a step function; that is, for a given compact subinterval $[a,b]$ of I there is a finite sequence of values $a = t_o < t_1 < \dots < t_k < t_{k+1} = b$ such that $G(t)$ is constant on each subinterval $(t_\alpha, t_{\alpha+1})$, $(\alpha = 0,1,\dots,k)$. In this case $(y;z)$ is a solution of (3.1) if and only if this pair is a solution of (3.2) on each subinterval $(t_\alpha, t_{\alpha+1})$, while the right- and left-hand limits of these functions satisfy the interface conditions

$$\begin{aligned} y(t_\beta) &= y(t_\beta^-) = y(t_\beta^+), \\ v_z(t_\beta) - v_z(t_\beta^-) &= [G(t_\beta) - G(t_\beta^-)]y(t_\beta), \\ v_z(t_{\beta-1}^+) - v_z(t_{\beta-1}) &= [G(t_{\beta-1}^+) - G(t_{\beta-1})]y(t_{\beta-1}), \\ &(\beta = 1,\dots,k+1). \end{aligned} \tag{3.3}$$

In particular, suppose that $A_1(t) \equiv B(t) \equiv E_n$, $A_0(t) \equiv 0$, and $M(t)$ is a step function matrix function $G(t)$ which is constant on each of the open subintervals $(t_{\beta-1}, t_\beta)$, $(\beta = 1,\ldots,k+1)$. If $G_0 = G(a^+) - G(a)$, $G_{k+1} = G(b) - G(b^-)$, and $G_i = G(t_i^+) - G(t_i^-)$, $(i = 1,\ldots,k)$, then $(y;z)$ is a solution of (3.1) on $[a,b]$ if and only if y is the polygonal vector function whose graph joins the successive points $(t_j, y(t_j))$, $(j = 0,1,\ldots,k+1)$, and the values $y(t_j)$ satisfy the linear second order difference system

$$\frac{y(t_{i+1}) - y(t_i)}{t_{i+1} - t_i} - \frac{y(t_i) - y(t_{i-1})}{t_i - t_{i-1}} - G_i y(t_i) = 0, \quad (i = 1,\ldots,k) \tag{3.4}$$

while $z(t) = y'(t) = [t_{\alpha+1} - t_\alpha]^{-1}[y(t_{\alpha+1}) - y(t_\alpha)]$ on $(t_\alpha, t_{\alpha+1})$, $(\alpha = 0,1,\ldots,k)$, $z(a) = [t_1 - t_0]^{-1}[y(t_1) - y(t_0)] - G_0 y(a)$, and $z(t_{\alpha+1}) = [t_{\alpha+1} - t_\alpha]^{-1}[y(t_{\alpha+1}) - y(t_\alpha)] + [G(t_{\alpha+1}) - G(t_\alpha)]y(t_{\alpha+1})$ for $\alpha = 0,1,\ldots,k$.

The fact that the theory of generalized differential systems is related to that of ordinary differential systems considered in Chapters V and VI is a consequence of the fact that for a given hermitian $2n \times 2n$ matrix Q and subspace S of $\mathbf{C}_{2n}$ the generalized differential system

$$\Delta[y,z](t) = 0, \quad L[y,z](t) = 0, \quad t \in [a,b]$$
$$\hat{y} \in S, \quad T[y,z] \equiv Q\hat{y} + [\operatorname{diag}\{-E_n, E_n\}]\hat{v}_z \in S^\perp$$

is the "Euler-Lagrange system" for the hermitian functional

$$J[y;a,b] = \hat{y}^*Q\hat{y} + \int_a^b \{z^*(t)B(t)z(t) + y^*(t)C(t)y(t)\}dt + \int_a^b y^*(t)[dM(t)]y(t)$$

subject to the restraints $y \in D[a,b]:z$, $\hat{y} \in S$. An even more formal reason is that under the stated hypotheses one may show that if $f \in \mathscr{L}[a,b]$ then $(y;z)$ is a solution of the non-homogeneous generalized differential system

$$\Delta[y,z](t) = f(t)dt, \quad L[y,z](t) = 0, \quad t \in [a,b]$$
$$y \in \hat{S}, \quad T[y,z] \in S^{\perp}$$

if and only if $(u^o;v^o) = (y;v_z - My)$ is a solution of the ordinary differential system

$$L_1^o[u^o,v^o](t) \equiv -v^{o\prime}(t) + C^o(t)u^o(t) - A^{o*}(t)v^o(t) = f(t)$$
$$L_2^o[u^o,v^o](t) \equiv u^{o\prime}(t) - A^o(t)u^o(t) - B^o(t)v^o(t) = 0$$
$$\hat{u}^o \in S, \ T^o[u^o,v^o] \equiv Q^o\hat{u}^o + [\operatorname{diag}\{-E_n,E_n\}]\hat{v}^o \in S^{\perp}$$

where $Q^o = Q + \operatorname{diag}\{-M(a),M(b)\}$. In spite of this possible reduction, however, there are distinct advantages in considering a generalized differential system in the context described above.

For a generalized differential system that is identically normal one may establish results of the Morse theory corresponding to those of Section V.8, (see Reid [18]), although at this point appears one fundamental difference in the treatment of ordinary differential systems and the generalized differential systems considered here. In the former case the solutions $(u^{(j)}(t);v^{(j)}(t))$ of (V.3.1) determined by the boundary conditions (V.8.2) were such that the $u^{(j)}$ and $v^{(j)}$ were continuous functions of $(t,t_{j-1},x^{j-1},t_j,x^j)$ as long as t_{j-1} and t_j varied in such a manner that the system (V.3.1) remained disconjugate on $[t_{j-1},t_j]$. For the generalized system, however, the component vector function

$y^{(j)}(t)$ of a solution pair $(y^{(j)}(t);z^{(j)}(t))$ possesses a continuity property of the same sort, but the associated vector function $v_z^{(j)}(t)$ does not. However, the coefficient matrices $Q^o_{\alpha\beta}\{\Pi\}$, $Q^{*o}_{\alpha\beta}\{\Pi\}$ and $Q^{o*}_{\alpha\beta}\{\Pi\}$ in the corresponding functional representations still possess the property of being continuous functions of $(t_o,t_1,\ldots,t_{k+1})$ on the set of such values belonging to fundamental partitions of $[a,b]$.

For basic results on disconjugacy for systems (3.1) and associated boundary problems the reader is referred to the papers [18, 22, 25, 43, 44] of the author; also, in papers [41, 43] such systems are employed to derive generalization of the classic Liapunov inequality. Also, since in a suitably defined metric space the solutions of (3.1) possess a completeness property that is not present when the matrix function $M(t)$ is restricted to be locally absolutely continuous, (see Reid [24]), the extremal solutions of certain problems of the sort considered by Krein [1], and Banks [1,3] are solutions of generalized differential systems.

Recently Denny [1] has shown that the major portion of basic results established by the author for generalized differential systems (3.1) remain valid for a class of similar systems that are not reducible to a system of ordinary differential equation. In particular, for $n = 1$ his results provide such extension to systems

$$-dv(t) + p(t)u(t)dt + [dm(t)]u(t) = 0, \quad du(t)-[dh(t)]v(t) = 0,$$

where on $[a,b]$ the functions p, m, h are real-valued with p of class $\mathscr{L}^\infty$, m of bounded variation, and h continuous and monotone.

4. Hestenes quadratic form theory in a Hilbert space

For certain types of quadratic forms in a Hilbert space Hestenes [2,3] has developed a theory of indices that may be applied to the second variation functional of a calculus of variations problem. For a linear space $\mathscr{A}$ over the field of reals, let $Q(x,y)$ denote a symmetric bilinear functional on $\mathscr{A} \times \mathscr{A}$, and denote by $Q(x)$ the corresponding quadratic form $Q(x,x)$. Corresponding to the terminology already used in Section VI.6, two vectors x and y in $\mathscr{A}$ are called Q-*orthogonal* whenever $Q(x,y) = 0$. If $\mathscr{B}$ is a linear subspace of $\mathscr{A}$ then the Q-*orthogonal complement* of $\mathscr{B}$, consisting of all $x \in \mathscr{A}$ that are Q-orthogonal to every $y \in \mathscr{B}$, is denoted by $\mathscr{B}^Q$. Elements of the subspace $\mathscr{B}_0 = \mathscr{B} \cap \mathscr{B}^Q$ are termed Q-*null vectors* of $\mathscr{B}$, and the dimension of $\mathscr{B}_0$ is called the *nullity*, $n(\mathscr{B})$, of Q on $\mathscr{B}$, or the Q-nullity of $\mathscr{B}$. If $n(\mathscr{B}) = 0$ the quadratic form $Q(x)$ is said to be *non-degenerate* on $\mathscr{B}$. The dimension of the minimal subspace $\mathscr{C}$ of $\mathscr{B}_0$ such that $\mathscr{B}_o = \mathscr{C} + \mathscr{B} \cap \mathscr{A}^Q$ is called the relative Q-nullity of $\mathscr{B}$, and denoted by $rn(\mathscr{B})$. In the most interesting cases $rn(\mathscr{B}) = rn(\mathscr{B}^Q)$, $n(\mathscr{B}^Q) = rn(\mathscr{B}) + n(\mathscr{A})$, so that if $n(\mathscr{A}) = 0$ then $n(\mathscr{B}) = rn(\mathscr{B}) = rn(\mathscr{B}^Q) = n(\mathscr{B}^Q)$. For a linear subspace $\mathscr{B}$ of $\mathscr{A}$ the dimension of the maximal subspace $\mathscr{C}$ of $\mathscr{B}$ on which Q is negative for all non-zero elements is called the *(negative) index* or *signature* of Q on $\mathscr{B}$, and denoted by $s(\mathscr{B})$. The quantity $s(\mathscr{B})$ may also be defined as the dimension of a maximal linear subspace $\mathscr{C}$ of $\mathscr{B}$ such that $Q(x) \leq 0$ for $x \in \mathscr{C}$ and $\mathscr{C} \cap \mathscr{C}^Q = 0$, or as the dimension of the minimal subspace $\mathscr{C}$ of $\mathscr{B}$ such that $Q(x) \geq 0$ for $x \in \mathscr{B} \cap \mathscr{C}^Q$. In particular, if

$s(\mathscr{B}) = 0$ then $Q(x) \geq 0$ for all $x \in \mathscr{B}$, and a vector $x_0 \in \mathscr{B}_0$ affords a minimum to $Q(x)$ on $\mathscr{B}$ if and only if $s(\mathscr{B}) = 0$.

Of particular significance in Hestenes' treatment of quadratic forms on a Hilbert space are forms $J(x)$ which for some closed subspace $\mathscr{B}$ of $\mathscr{A}$ are weakly lower-semicontinuous on $\mathscr{B}$ and such that if (x_n), $(n = 1,2,\ldots)$ is a sequence of vectors of $\mathscr{B}$ which is weakly convergent in $\mathscr{B}$, and the corresponding real sequence of functional values $(J(x_n))$ is convergent, then the sequence (x_n) converges strongly in $\mathscr{B}$. Such forms are called *(positively) elliptic*, or *Legendre*, on $\mathscr{B}$ since in the application of the general results to a second variation functional of the sort mentioned in Section VI.6 the above described condition holds if and only if the strengthened Legendre condition holds. As noted by Hestenes in [3, Sec. 7] the condition of ellipticity of $Q(x)$ admits many equivalent forms, two of which are as follows: (i) there exist positive constants m, M and a compact form K on $\mathscr{B}$ such that $m\|x\|^2 \leq J(x) + K(x) \leq M\|x\|^2$ for all $x \in \mathscr{B}$; (ii) there is a finite dimensional subspace $\mathscr{C}$ of $\mathscr{B}$ and a positive constant m such that $J(x) \geq m\|x\|^2$ for all $x \in \mathscr{B} \cap \mathscr{C}^{\perp}$, where $\mathscr{C}^{\perp}$ denotes the orthogonal complement of the subspace $\mathscr{C}$. In particular, this latter condition implies that $n(\mathscr{B})$ and $s(\mathscr{B})$ are finite.

Now suppose that $\mathscr{B}$ is a closed subspace of a Hilbert space $\mathscr{A}$, and the quadratic form $J(x)$ is positively elliptic on $\mathscr{B}$, while $K(x)$ is a compact form on $\mathscr{B}$; moreover, suppose that $J(x) > 0$ for all $x \neq 0$ of $\mathscr{B}$ satisfying $K(x) = 0$. Then for the pencil of quadratic forms $J_\lambda(x) = J(x) - \lambda K(x)$ there is a value μ such that $J_\mu(x) > 0$ for all $x \in \mathscr{B}$. If $n(\lambda)$ and $s(\lambda)$ denote the nullity and signature of J_λ on $\mathscr{B}$, then one has the unilateral conditions $s(\lambda^-) = s(\lambda)$, $s(\lambda^+) = s(\lambda) + n(\lambda)$ for $\lambda > \mu$, while $s(\lambda^-) = s(\lambda) + n(\lambda)$ and $s(\lambda^+) = s(\lambda)$ for $\lambda \leq \mu$. If λ is a point of discontinuity of $s(\lambda)$, then λ is called an eigenvalue of J relative to K on $\mathscr{B}$ of order $n(\lambda)$; the non-zero vectors of $\mathscr{B}^{J\lambda}$ are termed the corresponding eigenvectors.

In order to illustrate the abstract conjugate point theory considered by Hestenes, again let $\mathscr{B}$ be a closed subspace of $\mathscr{A}$, and let $\mathscr{B}(\lambda)$, $\lambda' \leq \lambda \leq \lambda''$, denote a family of subspaces of $\mathscr{B}$ and quadratic form $J(x)$ possessing the following properties: (i) $\mathscr{B}(\lambda') = \mathscr{B}(\lambda'^+) = 0$, $\mathscr{B}(\lambda'') = \mathscr{B}(\lambda''^-) = \mathscr{B}$; (ii) $\mathscr{B}(\lambda) \subset \mathscr{B}(\mu)$ for $\lambda' \leq \lambda < \mu \leq \lambda''$; (iii) $\bigcup_{\lambda<\mu} \mathscr{B}(\lambda) = \mathscr{B}(\mu) = \bigcap_{\lambda>\mu} \mathscr{B}(\lambda)$ for $\lambda' < \mu < \lambda''$; (iv) $J(x)$ is positively elliptic and $\mathscr{B}(\mu) \cap [\mathscr{B}(\mu)]^J = 0$ for $\lambda' \leq \mu < \lambda''$. Under these assumptions the nullity $n(\lambda)$ and signature $s(\lambda)$ of J on $\mathscr{B}(\lambda)$ are such that $n(\lambda') = s(\lambda') = 0$, while $s(\lambda^-) = s(\lambda)$ for $\lambda' < \lambda \leq \lambda''$ and $s(\lambda^+) = s(\lambda) + n(\lambda)$ for $\lambda' < \lambda < \lambda''$. The points of discontinuity of $s(\lambda)$ are called *focal points* and the jump $n(\lambda)$ the *order* of the focal point. Clearly, the signature $s(\lambda'')$ of J on $\mathscr{B}$ is equal to the sum of the orders of the focal points on $\lambda' < \lambda < \lambda''$.

For an application of these results to the case of a differential system of the sort considered in the preceding chapters, let $\mathscr{A}$ denote the class of absolutely continuous n-dimensional vector functions $x:x(t)$, $t \in [a,b]$, with $x'(t) \in \mathscr{L}^2[a,b]$. If Q is a $2n \times 2n$ real symmetric matrix, and P, Q, R real-valued continuous $n \times n$ matrix functions with R and P symmetric on this interval, then

$$J(x) = \hat{x}^*Q\hat{x} + \int_a^b \{x^{*\prime}[Rx' + Qx] + x^*[Q^*x' + Px]\}dt$$

is a form on $\mathscr{A}$ which is positive elliptic if and only if there exists a positive constant h such that $\xi^*R(t)\xi \geq h\xi^*\xi$ for $t \in [a,b]$ and arbitrary n-dimensional vectors ξ, which is the classical Legendre condition for a variational problem having the above form its second variation functional. If S is a linear subspace of $\mathbf{C}_{2n}$ and $\mathscr{B}$ is the closed subspace of elements $x \in \mathscr{A}$ with boundary vector $\hat{x} = (x(a);x(b))$ in S, then the above definitions of eigenvalues and eigenvectors of J relative to K on $\mathscr{B}$ are clearly in agreement with those concepts as used in Chapters V and VI. Also, if $\mathscr{B}$ denotes the class of elements x of $\mathscr{A}$, with $x(a) = 0 = x(b)$, and $\mathscr{B}(\lambda)$, $a \leq \lambda \leq b$, denotes the class of all arcs in $\mathscr{B}$ for which $x(t) \equiv 0$ on $[\lambda,b]$, then this class possesses the above properties (i)-(iv) and the above defined discontinuities of $s(\lambda)$ are the points conjugate to $t = a$ in the sense introduced in Chapter V, and $n(\lambda)$ is the order of the conjugate point in the usual sense. Other choices of $\mathscr{B}$ and J(x) lead to the classical theory of focal points.

As Hestenes [3-p. 27] notes, the above condition (iii) is stronger than necessary for his considerations of eigenvalue theory in the most general sense. Moreover, condition (iv) needs modification for application to accessory problems associated with variational problems of Lagrange or Bolza type wherein the conditions of identical normality is not satisfied.

5. The Weinstein method of intermediate problems

In the latter half of the 1930's, A. Weinstein developed a method for the determination of lower bounds for the eigenvalues of boundary problems involving partial differential equations. Shortly thereafter N. Aronszajn pointed out that the Weinstein procedure was a counterpart of an extension of the Rayleigh-Ritz method, which Aronszajn himself had developed under the name of a "generalized Rayleigh-Ritz method". Moreover Aronszajn provided generality and clarity of central aspects of the method by considering the corresponding problem for a compact, (completely continuous) symmetric linear operator in Hilbert space. In addition to the cited papers in the Bibliography of Aronszajn and Weinstein [1], Aronszajn [1], and Weinstein [2], the reader is referred to the book of Gould [1] for a comprehensive discussion of the method and applications to the time of its publication in 1967. In certain later considerations, (see Weinstein and Stenger [1]), the type of considered operators in a Hilbert space was extended to the class $\mathscr{S}$ of selfadjoint linear operators that are bounded below and whose spectrum has a lower part consisting of a denumerable sequence of isolated eigenvalues, each of which is of finite multiplicity. An application of this method to establish the classical separation theorem for a simple Sturm-Loiuville problem is given in Weinstein [1]. In turn, this paper led Weinberger [1] to employ the Weinstein procedure to establish for self-adjoint boundary problems involving higher order self-adjoint ordinary linear differential equations separation theorems that include some of those of Chapter VI involving a given problem and a subproblem.

Stated briefly, the so-called "first method of Weinstein" consists of associating with an initial problem a base, or auxiliary problem whose eigenvalues and eigenvectors are assumed known, and to link the base problem with the initial problem by a finite or infinite sequence of intermediate problems, each of which is solvable explicitly in terms of the base problem. In most cases, the domain of the base problem includes that of the given problem and the domains of the intermediate problems form a shrinking sequence. A second type of intermediate problem is occasioned by successively changing the base operator into a sequence of different operators, usually with the same domain of definition.

For $\mathscr{H}$ a Hilbert space, and A a self-adjoint linear operator of the class $\mathscr{S}$ defined above, let λ_∞ denote the smallest value in the essential spectrum of A, if existent. All further comments will refer to the sequence of isolated eigenvalues λ_j that are below λ_∞; for simplicity, this sequence will be assumed to be infinite. Also, $\{\lambda_j, u_j(t)\}$, $(j = 1,2,\ldots)$, will denote these eigenvalues and associated eigenvectors indexed so that $\lambda_1 \leq \lambda_2 \leq \ldots$, with each eigenvalue repeated a number of times equal to its multiplicity and the associated eigenvectors forming an orthonormal set. If λ is not a number of the spectrum of A then the resolvent operator $R_\lambda = [A-\lambda I]^{-1}$ has for $x \in H$ the evaluation

$$R_\lambda x = \textstyle\sum_j (\lambda_j - \lambda)^{-1}(x,u_j)u_j + \displaystyle\int_{\lambda_\infty^-}^{\infty} (\mu - \lambda)^{-1} dE_\mu x, \qquad (5.1)$$

where E_μ is an appropriate family of projection operators on $\mathscr{H}$. In particular, if the essential spectrum of A is empty the integral term in (5.1) does not appear. If λ is one of the eigenvalues of A appearing in the sequence of such values defined above, then R'_λ will denote the corresponding generalized resolvent function defined by

$$R'_\lambda x = \sum_{\lambda_j \neq \lambda} (\lambda_j - \lambda)^{-1}(x,u_j)u_j + \int_{\lambda_\infty^-}^{\infty} (\mu-\lambda)^{-1} dE_\mu x. \tag{5.2}$$

Now let $\mathscr{P}$ be a closed proper subspace of $\mathscr{H}$, $\mathscr{Q}$ the orthogonal complement of $\mathscr{P}$ in $\mathscr{H}$, and denote by P and $Q = I - P$ the orthogonal projection operators onto $\mathscr{P}$ and $\mathscr{Q}$, respectively. For $A \in \mathscr{S}$, consider eigenvalue problems of the form

$$Au - PAu = \lambda u, \quad Pu = 0, \tag{5.3}$$

which may be written also as

$$QAu = \lambda u, \qquad u = Qu, \tag{5.3'}$$

where it is assumed that A and QA are of class $\mathscr{S}$ in their respective domains $D(A)$ and $D(A) \cap Q$. Also, in a certain sense (see Weinstein and Stenger [1-Secs. 3.2, .4.2]), for non-trivial eigenvalues (5.3') is equivalent to

$$QAQu = \lambda u, \tag{5.3''}$$

where QAQ has domain $\mathscr{P} \oplus D(A) \cap \mathscr{Q}$. For the base problem $\mathscr{A}^{o}$ defined by

$$Au = \lambda u, \qquad u \in D(A), \tag{5.4}$$

the above described set of eigenvalues and corresponding eigenvectors is denoted by $\{\lambda_j^{(o)}, u_j^{(o)}\}$, with the corresponding set for the problem $\mathscr{A}^{\infty}$ defined by (5.3") denoted by $\{\lambda_j^{(\infty)}, u_j^{(\infty)}\}$.

Now let $\{p_j\}$, be a sequence of linearly independent vectors in $\mathscr{P}$; this sequence may be finite or infinite, but for simplicity it is assumed to be infinite since in the contrary case the described sequence of procedures terminates.

Then for each positive integer n let $\mathscr{P}_n$ denote the n-dimensional subspace spanned by $p_1,\dots,p_n$ and denote by P_n the orthogonal projection onto $\mathscr{P}_n$ and $Q_n = I - P_n$. Then the corresponding *intermediate problem* $\mathscr{A}^n$ is defined by

$$Au - P_n Au = \lambda u, \qquad P_n u = 0, \tag{5.5}$$

or, correspondingly, by

$$Q_n A Q_n = \lambda u, \qquad Q_n u = u. \tag{5.5'}$$

If $\{\lambda_j^{(n)}, u_j^{(n)}\}$ denotes the associated sequence of eigenvalues and eigenvectors of $\mathscr{A}^n$, then extremum properties of the eigenvalues of the involved problems yield the inequalities (Weinstein-Stenger [1-Chs. 3,4]):

$$\lambda_i^{(m)} \le \lambda_i^{(m+n)} \le \lambda_{n+i}^{(m)} \qquad (i,m,n = 1,2,\dots). \tag{5.6}$$

Also, further basic properties of the eigenvalues are as follows.

(i) {Weinstein and Stenger [1-Sec. 4.3]}. *If* λ *is an eigenvalue of* $\mathscr{A}^n$ *that is not an eigenvalue of* $\mathscr{A}^0$, *then the* $n \times n$ *matrix* $M^{on}(\lambda) = [(R_\lambda p_i, p_j)]$, $(i,j = 1,\dots,n)$ *is singular, and the multiplicity of* λ *as a zero of the function*

$$W^{on}(\lambda) = \det[(R_\lambda p_i, p_j)] \tag{5.7}$$

is equal to the nullity $\nu(\lambda)$ *of the matrix* $[(R_\lambda p_i, p_j)]$. {The function $W^{on}(\lambda)$ of (5.7) is known as the "Weinstein determinant"}.

(ii) {Weinstein and Stenger [1-Sec. 4.7]}. *If* λ_* *is an eigenvalue of* $\mathscr{A}^0$ *of index* μ *and* $u^{(\alpha)}$, $(\alpha = 1,\dots,\mu)$, *is a basis for the corresponding set of solutions of* (5.4) *for* $\lambda = \lambda_*$, *then* $\lambda = \lambda_*$ *is an eigenvalue of the intermediate* *problem* $\mathscr{A}^n$ *iff the* $(n+\mu) \times (n+\mu)$ *matrix*

$$\begin{vmatrix} (R'_{\lambda_*}p_i, p_j) & (p_i, u^{(\beta)}) \\ (u^{(\alpha)}, p_j) & 0_{\alpha\beta} \end{vmatrix}, \quad \begin{matrix} (i,j = 1,\ldots,n; \\ \alpha,\beta = 1,\ldots,\mu) \end{matrix} \tag{5.8}$$

is singular, and the multiplicity of λ_* *as an eigenvalue of* $\mathscr{A}^n$ *is equal to the nullity of this matrix.*

Now for $j > i \geq 0$ the problem $\mathscr{A}^j$ may be considered as an intermediate problem to $\mathscr{A}^i$ of index $j - i$, and there is a $(j-i) \times (j-i)$ matrix $M^{ij}(\lambda)$ and corresponding Weinstein determinant $W^{ij}(\lambda) = \det M^{ij}(\lambda)$ linking $\mathscr{A}^i$ to $\mathscr{A}^j$ in the same manner that $M^{on}(\lambda)$ and $W^{on}(\lambda)$ link $\mathscr{A}^o$ to $\mathscr{A}^n$. Now $W^{ij}(\lambda)$ is a meromorphic function of λ on the lower portion of the spectrum of $\mathscr{A}^i$, and thus for each such λ the order $\omega^{ij}(\lambda)$ of $W^{ij}(\lambda)$ is well-defined as follows: $\omega^{ij}(\lambda) = 0$ if $W^{ij}(\lambda)$ is finite and non-zero; $\omega^{ij}(\lambda) = +k$ if $W^{ij}(\lambda)$ has a zero of order k at $\lambda = \lambda_*$; $\omega^{ij}(\lambda) = -k$ if $W^{ij}(\lambda)$ has a pole or order k at $\lambda = \lambda_*$. The following result is known as *Aronszajn's Rule*, and is a consequence of the decomposition $W^{on}(\lambda) = W^{o1}(\lambda)W^{12}(\lambda)\ldots W^{n-1,n}(\lambda)$.

(iii) {Weinstein and Stenger [1-Sec. 4.8]}. *If a given value* λ *is an eigenvalue of* $\mathscr{A}^o$ *and* $\mathscr{A}^n$ *of respectively multiplicities* $\mu_o(\lambda)$ *and* $\mu_n(\lambda)$, *then*

$$\mu_n(\lambda) = \mu_o(\lambda) + \omega^{on}(\lambda). \tag{5.9}$$

(iv) {Weinstein and Stenger [1-Sec. 7.8]}. *For each positive integer* n *the inequality* $\lambda_i^{(n)} \leq \lambda_{n+i}^o$ *holds for* $i = 1,2,\ldots$. *Moreover, for a given value* i *the equality* $\lambda_i^{(n)} = \lambda_{n+i}^o$ *holds iff for* ε *a sufficiently small positive value the hermitian form with matrix* $[(R_\lambda p_i, p_j)]$, (i,j = 1,...,n) *has for* $\lambda_{n+i}^{(o)} - \varepsilon < \lambda < \lambda_{n+i}^{(o)}$ *negative index at least* m - i *and at most* m - 1, *where* $m = \text{Min}\{j : \lambda_j^{(o)} = \lambda_{n+i}^{(o)}\}$; *equivalently, for* $\lambda_{n+i}^{(o)} - \varepsilon < \lambda < \lambda_{n+i}^{(o)}$ *the number of changes of sign in the sequence* $1, W^{01}(\lambda), W^{12}(\lambda),\ldots, W^{n-1,n}(\lambda)$ *is at least* m - i *and at most* m - 1.

Intermediate problems of the second type are provided by a base problem $\mathscr{A}^0$ defined by

$$Au = \lambda u, \quad u \in D(A),$$

where A is a self-adjoint linear operator of class $\mathscr{S}$, and for n = 1,2,... the intermediate problem $\mathscr{A}^n$ is defined by

$$Au + \sum_{j=1}^{n} (u,Bp_j)p_j = \lambda u, \quad u \in D(A) \cap D(B)$$

where B is a positive self-adjoint linear operator, $\{p_j\}$, (j = 1,2,...), is a linearly independent set of elements in D(B) which for simplicity will be chosen B-orthonormal in the sense that $(p_i,Bp_j) = \delta_{ij}$. For intermediate problems of this second type there exist results that are quite analogous to those for intermediate problems of the first type as stated above, (see Weinstein and Stenger [1-Ch. 5]). Precise statements of such results will not be given here, however, and further comments on such problems are limited to the following.

(a) There are interrelations between the Hestenes theory of quadratic forms in Hilbert space as discussed in the preceding section and the area of Weinstein intermediate problems. In particular, if $\mathscr{H}$ is a Hilbert space over the field of reals, and $A:\mathscr{H} \to \mathscr{H}$ is a self-adjoint linear operator belonging to the above defined class $\mathscr{S}$, then A is a bounded self-adjoint operator and for $\lambda < \lambda_\infty$ the form $([A+\lambda I]x,x)$ is a Legendre form in the sense of Hestenes.

(b) In the complex Hilbert space of n-dimensional vector functions that are of class $\mathcal{L}^2$ on a compact interval $[a,b]$ a base problem $\mathcal{A}$ may be defined by a boundary problem of the form $(\mathcal{B})$ of Chapter VI satisfying hypotheses $(\mathcal{A}_{\mathcal{B}})$ and $(\mathcal{A}'_K\text{i,ii,iii})$. If $F = [f_1,\ldots,f_k]$, $1 \le k \le \infty$, is a set of n-dimensional vector functions satisfying for each $r = 1,2,\ldots,k$ the conditions of Exercise VII.7.6, then the corresponding system (7.6) defines a problem $\mathcal{A}^r$ such that $\mathcal{A},\mathcal{A}^1,\ldots,\mathcal{A}^k$ forms a sequence of intermediate problems of the first type.

(c) Suppose that $K(t,s)$ is a continuous positive definite hermitian $n \times n$ matrix kernel on $[a,b] \times [a,b]$, and $\{\mu_j,\phi_j(t)\}$, $(j = 1,2,\ldots)$, is the sequence of eigenvalues and corresponding eigensolutions of the integral equation

$$\phi(t) = \mu \int_a^b K(t,s)\phi(s)ds, \quad t \in [a,b],$$

where each eigenvalue is repeated a number of times equal to its multiplicity and the $\phi_j(t)$ form an orthonormal set. Then for $m = 1,2,\ldots$ the degenerate kernel $K^{(m)}(t,s) = \sum_{j=1}^{m} \mu_j^{-1}\phi_j(t)\phi_j^*(s)$ is such that a differential boundary problem $(\mathcal{B})$ satisfying the conditions of comment (b) above, together with the sequence of integro-differential boundary problems of Section 1 with $N(t,s) = K^{(m)}(t,s)$, $(m = 1,2,\ldots)$, forms a base problem $\mathcal{A}$ and a sequence $\mathcal{A}^m$ of intermediate problems of the second type.

6. Oscillation phenomena for Hamiltonian systems in a B*-algebra

Within recent years there has been considerable study of differential equations in abstract spaces, and in such settings there have been established generalizations of some of the oscillation phenomena of the classical Sturmian theory.

Let X be a B-space (Banach space), and denote by $[X]$ the algebra of bounded linear operators $T:X \to X$, and for I a subinterval of the real line $\mathbb{R}$ let the concepts of continuity, derivative, and integral of $f:I \to X$ and $F:I \to [X]$ be taken in the norm topologies. For continuous $A:[0,\infty) \to [X]$ and $B:[0,\infty) \to [X]$, Heimes [1] derived conditions on A, B and the length $\tau > 0$ which imply the existence of a unique solution for the two-point boundary problem

$$\begin{aligned} &\text{(a)} \quad y''(t) = A(t)y'(t) + B(t)y(t), \\ &\text{(b)} \quad y(0) = \alpha, \quad y(\tau) = \beta, \end{aligned} \tag{6.1}$$

for arbitrary elements α,β of X. In particular, he established the "Liapunov-type inequality" that if $A(t) \equiv 0$ and equation (6.1a) has a non-identically vanishing solution $y(t)$ which is equal to 0 at two distinct values on $[0,\tau]$ then $\int_0^\tau \|B(t)\|dt > 4/\tau$.

For the treatment of abstract oscillation phenomena one of the most fruitful domains of consideration has proved to be that of a B*-algebra. Hayden and Howard [1] obtained some preliminary results in this context. Hille [2-Sec. 9.6] has provided an excellent discussion of problems in this setting. In particular, he established certain criteria of non-oscillation for large t that are generalizations of central results of his earlier paper [1] on scalar linear second order differential equations, and also provided an extension of the matrix generalizations of the trigonometric sine and cosine functions due to Barrett [3]. In this latter regard, Benson and Kreith [1] presented a modification of Barrett's original argument that yields an extension of Hille's results. Williams [1] has elaborated upon the results of Hille, and showed various relationships between such results and their analogues, for matrix differential systems as discussed in Chapter V. Further detailed comments on problems in this area will be presented in the context occurring in Hille [2] and Williams [1].

In order that there be no ambiguity in some of the future statements, by a B*-algebra $\mathscr{B}$ we shall mean a Banach space of elements $x, y, \ldots,$ with complex scalars, norm function $\| \quad \|$, a unit element e, an associative multiplication xy satisfying $\|xy\| \leq \|x\| \, \|y\|$, and an operation $(\ )^*$ with the following properties for x, y arbitrary elements of $\mathscr{B}$ and $\gamma \in \mathbf{C}$:

(a) for each $x \in \mathscr{B}$, there exists a unique $x^* \in \mathscr{B}$, and $(x^*)^* = x$;

(b) $(x + y)^* = x^* + y^*$;

(c) $(\gamma x)^* = \bar{\gamma} x^*$;

(d) $(xy)^* = y^* x^*$;

(e) $\|x^* x\| = \|x\|^2$.

An element $x \in \mathscr{B}$ is called *non-singular* or *regular* in case there exists an element $x^{-1} \in \mathscr{B}$ such that $xx^{-1} = x^{-1}x = e$, and *singular* in the contrary case. Also, for $x \in \mathscr{B}$ the spectrum $\sigma(x)$ is the set of all complex λ such that $\lambda e - x$ is singular. An element x is called *symmetric*, or *hermitian*, if $x = x^*$, and the further additional properties are required.

(f) if $x \in \mathscr{B}$ is symmetric, then $\sigma(x)$ is a subset of the real line;

(g) the set $\mathscr{B}^+$ of all symmetric elements with non-negative real spectra is a positive cone; i.e., $\mathscr{B}^+$ is closed under addition, multiplication by positive scalars, and passage to the limit;

(h) each element of the form x^*x belongs to $\mathscr{B}^+$.

Whenever x is symmetric and $\sigma(x) \subset (0,\infty)$, $\{\sigma(x) \subset [0,\infty)\}$, the element x is said to be positive, {non-negative}, and we write $x > 0$, $\{x \geq 0\}$.

One example of such a B*-algebra is the algebra of $n \times n$ matrices A with complex elements, with A^* the corresponding conjugate transpose matrix, and $\|A\| = \sup\{|A\xi| : \xi \in \mathbf{C}_n\}$. Another example is the algebra of all bounded linear operators $T: \mathscr{H} \to \mathscr{H}$, where $\mathscr{H}$ is a complex Hilbert space and T^* denotes the operator adjoint to T. Indeed, it is known that any B*-algebra is isometrically *-isomorphic to an algebra of bounded linear operators over a suitable complex Hilbert space.

Similar to the convention mentioned above in connection with reference to Heimes' paper, the concepts of continuity, derivative and integral will be taken in the norm topologies. Moreover, integration will be restricted to the case of continuous functions, and the ordinary Riemann-type integral, such as treated in Hille-Phillips [1, pp. 62-71] suffices to for the present discussion.

For I a non-degenerate interval on the real line, let a,b,c be continuous $\mathscr{B}$-valued functions on I such that $b(t)$ and $c(t)$ are symmetric for each $t \in I$. Then corresponding to the vector system (V.2.5) or (V.3.1) we may consider the (Hamiltonian) system

$$\begin{aligned} L_1[u,v](t) &\equiv -v'(t) + c(t)u(t) - a^*(t)v(t) = 0, \\ L_2[u,v](t) &\equiv u'(t) - a(t)u(t) - b(t)v(t) = 0. \end{aligned} \tag{6.2}$$

Also, corresponding to the matrix Riccati differential equation (V.4.1) we have the non-linear equation

$$\begin{aligned} k[w](t) \equiv w'(t) + a(t)w(t) + w(t)a^*(t) \\ + w(t)b(t)w(t) - c(t) = 0. \end{aligned} \tag{6.3}$$

Well-known existence theorems for such differential equations, (see, for example, Hille-Phillips [1, Sec. 3.4]), yield global existence and uniqueness of solutions of (6.2) and corresponding local results for (6.3). Moreover, as in the case of matrix equations considered in Chapter V, if $y_1(t) = (u_1(t);v_1(t))$ and $y_2(t) = (u_2(t);v_2(t))$ are solutions of (6.2) on I then $\{y_1,y_2\}(t) = v_2^*(t)u_1(t) - u_2^*(t)v_1(t)$ is an element of $\mathscr{B}$ which is constant on I. If the value of this constant is 0, then $y_1(t)$ and $y_2(t)$ are said to be *conjugate* or *conjoined* solutions of (6.2). Also, if $(u(t);v(t))$ is a solution of (6.2) with $u(t)$ non-singular on a subinterval I_0 of I then $w(t) = v(t)u^{-1}(t)$ is a solution of the Riccati differential equation (6.3) on I_0; conversely, if $w(t)$ is a solution of (6.3) on a subinterval I_0 of I then there exists a solution $(u(t);v(t))$ of (6.2) with $u(t)$ non-singular and $w(t) = v(t)u^{-1}(t)$ on I_0.

Furthermore, $u^*(t)[w(t) - w^*(t)]u(t) = u^*(t)v(t) - v^*(t)u(t)$ on I_o so that $w(t)$ is a symmetric solution of (6.3) on this subinterval if and only if $(u(t);v(t))$ is a *self-conjoined* or *isotropic* solution of (6.2).

Corresponding to the terminology of Chapter V for matrix systems, two distinct points t_1, t_2 of I are said to be *(mutually) conjugate* with respect to (6.2) provided there exists a solution $(u(t);v(t))$ of this system with $u(t_1) = 0 = u(t_2)$ and $u(t) \not\equiv 0$ on the subinterval with endpoints t_1 and t_2. If no two distinct points of a subinterval I_o of I are conjugate with respect to (6.2), then this system is said to be *disconjugate* on I_o. For $\tau \in I$ the solution $y_\tau(t) = (u_\tau(t);v_\tau(t))$ determined by the initial conditions

$$u_\tau(\tau) = 0, \quad v_\tau(\tau) = e \tag{6.4}$$

clearly plays a central role in the determination of points conjugate to τ.

In order to present in a concise manner results for (6.2) which correspond to, or contrast with, certain results for the matrix systems of Chapter V, there are introduced the following abbreviations of certain concepts for (6.2) relative to a compact subinterval $[\alpha,\beta]$ of I.

NCP$[\tau;\alpha,\beta]$: $\tau \in [\alpha,\beta]$, *and there exists no point on* $[\alpha,\beta]$ *distinct from* τ *and conjugate to* τ.

NCP$^+[\tau;\alpha,\beta]$: $\tau \in [\alpha,\beta]$, *and for* $y_\tau(t) = (u_\tau(t);v_\tau(t))$ *the solution of* (6.2) *determined by* (6.4) *we have* $u_\tau(t)$ *non-singular for* $t \neq \tau$ *and* $t \in [\alpha,\beta]$.

DC$[\alpha,\beta]$: *Condition* NCP$[\tau;\alpha,\beta]$ *holds for all* $\tau \in [\alpha,\beta]$.

US$[\alpha,\beta]$: *for arbitrary distinct values* t_1, t_2 *on* $[\alpha,\beta]$ *and arbitrary elements* u_1, u_2 *of* $\mathcal{B}$ *there is a unique solution* $y(t) = (u(t);v(t))$ *of* (6.2) *satisfying*

$$u(t_1) = u_1, \quad u(t_2) = u_2, \tag{6.5}$$

$IN[\alpha,\beta]$: *for each* $\tau \in [\alpha,\beta]$ *and* $y_\tau(t) = (u_\tau(t);v_\tau(t))$ *the solution of* (6.2) *determined by* (6.4) *there is a corresponding positive* $\delta = \delta_\tau$ *such that* $u_\tau(t)$ *is non-singular for* $t \in (\tau-\delta,\tau+\delta) \cap [\alpha,\beta]$.

$NO[\alpha,\beta]$: *there exists a self-conjoined solution* $y(t) = (u(t);v(t))$ *of* (6.2) *with* $u(t)$ *non-singular on* $[\alpha,\beta]$.

$H_b[\alpha,\beta]$: $b(t) \geq 0$ *for* $t \in [\alpha,\beta]$.

$H_b^+[\alpha,\beta]$: $b(t) > 0$ *for* $t \in [\alpha,\beta]$.

The condition $IN[\alpha,\beta]$ corresponds to the condition of identical normality for matrix systems (V.3.1), and whenever this condition holds condition $NCP[\alpha,\beta]$ is clearly equivalent to $NCP^+[\alpha,\beta]$.

The symbol $PC[\alpha,\beta]$ will denote the class of functions $\xi:[\alpha,\beta] \to \mathscr{B}$ which are piecewise continuous on $[\alpha,\beta]$ in the sense that $\xi(t)$ is continuous on this interval except for at most a finite number of points, and the right- and left-hand limits exist at the points of discontinuity. Correspondingly, $PS[\alpha,\beta]$ denotes the class of functions $\eta:[\alpha,\beta] \to \mathscr{B}$ which are piecewise smooth on $[\alpha,\beta]$ in the sense that $\eta(t)$ is continuous, its derivative function $\eta'(t)$ exists on $[\alpha,\beta]$ except for at most a finite number of points and is piecewise continuous. The set of $\eta \in PS[\alpha,\beta]$ for which there exists a corresponding $\xi \in PC[\alpha,\beta]$ satisfying

$$L_2^*[\eta,\zeta](t) \equiv \eta'(t) - a(t)\eta(t) - b(t)\xi(t) = 0 \tag{6.6}$$

whenever $\eta'(t)$ exists is denoted by $D[\alpha,\beta]$ and the association of ξ with η is signified by $\eta \in D[\alpha,\beta]:\xi$. The subset of $D[\alpha,\beta]$ on which $\eta(\alpha) = 0 = \eta(\beta)$ is denoted by $D_0[\alpha,\beta]$. Also, if $\eta_j \in D[\alpha,\beta]:\xi_j$, $(j = 1,2,)$, the integral

$$\int_\alpha^\beta \{\xi_2^*(s)b(s)\xi_1(s) + \eta_2^*(s)c(s)\eta_1(s)\}ds \tag{6.6}$$

defines a symmetric or hermitian mapping on $D[\alpha,\beta] \times D[\alpha,\beta]$ into $\mathscr{B}$, whose value is independent of the particular ξ_j associated with η_j by $\eta_j \in D[\alpha,\beta]:\xi_j$. Consequently, (6.6) is denoted by $J[\eta_1,\eta_2;\alpha,\beta]$, with the symbol $J[\eta_1,\eta_1;\alpha,\beta]$ contracted to $J[\eta_1;\alpha,\beta]$. Also, for real values λ we are concerned with the differential system

$$\begin{aligned} -v'(t) + [c(t) - \lambda e]u(t) - a^*(t)u(t) &= 0, \\ u'(t) - a(t)u(t) - b(t)v(t) &= 0 \end{aligned} \tag{6.2_λ}$$

and the associated functional

$$J[\eta|\lambda;\alpha,\beta] = J[\eta;\alpha,\beta] - \lambda\int_a^b \eta^*(s)\eta(s)ds \tag{6.7}$$

for $\lambda \in D[\alpha,\beta]$. The symbol $J_\lambda[\alpha,\beta]$ is used to denote the condition that λ is a real number and $J[\eta|\lambda;\alpha,\beta] \geq 0$ for all $\eta \in D_0[\alpha,\beta]$, and $J_\lambda^+[\alpha,\beta]$ is used to signify that condition $J_\lambda[\alpha,\beta]$ holds with equality only in case $\eta(t) \equiv 0$ on $[\alpha,\beta]$. In particular, $J_0^+[\alpha,\beta]$ is the condition that $J[\eta;\alpha,\beta] \geq 0$ for $\eta \in D_0[\alpha,\beta]$, and equality holds only if $\eta(t) \equiv 0$ on $[\alpha,\beta]$.

In terms of these notations we have the following results, wherein $[\alpha,\beta]$ denotes an arbitrary subinterval of I and the cited theorems refer to Williams [1].

(i) {Th. 4.1}. $NO[\alpha,\beta]$, $H_b^+[\alpha,\beta] \rightarrow NCP^+[\alpha;\alpha,\beta]$.

(ii) {Th. 4.4}. $H_b[\alpha,\beta]$, $NCP^+[\alpha,\beta] \rightarrow NO[\alpha,\beta]$.

(iii) {Th. 4.5}. $H_b[\alpha,\beta]$, $IN[\alpha,\beta]$, $NO[\alpha,\beta] \to NCP^+[\tau;\alpha,\beta]$ *for all* $\tau \in [\alpha,\beta]$.

(iv) {Th. 4.6}. $US[\alpha,\beta] \to NCP^+[\alpha,\beta]$.

(v) {Th. 4.7}. $H_b[\alpha,\beta]$, $US[\alpha,\beta] \to NO[\alpha,\beta]$.

(vi) {Th. 4.8}. $H_b[\alpha,\beta]$, $IN[\alpha,\beta]$, $NO[\alpha,\beta] \to US[\alpha,\beta]$.

(vii) {Th. 5.1}. $J_o^+[\alpha,\beta] \to DC[\alpha,\beta]$.

(viii) {Th. 5.2}. $H_b[\alpha,\beta]$, $NO[\alpha,\beta] \to J_o^+[\alpha,\beta]$.

(ix) {Th. 5.3}. $H_b[\alpha,\beta]$, $IN[\alpha,\beta]$, $DC[\alpha,\beta] \to NO[\alpha,\beta]$ *iff for* $\tau \in [\alpha,\beta]$ *and* $(u_\tau(t);v_\tau(t))$ *the solution of* (6.2) *determined by* (6.4) *then for arbitrary* $t \in [\alpha,\beta]$ *either* $u_\tau(t)$ *is non-singular or there exists a non-zero* $x \in \mathscr{B}$ *such that* $xu_\tau(t) = 0$.

(x) {Th. 6.1}. $H_b[\alpha,\beta]$, $NO[\alpha,\beta] \to$ *there exists a* $\mu > 0$ *such that condition* $J_\mu[\alpha,\beta]$ *holds.*

(xi) {Th. 6.2}. $H_b[\alpha,\beta]$, $IN[\alpha,\beta]$, $J_\mu[\alpha,\beta]$ *holds with* $\mu > 0 \to NO[\alpha,\beta]$.

(xii) {Th. 6.3}. $H_b^+[\alpha,\beta]$, $\mu = \sup\{\lambda : J_\lambda[\alpha,\beta]$ *holds*$\} \to$ (6.2_λ) *satisfies condition* $NO[\alpha,\beta]$ *for each* $\lambda < \mu$.

With the aid of the results (xi) and (xii), Williams [1-Ths. 6.6, 6.7] obtained some simple comparison theorems of Sturmian type. Also, for systems of the form (6.2) on a non-compact interval $[\alpha,\infty)$, and with $b(t) > 0$ and $c(t) \leq 0$, in his Theorems 8.1 and 8.2 Williams obtained slight generalizations of Hille's Theorems 9.6.2 and 9.6.3. Finally, Williams [1-Sec. 9] obtained sufficient conditions for the non-existence of self-conjoined solutions $(u(t);v(t))$ of (6.2) with $u(t)$ non-singular on some subinterval $[\alpha_0,\infty)$ of $[\alpha,\infty)$, which are generalizations of the results of Theorems 2, 3, 5 of Howard [4].

In contrasting the above cited results for systems (6.2) in a B*-algebra with those for matrix systems as considered in Chapter V, it is to be noted that in the matrix case Theorem V.6.3 implies that whenever $H_b[\alpha,\beta]$ holds the three conditions $NO[\alpha,\beta]$, $J_o^+[\alpha,\beta]$ and $DC[\alpha,\beta]$ are equivalent. In the general B*-algebra case considered above, these three conditions are no longer equivalent, even when $H_b[\alpha,\beta]$ is strengthened to $H_b^+[\alpha,\beta]$. This phenomenon is illustrated by an example of Heimes [1, p. 217], involving the B*-algebra of bounded linear operators on the Hilbert space ℓ^2, wherein $a(t) \equiv 0$, $b(t) \equiv e$, and $c(t)$ is the constant operator defined by $c(\underline{e}^n) = -k_n^2 \underline{e}^n$ with $k_n = n\pi/(n+1)$ and $\underline{e}^n$ the unit vector $(\underline{e}_j^n)$ with $\underline{e}_n^n = 1$, $\underline{e}_j^n = 0$ for $j \neq n$. In this case, condition $DC[0,1]$ holds, but condition $NO[0,1]$ does not hold. An element $y \in \mathscr{B}$ is said to be *compact* if for each bounded sequence $\{x_n\}$ of elements in $\mathscr{B}$ the sequence $\{yx_n\}$ contains a convergent subsequence, or, equivalently, if T_y is the bounded linear operator on $\mathscr{B}$ defined by $T_y(x) = yx$ then $T_y : \mathscr{B} \to \mathscr{B}$ is a compact operator. Corresponding to Lemma 2 of Heimes [1], Theorem 5.4 of Williams [1] yields the result that if $b(t) > 0$ on $[\alpha,\beta]$ and the $\mathscr{B}$-valued continuous functions $a(t)$, $a^*(t)$, $c(t)$ are compact for each $t \in [\alpha,\beta]$, then the conditions $NO[\alpha,\beta]$, $J_o^+[\alpha,\beta]$ and $DC[\alpha,\beta]$ are equivalent.

In regard to other differences, it is to be noted that for the matrix case of Chapter V the existence of a value λ such that condition $J_\lambda[\alpha,\beta]$ holds implies that $b(t) \geq 0$ for $t \in [\alpha,\beta]$, and no counterpart of this result has been established for systems (6.2). In particular, the above properties of solutions of systems (6.2) in a B^*-algebra provide only meager results in areas analogous to those of separation theorems and boundary problems involving matrix systems of the sort considered in Chapters V and VI.

7. Topological interpretations of the Sturmian theorems

The basic work of Morse [1, 4-Ch. IV] on the Sturmian theory for general self-adjoint differential systems was actually subsidiary to the development of his critical point theory and variational theory in the large, dealing in particular with closed geodesics on a Riemannian manifold [4-Chs. VIII, IX]. In the years since Morse's initial contributions there have been considerable extensions of his critical point theory, and generalizations of his theory of geodesics. In particular, any result dealing with generalizations of the concepts of conjugate and focal point might be considered as belonging to "Sturmian theory". However, this section is limited to brief statements on certain generalizations and interpretations that are more readily connected with the classical analytic theory.

In the course of his study of the Morse index and nullity of the iterates of closed geodesics on a Riemannian manifold, R. Bott [1] presented a topological intersection theory for self-adjoint linear differential systems which yields the Morse comparison and oscillation theorems. In notation corresponding to that used for similar situations in Chapters V and VI, for an interval $[0,b]$ on the real line R the problem considered by Bott may be written as the first order system

$$\begin{aligned} &\text{(i)} \quad \mathcal{J}y'(t) + \mathcal{A}_\lambda(t)y(t) = 0, \quad t \in [0,b], \\ &\text{(ii)} \quad y(0) = \mathbf{P}\xi, \quad y(b) = \mathbf{Q}\xi, \end{aligned} \qquad (\mathcal{B}\text{-b})$$

in the $N = 2n$ dimensional vector function $y(t)$, where $\mathcal{J}$ and $\mathcal{A}(t)$ are defined by (V.3.2) with the component $n \times n$ matrix functions A, B, C of class $\mathcal{C}'$ and $B(t) > 0$ on an interval I containing $[0,b]$, $\mathcal{A}_\lambda(t) = \mathcal{A}(t) - \lambda T$ with $T = \operatorname{diag}\{E_n, 0\}$, while $\mathbf{P}$, $\mathbf{Q}$ are $N \times N$ matrices with $(\mathbf{P};\mathbf{Q})$ of rank N and satisfying $\mathbf{P}^*\mathcal{J}\mathbf{P} - \mathbf{Q}^*\mathcal{J}\mathbf{Q} = 0$.

Let $Y_\lambda(t)$ be the fundamental matrix solution of ($\mathcal{B}$-bi) satisfying $Y_\lambda(0) = E_N$. Then for $(t,\lambda) \in \Delta = I \times \mathbf{R}$ the matrix $Y_\lambda(t)$ is non-singular, and $(t,\lambda) \to Y_\lambda(t)$ defines a differentiable map of Δ into the full linear group $GL(N;\mathbf{C})$ which for fixed t is real analytic in λ on $\mathbf{R}$. The fact that the corresponding $N \times N$ matrix function $\{Y_\lambda, Y_\lambda\}$ of (V.3.3) is constant with respect to t for fixed λ then yields the result that for arbitrary $(t,\lambda) \in \Delta$ we have that $Y_\lambda(t)$ lies in the Lie subgroup $\mathcal{U}$ of $GL(N;\mathbf{C})$ characterized by the condition $Y^*\mathcal{J}Y = \mathcal{J}$. Moreover, corresponding to the discussion of Section VI.5, for a given $b > 0$ a value λ_0 is an eigenvalue of ($\mathcal{B}$-b) of multiplicity k iff the matrix $Y_{\lambda_0}(b)\mathbf{P} - \mathbf{Q}$ is of rank $N - k$. Consequently, if $\mathcal{B}^0 = \{Y : Y \in \mathbf{H},\ Y\mathbf{P} - \mathbf{Q} \text{ singular}\}$, then λ_0 is an eigenvalue of ($\mathcal{B}$-b) iff $Y_{\lambda_0}(b) \in \mathcal{B}^0$, and thus the study of the boundary problem ($\mathcal{B}$-b) is transformed into the consideration of the intersections of the curve $\lambda \to Y_\lambda(b)$ and the set $\mathcal{B}^0$. Bott showed that $\mathcal{B}^0$ is covered by a $(N^2 - 1)$-dimensional locally finite cycle $\gamma_{\mathcal{B}}$, and that for any compact subinterval $[\lambda', \lambda'']$ with neither λ' nor λ'' an eigenvalue the spectral multiplicity of ($\mathcal{B}$-b) on $[\lambda', \lambda'']$, (i.e., the sum of the eigenvalues of ($\mathscr{B}$-b) on this interval, each counted according to its multiplicity), equals the topological intersection number of γ_B and the curve $\lambda \to Y_\lambda(b)$. This basic result was then employed to present topological proofs of the comparison, oscillation and index theorems of Morse for boundary problems of the form ($\mathscr{B}$-b), these individual theorems appearing as statements about the behavior of the intersection number of two cycles under deformation of the cycles.

The proof of Bott's fundamental result involved a number of detailed topological considerations. In view of the fact that points of the spectrum of ($\mathscr{B}$-b) might have a multiplicity greater than one, the construction of $\gamma_{\mathscr{B}}$ was by way of an auxiliary "resolution of $\mathscr{B}^{(o)}$", a sub-manifold $\mathscr{B}^{(1)}$ in the cartesian product of $\mathscr{A}$ and a complex projective space $G^{(1)}$; under the canonical projection $f^{(1)}: \mathscr{A} \times G^{(1)} \to \mathscr{A}$, $\mathscr{B}^{(1)}$ mapped onto $\mathscr{B}^{(o)}$ and $\gamma_{\mathscr{B}}$ was defined as the image of the fundamental class of $B^{(1)}$ under $f^{(1)}$. Another subtle topological concept was that of "clean intersection of manifolds".

Following Bott's intersection-theoretic method, Edwards [1] made further generalizations of the theory. In contrast to Bott's consideration of operators, Edwards dealt with hermitian sesquilinear forms, corresponding to the $J[\eta;\lambda:\mathscr{B}]$ associated with $J[\eta:\mathscr{B}]$ of (VI.1.10") for systems (VI.3.1) equivalent to a boundary problem involving a self-adjoint scalar differential equation of order $2n$ as described at the end of Section V.2. Many of his analytic definitions and details are reminiscent of the applications of Hestenes [2] to the second variation functional of a calculus of variations problem, although no mention is made of Hestenes' work. Specifically, Edwards' concept of a "Sturm form" corresponds to Hestenes' "Legendre form". Topologically, Edwards studied the structure of U-manifolds, where by definition a U-manifold is a set $U(\mathscr{L},\psi)$ obtained from an even-dimension complex space $\mathscr{L}$ and a non-degenerate hermitian form ψ on $\mathscr{L}$ of signature zero, by setting $U(\mathscr{L},\psi)$ equal to the collection of all subspaces $\mathscr{P}$ of $\mathscr{L}$ with $2 \dim \mathscr{P} = \dim \mathscr{L}$, and such that ψ restricted to $\mathscr{P}$ is the zero form. The final portion of the paper dealt with a discussion of the relevance of his general theory for the second variation functional of a calculus of variational problem, in which context the Sturmian theorems appeared as results in his theory of multiplicities of intersections of curves in U-manifolds with certain subvarieties of codimension one.

Further results in this area appear in the recent papers of Duistermaat [1] and Cushman and Duistermaat [1]. Let $\mathscr{L}$ be a finite dimensional real vector space, and σ a symplectic form on $\mathscr{L}$; that is, σ is a non-degenerate antisymmetric bilinear form on $\mathscr{L}$. For a linear subspace α of $\mathscr{L}$, let α^σ denote the orthogonal complement of α with respect to σ; that is, $\alpha^\sigma = \{v : v \in \mathscr{L},\ \sigma(u,v) = 0 \text{ for } u \in \alpha\}$. Duistermaat [1] terms α *isotropic* if $\alpha \subset \alpha^\sigma$ and α is called a Lagrange subspace of the symplectic vector space $(\mathscr{L},\sigma)$ if it is maximal with this property, in which case $\alpha = \alpha^\sigma$ and $2 \dim \alpha = \dim \mathscr{L}$. He developed an intersection theory for curves of Lagrange spaces, which upon application to such curves arising via a fundamental matrix solution of a Hamiltonian system of the above form ($\mathscr{B}$-b) yield the Morse theorems on conjugate and focal points. In Cushman and Duistermaat [1] this intersection theory is applied to periodic linear Hamiltonian systems. In particular, there is obtained a generalization of the results of Bott [1] for the Morse index of iterated periodic geodesics to the case of curves $t \to \Phi(t)$ of h-unitary transformations, where h is an hermitian form of arbitrary signature, and which are not necessarily related to a variational problem with a positivity condition insuring the finiteness of its Morse index.

In the above cited papers there are of course numerous points of contact with the material of the preceding chapters, especially in regard to such specific results as those presented in Topics and Exercises V.15.3, V.15.9, VI.7.1, and the methods of treatment appearing in many places, notably in Sections V.10 and VI.5. Indeed, in these latter instances the final basic step in the argument is in essence topological, being interpretable as certain cycles homologous to zero in view of the uniqueness and continuity of certain functions. In the study of differential systems of the sort forming the focus of attention in the earlier chapters, for the presented results the type of treatment there given is simpler in detail than the development of more general topological theories that yields these results as direct applications. However, one is ever mindful of the fact that mathematical insight and development is often fostered by the discovery of a relatively sophisticated manner in which to describe a phenomenon considered simple or ordinary, and, in particular, the individual method to be employed in a given case is frequently influenced by the envisioned goal of the investigation. In this regard, one is reminded of the following quotation from the preface to the Colloquium Publication [4] of Morse.

"Any problem which is non-linear in character, which involves more than one coordinate system or more than one variable, or whose structure is initially defined in the large, is likely to require considerations of topology and group theory in order to arrive at its meaning and its solution. In the solution of such problems classical analysis will frequently appear as an instrument in the small, integrated over the whole problem with the aid of group theory of topology. Such conceptions are not due to the author. It will be sufficient to say that Henri Poincaré was among the first to have a conscious theory of macro-analysis, and of all mathematicians was doubtless the one who most effectively put such a theory into practice."

ABBREVIATIONS FOR MATHEMATICAL PUBLICATIONS MOST FREQUENTLY USED

AAST	Atti della Accademia delle Scienze di Torino. Classe di Scienze Fisiche, Mathematiche e Naturali
ACMT	Acta Mathematica
AMJM	American Journal of Mathematics
AMMM	American Mathematical Monthly
ANPM	Annales Polonici Mathematici
AMPA	Annali di Matematica Pura ed Applicata
AMST	American Mathematical Society, Translations
ANNM	Annals of Mathematics
ANLR	Atti della Accademia Nazionale dei Lincei. Rendiconti. Classe di Scienze, Fisiche, Mathematiche e Naturali
APLA	Applicable Analysis
ARMA	Archive for Rational Mechanics and Analysis
ARMT	Archiv der Mathematik
ARVM	Archivum Mathematicum
ASEN	Annales Sciettifiques de l'École Normale Superieure
ASNP	Annali della Scuola Normale Superiore di Pisa
BAMS	Bulletin of the American Mathematical Society
BAUS	Bulletin of the Australian Mathematical Society
BAPS	Bulletin de l'Académie Polonaise des Sciences. Série des Sciences Mathématiques, Astronomiques et Physiques
BCMS	Bulletin of the Calcutta Mathematical Society
BMSR	Bulletin Mathématique de la Société des Sciences Mathématiques R. S. Roumanie
BSMF	Bulletin de la Societé Mathématique de France
BUMI	Bolletino della Unione Matematica Italiana
CAPM	Časopis pro Pěstování Makematiky
CDJM	Canadian Journal of Mathematics
CDMB	Canadian Mathematical Bulletin
CLQM	Colloquium Mathematicum (Warsaw)
CMMH	Commentarii.Mathematici Helvetici
COMT	Compositio Mathematica
CPAP	Communications on Pure and Applied Mathematics
CTCV	Contributions to the Calculus of Variations
CZMJ	Czechoslovak Mathematical Journal

DFUJ	Differencial'nye Uravnenija (Minsk); translated as Differential Equations
DKMJ	Duke Mathematical Journal
DOKL	Doklady Akademiǐ Nauk SSSR
EMTG	Ergebnisse der Mathematik und ihrer Granzgebiete
EMTW	Enzyklopädie der Mathematischen Wissenschaften
ESMT	Encyclopedie des Sciences Mathématiques
GMTW	Die Grundlehren der Mathematischen Wissenschaften
ILJM	Illinois Journal of Mathematics
INDM	Indigationes Mathematicae
IVZM	Izvestija Vysših Učebnyh Zavedeniǐ. Matematika (Kazan)
JAMT	Journal d'Analyse Mathématique
JDEQ	Journal of Differential Equations
JFSH	Journal of the Faculty of Science, Hokkaido University, Series I. Mathematics
JIMS	The Journal of the Indian Mathematics Society
JJMT	Japanese Journal of Mathematics
JLFA	Journal of Functional Analysis
JLMS	The Journal of the London Mathematical Society
JMAA	Journal of Mathematical Analysis and Application
JMMC	Journal of Mathematics and Mechanics
JMPA	Journal de Mathématiques Pures et Appliquées
JMPH	Journal of Mathematics and Physics
JRAM	Journal für die reine und angewandte Mathematik
MCMJ	Michigan Mathematical Journal
MCSA	Matematický Časopis Slovenskej Akadémie Vied. (Bratislava)
MOMT	Monatshefte für Mathematik
MTAN	Mathematische Annalen
MTHN	Mathematical Notes
MTNR	Mathematische Nachrichten
MTSA	Mathematica Scandinavica
MTSK	Matematičeskiǐ Sbornik
MTZT	Mathematische Zeitschrift
NAWK	Nieuw Archief voor Wiskunde
PAMS	Proceedings of the American Mathematical Society
PCPS	Proceedings of the Cambridge Philosophical Society
PEMS	Proceedings of the Edinburgh Mathematical Society
PFJM	Pacific Journal of Mathematics
PNAS	Proceedings of the National Academy of Sciences. USA

PRMF	Prace Matematyczno-Fizyczne
PTGM	Portugaliae Mathematica
PUZM	Permskiǐ Gos. Universitet. Učenye Zapiski
QAMT	Quarterly of Applied Mathematics
QJMO	Quarterly Journal of Mathematics, Oxford
RCMP	Rendiconti del Circolo Matematico di Palermo
RMJM	Rocky Mountain Journal of Mathematics
RMTS	Russian Mathematical Surveys (English translation of Uspehi Matematičeskih Nauk)
RMUP	Rivista di Matematica della Università di Parma
SIJC	SIAM Journal on Control
SJAM	SIAM Journal on Applied Mathematics
SJMA	SIAM Journal on Mathematical Analysis
SMDK	Soviet Mathematics. Doklady (English translation of mathematics section of Doklady Akademii Nauk SSSR)
STMT	Studia Mathematica
TAMS	Transactions of the American Mathematical Society
TRSC	Translations of the Royal Society, Canada
UKMZ	Ukrainian Mathematical Journal (English translation of Ukrainskii Matematičeskii Zurnal)
USMN	Uspehi Matematičeskiȟ Nauk

BIBLIOGRAPHY

Abramovich, S.

1. *Monotonicity of eigenvalues under symmetrization*, SJAM, 28(1975), 350-361.

Achieser, N. I. and I. M. Glazman

1. *Theorie der linearen Operatoren im Hilbert-Raum*, Akademic-Verlag, Berlin, 1954. MR 16-596.

Ahlbrandt, C. D.

1. *Disconjugacy criteria for self-adjoint differential systems*, (Dissertation, Univ. of Oklahoma, 1968), JDEQ 6(1969), 271-295. MR 39 #5855.
2. *Equivalent boundary value problems for self-adjoint differential systems*, JDEQ, 9(1971), 420-435. MR 44 #1860.
3. *Principal and antiprincipal solutions of self-adjoint differential systems and their reciprocals*, RMJM, 2(1972), 169-182. MR 45 #5448.

Anderson, N., A. M. Arthurs and R. R. Hall

1. *Extremum principle for a nonlinear problem in magneto-elasticity*, PCPS, 72(1972), 315-318. MR 45 #7151.

Aronszajn, N., and A. Weinstein

1. *On the unified theory of eigenvalues of plates and membranes*, AMJM, 64(1942), 623-645. (This paper contains a bibliography of other joint papers by these two authors). MR 4-101.

Aronszajn, N.

1. *Rayleigh-Ritz and A. Weinstein methods for approximation of eigenvalues*, I. *Operators in a Hilbert space*, II. *Differential Operators*, PNAS, 34(1948), 474-480 and 594-601. MR 10-382.

Atkinson, F. V.

1. *On second-order linear oscillators*, Univ. Nac. Tucuman, Revista A. 8(1951), 71-87. MR 14-50.

2. *Discrete and Continuous Boundary Problems*, Academic Press, New York, 1964. MR 31 #416.

Bailey, P., and P. Waltman

1. *On the distance between consecutive zeros for second order differential equations*, JMAA 14(1966), 23-30. MR 33 #6009a.

Bamforth, F. R.

1. *A classification of boundary problems for a system of ordinary differential equations containing a parameter*, (Dissertation, Chicago, 1928).

Banks, D. O.

1. *Bounds for the eigenvalues of some vibrating systems*, PFJM, 10(1960), 439-474. MR 22 #8158.

2. *Upper bounds for the eigenvalues of some vibrating systems*, PFJM, 11(1961), 1183-1203. MR 26 #2681.

3. *Bounds for eigenvalues and generalized convexity*, PFJM, 13(1963), 1031-1052. MR 27 #5967.

4. *An integral inequality*, PAMS, 14(1963), 823-828. MR 27 #3767.

5. *Bounds for eigenvalues and conditions for existence of conjugate points*, SJAM, 27(1974), 365-375.

Banks, D. O. and G. J. Kurowski

1. *A Prüfer transformation for the equation of the vibrating beam*, TAMS, 199(1974), 203-222.

Bargmann, V.

1. *On the number of bound states in a central field of force*, PNAS, 38(1952). MR 14-875.

Barnes, D. C.

1. *Positivity conditions for quadratic forms*, JMAP, 37(1972), 607-616. MR 46 #3894.

Barrett, J. H.

1. *Behavior of solutions of second order self-adjoint differential equations*, PAMS, 6(1955), 247-251. MR 17-37.

2. *Matrix systems of second order differential equations*, PGLM 14(1956), 79-89. MR 18-211.

3. *A Prüfer transformation for matrix differential equations*, PAMS, 8(1957), 510-518. MR 19-415.

4. *Second order complex differential equations with a real independent variable*, PFJM 8(1958), 187-200. MR 20 #4675.

5. *Disconjugacy of second order linear differential equations with non-negative coefficients*, PAMS 10(1959), 552-561. MR 21 #7329.

6. *Disconjugacy of a self-adjoint differential equation of the fourth order*, PFJM 11(1961), 25-37. MR 23 #A2594.

7. *Systems-disconjugacy of a fourth order differential equation*, PAMS 12(1961), 205-213. MR 24 #A304.

8. *Fourth order boundary value problems and comparison theorems*, CDJM 13(1961), 625-638. MR 24 #A3350.

9. *Two-point boundary problems for linear self-adjoint differential equations of the fourth order with middle term*, DKMJ 29(1962), 543-544. MR 26 #6477.

10. *Oscillation Theory of Ordinary Linear Differential Equations*, Advances in Mathematics, 3(1969), 415-509. (reprinted in *Lectures on Ordinary Differential Equations*, edited by R. McKelvey, Academic Press, 1970). MR 41 #2113.

Beesack. P. R.

0. *Nonoscillation and disconjugacy in the complex domain*, TAMS 81(1956), 211-242.

1. *A note on an integral inequality*, PAMS, 8(1957), 875-879. MR 19-947.

2. *Integral inequalities of the Wirtinger type*, DKMJ, 25(1958), 477-498. MR 20 #3947.

3. *Extensions of Wirtinger's inequality*, TRSC, 53(1959), 21-30.

4. *Integral inequalities involving a function and its derivative*, AMMM, 78(1971), 705-741.

5. *On Sturm's separation theorem*, CDMB, 15(1972), 481-487. MR 48 #2489.

Beesack, P. R., and Schwarz, B.

1. *On the zeros of solutions of second-order linear differential equations*, CDJM, 8(1956), 504-515. MR 18-211.

Bellman, R.

1. *The boundedness of solutions of linear differential equations*, DKMJ 14(1947), 83-97. MR 9-35.

Benson, D. C., and K. Kreith

1. *On abstract Prüfer transformations*, PAMS, 26(1970), 137-140. MR 41 #7243.

Bhatia, N. P.

1. *Some oscillation theorems for second order differential equations*, JMAA 15(1966), 442-446. MR 34 #3017.

Bieberbach, L.

1. *Theorie der Differentialgleichungen*, 3rd ed., New York, Dover, 1944. MR 6-153.

Birkhoff, G. D.

1. *On the asymptotic character of the solutions of certain linear differential equations containing a parameter,* TAMS 9(1908), 219-231.
2. *Boundary value and expansion problems of ordinary linear differential equations,* TAMS 9(1908), 373-395.
3. *Existence and oscillation theorems for a certain boundary value problem,* TAMS 10(1909), 259-270.
4. *The scientific work of Maxime Bôcher,* BAMS, 25(1918-19), 197-215.

Birkhoff, G. D., and M. R. Hestenes

1. *Natural isoperimetric conditions in the calculus of variations,* DKMJ, 1(1935), 198-286.

Birkhoff, Garrett and G.-C. Rota

1. *Ordinary Differential Equations,* Ginn and Co., 1959; 2nd ed., Blaisdell Pub. Co., 1962. MR 25 #2253.
2. *On the completeness of Sturm-Liouville expansions,* AMMM, 67(1960), 835-841. MR 23 #A2577.

Bliss, G. A.

1. *Some recent developments in the calculus of variations,* BAMS 26(1919-20), 343-361.
2. *A boundary value problem in the calculus of variations,* BAMS 32(1926), 317-331.
3. *The transformation of Clebsch in the calculus of variations,* Proc. Int. Congress held in Toronto, 1922, 1(1924), 589-603.
4. *A boundary value problem for a system of ordinary differential equations of the first order,* TAMS 28(1926), 561-584.
5. *The problem of Lagrange in the calculus of variations,* AMJM 52(1930), 673-744.
6. *Definitely self-adjoint boundary value problems,* TAMS 44(1938), 413-428.
7. *Lectures on the Calculus of Variations,* Univ. of Chicago Press, Chicago, 1946. MR 8-212.

Bobonis, A.

1. *Differential systems with boundary conditions involving the characteristic parameter,* (Dissertation, Univ. of Chicago, 1939), CTCV, (1938-41), 99-138. MR 4-200.

Bôcher, M.

1. *Ueber die Reihenentwickelungen der Potentialtheorie,* (Dissertation, Göttingen, 1891). An expanded version also published in book form by Teubner, 1894.
2. *The theorems of oscillation of Sturm and Klein,* BAMS, 4(1897-98), 295-313, 365-376; 5(1898-99), 22-43.

3. *Applications of a method of d'Alembert to the proof of Sturm's theorem of comparison*, TAMS 1(1900), 414-420.

4. *Randwertaufgaben bei gewohnlichen Differentialgleichungen*, Enzyklopädie der Mathematischen Wissenschaften, II. A. 7a(1900), 437-463.

5. *Boundary problems in one dimension*, Proc. of the Fifth International Congress of Mathematicians, I, Cambridge, 1912, 163-195.

6. *Applications and generalizations of the conception of adjoint systems*, TAMS, 14(1913), 403-420.

7. *Lecons sur les méthodes de Sturm dans la theorie des équations différentielles linéaires, et leurs développements modernes*, Gauthier-Villars, Paris, 1917. (See also G. D. Birkhoff [4], for a complete bibliography of Bôcher's publications).

Boerner, H.

1. *Das Eigenwertproblem der selbstadjungierten linearen Differentialgleichung vierter Ordnung*, MTZT, 34(1931), 293-319.

2. *Über einige Variationsprobleme*, MTZT, 35(1932), 161-189.

3. *Zur Theorie der zweiten Variation*, MTZT, 39(1935), 492-500.

Bohl, P.

1. *Über eine Differentialgleichung der Störungstheorie*, JRAM, 131(1906), 268-321.

2. *Sur certaines équations différentielles d'une type général utilisable en mecanique*, BSMF, 38(1910), 5-138. (French translation of the author's 1900 doctoral dissertation.)

Bolza, O.

1. *Lectures on the Calculus of Variations*, Univ. of Chicago Press, 1904; reprinted by G. E. Stechert and Co., New York, 1931, Chelsea, New York.

2. *Vorlesungen über Variationsrechnung*, Teubner, Leipzig 1909; reprinted by Koehler's Antiquarium, Leipzig, 1933, Chelsea, New York.

Borg, G.

1. *Eine Umkehrung der Sturm-Liouvilleschen Eigenwertaufgabe Bestimmung der Differentialgleichungen durch die Eigenwerte*, ACTM, 78(1946), 1-96. MR 7-382.

2. *On a Liapunoff criterion of stability*, AMJM, 71(1949), 67-70. MR 10-456.

Borůvka, O.

1. *Théorie analytique et constructive des transformations différentielles linéaires du second ordre*, BMSR, 1(49), (1957), 125-130. MR 21 #3608.

2. *Transformation of ordinary second-order linear differential equations and their applications*, Proceedings of Conference, Prague, 1962, Academic Press, New York (1963), 27-38. MR 30 #295.

3. *Sur quelques applications des dispersions centrales dans la théorie des équations différentielles linéaries du deuxieme ordre*, ARVM, 1(1965), 1-20. MR 33 #5984.

4. *Lineare Differentialtransformationen zweiter Ordnung*, Veb. Deutsche Verlag der Wissenschaften, Berlin (1967), MR 38 #4743. (Expanded and translated to English by F. M. Arscott, Published under the title "Linear Differential Transformations of the Second Order," by the English University Press Ltd. - Pittman, 1971.)

Bott, R.

1. *On the iteration of closed geodesics and the Sturm intersection theory*, CPAP, 9(1956), 171-206. MR 19-859.

Bradley, J. S.

1. *Adjoint quasi-differential operators of Euler type*, (Dissertation Univ. of Iowa, 1964), PFJM, 16(1966), 213-237; Correction to: PFJM, 19(1966), 587-588. MR 34 #409(E); 34 #1601.

2. *Conditions for the existence of conjugate points for a fourth order linear differential equation*, SJAM, 17(1969), 984-991. MR 40 #5970.

Bradley, J. S., and W. N. Everitt

1. *Inequalities associated with regular and singular problems in the calculus of variations*, TAMS, 182(1973), 303-321. MR 48 #8943.

Breuer, S. and Gottlieb, D.

1. *Upper and lower bounds on solutions of initial value problems*, JMAA, 36(1971), 283-300. MR 46 #424.

2. *Upper and lower bounds on eigenvalues of Sturm-Liouville systems*, JMAA, 36(1971), 465-476. MR 44 #2973.

Buckley, E. D. J.

1. *A bibliography of publications concerned with the oscillation of solutions to the equation* (p(t)y')' + q(t)y = 0. Publication of the Department of Mathematics of the University of Alberta, Edmonton, Alberta, 1972.

Burkhardt, M. H.

1. *Sur les fonctions de Green relative à un domaine d'une dimension*, BSMF, 22(1894), 71-73.

Butler, G. and J. W. Macki

1. *Oscillation and comparison theorems for second order linear differential equations with integrable coefficients*, CNJM, 26(1974), 294-301.

Carathéodory, C.

1. *Vorlesungen über reelle Funktionen*, Teubner, Leipzig, 1918; 2nd. edition, 1927, Chelsea, 1968.
2. *Die Theorie der zweiten Variation beim Problem von Lagrange*, Munchner Berichte, 1932, 99-144.
3. *Über die Einteilung der Variationsprobleme von Lagrange nach Klassen*, CMMH, 5(1933), 1-19.

Carmichael, R. D.

1. *Comparison theorems for homogeneous linear differential equations of general order*, ANNM, 19(1918), 159-171.
2. *Boundary value and expansion problems*, AMJM, 43(1921), 69-101 and 232-70; 44(1922), 129-152.
3. *Algebraic guides to transcendental problems*, BAMS, 28(1922), 179-210.

Chellevold, J. O.

1. *Conjugate points of singular quadratic functionals for N dependent variables*, Proc. Iowa Acad. Sci., 59(1952), 331-337. MR 14-769.

Cheng, Sui-Sun

1. *Systems - conjugate and focal points of fourth order nonselfadjoint differential equations*, TAMS, 223(1976), 155-165. MR 57 #13005.

Cimmino, G.

1. *Autosoluzione e autovalore nelle equazioni differenziali lineari ordinarie autoaggiunte di ordine superiore*, MTZT, 32(1930), 4-58.
2. *Extensione dell'identita di Picone alla piu generale equazione differenziale lineare ordinaria autoaggiunta*, ANLR, 28(1939), 354-364.

Clebsch, A.

1. *Ueber die Reduktion der zweiten Variation auf ihre einfachste Form*, JRAM, 55(1858), 254-270.
2. *Ueber diejenigen Probleme der Variationsrechnung, welche nur eine unabhangige Variable enthalten*, JRAM, 55(1858), 335-355.

Coddington, E. A., and Levinson, N.

1. *Theory of Ordinary Differential Equations*, McGraw-Hill, New York, 1955. MR 16-1022.

Cohn, J. H. E.

1. *On the number of negative eigen-values of a singular boundary value problem*, JLMS, 40(1965), 523-525. MR 32 #7837.

2. *A theorem of Bargman's*, QJMO, 17(1966), 51-52. MR 33 #7617.

3. *Consecutive zeros of solutions of ordinary second order differential equations*, JLMS, 5(1972), 465-468. MR 47 #7127.

Colautti, Maria P.

1. *Sul calcolo degli autovalori di un problema ai limiti* I, II, III, ANLR (8) 51(1971), 477-485 (1972); (8) 52(1972), 24-35; (8) 52(1972), 141-149. MR 49 #1791.

Coles, W. J.

1. *Linear and Riccati systems*, DKMJ, 22(1955), 333-338. MR 17-482.

2. *A general Wirtinger-type inequality*, DKMJ, 27(1960), 133-138. MR 22 #1638.

3. *A note on matrix Riccati systems*, PAMS, 12(1961), 557-559. MR 26 #1522.

4. *Wirtinger-type integral inequalities*, PJMT, 11(1961), 871-877. MR 25 #4049.

5. *Some boundary value problems for linear differential systems*, PAMS, 14(1963), 956-960. MR 28 #4168.

6. *Matrix Riccati differential equations*, SJAM, 13(1965), 627-634. MR 32 #2666.

7. *An oscillation criterion for second-order linear differential equations*, PAMS, 19(1968), 755-759.

8. *A simple proof of a well-known oscillation theorem*, PAMS, 19(1968), 507. MR 36 #6692.

Coles, W. J. and D. Willett

1. *Summability criteria for oscillation of second order linear differential equations*, AMPA, 79(1968), 391-398. MR 38 #4757.

Collatz, L.

1. *Eigenwertprobleme und ihre numerische Behandlung*, Akademische Verlagsgesellschaft, 1945; Reprinted by Chelsea, New York, 1948. MR 8-574.

Cope, T.

1. *An analogue of Jacobi's condition for the problem of Mayer with variable end points*, (Dissertation, Univ. of Chicago, 1927), AMJM, 59(1937), 655-672.

Coppel, W. A.

1. *Comparison theorems for canonical systems of differential equations*, JMAA, 12(1965), 306-315. MR 32 #7825.

2. *Disconjugacy*, Lecture Notes in Mathematics, No. 220, Springer Verlag, 1971.

3. *Matrix quadratic equations*, BAUS, 10(1974), 377-401.

Courant, R. and Hilbert, D.

1. *Methoden der Mathematischen Physik*, I, II, Springer, Berlin, 1924.
2. *Methods of Mathematical Physics*, I, II, Interscience, New York, 1953.

Courant, R.

1. *Uber die Eigenwerte bei den Differentialgleichungen der Mathematischen Physik*, MTZT, 7(1920), 1-57.
2. *Uber die Anwendung der Variationsrechnung in der Theorie der Eigenschwingungen und uber neue Klassen von Funktionalgleichungen*, ACMT, 49(1926), 1-68.

Crum, M. M.

1. *Associated Sturm-Liouville systems*, QJMO, (2) 6(1955), 121-127. MR 17-266.

Cushman, R., and J. J. Duistermaat

1. *The behavior of the index of a periodic linear Hamiltonian system under iteration*, Advances in Math., 23(1977), 1-21.

Denny, W. F.

1. *Oscillation criteria for a linear Riemann-Stieltjes integral equation system*, JDEQ, 22(1976), 14-27.

Diaz, J. B., and J. R. McLaughlin

1. *Sturm separation and comparison theorems for ordinary and partial differential equations*, ANLR (VIII) 9(1969), 134-194.
2. *Sturm comparison and separation theorems for linear, second order, self-adjoint ordinary differential equations and for first order systems*, APLA, 2(1972), 355-376.

Drahlin, M. E.

1. *On the zeros of solutions of a Riccati equation* (Russian), IVZM, (1965), No. 5 (48), 54-64. MR 32 #7847.
2. *On the existence of a denumerable set of zeros for the solutions of the Riccati equation*, PUZM, No. 103 (1963), 164-172. MR 31 #2454.
3. *A comparison principle for differential equations of the second-order on an infinite interval* (Russian), IVZM, (1967), No. 9 (64), 26-30. MR 36 #467.

Duistermaat, J. J.

1. *On the Morse index in variational calculus*, Advances in Math., 21(1976), 173-195.

Dunford, N., and J. T. Schwartz

1. *Linear Operators*, I, II, III. Interscience, New York, 1958, 1963, 1971.

Eastham, M. S. P.

1. *The Picone identity for self-adjoint differential equations of even order*, Mathematika, 20(1973), 197-200.

Edelson, A., and K. Kreith

1. *Upper bounds for conjugate points for nonselfadjoint fourth order differential equations*, ANLR 58(1975), 686-695. MR 55 #767.

Edwards, H. M.

1. *A generalized Sturm theorem*, ANNM, 80(1964), 22-57. MR 29 #1652.

Eisenfeld, J.

1. *On the number of interior zeros of a one parameter family of solutions to a second order differential equation satisfying boundary conditions at one end-point*, JDEQ, 11(1972), 202-206. MR 45 #7155.

Eliason, S. B.

1. *The integral* $T\int_{-T/2}^{T/2} p(t)dt$ *and the boundary value problem* $x'' + p(t)x = 0$, $x(-T/2) = x(T/2) = 0$, JDEQ, 4(1968), 646-660. MR 38 #1313.

2. *A Lyapunov inequality for a certain second order non-linear differential equation*, JLMS (2), 2(1970), 461-466. MR 42 #2093.

Eliason, S. B. and D. F. St. Mary

1. *Upper bounds of* $T\int_{-T/2}^{T/2} p(t)dt$ *and the differential equation* $x'' + p(t)x = 0$, JDEQ, 6(1969), 154-160. MR 39 #1742.

El'šin, M. I.

1. *Qualitative problems on the linear differential equation of the second order*, (Russian), DOKL, 68(1949), 221-224. MR 11-110.

2. *The phase method and the classical method of comparison*, (Russian), DOKL, 68(1949), 813-816. MR 11-247.

3. *Qualitative solution of a linear differential equation of the second order*, (Russian), USMN, 5(1950), 155-158. MR 12-27.

4. *Qualitative investigation of a system of two linear homogeneous equations of the first order*, (Russian), DOKL, 94(1954), 5-8. MR 15-957.

5. *On a solution of a classical oscillation problem*, (Russian), DOKL, 147(1962), 1013-1016 = SMDK, 3(1962), 1752-1755. MR 26 #1581.

von Escherich, G.

1. *Die zweite Variation der einfachen Integral*, Wiener Sitzungsberichte, (8) 107, (1898), 1191-1250, 1267-1326, 1383-1430; 108(1899), 1269-1340; 110(1901), 1355-1421.

Etgen, G. J.

1. *Oscillation properties of certain nonlinear matrix equations of second order*, TAMS, 122(1966), 289-310. MR 32 #7834.

2. *A note on trigonometric matrices*, PAMS, 17(1966), 1226-1232. MR 35 #4504.

3. *On the determinants of second order matrix differential systems*, JMAA, 18(1967), 585-598. MR 35 #4505.

4. *On the oscillation of solutions of second order self-adjoint matrix differential equations*, JDEQ, 6(1969), 187-195. MR 39 #3091.

5. *Two point boundary problems for second order matrix differential systems*, TAMS, 149(1970), 119-267. MR 42 #7977.

6. *Oscillation criteria for nonlinear second order matrix differential equations*, PAMS, 27(1971), 259-267. MR 43 #617.

Etgen, G. J., and J. B. Scott

1. *On the conjugate points of fourth order, self-adjoint linear differential equations*, PAMS, 29(1971), 349-350. MR 43 #6514.

Ettlinger, H. J.

1. *Existence theorems for the general real self-adjoint linear systems of the second order*, TAMS, 19(1918), 79-96.

2. *Existence theorem for the non-self-adjoint linear system of the second order*, ANNM, 21(1919-20), 278-290.

3. *Oscillation theorems for the real self-adjoint linear system of the second order*, TAMS, 22(1921), 136-143.

Fair, W. G.

1. *Continued fraction solution to the Riccati equation in a Banach algebra*, JMAA, 39(1972), 318-323. MR 46 #5784.

Fan, K., Taussky, O., and Todd, J.

1. *Discrete analogs of inequalities of Wirtinger*, MOMT, 59(1955), 73-90. MR 17-19.

Feller, W.

1. *Generalized second order differential operators and their lateral conditions*, ILJM, 1(1957), 459-504. MR 19-1052.

Fink, A. M.

1. *Maximum amplitude of controlled oscillations*, JMAA, 14(1966), 253-262. MR 34 #4614.

2. *The functional* $T\int_0^T R$ *and the zeros of a second order linear differential equation*, JMPA, 45(1966), 387-394. MR 34 #7863.

3. *On the zeros of* y" + py = 0 *with linear, convex and concave* p, JMPA, 46(1967), 1-10. MR 35 #4509.

4. *Comparison theorems for* $\int_a^b p$ *with* p *an admissible sub or superfunction*, JDEQ, 5(1969), 49-54. MR 38 #1315.

5. *Eigenvalue of the square of a function*, PAMS, 20(1969), 73-74. MR 38 #2365.

6. *Comparison theorems for eigenvalues*, QAMT, 28(1970), 289-292. MR 41 #8749.

Fink, A. M., P. Hartman and D. F. St. Mary

1. *On disconjugacy criteria*, manuscript.

Fink, A. M. and D. F. St. Mary

1. *A generalized Sturm comparison theorem and oscillation coefficients*, MOMT, 73(1969), 207-212. MR 39 #5875.

2. *On an inequality of Nehari*, PAMS, 21(1969), 640-642. MR 39 #1737.

Fite, W. B.

1. *Concerning the zeros of the solutions of certain differential equations*, TAMS, 19(1918), 341-352.

Fort, T.

1. *Linear difference and differential equations*, AMJM, 39(1917), 1-26.

2. *Some theorems of comparison and oscillation*, BAMS, 24(1918), 330-334.

3. *Finite Differences and Difference Equations in the Real Domain*, Oxford, Clarendon Press, 1948. MR 9-514.

Fubini, G.

1. *Alcuni nuovi problemi di calcolo di variazioni, con applicazioni alla teoria delle equazioni integro-differenziali*, AMPA, 20(1913), 217-244.

2. *Su un teorema di confronto per le equizioni del secondo ordine alle derivate ordine*, ASNP, II, 2(1933), 283-284.

Gagliardo, E.

1. *Sul comportamento asintotico degli integrali dell' equazione differenziale* y" + A(x)y = 0 *con* A(x) $\geq$ 0, BUMI (3) 8(1953), 177-185. MR 15-126.

2. *Sui criteri di oscillazione per gli integrali di un'equazione differenziale lineare del secondo ordine*, BUMI (3) 9(1954), 177-189. MR 16-247.

Galbraith, A. S.

1. *On the zeros of solutions of ordinary differential equations of the second order*, PAMS, 17(1966), 333-337. MR 32 #7848.

Giuliano, L.

1. *Sul teorema di confronto di Sturm*, BUMI, (3) 2(1947), 16-19. MR 9-36.

Glazman, I. M.

1. *Direct Methods of Qualitative Spectral Analysis of Singular Differential Operators*, Translated from the Russian by the Israel Program for Scientific Translations, Jerusalem, 1965. MR 32 #8210.

Gottlieb, M. J.

1. *Oscillation theorems for self-adjoint boundary value problems*, DKMJ, 15(1948), 1073-1091. MR 10-537.

Gould, S. H.

1. *Variational Methods for Eigenvalue Problems*, 2nd edition, Mathematical Expositions, No. 10, Univ. of Toronto Press, Toronto, 1966. MR 35 #559.

Gregory, J.

1. *A theory of focal points and focal intervals for an elliptic quadratic form on a Hilbert space*, TAMS, 157(1971), 119-128. MR 43 #3878.
2. *An approximation theory for elliptic quadratic forms on Hilbert spaces: application to the eigenvalue problem for compact quadratic forms*, PFJM, 37(1970), 383-395.

Guggenheimer, H. W.

1. *Geometric theory of differential equations*, I. *Second order linear equations*, SJMA, 2(1971), 233-241. MR 44 #2977.
2. *Geometric theory of differential equations*, III. *Second order equations on the reals*, ARMA, 41(1971), 219-240.
3. *Inequalities for eigenvalues of homogeneous boundary value problems*, AMPA,(4) 98(1974), 281-296.

Gustafson, G. B.

1. *The nonequivalence of oscillation and disconjugacy*, PAMS, 25(1970), 254-260. MR 44 #1872.
2. *Eventual disconjugacy of self-adjoint fourth-order linear differential equations*, PAMS, 35(1972), 187-192. MR 45 #7178.
3. *Interpolation between consecutive conjugate points of an* n-*th order linear differential equations*, TAMS, 177(1973), 237-255. MR 47 #8958.
4. *Higher order separation and comparison theorems with applications to solution space problems*, AMPA, 95(1973), 245-254. MR 48 #4407.

Hahn, H.

1. *Über räumliche Variationsprobleme*, MTAN, 70(1911), 110-142.
2. *Über Variationsprobleme mit variablen Endpunkten*, MOMT, 22(1911), 127-136.

Halanay, A. and S. Sandor

1. *Sturm-type theorems for self-conjugate systems of linear differential equations of higher order*, BMSR, 1(49), (1957), 401-431. (Russian). MR 21 #5739.

Hardy, G. H., J. E. Littlewood, and G. Pólya

1. *Inequalities*, Cambridge Univ. Press, New York, 1934.

Harris, V. C.

1. *A system of linear difference equations and an associated boundary value problem*, (Dissertation, Northwestern Univ., 1950).

Hartman, P.

1. *On a theorem of Milloux*, AMJM, 70(1948), 395-399. MR 10-120.

2. *On the linear logarithmico-exponential differential equation of the second order*, AMJM, 70(1948), 764-779. MR 10-376.

3. *Differential equations with non-oscillatory eigenfunctions*, DKMJ, 15(1948), 697-709. MR 10-376.

4. *The number of* L^2- *solutions of* $x'' + q(t)x = 0$, AMJM, 73(1951), 635-645. MR 13-462.

5. *On the eigenvalues of differential equations*, AMJM, 73(1951), 657-662. MR 13-463.

6. *On linear second order differential equations with small coefficients*, AMJM, 73(1951), 955-962. MR 13-652.

7. *On non-oscillatory linear differential equations of the second order*, AMJM, 74(1952), 389-400. MR 14-50.

8. *On the zeros of solutions of second order linear differential equations*, JLMS, 27(1952), 492-496. MR 14-278.

9. *On the derivatives of solutions of linear, second order, ordinary differential equations*, AMJM, 75(1953), 173-177. MR 14-754.

10. *Self-adjoint, non-oscillatory systems of ordinary, second order, linear differential equations*, DKMJ, 24(1957), 25-35. MR 18-576.

11. *On oscillators with large frequencies*, BUMI, 14(1959), 62-65. MR 21 #3610.

12. *The existence of large or small solutions of linear differential equations*, DKMJ, 28(1961), 421-430. MR 24 #A293.

13. *Ordinary Differential Equations*, John Wiley and Sons, Inc., New York, (1964), MR 30 #1270.

14. *Disconjugate nth order differential equations and principal solutions*, BAMS, 74(1968), 125-129. MR 36 #5440.

15. *On disconjugate differential equations*, TAMS, 134(1968), 53-70. MR 37 #1728.

16. *Principal solutions of disconjugate* n-*th order differential equations,* AMJM, 91(1969), 306-362; Corrigendum and Addendum, AMJM, 93(1971), 439-451. MR 40 #450 (E).

17. *On an ordinary differential equation involving a convex function,* TAMS, 146(1969), 179-202. MR 43 #2283.

18. *On disconjugacy criteria,* PAMS, 24(1970), 374-381. MR 40 #4535.

19. *Boundary value problems for second order, ordinary differential equations involving a parameter,* JDEQ, 12(1972), 194-212. MR 49 #705.

Hartman, P. and A. Wintner

1. *The asymptotic arcus variation of solutions of real linear differential equations of second order,* AMJM, 70(1948), 1-10. MR 9-435.

2. *On the asymptotic problem of the zeros in wave mechanics,* AMJM, 70(1948), 461-480. MR 10-194.

3. *On non-conservative linear oscillators of low frequency,* AMJM, 70(1948), 529-539. MR 10-194.

4. *Oscillatory and non-oscillatory linear differential equations,* AMJM, 71(1949), 627-649. MR 11-109.

5. *On an oscillation criterion of Liapunoff,* AMJM, 73(1951), 885-900. MR 13-652.

6. *On non-oscillatory linear differential equations,* AMJM, 75(1953), 717-730. MR 15-527.

7. *On non-oscillatory linear differential equations with monotone coefficients,* AMJM, 76(1954), 207-219. MR 15-527.

8. *On the assignment of asymptotic values for the solutions of linear differential equations of the second order,* AMJM, 77(1955), 475-483. MR 17-36.

9. *An inequality for the first eigenvalue of an ordinary boundary value problem,* QAPM, 13(1955), 324-326. MR 17-619.

10. *On an oscillation criterion of de la Vallée Poussin,* QAPM, 13(1955), 330-332. MR 17-484.

11. *On disconjugate differential systems,* CDJM, 8(1956), 72-81. MR 17-611.

Hayden, T. L., and H. C. Howard

1. *Oscillation of differential equations in Banach spaces,* AMPA, 85(1970), 383-394. MR 42 #2137.

Hazard, K.

1. *Index theorems for the problem of Bolza in the calculus of variation,* CTCV, (1938-1941), Univ. of Chicago Press, 293-356. MR 4-47.

Heimes, K. A.

1. *Two-point boundary problems in Banach spaces,* JDEQ, 5(1969), 215-225. MR 39 #1779.

Heinz, E.

1. *Halbbeschränktheit gewöhnlicher Differentialoperatoren höherer Ordnung,* MTAN, 135(1958), 1-49. MR 21 #743.

Hestenes, M. R.

1. *The problem of Bolza in the calculus of variations,* BAMS, 48(1942), 57-75. MR 3 #248.
2. *Applications of the theory of quadratic forms in Hilbert space to the calculus of variations,* PFJM, 1(1951), 525-581. MR 13-759.
3. *Quadratic variational theory,* Control Theory and the Calculus of Variations, Academic Press, New York, 1969, 1-37.

Hickson, A. O.

1. *An application of the calculus of variations to boundary problems,* (Dissertation, Univ. of Chicago, 1928), TAMS, 31(1929), 563-597.

Hilbert, D.

1. *Grundzüge einer allgemeiner Theorie der linearen Integralgleichungen,* Teubner, Leipzig, 1912; reprinted by Chelsea, New York, 1952.

Hille, E.

1. *Non-oscillation theorems,* TAMS, 64(1948), 234-252. MR 10-376.
2. *Lectures of Ordinary Differential Equations,* Addison-Wesley, Reading, Mass., 1969. MR 40 #2939.

Hille, E., and R. S. Phillips

1. *Functional Analysis and Semi-groups,* AMS Colloquium Publications, 31(1957). MR 19 #664.

Hinton, D. B.

1. *Clamped end boundary conditions for fourth order self-adjoint differential equations,* DKJM, 34(1967), 131-138. MR 34 #7864.
2. *A criterion for* n-n *oscillation in differential equations of order* 2n, PAMS, 19(1968), 511-518. MR 37 #1701.
3. *Disconjugate properties of a system of differential equations,* JDEQ, 2(1966), 420-437. MR 34 #7856.
4. *Limit point criteria for positive definite fourth-order differential operators,* QTJM, 24(1973), 367-376. MR 48 #6529.

Hockstadt, H.,

1. *On an inequality of Lyapunov,* PAMS, 22(1969), 282-284.

Hölder, Ernst

1. *Die Lichtensteinsche Methoden fur die Entwicklung der zweiten Variation, angewandt auf das Problem von Lagrange,* PRMF, 43(1935), 307-346.

2. *Entwicklungssätze aus der Theorie der zweiten Variation, allgemeine Randbedigungen*, ACMT, 70(1939), 193-242.

3. *Einordnung besonderer Eigenwertprobleme in die Eigenwerttheorie kanonischer Differentialgleichungssysteme*, MTAN, 119(1943), 21-66. MR 5-265.

4. *Über den aufbau eines erweiterten Greenschen Tensore kanonische Differentialgleichungen aus assozierten Lösungssystemen*, ASPM, 25(1952), 115-121. MR 15-224.

Horgan, C. O.

1. *A note on a class of integral inequalities*, PCPS, 74(1973), 127-131. MR 48 #9486.

Howard, H. C.

1. *Oscillation criteria for fourth order linear differential equations*, TAMS, 96(1960), 296-311. MR 22 #8159.

2. *Oscillation and non-oscillation criteria for* $y'' + f(y(x))p(x) = 0$, PFJM, 12(1962), 243-251. MR 26 #2685.

3. *Oscillation criteria for even order linear differential equations*, AMPA, (IV) 66(1964), 221-231. MR 30 #2195.

4. *Oscillation criteria for matrix differential equations*, CDJM, 19(1967), 184-199; MR 35 #3126. 21(1969), 1279-1280.

5. *Some oscillation criteria for general self-adjoint differential equations*, JLMS, 43(1968), 401-406; corrigendum (2) 1(1969), 660. MR 40 #1649 (E).

6. *Oscillation criteria for matrix differential equations with oscillatory coefficients*, AMPA, (IV) 85(1970), 83-91. MR 42 #2095.

Hu, K.-S.

1. *The problem of Bolza and its accessory boundary value problem*, (Dissertation, Univ. of Chicago, 1932), CTCV, (1931-32), 361-443.

Hunt, R. W.

1. *The behavior of solutions of ordinary, self-adjoint differential equations of arbitrary even order*, PFJM, 12(1962), 945-961. MR 26 #5239.

2. *Oscillation properties of even-order linear differential equations*, TAMS, 115(1965), 54-61. MR 34 #2985.

Hunt, R. W., and M. S. T. Namboodiri

1. *Solution behavior for general self-adjoint differential equations*, PLMS, 21(1970), 637-650. MR 43 #7710.

Ince, E. L.

1. *Ordinary Differential Equations*, Longmans, Green and Co., London, 1927.

Jackson, Rosa L.

1. *The boundary value problem of the second variation for parametric problems in the calculus of variations*, (Dissertation, Univ. of Chicago, 1928).

Jacobi, C. G. J.

1. *Zur Theorie der Variationsrechnung und die Differentialgleichungen*, JRAM, 17(1837), 68-82. A French translation of this paper appeared in JMPA, 3(1838), 44-59.

Jakubovic, V. A.

1. *A two-sided estimate for the solution of a homogeneous differential equation of order two*, (Russian) Metody Vycisl. 1(1963), 30-44. MR 32 #4305.

2. *Oscillatory properties of the solutions of canonical equations*, (Russian), Mat. Sb. 56(98), (1962), 3-42-AMST (2) 42(1964), 247-288. MR 25 #2303.

Kac, I. S., and M. G. Krein

1. *On the spectral function of the string*, AMST, (2) 103(1974), 19-102.

Kamenev, I. V.

1. *Integral nonoscillation criteria*, Mat. Zametki, 13(1973), 51-74 = MTHN, 13(1973), 31-32. MR 48 #2492.

2. *Oscillation criteria, connected with averaging, for the solutions of second order ordinary differential equations*, DFUJ, 10(1974), 246-252, 371. MR 49 #718.

Kamke, E.

1. *Differentialgleichungen reeller Funktionen*, Academische Verlagsgesellschaft, Leipzig, 1930; reprinted by Chelsea, New York, 1947. MR 8-514.

2. *Zur Theorie der Systeme gewöhnlicher Differentialgleichungen* II, ACMT, 58(1932), 57-85.

3. *Über die Existenz von Eigenwerten bei Randwertaufgaben zweiter Ordnung*, MTZT, 44(1939), 619-634.

4. *Neue Herleitung der Oszillationssätze für die linearen selbstadjungierten Randwertaufgaben zweiter Ordnung*, MTZT, 44(1939), 635-658.

5. *A new proof of Sturm's comparison theorems*, AMMM, 46(1939), 417-421; a slightly revised version of this paper appeared as *Über Sturms Vergleichssätze für homogene lineare Differentialgleichungen zweiter Ordnung und Systeme von zwei Differentialgleichungen erster Ordnung*, MTZT, 47(1942), 788-795. MR 7-297.

6. *Über die definiten selbstadjungierten Eigenwertaufgaben bei gewöhnlichen linearen Differentialgleichungen*, I, II, III, IV, MTZT, 45(1939), 759-787; 46(1940), 231-250; 251-286; 48(1942), 67-100.

7. *Differentialgleichungen Lösungsmethoden und Lösungen*, I (Gewöhnliche Differentialgleichungen), 6th ed., Akademische Verlagsgesellschaft, Leipzig, 1959.

Karush, W.

1. *Isoperimetric problems and index theorems in the calculus of variations*, (Dissertation, Univ. of Chicago, 1942).

Kaufman, H., and R. L. Sternberg

1. *A two-point boundary problem for ordinary self-adjoint differential equations of even order*, DKMJ, 20(1953), 527-531. MR 15-530.

Keener, M. S.

1. *On the equivalence of oscillation and the existence of infinitely many conjugate points*, RMJM, 5(1975), 125-134.

Kegley, J. Colby

1. *Convexity with respect to Euler-Lagrange differential operators*, PFJM, 16(1966), 87-111. MR 32 #7889.

Kestens, J.

1. *Le problème aux valeurs propres normal et bornes supérieures et inférieures par la méthode des itérations*, Acad. Roy. Belg. Cl. Sci. Mem. Coll. in 8°, 29(1956), No. 4, MR 19-961.

Kim, W. J.

1. *Disconjugacy and non-oscillation criteria for linear differential equations*, JDEQ, 8(1970), 163-172. MR 42 #2096.

2. *On the extremal solutions of the nth order linear differential equation*, PAMS, 33(1972), 62-68. MR 45 #3848.

Kneser, A.

1. *Untersuchungen über die reellen Nullstellen der Integrale linearer Differentiagleichungen*, MTAN, 42(1893), 409-435.

2. *Untersuchungen über die Darstellung Willkürlicher Funktionen in der Mathematischen Physik*, MTAN, 58(1904), 81-147.

3. *Beiträge zur Theorie der Sturm-Liouvilleschen Darstellung willkürlicher Funktionen*, MTAN, 60(1905), 402-423.

4. *Die Theorie der Integralgleichungen und die Darstellung willkürlicher Funktionen in der Mathematischen Physik*, MTAN, 63(1907), 477-524.

Komkov, V.

1. *A technique for the detection of oscillation of second order ordinary differential equations*, PFJM, 42(1972), 105-115. MR 47 #2134.

2. *A note on the semi-inverse approach to oscillation theory for ordinary second-order linear differential equations*, CLQM, 28(1973), 141-144. MR 49 #719.

Kondrat'ev, V. A.

1. *Elementary derivation of a necessary and sufficient condition for non-oscillation of the solutions of a linear differential equation of second order* (Russian), USMN, 12(1957), 159-160. MR 19-856.

2. *Sufficient conditions for non-oscillatory or oscillatory nature of solutions of second order equations* $y'' + p(x)y = 0$. (Russian), DOKL, 113(1957), 742-745. MR 19-960.

König, R.

1. *Die Oszillationseigenschaften der Eigenfunktionen der Integralgleichungen mit definiten kern und das Jacobische Kriterium der Variationsrechnung*, (Dissertation, Göttingen, 1907).

Krbilja, J.

1. *Structure formulas of quadratic functionals*, (Russian), MCSA, 23(1973), 353-363. MR 49 #692.

Krein, M. G.

1. *On certain problems on the maximum and minimum of characteristic values and on the Lyapunov zones of stability*, AMST, Series 2, 1(1955), 163-187. MR 17-484.

Kreith, K.

1. *An abstract oscillation theorem*, PAMS, 19(1968), 17-20. MR 36 #5745.

2. *A direct method for self-adjoint systems of second order differential equations*, ANLR, 46(1969), 330-332. MR 41 #5706.

3. *A comparison theorem for fourth order differential equations*, ANLR, 46(1969), 664-666. MR 42 #591.

4. *A comparison theorem for conjugate points of general selfadjoint differential equations*, PAMS, 25(1970), 656-661. MR 41 #3888.

5. *A Picone identity for first order differential systems*, JMAA, 31(1970), 297-308. MR 41 #5707.

6. *Oscillation criteria for nonlinear matrix differential equations*, PAMS, 26(1970), 270-272. MR 41 #8759.

7. *Comparison theorems for non-selfadjoint equations based on integral inequalities*, PAMS, 34(1972), 105-109. MR 46 #3902.

8. *A Prüfer transformation for nonselfadjoint systems*, PAMS, 31(1972), 147-151. MR 45 #2295.

9. *Oscillation Theory*. Lecture Notes in Mathematics, No. 324, Springer-Verlag, 1973.

10. *A non-trigonometric Prüfer transformation*, JLMS, (2), 7(1974), 728-732. MR 49 #769.

11. *A nonselfadjoint dynamical system*, PEMS, 19(1974), 77-87. MR 49 #7513.

12. *A dynamical approach to fourth order oscillation*, Ordinary and Partial Differential Equations, (Proceedings of the Conference held at Dundee, Scotland, 26-29 March, 1974), Lectures Notes in Mathematics, No. 415, Springer-Verlag, 1974, 188-192.

13. *Nonselfadjoint fourth order differential equations with conjugate points*, BAMS, 80(1974), 1190-1192. MR 50 #13691.

14. *Rotation properties of a class of second order differential systems*, JDEQ, 17(1975), 395-405.

15. *A dynamical criterion for conjugate points*, PFJM, 58(1975), 123-132. MR 52 #888.

Kryloff, N.

1. *Les méthodes de solution approchée des problèmes de la physique mathématique*, Memorial des Sciences Mathématiques, fasc. 49, Gauthier-Villars et Cie, Paris, 1931.

Kuks, L. M.

1. *A Sturm type comparison theorem for systems of fourth order ordinary differential equations*, (Russian), DFUJ, 10(1974), 751-754, 768. MR 49 #720.

Kummer, E. E.

1. *De generali quadam aequatione differentiali tertii ordinis*, Programm des evangelischen Königl . und Stadtgym. in Liegnitz vom Jahre 1823; reprinted JRAM, 100(1887), 1-9.

Legendre, A. M.

1. *Mémoire sur la manière de distinguer les maxima des minima dans le calcul des variations*, Mémoires de l'Academie des Sciences, 1786.

Leighton, W.

1. *A substitute for the Picone formula*, BAMS, 55(1949), 325-328. MR 11-109.

2. *Principal quadratic functionals and self-adjoint second-order differential equations*, PNAS, 35(1949), 192-193. MR 11-33.

3. *Principal quadratic functionals*, TAMS, 67(1949), 253-274. MR 11-603.

4. *The detection of the oscillation of solutions of a second order linear differential equation*, DKMJ, 17(1950), 57-62. MR 11-248, 871.

5. *On self-adjoint differential equations of the second order*, JLMS, 27(1952), 37-47. MR 13-745.

6. *Comparison theorems for linear differential equations of the second order*, PAMS, 13(1962), 603-610. MR 25 #4173.

7. *On the zeros of solutions of a second-order linear differential equation*, JMPA, 44(1965), 297-310; *Erratum*, JMPA, 46(1967), 10. MR 32 #4323.

8. *Some elementary Sturm theory*, JDEQ, 4(1968), 187-193. MR 37 #506.

9. *Regular singular points in the nonanalytic case; singular functionals*, JMAA, 28(1969), 59-76. MR 40 #2960.

10. *The conjugacy function*, PAMS, 24(1970), 820-823. MR 41 #2115.

11. *Quadratic functions of second order*, TAMS, 151(1970), 309-322. MR 41 #9078.

12. *Upper and lower bounds for eigenvalues*, JMAA, 35(1971), 381-388. MR 43 #7052.

13. *Computing bounds for focal points and for σ-points for second-order linear differential equations*, *Ordinary Differential Equations*, 1971 NRL-MRC Conference, Academic Press, New York, (1972), 497-503.

14. *More elementary Sturm theory*, APLA, 3(1973), 187-203.

Leighton, W., and William Oo Kian Ke

1. *A comparison theorem*, PAMS, 28(1971), 185-188. MR. 42 #8002.

2. *Determining bounds for the first conjugate point*, AMPA (4), 86(1970), 99-114. MR 43 #2292.

Leighton, W., and A. D. Martin

1. *Quadratic functionals with a singular end point*, TAMS, 78(1955), 98-128. MR 16-598.

Leighton, W., and Z. Nehari

1. *On the oscillation of solutions of self-adjoint linear differential equations of the fourth order*, TAMS, 89(1958), 325-377. MR 21 #1429.

Leighton, W., and A. S. Skidmore

1. *On the oscillation of solutions of a second-order linear differential equation*, RCMP; (II), 14(1965), 327-334. MR 35 #5706.

2. *On the differential equation* $y'' + p(x)y = f(x)$, JMAA, 43(1973), 46-55. MR 47 #3762.

Levi-Civita, T.

1. *Sur les équations linéaires a coefficients périodiques et sur le moyen mouvement du noeud lunaire*, ASEN, (3), 28(1911), 325-376.

Levin, A. Ju.

1. *A comparison principle for second-order differential equations*, (Russian), DOKL, 135(1960), 783-786 = SMDK, 1(1961), 1313-1316. MR 23 #A1875.
2. *On linear second-order differential equations*, (Russian), DOKL, 153(1963), 1257-1260 = SMDK, 4(1963), 1814-1817. MR 28 #2278.
3. *Integral criteria for the equation* $x'' + q(t)x = 0$ *to be non-oscillatory*, (Russian), USMN, 20(1965), 244-246. MR 32 #248.
4. *Non-oscillation of solutions of the equation* $x^{[n]} + p_1(t)x^{[n-1]} + \ldots + p_n(t)x = 0$, USMN, 24(1969), No. 2 (146), 43-96 = RMTS, 24(1969), 43-99. MR 40 #7537.

Levin, J. J.

1. *On the matrix Riccati equation*, PAMS, 10(1959), 519-524. MR 21 #7344.

Levinson, N.

1. *The inverse Sturm-Liouville problem*, Mat. Tidsskr. B. 1949, 25-30, (1949). MR 11-248.
2. *The $\mathscr{L}$-closure of eigenfunctions associated with self-adjoint boundary value problems*, DKMJ, 19(1952), 23-26. MR 13-654.

Lewis, R. T.

1. *Oscillation and nonoscillation criteria for some self-adjoint even order linear differential operators*, PFJM, 51(1974), 221-234.
2. *The oscillation of fourth order linear differential equations*, CDJM, 27(1975), 138-145.
3. *The existence of conjugate points for self-adjoint differential equations of even order*, PAMS, 56(1976), 162-166.

Liapunov, A. M.

1. *Problème Général de la Stabilité du Mouvement*, (French translation of a Russian paper dated 1893), Ann. Fac. Sci. Univ. Toulouse, 2(1907), 27-247; reprinted as Ann. Math. Studies, No. 17, Princeton, 1949.

Lichtenstein, L.

1. *Zur Analysis der unendlich vielen Variablen*, RCMP, 38(1914), 113-166.

2. *Über eine Integro-Differentialgleichungen und die Entwicklung willkürlicher Funktionen nach deren Eigenfunktionen*, Schwarz Festschrift, Berlin (1914), 274-285.

3. *Zur Variationsrechnung*, I, Göttinger Nachrichten, (1919), 161-192.

4. *Zur Variationsrechnung*, II,-Das isoperimetrische Problem, JRAM, 165(1931), 194-216.

5. *Zum Sturm-Liouville Problem*, MTZT, 31(1929), 346-349.

Lidskiĭ, V. B.

1. *Oscillation theorems for canonical systems of differential equations*, (Russian), DOKL, 102(1955), 877-888. MR 17-483.

Liouville, J.

1. *Sur le développement des fonctions ou parties de fonctions en séries dont les divers termes sont assujettis à satisfaire a une même équation différentielles du second ordre contenant un paramètre variable*, JMPA, 1(1836), 253-265; 2(1837), 16-35; 418-436.

2. *D'un problème d'analyse, relatif aux phénomènes thermo-mécaniques*, JMPA, 2(1837), 439-456.

Lorch, L., and D. J. Newman

1. *A supplement to the Sturm separation theorem*, AMMM, 72(1965), 359-366. MR 31 #422.

Lorch, L., and P. Szego

1. *Higher monotonicity properties of certain Sturm-Liouville functions*, ACMT, 109(1963), 55-73, MR 26 #5209; II. BAPS, 11(1963), 455-457. MR 28 #1340; III (with M. E. Muldoon) CDJM, 22(1970), 1238-1265, MR 43 #603; IV (with M. E. Muldoon) CDJM, 24(1972), 349-368.

Lovitt, W. V.

1. *Linear Integral Equations*, McGraw-Hill, New York, 1924.

Macki, J. W., and J. S. W. Wong

1. *Oscillation theorems for linear second order ordinary differential equations*, PAMS, 20(1969), 67-72. MR 38 #3513.

Makai, E.

1. *Ueber eine Eigenwertabschätzung bei gewissen homogenen linearen Differentialgleichungen zweiter Ordnung*, COMT, 6(1938-39), 368-374.

2. *Eine Eigenwertabschätzung bei gewissen Differentialgleichungen zweiter Ordnung*, Mat. Fiz. Lapok, 48(1941), 510-532. MR 8-208.

3. *Ueber die Nullstellen von Funktionen, die Lösungen Sturm-Liouville'scher Differentialgleichungen sind*, CMMH, 16(1944), 153-199. MR 6-2.

Marik, J., and M. Rab

1. *Asymptotische Eigenschaften von Lösungen der Differentialgleichung* y" = A(x)y *in nichtoszillatorischen Fall*, CZMJ, 10(85) (1960), 501-522. MR 25 #2283.
2. *Nichtoszillatorische Lineare Differentialgleichungen 2 Ordnung*, CZMJ, 13(88), 1963, 209-225. MR 28 #3206.

Markus, L., and R. A. Moore

1. *Oscillation and disconjugacy for linear differential equations with almost periodic coefficients*, ACMT, 96(1956), 99-123. MR 18-306.

Martin, A. D.

1. *A regular singular functional*, CDJM, 8(1956), 53-68. MR 17-633.
2. *A singular functional*, PAMS, 7(1956), 1031-1035. MR 18-809.

Mason, M.

1. *Zur Theorie der Randwertaufgaben*, MTAN, 58(1904), 528-544.
2. *On the boundary problems of linear ordinary differential equations of second order*, TAMS, 7(1906), 337-360.
3. *The expansion of a function in terms of normal functions*, TAMS, 8(1907), 427-432.
4. *Selected topics in the theory of boundary value problems of differential equations*, New Haven, Colloquium of AMS, (1910), 173-222.

Mayer, A.

1. *Ueber die Kriterien des Maximums und Minimums der einfachen Integrale*, JRAM, 69(1868), 238-263.

McCarthy, P. J.

1. *On the disconjugacy of second order linear differential equations*, AMMM, 66(1959), 892-894. MR 22 #793.

Miller, J. C. P.

1. *On a criterion for oscillatory solutions of a linear differential equation of the second order*, PCPS, 36(1940), 283-287. MR 2-50.

Miller, W. B.

1. *The behavior of solutions of self-adjoint linear differential equations of the fourth order*, (Dissertation, Lehigh University, 1962). An abbreviated version appeared as *Separation theorems for self-adjoint linear differential equations of the fourth order*, SJMA, 6(1975), 742-759.

Milloux, H.

1. *Sur l'équation différentielle* x" + A(x)x = 0, PRMF, 41(1934), 39-53.

Milne, W. E.

1. *The behavior of a boundary value problem as the interval becomes infinite*, TAMS, 30(1928), 797-802.
2. *On the degree of convergence of expansions in an infinite interval*, TAMS, 31(1929), 907-918.

Molinari, B. P.

1. *The stabilizing solution of the algebraic Riccati equation*, SIJC, 11(1973), 262-271.
2. *Equivalence relations for the algebraic Riccati equation*, SIJC, 11(1973), 272-285.

Moore, R. A.

1. *The behavior of solutions of a linear differential equation of the second order*, PFJM, 5(1955), 125-145. MR 16-925.

Morse, M.

1. *A generalization of the Sturm separation and comparison theorems*, MTAN, 103(1930), 52-69.
2. *The order of vanishing of the determinant of a conjugate base*, PNAS, 17(1931), 319-320.
3. *Sufficient conditions in the problem of Lagrange with variable end conditions*, AMJM, 53(1931), 517-546.
4. *"The Calculus of Variations in the Large"*, AMS Colloquium Publications, 18(1934); fourth printing, 1965.
5. *Sufficient conditions in the problem of Lagrange without assumptions of normalcy*, TAMS, 37(1935), 147-160.
6. *Recent advances in variational theory in the large*, Proc. Int. Cong. Math., Cambridge, Mass., 1950, 2(1950), 143-156. MR 13-474.
7. *Subordinate quadratic forms and their complementary forms*, Rev. Roumaine Math., pures appl. 16(1971), 559-569. MR 44 #4762.
8. *Singular quadratic functionals*, MTAN, 201(1973), 315-340.
9. *Variational Analysis: Critical Extremals and Sturmian Extensions*, Wiley-Interscience, New York, 1973.

Morse, M., and W. Leighton

1. *Singular quadratic functionals*, TAMS, 40(1936), 252-286.

Myers, S. B.

1. *Sufficient conditions in the problem of the calculus of variations in n-space in parametric form and under general end conditions*, TAMS, 35(1933), 746-760.

Naimark, M. A.

1. *Linear Differential Operators*, I, II, Harrap, London, 1968.

Nehari, Z.

1. *On the zeros of solutions of second order linear differential equations*, AMJM, 76(1954), 689-697. MR 16-131.
2. *Oscillation criteria for second-order linear differential equations*, TAMS, 85(1957), 428-445. MR 19-415.
3. *On an inequality of Lyapunov*, Studies in Mathematical Analysis and Related Topics, Stanford Univ. Press, Stanford (1962), 256-261. MR 26 #2684.
4. *Disconjugate linear differential operators*, TAMS, 129(1967), 500-516. MR 36 #2860.
5. *Disconjugacy criteria for linear differential equations*, JDEQ, 4(1968), 604-611. MR 38 #1329.
6. *A disconjugacy criterion for self-adjoint linear differential equations*, JMAA, 35(1971), 591-599. MR 43 #7694.
7. *Conjugate points, triangular matrices, and Riccati equations*, TAMS, 199(1974), 181-198.

Nikolenko, L. D.

1. *On oscillation of solutions of the differential equation* $y'' + p(x)y = 0$, (Russian), UKMZ, 7(1955), 124-127. MR 17-263.
2. *Some criteria for non-oscillation of a fourth order differential equation*, (Russian), DOKL, 114(1957), 483-485. MR 19-960.

Noussair, E. S., and C. A. Swanson

1. *Oscillation criteria for differential systems*, JMAA, 36(1971), 575-580. MR 45 #5477.

Oakley, C. O.

1. *A note on the methods of Sturm*, ANNM, II, 31(1930), 660-662.

Olech, C., Z. Opial, and T. Wazewski

1. *Sur le problème d'oscillation des intégrales de l'équation* $y'' + g(t)y = 0$, BAPS, 5(1957), 621-626. MR 19-650.

Opial, Z.

1. *Sur les intégrales oscillantes de l'équation différentialle* $u''+f(t)u = 0$, ANPM, 4(1958), 308-313. MR 20 #4051.
2. *Sur un inégalité de C. de la Vallée Poussin dans la theorie de l'équation différentielle linéaire du second ordre*, ANPM, 6(1959-60), 87-91, MR 21 #3626.
3. *Sur une critère l'oscillation des intégrales de l'équation différentielle* $(Q(t)x')' + f(t)x = 0$, ANPM, 6(1959-60), 99-104. MR 21 #3627.
4. *Sur la répartition asymptotique des zeros des fonctions caractéristiques du problème de Sturm*, ANPM, 6(1959-60), 105-110. MR 21 #3614.

Patula, W. T., and P. Waltman

1. *Limit point classification of second order linear differential equations*, JLMS, (2), 8(1974), 209-216.

Peterson, A. C.

1. *Distribution of zeros of solutions of linear differential equations of order four*, (Ph.D. Dissertation, University of Tennessee), 1968.
2. *The distribution of zeros of extremal solutions of fourth order differential equations for the N-th conjugate point*, JDEQ, 8(1970), 502-511. MR 42 #4821.
3. *A theorem of Aliev*, PAMS, 23(1969), 364-366. MR 40 #2961.

Petty, C. M., and J. E. Barry

1. *A geometric approach to the second order linear differential equation*, CDJM, 14(1962), 349-358. MR 26 #3973.

Picard, E'.

1. *Sur l'application des méthodes d'approximations successives à l'étude de certaines équations différentielles ordinaires*, JMPA, (4) 9(1893), 217-271.
2. *Traité d'Analyse*, Tome III, Gauthier-Villars, Paris, 1896.

Picone, M.

1. *Su un problema al contorno nelle equazioni differenziali lineari ordinarie del secondo ordine*, ASNP, 10(1909), 1-92.
2. *Sui valori eccezionali di un parametro da cui dipende un'equazione differenziale lineare ordinaria del second ordine*, ASNP, 11(1910), 1-141.
3. *Sulle autosoluzione e sulle formule di maggiorazione per gli integrali delle equazioni differenziali lineari ordinarie autoaggiunte*, MTZT, 28(1928), 519-555.

Pokornyi, V. V.

1. *On some sufficient conditions for univalence*, DOKL, 79(1951), 743-746.

Pólya, G.

1. *On the mean-value theorem corresponding to a given linear homogeneous differential equation*, TAMS, 24(1922), 312-324.

Pólya, G. and G. Szegö

1. *Isoperimetric Inequalities in Mathematical Physics*, Annals of Math. Studies, No. 27, Princeton Univ. Press, 1951. MR 13-270.

Porter, M. B.

1. *On the roots of functions connected with a linear recurrent relation of the second order*, ANNM, 3(1902), 55-70.

Potter, Ruth L.

1. *On self-adjoint differential equations of second order*, PFJM, 3(1953), 467-491. MR 15-32.

Prüfer, H.

1. *Neue Herleitung der Sturm-Liouvilleschen Reihenentwicklung stetiger Funktionen*, MTAN, 95(1926), 499-518.

Pudei, V.

1. *On the properties of solutions of the differential equation* $y^{(iv)} + p(x)y'' + q(x)y = 0$, CAPM, 93(1968), 201-216. MR 38 #3515.

2. *Über die Eigenschaften der Lösungen linearen Differentialgleichungen gerader Ordnung*, CAPM, 94(1969), 401-425. MR 42 #7993.

Putnam, C. R.

1. *An oscillation criterion involving a minimum principle*, DKMJ, 16(1949), 633-636. MR 11-437.

2. *Note on some oscillation criteria*, PAMS, 6(1955), 950-952. MR 17-615.

3. *Necessary and sufficient conditions for the existence of negative spectra*, QAMT, 13(1955), 335-337. MR 17-370.

Rab, M.

1. *Eine Bemerkung zu der Frage über die oszillatorischen Eigenschaften der Lösungen der Differentialgleichung* $y'' + A(x)y = 0$, CAPM, 82(1957), 342-348. MR 20 #1023.

2. *Kriterien für die Oszillation der Lösungen der Differentialgleichung* $(p(x)y')' + q(x)y = 0$, CAPM, 84(1959), 335-370; erratum, ibid 85(1960), 91. MR 22 #5773.

Radon, J.

1. *Über die Oszillationstheoreme der konjugierten Punkte biem Probleme von Lagrange*, Munchener Sitzungsberichte, 57(1927), 243-257.

2. *Zum Probleme von Lagrange*, Abh. Math. Sem. Univ. Hamburg, 6(1929), 273-299.

Reid, W. T.

1. *Expansion problems associated with a system of integral equations*, TAMS, 33(1931), 475-485.

2. *A boundary problem associated with the calculus of variations*, AMJM, 54(1932), 769-790.

3. *Analogues of the Jacobi condition for the problem of Mayer in the calculus of variations*, ANNM, 35(1934), 836-848.

4. *Discontinuous solutions in the non-parametric problem of Mayer in the calculus of variations*, AMJM, 57(1935), 69-93.

5. *The theory of the second variation for the non-parametric problem of Bolza*, AMJM, 57(1935), 573-586.

6. *Boundary value problems of the calculus of variations*, BAMS, 43(1937), 633-666.

7. *An integro-differential boundary problem*, AMJM, 60(1938), 257-292.

8. *A system of ordinary linear differential equations with two-point boundary conditions*, TAMS, 44(1938), 508-521.

9. *A new class of self-adjoint boundary value problems*, TAMS, 52(1942), 381-425. MR 4-100.

10. *A matrix differential equation of Riccati type*, AMJM, 68(1946), 237-246; "Addendum", 70(1948), 250. MR 7-446.

11. *Symmetrizable completely continuous linear transformations in Hilbert space*, DKMJ, 18(1951), 41-56. MR 13-564.

12. *Oscillation criteria for linear differential systems with complex coefficients*, PFJM, 6(1956), 733-751. MR 18-898.

13. *A comparison theorem for self-adjoint differential equations of second order*, ANNM, 65(1957), 197-202. MR 19-1052.

14. *Adjoint linear differential operators*, TAMS, 85(1957), 446-461. MR 19-550.

15. *Principal solutions of non-oscillatory self-adjoint linear differential systems*, PFJM, 8(1958), 147-169. MR 20 #4682.

16. *A class of two-point boundary problems*, ILJM, 2(1958), 434-453.

17. *A Prüfer transformation for differential systems*, PFJM, 8(1958), 575-584. MR 20 #5913.

18. *Generalized linear differential systems*, JMMC, 8(1959), 705-726. MR 21 #5777.

19. *Oscillation criteria for self-adjoint differential systems*, TAMS, 101(1961), 91-106. MR 24 #A3349.

20. *Riccati matrix differential equations and non-oscillation criteria for associated linear differential systems*, PFJM, 13(1963), 665-685. MR 27 #4991.

21. *Principal solutions of non-oscillatory linear differential systems*, JMAA, 9(1964), 397-423. MR 29 #6110.

22. *Generalized linear differential systems and related Riccati matrix integral equations*, ILJM, 10(1966), 701-722. MR 37 #1682.

23. *Generalized Green's matrices for two-point boundary problems*, SJAM, 15(1967), 856-870. MR 36 #2866.

24. *Some limit theorems for ordinary differential systems*, JDEQ, 3(1967), 423-439. MR 37 #1683.

25. *Variational methods and boundary problems for ordinary linear differential systems*, Proc. US-Japan Sem. on Differential and Functional Equations, Univ. of Minnesota, Minneapolis, Minn., June 26-30, 1967, W. A. Benjamin, Inc., 267-299. MR 37 #4322.

26. *Generalized inverses of differential and integral operators*, Proc. Symposium on Theory and Application of Generalized Inverses of Matrices, Texas Technological College, Lubbock, Texas, March, 1968, 1-25. MR 41 #2470.

27. *Remarks on the Morse index theorem*, PAMS, 20(1969), 339-341. MR 38 #2814.

28. *A maximum problem involving generalized linear differential equations of the second order*, JDEQ, 8(1970), 283-293. MR 42 #3643.

29. *Monotoneity properties of solutions of hermitian Riccati matrix differential equations*, SJMA, 1(1970), 195-213. MR 41 #7202.

30. *A matrix Liapunov inequality*, JMAA, 32(1970), 424-434. MR 42 #3354.

31. *Some remarks on special disconjugacy criteria for differential systems*, PFJM, 35(1970), 763-772. MR 43 #7712.

32. *Generalized polar coordinate transformations for differential systems*, RMJM, 1(1971), 383-406. MR 43 #6488.

33. *Discontinuous solutions for a non-parametric variational problem*, APLA, 1(1971), 161-182. MR 44 #4605.

34. *A disconjugacy criterion for higher order linear vector differential equations*, PFJM, 39(1971), 795-806. MR 46 #7632.

35. *Ordinary Differential Equations*, John Wiley and Sons, New York, 1971. MR 42 #7963.

36. *Variational aspects of oscillation phenomena for higher order differential equations*, JMAA, 40(1972), 446-470. MR 47 #8977.

37. *Involutory matrix differential equations*, Ordinary Differential Equations: 1971 NRL-MRC Conference, Academic Press, New York (1972), 221-240.

38. *Riccati Differential Equations*, Academic Press, 1972.

39. *A continuity property of principal solutions of linear Hamiltonian differential systems*, Scripta Mathematica, 29(1973), 337-350.

40. *Boundary problems of Sturmian type on an infinite interval*, SJMA, 4(1973), 185-197.

41. *A generalized Liapunov inequality*, JDEQ, 13(1973), 182-196. MR 48 #8921.

42. *A supplement to oscillation and comparison theory for Hermitian differential systems*, JDEQ, 16(1974), 550-573.

43. *Variational methods and quadratic functional inequalities*, SJMA, 6(1975), 404-416.

44. *Generalized linear differential systems and associated boundary problems*, Proc. International Symposium on Dynamical Systems, Brown Univ., Providence, Rhode Island, Aug. 12-16, 1974, (1976), 651-673, Academic Press.

45. *A historical note on Sturmian theory*, JDEQ, 20(1976), 316-3

46. *Related self-adjoint differential and integro-differential systems*, JMAA, 54(1976), 89-114.

47. *Interrelations between a trace formula and Liapunov type inequalities*, JDEQ, 23(1977), 448-458.

Richardson, R. G. D.

1. *Das Jacobische Kriterium der Variationsrechnung und die Oscillationseigenschaften linearer Differentialgleichungen, 2. Ordnung*, MTAN, 68(1910), 279-304; (*Zweite Mitteilung*), MTAN, 71(1911), 214-232.

2. *Theorems of oscillation for two linear differential equations of the second order with two parameters*, TAMS, 13(1912), 22-34.

3. *Über die notwendige und hinreichenden Bedingungen für das Bestehen eines Kleinschen Oszillationstheorems*, MTAN, 73(1913), 289-304; *Berichtigung*, 74(1913), 312.

4. *Contributions to the study of oscillation properties of the solutions of linear differential equations of the second order*, AMJM, 40(1918), 283-316.

5. *Bôcher's boundary problems for differential equations* [Review of Bôcher's Lecons sur les méthodes de Sturm, Paris, 1917], BAMS, 26(1920), 108-124.

Ridenhour, J. R.

1. *On* (n,n)*-zeros of solutions of linear equations of order* 2n, PAMS, 44(1974), 135-140.

2. *On the zeros of solutions of* n*-th order linear differential equations*, JDEQ, 16(1974), 45-71.

3. *Linear differential equations where nonoscillation is equivalent to eventual disconjugacy*, PAMS, 49(1975), 366-372.

Ridenhour, J. R., and T. L. Sherman

1. *Conjugate points for fourth order linear differential equations*, SJAM, 22(1972), 599-603. MR 46 #2151.

Ritcey, L.

1. *Index theorems for discontinuous problems in the calculus of variations*, (Dissertation, Univ. of Chicago, 1945).

Rozenberg, J.

1. *Über das Verhalten von Extremalenbogen die den zum Anfangspunkt konjugierten Punkte enthalten beim Lagrangschen Problem der Variationsrechnung*, MOMT, 24(1913), 65-86.

Sandor, S.

1. *Quelques critériums de non-oscillation*, Com. Acad. R. P. Romîne 6(1956), 753-756. (Romanian. Russian and French summaries). MR 20 #4054.

2. *Sur l'équation différentielle matricielle de type Riccati*, BMSR, 3(51) (1959), 229-249. MR 23 #A1863.

Sansone, G.

1. *Equazioni differenziali nel campo reale*, 2nd ed., Zanichelli, Bologna, Vol. 1 (1948), MR 10-193; Vol. 2 (1949), MR 11-32.

Schäfke, F. W., and A. Schneider

1. S-*hermitesche Rand-Eigenwertprobleme* I, II, III. MTAN, 162(1965), 9-26, MR 32 #5968; 165(1966), 236-260, MR 33 #7664; 177(1968), 67-94, MR 37 #4315.

Scheeffer, L.

1. *Die Maxima und Minima der einfachen Integrale zwischen festen Grenzen*, MTAN, 25(1885), 522-593.

Schmidt, E.

1. *Entwicklung willkürlicher Funktionen nach Systemen vorgeschreibener*, (Dissertation, Göttingen), MTAN, 63(1907), 433-476.

Schneider, A.

1. *Die Greensche Matrix* S-*hermitescher Rand-Eigenwertprobleme im Normalfall*, MTAN, 180(1969), 307-312. MR 39 #3086.

Schneider, L. J.

1. *Oscillation properties of the* 2-2 *disconjugate fourth order self-adjoint differential equation*, PAMS, 28(1971), 545-550. MR 43 #7713.

Schubert, H.

1. *Über die Entwicklung zulässiger Funktionen nach den Eigenfunktionen bei definiten selbstadjungierten Eigenwertaufgaben*, Sitzungsberichte Heidelberger Acad. Wiss., (1948), 178-192. MR 11-173.

Schwarz, H. A.

1. *Ueber ein die Flachen kleinsten Inhalts betreffendes Problem der Variationsrechnung*, Acta Societatis Scientiarum Fennicae, 15(1885), 315-362.

Sherman, T. L.

1. *Properties of solutions of* n-*th order linear differential equations*, PFJM, 15(1965), 1045-1060. MR 32 #2654.

2. *Properties of solutions of quasi-differential equations*, DKMJ, 32(1965), 297-304. MR 31 #3659.

3. *On solutions of nth order linear differential equations with* N *zeros*, BAMS, 74(1968), 923-925. MR 38 #1331.

4. *Conjugate points and simple zeros for ordinary linear differential equations*, TAMS, 146(1969), 397-411. MR 41 #572.

Simons, W.

1. *Some disconjugacy criteria for self-adjoint linear differential equations*, JMAA, 34(1971), 445-464. MR 44 #521.

2. *Disconjugacy criteria for systems of self-adjoint differential equations*, JLMS, 6(1972-73), 373-381. MR 47 #5372.

3. *Monotoneity in some non-oscillation criteria for differential equations*, JDEQ, 13(1973), 124-126.

Sloss, F. B.

1. *A self-adjoint boundary value problem with end conditions involving the characteristic parameter*, (Dissertation, Northwestern University, 1955).

Stafford, R. A., and J. W. Heidel

1. *A new comparison theorem for scalar Riccati equations*, BAMS, 80(1974), 754-757.

Stark, M. E.

1. *A self-adjoint boundary value problem associated with a problem of the calculus of variations*, (Dissertation, Chicago, 1926).

Stein, J.

1. *Singular quadratic functionals*, (Dissertation, Univ. of California, Los Angeles, 1971). Results of this dissertation have been reported in the following:
A. *Index theory for singular quadratic functionals in the calculus of variations*, BAMS, 79(1973), 1189-1192. MR 48 #7068.
B. *Hilbert space and variational methods for singular self-adjoint systems of differential equations*, BAMS, 80(1974), 744-747.

Stenger, W.

1. *Some extensions and applications of the new maximum-minimum theory of eigenvalues*, JMMC, 19(1970), 931-949.

2. *On two complementary variational characterizations of eigenvalues*, Inequalities Vol. II, Academic Press, 1970, 375-387. MR 42 #878.

3. *Non-classical choices in variational principles for eigenvalues*, JLFA, 6(1970), 157-164. MR 41 #4281.

Sternberg, Helen M. and R. L. Sternberg

1. *A two-point boundary problem for ordinary self-adjoint differential equations of fourth order*, CDJM, 6(1954), 416-419. MR 15-874.

Sternberg, R. L.

1. *Variational methods and non-oscillation theorems for systems of differential equations*, (Dissertation, Northwestern Univ., 1951), DKMJ, 19(1952), 311-322. MR 14-50.

2. *A theorem on hermitian solutions for related matrix differential and integral equations*, PTGM, 12(1953), 135-139. MR 15-706.

Stevens, R. R.

1. *A conjugacy criterion*, PAMS, 33(1972), 75-80. MR 44 #7047.

Stickler, D. C.

1. *Bounds for the solution of the linear self-adjoint second order differential equation*, JMAA, 10(1965), 419-423. MR 31 #432.

St. Mary, D. F.

1. *Some oscillation and comparison theorems for* $(r(t)y')' + p(t)y = 0$, JDEQ, 5(1969), 314-323. MR 38 #1332.

Stokes, A. N.

1. *Differential inequalities and the matrix Riccati equation*, (Dissertation, Australian National University, Canberra, 1972).

Sturdivant, J. H.

1. *Second order linear systems with summable coefficients*, TAMS, 30(1928), 560-566.

Sturm, J. C. F.

1. *Mêmoire sur les équations différentielles linéaires du second ordre*, JMPA, 1(1836), 106-186.

Swanson, C. A.

1. *Comparison and Oscillation Theory of Linear Differential Equations*, Academic Press, New York, 1968.

2. *Oscillation Criteria for nonlinear matrix differential inequalities*, PAMS, 24(1970), 824-827. MR 41 #3890.

Sz.-Nagy, B.

1. *Vibrations d'une corde non-homogène*, BSMF, 75(1947), 193-209. MR 10-269.

Taam, C.-T.

1. *Non-oscillatory differential equations*, DKMJ, 19(1952), 493-497. MR 14-557.

2. *Non-oscillation and comparison theorems of linear differential equations with complex-valued coefficients*, PTGM, 12(1953), 57-72. MR 14-873.

3. *On the solution of second order linear differential equations*, PAMS, 4(1953), 876-879. MR 15-625.

Titchmarsh, E. C.

1. *Eigenfunction Expansions Associated with Second-Order Differential Equations*, Clarendon Press, Oxford, 1st. ed., 1946; Part I, 2nd ed., 1962.

Tomastik, E. C.

1. *Singular quadratic functionals of* n *dependent variables*, TAMS, 124(1966), 60-76. MR 33 #4743.

2. *Oscillation of nonlinear matrix differential equations of second order*, PAMS, 19(1968), 1427-1431. MR 38 #372.

Tonelli, L.

1. *Fondamenti di calcolo delle Variazioni*, I, II, Zanichelli, Bologna, 1923.

Travis, C. C.

1. *Remarks on a comparison theorem for scalar Riccati equations*, PAMS, 52(1975), 311-314.

Vosmansky, J.

1. *The monotoneity of extremants of integrals of the differential equation* $y'' + q(t)y = 0$, ARVM, 2(1966), 105-111. MR 35 #6907.

de la Vallée Poussin, Ch. J.

1. *Sur l'équation différentielle linéaire du second ordre. Determination d'une integrale par deux valeurs assigneés. Extension aux équation d'ordre* n., JMPA, (9) 8(1929), 125-144.

Walter, J.

1. *Regular eigenvalue problems with eigenvalue parameter in the boundary conditions*, MTZT, 133(1973), 301-312. MR 49 #713.

Weinberger, H.

1. *An extension of the classical Sturm-Liouville theory*, DKMJ, 22(1955), 1-14. MR 16-824.

Weinstein, A.

1. *Quantitative methods in Sturm-Liouville theory*, Proceedings of Symposium on Spectral Theory and Differential Equations, held at Oklahoma A. and M. College, 1951, pp. 345-352. MR 13-240.

2. *On the new maximum-minimum theory of eigenvalues*, Inequalities, Academic Press, 1967, 329-338. MR 36 #4369.

Weinstein, A., and W. Stenger

1. *Methods of Intermediate Problems for Eigenvalues*, Academic Press, New York, 1972.

Weyl, H.

1. *Ueber gewöhnliche Differentialgleichungen mit Singularitäten und die zugehörigen Entwicklungen willkürlicher Funktionen*, MTAN, 68(1910), 220-269.
2. *Ueber gewöhnliche Differentialgleichungen mit singulären Stellen und ihre Eigenfunktionen*, Göttinger Nachrichten, (1910), 442-467.

Whyburn, W. M.

1. *Second order differential systems with integral and* k-*point boundary conditions*, TAMS, 30(1928), 630-640.
2. *Existence and oscillation theorems for non-linear differential systems of the second order*, TAMS, 30(1928), 848-854.
3. *On related difference and differential systems*, AMJM, 51(1929), 265-286.
4. *On related difference and differential systems*, BAMS, 36(1930), 94-98.
5. *On self-adjoint ordinary differential equations of the fourth order*, AMJM, 52(1930), 171-196.
6. *Matrix differential equations*, AMJM, 56(1934), 587-592.
7. *Differential equations with general boundary conditions*, BAMS, 48(1942), 692-704. MR 4-100.
8. *Differential systems with general boundary conditions*, Seminar Reports in Math., (Los Angeles) Univ. Calif. Publ. Math., 2(1944), 45-61. MR 5-265.
9. *Differential systems with boundary conditions at more than two points*, Proc. Conference on Differential Equations, Univ. of Maryland, College Park, MD., (1955), 1-21. MR 18-481.
10. *A nonlinear boundary value problem for second order differential equations*, PFJM, 5(1955), 147-160. MR 16-1027.
11. *On a class of linear differential systems*, Rev. Ci. (Lima), 60(1958), 43-59. MR 23 #A2568.

Whyburn, W. M. and T. J. Pignani

1. *Differential equations with interface and general boundary conditions*, J. Elisha Mitchell Scientific Soc., 72(1956), 1-14. MR 18-42.

Wielandt, H.

1. *Über die Eigenwertaufgaben mit reellen diskreten Eigenwerten*, MTNR, 4(1950-51), 308-314. MR 12-717.

Wiggin, E. P.

1. *A boundary value problem of the calculus of variations*, CTCV, (1933-37), Univ. of Chicago Press, 243-275.

Wilkins, J. E., Jr.

1. *Definitely self-conjugate adjoint integral equations*, DKMJ, 11(1944), 155-166. MR 5-267.

Willett, D.

1. *On the oscillatory behavior of the solutions of second order linear differential equations*, ANPM, 21(1969), 175-194. MR 40 #2964.

2. *Classification of second order linear differential equations with respect to oscillation*, Advances in Math. 3(1969), 594-623. (reprinted in *Lectures on Ordinary Differential Equations*, edited by R. McKelvey, Academic Press, 1971). MR 43 #6519.

3. *A necessary and sufficient condition for the oscillation of some linear second order differential equations*, RMJM, 1(1971), 357-365. MR 44 #7048.

Williams, C. M.

1. *Oscillation phenomena for linear differential equations in a* B**-algebra*, (Dissertation, University of Oklahoma, 1971).

Wiman, A.

1. *Über die reellen Lösungen der linearen Differentialgleichungen zweiter Ordnung*, Arkiv för Matematik, Astronomi och fysik, 12(1917), No. 14.

2. *Über eine Stabilitätsfrage in der Theorie der linearen Differentialgleichungen*, ACMT, 66(1936), 121-145.

Wintner, A.

1. *The adiabatic linear oscillator*, AMJM, 68(1946), 385-397. MR 8-71.

2. *Asymptotic integration constants*, AMJM, 68(1946), 553-559. MR 8-272.

3. *Asymptotic integrations of the adiabatic oscillator*, AMJM, 69(1947), 251-272. MR 9-35.

4. *On the Laplace-Fourier transcendents occurring in mathematical physics*, AMJM, 69(1947), 87-98. MR 8-381.

5. *A norm criterion for non-oscillatory differential equations*, QAMT, 6(1948), 183-185. MR 9-589.

6. *A criterion of oscillatory stability*, QAMT, 7(1949), 115-117. MR 10-456.

7. *On the non-existence of conjugate points*, AMJM, 73(1951), 368-380. MR 13-37.

8. *On a theorem of Bôcher in the theory of ordinary differential equations*, AMJM, 76(1954), 183-190. MR 15-426.

9. *On disconjugate linear differential equations*, Arch. Math., 8(1957), 290-293. MR 19-855.

10. *On the comparison theorem of Kneser-Hille*, MTSA, 5(1957), 255-256. MR 20 #3349.

11. *Comments on 'flat' oscillations of low frequency*, DKMJ, 24(1957), 365-366. MR 19-855.

12. *On stable oscillations of high frequency*, BUMI, (3) 12(1957), 9-11. MR 19-416.

13. *A comparison theorem for Sturmian oscillation numbers of linear systems of second order*, DKMJ, 25(1958), 515-518. MR 20 #7129.

14. *On an inequality of Liapunoff*, QAPM, 16(1958), 175-178. MR 20 #1815.

15. *A stability criterion for quasi-harmonic vibrations*, QAPM, 16(1959), 423-426. MR 20 #5316.

Wolfson, K. G.

1. *On the spectrum of a boundary problem with two singular endpoints*, AMJM, 72(1950), 713-719. MR 10-946.

Wong, J. S. W.

1. *Second order linear oscillation with integrable coefficients*, BAMS, 74(1968), 909-911. MR 38 #1333.

2. *Oscillation and nonoscillation of solutions of second order linear differential equations with integrable coefficients*, TAMS, 144(1969), 197-215. MR 40 #4536.

Wray, S. D.

1. *Integral comparison theorems in oscillation theory*, JLMS, 8(1974), 595-606.

Zaanen, A. C.

1. *Ueber vollstetige symmetrische und symmetrisierbare Operatoren*, NAWK, (2) 22(1943), 57-80. MR 7-453, 621.

2. *Normalizable transformations in Hilbert space and systems of linear integral equations*, ACMT, 83(1950), 197-248. MR 13-564.

3. *Linear Analysis*, Interscience, New York, 1953. MR 15-878.

Zettl, A.

1. *Factorization of differential operators*, PAMS, 27(1971), 425-426. MR 42 #7966.

2. *Factorization and disconjugacy of third order differential equations*, PAMS, 31(1972), 203-208. MR 45 #5481.

3. *Explicit conditions for the factorization of* nth *order linear differential operators*, PAMS, 41(1973), 137-145. MR 47 #8952.

Zimmerberg, H. J.

1. *A class of definite boundary value problems,* (Dissertation, Univ. of Chicago, 1945).
2. *A self-adjoint differential system of even order,* DKMJ, 13(1946), 411-417. MR 8-213.
3. *Definite integral systems,* DKMJ, 15(1948), 371-388. MR 11-37.
4. *Two-point boundary problems involving a parameter linearly,* ILJM, 4(1960), 593-608. MR 22 #12267.
5. *Two-point boundary conditions linear in a parameter,* PFJM, 12(1962), 385-393. MR 25 #3210.
6. *Reduction of symmetrizable problems with integral boundary conditions,* JDEQ, 14(1973), 568-580. MR 48 #11650.
7. *Symmetric integro-differential-boundary problems,* TAMS, 188(1974), 407-417. MR 49 #3481.
8. *Linear integro-differential boundary parameter problems,* AMPA, (4), 105(1975), 241-256.

Zlámal, M.

1. *Oscillation criterions,* CAPM, 75(1950), 213-218. MR 13-132.

Zubova, A. F.

1. *Concerning oscillation of the solutions of an equation of the second order,* (Russian), Vestnik Leningrad Univ., 12(211), (1957), 168-174. MR 19-1177.
2. *Oscillations and stability of solutions of second-order equations,* (Russian), Sibirsk. Mat. Ž, 4(1963), 1060-1070. MR 27 #5985.

SPECIAL SYMBOLS

SUBJECT INDEX

AUTHOR INDEX

F